Sporulation and Germination

Sporulation and Germination

Proceedings of the Eighth International Spore Conference
Woods Hole, Massachusetts
9–12 October 1980

Edited by

Hillel S. Levinson
Science and Advanced Technology
 Laboratory
U. S. Army Natick Research and
 Development Laboratories
Natick, Massachusetts 01760

A. L. Sonenshein
Department of Molecular Biology and
 Microbiology
Tufts University School of Medicine
Boston, Massachusetts 02111

Donald J. Tipper
Department of Molecular Genetics
 and Microbiology
University of Massachusetts Medical
 School
Worcester, Massachusetts 01605

American Society for Microbiology
Washington, D. C.
1981

Copyright ©1981 American Society for Microbiology
1913 I St., N.W.
Washington, DC 20006

Library of Congress Cataloging in Publication Data

International Spore Conference (8th:1980:
  Woods Hole, Mass.)
Sporulation and germination.

  Includes index.
  1. Spores (Bacteria)—Congresses.
2. Germination—Congresses. I. Levinson, Hillel, 1918–   . II. Sonenshein, A. L., 1944–   . III. Tipper, Donald J., 1935–   . IV. Title.
QR79.I57 1980   589′.0165   81-3670

ISBN 0-914826-35-2   AACR2

Dedication

This volume is dedicated to the memory of two outstanding microbiologists whose untimely deaths prevented their participation in the Eighth International Spore Conference. Z. John Ordal and William Steinberg were representatives of different generations of spore researchers but shared an enthusiasm for their subject and an exuberance for life that made them both beloved and respected.

Z. John Ordal (1913–1979) was Professor of Food Microbiology in the Departments of Food Science and Microbiology, University of Illinois. His distinguished research career spanned four decades. While he had many interests, his major and abiding interest was directed towards the physiology of the bacterial spore and related studies on injury and repair in bacteria. He did much to contribute to the success of the First International Spore Conference held in 1956 at Allerton Park, University of Illinois, and to many of the subsequent spore conferences. His university, his students, and spore science were all enriched by his presence.

Bill Steinberg (1941–1979) was Associate Professor of Microbiology at the University of Virginia. His tragic death at the age of 38 cut short a career that was already highly productive. His intelligence was evident from the elegance of his experimental approaches. One of the saddest aspects of his passing is that the spore field is just now reaching the stage in its development at which his biochemical and genetic skills would have been most effectively applied.

We mourn the passing of these two friends and hope that their families will find comfort in knowing how much they were loved and respected by their colleagues.

Contents

Review Articles

Research Articles

A. Genetics of Sporeforming Bacteria and Cloning of Sporulation Genes

B. Initiation of Sporulation and Regulation of Metabolism

C. Macromolecular Synthesis During Sporulation

D. Dormancy, Germination, and Outgrowth

E. Resistance of Spores

F. Resting Forms of Microorganisms Other than Bacillus and Clostridium

CONTENTS

Preface

The Eighth International Spore Conference was held in Woods Hole, Massachusetts, on 9–12 October 1980. Within the beautiful facilities of the Marine Biological Laboratories, 250 scientists from five continents gathered for intensive and informative discussions on sporeforming microorganisms. This book hopefully reflects the high caliber of the conference, the excitement generated by many of the presentations, and the extent to which the persistent and provocative questions concerning the description and control of sporulation and germination events are yielding to the application of modern molecular genetics and biochemical techniques.

The Local Organizing Committee (Arnold Demain, Harlyn Halvorson, Ron Labbe, Richard Losick [Chairman], Henry Paulus, Hillel Levinson, Linc Sonenshein, Don Tipper, and Annamaria Torriani) worked hard to create the structure of the conference and to make the difficult choices of topics to be covered and speakers to be invited. No one worked harder than Rich Losick. From incorporating (with the legal assistance of Larry Levinson) the International Spore Conference as a nonprofit corporation, to applying for support grants, to signing in registrants, our chairman was clearly the greatest contributor. Even his indefatigable efforts would have been much less effective, however, were it not for the intelligent organization of the incredibly efficient conference secretary, Thi Van Huynh.

The International Advisory Committee for this conference included Lyle Brown, Amikam Cohen, Jim Hoch, Yasuo Kobayashi, Pat Piggott, Hans Rhaese, Jacqueline Millet, Jim Vary, and Alan Warth. Their advice on matters of organization and substance was of great assistance.

Generous financial support was provided by the National Institutes of Health, the National Science Foundation, and the U.S. Army Research Office. In addition, unprecedented support was raised from the following industrial sources: Ajinomoto, AMSCO Medical Products, Bristol-Myers, Campbell Institute for Food Research, Cetus, Chemapac, Dow Chemical, DuPont, Eastman Kodak, Gist-Brocades, Hoffmann-LaRoche, Johnson and Johnson, Merck Sharp & Dohme, Miles Laboratories, New Brunswick Scientific, Novo, Pfizer, Rhône-Poulenc, Shionogi, Smith Kline & French, Squibb, and Sterling-Winthrop Research Institute. We are exceedingly grateful for this support, which not only permitted many participants from outside the United States to attend but also made possible publication of this book.

The editors of this volume are grateful to those who submitted manuscripts. We wish that we had been able to publish all the worthy papers received. Stringent space limitation prevented that and, moreover, forced the authors of accepted manuscripts to assist us by condensing their papers more than they would have liked. We are especially grateful to the authors of invited reviews who undertook difficult tasks with both care and diligence, to the many referees who provided the editors and authors with careful, critical, and helpful analyses of submitted manuscripts, and to Cheryl Cross of the ASM Publications Office who provided much needed editorial management and assistance in preparing this volume.

The Ninth International Spore Conference, under the chairmanship of John Spizizen, Department of Microbiology, University of Arizona School of Medicine, is scheduled for October 1983 in Phoenix, Arizona.

Hillel Levinson
Science and Advanced Technology Laboratory
U.S. Army Natick Research and Development Laboratories
A. L. Sonenshein
Department of Molecular Biology and Microbiology
Tufts University School of Medicine
Donald J. Tipper
Department of Molecular Genetics and Microbiology
University of Massachusetts Medical School

Initiation of Bacterial Sporulation

ERNST FREESE

Laboratory of Molecular Biology, National Institute of Neurological and Communicative Disorders and Stroke, Bethesda, Maryland 20205

INTRODUCTION

Many gram-positive bacilli (including members of the aerobic genus *Bacillus* and the anaerobic genus *Clostridium*) form endospores when they encounter certain adverse nutritional conditions which no longer allow rapid growth. The sporulation development starts when nutritional deprivation causes the cells to divide asymmetrically; the two membranes separating the smaller from the larger (mother) cell grow around the small cell compartment, producing, within the cytoplasm of the mother cell, a forespore enclosed by two membranes with opposite polarities. Because this membrane combination no longer allows normal active transport, the metabolic milieu of the forespore must differ from that of the mother cell so that the synthesis of spore-specific proteins can be derepressed. Whereas the asymmetric septation usually depends on a nutritional signal, the subsequent forespore development depends only on the ge-

netic constitution of the organism. I present here the essential ideas and results concerning the signal initiating sporulation.

EARLY STUDIES ON NUTRITIONAL AND BIOCHEMICAL CHANGES DURING INITIATION OF SPORULATION

Nutritional Conditions Which Allow Sporulation of Wild-Type Strains

As is typical for all differentiating organisms, bacilli do not differentiate as long as they can grow rapidly. In nature, they multiply until they (and competing organisms) have reduced the concentration of one of the several nutrients required for growth sufficiently that the cell recognizes the nutritional deficiency as a signal to start sporulation. Because most nutrients are actively transported by bacteria, their deprivation is recognized only at the last moment so that the remaining nutrient does not suffice for the synthesis of the macromolecules needed for differentiation. Therefore, organisms have evolved special mechanisms that enable them to continue the relatively slow differentiation process. These include (25, 33, 52, 82): (i) intracellular or extracellular accumulation of slowly metabolizable carbon sources, (ii) adaptation to the new environment by turnover of proteins by proteases and of RNA by RNases and phosphorylases, and (iii) stringent response to amino acid deprivation.

In the laboratory, massive sporulation can be initiated by exposing the cells to different starvation conditions: (i) at the end of growth in media in which one or more nutrients eventually run out (19, 22, 66, 71); (ii) by the transfer of cells from a growth medium to a "replacement medium" (46, 53, 75, 76) in which cells, if they would not sporulate, could grow only at a very low rate and only after a long adaptation time; or (iii) by "microcycle sporulation" (70). In the last case spores are germinated in a medium in which the outgrowing cells find conditions that allow them to pass through one round of DNA duplication during which sporulation is again initiated.

When Schaeffer et al. continuously diluted cultures of *B. subtilis* into the same medium (67), they found that cells could sporulate during exponential growth with a probability that increased greatly when the growth rate was low because the nitrogen or carbon source could be only slowly metabolized. They concluded that sporulation is repressed by a carbon- and nitrogen-containing catabolite which might control the synthesis of an enzyme needed for sporulation. Similar results were obtained in chemostat cultures by Dawes and Mandelstam (10).

It seems likely (and is experimentally more productive to assume) that there is ultimately one compound of low molecular weight which, in combination with a protein, suppresses sporulation; according to the above findings, this compound must contain carbon, nitrogen, and phosphorus atoms. Because we do not know whether the asymmetric prespore septation requires any change in transcription, I use "suppression" rather than the word "repression," which has been associated with the control of mRNA synthesis on the DNA level.

Cellular Changes

At the end of exponential growth or after cell transfer to a poor medium, the rate of cell expansion is considerably reduced as a consequence of a reduced rate of RNA and protein synthesis. Membrane synthesis continues, whereas wall synthesis decreases transiently, to resume during cortex formation (14, 23, 52, 65, 82).

Spores contain one or more chromosomes (69). Studies with a DNA initiation mutant of *B. subtilis* have shown that step-down conditions must exist before DNA synthesis has finished a round of replication; otherwise no sporulation ensues (41, 49, 82). This result has been interpreted by two quite different models for the initiation of sporulation, both of which postulate macromolecular changes that are difficult to examine experimentally. One model postulates that a certain gene required to trigger sporulation (i.e., asymmetric septation) can be activated only while it is being duplicated; the cell has to be under step-down conditions for this activation (9, 82). The other model postulates that the origins of the chromosome, which are attached to the membrane, have to move toward the cell poles and initiate on one cell end the growth of an asymmetric septum before duplication of the terminus of the chromosome enables symmetric septation; the slowdown of cell expansion is essential for the proper movement of the origin toward the cell pole and the triggering of the asymmetric septation (23).

ATTEMPTS TO ANALYZE INITIATION OF SPORULATION BY USE OF MUTANTS AND INHIBITORS

In the classical analysis of biochemical pathways, many individual reactions and their corresponding enzymes were identified by the use of mutants or specific inhibitors which affected the function of a specific enzyme. Initially, a similar approach for the identification of biochemical reactions needed for differentiation seemed promising. In fact, by use of mutants it has been shown that certain enzymes are not, or not always, needed for sporulation and that sporulation is prevented by accumulation of acidic compounds. However, these studies have not helped to identify the mechanism whereby sporulation is initiated.

spo0 Mutations

Many mutants unable to sporulate are also unable to produce the asymmetric prespore septa. If the biochemical deficiency of the relevant mutation was not known, the gene was called *spo0* and carefully mapped genetically (36). If the biochemical deficiency was known, the genotype was named accordingly (29); some of these mutations have also been mapped. Most strains containing these biochemically characterized mutations can sporulate under special conditions (23). It may be simply as a result of our ignorance about some of the biochemically uncharacterized *spo0* mutants that no conditions have yet been devised under which they can also sporulate. For details about *spo0* mutations, see references 36 and 52.

Mutants Blocked in Known Biochemical Reactions

When it became clear that it would be difficult to identify the biochemical lesions of *spo0* mutations, other mutations were isolated whose enzymatic or other protein lesion could be identified. Many of these mutations interfere with sporulation, at least under certain conditions.

The developmental effects of auxotrophic mutations of *B. subtilis* could be divided into the following five types. (i) The nutrient (e.g., leucine or biotin) required for growth does not have to be present during development because it is provided by turnover or accumulated during vegetative growth (13, 28). (ii) The nutrient is required during development, sometimes in larger concentrations than during growth (nicotinic acid, thiamine); apparently, certain enzymes (pyruvate dehydrogenase) are needed at higher activity for sporulation than for growth (28). (iii) For some mutants the nutrient has to be provided at a rather precise intracellular concentration (for glycerol and glucosamine auxotrophs). (iv) Mutants blocked at different steps of the Embden-Meyerhof pathway accumulate phosphorylated intermediates when even small amounts of a carbohydrate are present and then do not sporulate (28). (v) Mutants blocked in the citrate cycle or in phosphoenolpyruvate carboxykinase cannot maintain ATP or gluconeogenesis if sporulation is initiated by carbon deprivation; the citrate cycle mutants cannot sporulate even if ATP regeneration is enabled by addition of the proper reducing compounds, because acidic intermediates accumulate (28). However, under other conditions these mutants can sporulate (31).

As was shown by the introduction of mutations, certain enzymes are not necessary for sporulation (amylase and metal-dependent protease [65]; phosphoglucomutase controlling the production and utilization of carbohydrate polymers [24]; fructose bis-phosphatase, which can be circumvented by a bypass allowing glucosamine synthesis during gluconeogenesis [32]).

Some mutants resistant to inhibitors of RNA synthesis (rifampin, streptolydigin, streptovaricin) are altered in RNA polymerase and are unable to sporulate at 45°C; they sporulate only at lower temperatures (33°C) (73). As the genetic analysis of these mutants is limited, it is not always clear whether the alteration of RNA polymerase itself or a secondary mutation is responsible for the sporulation deficiency. This is not a problem for mutants temperature sensitive in both RNA polymerase and sporulation; they are blocked in the middle and late stages of sporulation (44). Although an alteration of RNA polymerase (new σ-factor) apparently is required for the transcription of a sporulation-specific gene (34), there is no evidence that it is needed for asymmetric septation.

Many mutants selected for resistance to antibiotics that bind to ribosomes are unable to sporulate at all, are unable to sporulate in the presence of the antibiotics, or are temperature sensitive in sporulation (7, 73). The relevant mutations mostly map in the cluster of ribosomal genes, suggesting that they are altered in a ribosomal constituent; but for only one temperature-sensitive erythromycin-resistant mutant has the alteration of a ribosomal protein been demonstrated (78). The results have generated the impression that certain ribosomal structures are specifically needed for the translation of certain sporulation messages. However, it is just as plausible that a slight change of a ribosomal structure affects growth only minimally and yet interferes with the delicate metabolic balance required for the initiation of sporulation, e.g., under sporulation conditions that depend on the stringent response (which is also associated with ribosomes) (see below).

Chemicals Interfering Predominantly with Sporulation

At low concentrations, many compounds inhibit massive sporulation without significantly affecting the rate or extent of vegetative growth. For most compounds the reasons for this effect are unknown. Some of them may interfere with transcription because they bind to DNA (Netropsin, tylorone) (40, 59) or intercalate into DNA (acridines, ethidium bromide, phenothiazines) (6, 62). Ethanol prevents sporulation, perhaps by altering membrane function (4). α-Picolinate and other chelators of iron (21) prevent activation of the iron-dependent aconitase.

POSSIBLE ROLE OF COMPOUNDS MADE EARLY IN SPORULATION

Although the usual sporulation conditions suggested that the concentration of a suppressor compound had to decrease for sporulation to be initiated, the paradigms of virus infection and enzyme induction enticed many investigators of sporulation to look for the appearance of new compounds that might cause the initiation of sporulation. I will discuss only small-molecular-weight compounds; the changes of RNA polymerase, tRNA's, ribosomes, and degradative enzymes have been covered before (34, 73, 82) or are analyzed by R. Losick in this volume.

Miscellaneous Compounds

A compound which is made as soon as sporulation of *B. megaterium* begins but which has not been proposed as being important for the initiation of sporulation is L(+)-*N*-succinylglutamate (2). Srinivasan and Halvorson reported that a compound (with spectral properties of a nucleoside) produced by *B. megaterium* could initiate sporulation of this organism (74). This compound may have been inosine, which has a very similar UV spectrum (Hageman, personal communication).

Nucleotides

Cyclic AMP is undetectable (less than 1 nM) during growth and sporulation of *B. megaterium* and *B. licheniformis* (3, 68). The concentration of cyclic GMP has been measured in *B. licheniformis*, where it is very low (6 nM) during exponential growth in a glucose medium and decreases 10-fold toward the end of growth (3). No information is available about the role of this compound in sporulation.

The highly phosphorylated guanine nucleotides ppGpp and pppGpp are typically made by bacteria when one or more types of tRNA molecules are uncharged (33). This increase is part of the stringent response, which also includes a reduction in the synthesis of RNA and other macromolecules. The initiation of sporulation often involves a step-down to nutrient limitation with a corresponding increase of (p)ppGpp (50, 56; J. M. Lopez, K. Ochi, and E. Freese, this volume). But the stringent response is not necessary for sporulation because relaxed (*relA*) mutants, which do not produce (p)ppGpp and the stringent response, can still sporulate in nutrient sporulation medium (50, 56). Nevertheless, for several frequently used sporulation conditions, stringent response is essential (see below).

Rhaese and co-workers have observed several highly phosphorylated adenine nucleotides which appeared after the end of exponential growth in a nutrient sporulation medium containing 0.1% glucose or after cell transfer from a glucose medium to one containing no glucose (56, 57, 59). Two of these compounds (ppApp and pppApp) were made only by ribosomes of sporulating cells but not by those of *spo0* mutants (59). The authors also claimed that ribosomes from sporulating cells could no longer produce (p)ppGpp; however, this was not confirmed by Smith et al. (72). Rhaese et al. postulated that another compound (HPNIV, tentatively identified as pppAppp) is required for sporulation (57, 59). The enzyme(s) producing pppAppp was membrane bound and present during growth, but it started producing pppAppp only after the cells were in a medium containing no sugar or phosphate. Whereas most *spo0* mutants produced pppAppp after cell transfer to a sugar-free medium, a *spo0* mutant produced less than one-tenth the normal amount and a *spo0E* mutant produced a reduced amount (59). The compound was also not made in mutants in which a sugar-phosphate accumulated, a condition which prevents sporulation. Whereas all this information could have been attributed to a pleiotropic effect of an early block in sporulation, the claim that a temperature-sensitive *spo0F* mutant was also temperature sensitive in the enzyme activity needed for pppAppp synthesis suggested that the *spo0F* gene may code for one of the two enzymes responsible for the synthesis of pppAppp (58). Several other groups have not detected the compound under their sporulation conditions (nutrient sporulation medium without glucose [50, 72]; less than 2 μM intracellular concentration during initiation of sporulation by decoyinine in synthetic medium [J. M. Lopez and E. Freese, unpublished data]). In nutrient sporulation medium, Nishino et al., using conditions under which they obtained 30% spores and labeling with $^{32}PO_4$ at a specific activity of about 300,000 cpm/nmol of PO_4, stated that they could have detected an intracellular concentration of about 3 μM pppAppp (50). This would correspond to about 1,500 molecules per cell. However, Rhaese et al. had added carrier-free $^{32}PO_4$ to their sporulation media, providing a high specific activity; they did not state how much pppAppp was produced under their conditions. Because pppAppp seems to be made only under certain conditions, its synthesis probably is not absolutely required for sporulation. Rhaese et al. proposed a "sensor model" according to which the pppAppp-synthesizing enzyme(s) may be inhibited by sugar-phosphates produced via the phosphoenolpyruvate:sugar-phosphoryl transferase system. When the sugar-phosphates run out, they argued, the enzyme may become active and phosphorylate ATP (58). This would limit the role of pppAppp to the exhaustion of sugars or phosphate and would not explain sporulation caused by nitrogen starvation or other conditions (GTP deprivation) discussed below.

Antibiotics Produced by Sporulating Cells

Toward the end of exponential growth in a sporulation medium, many (including nonsporulating) bacteria produce one or more antibiotics which not only inhibit other bacteria but usually also inhibit exponentially growing cells of the producer organism (39). Antibiotic production may be useful to the organism, under certain sporulation conditions, by slowing down the rate at which intra- or extracellular metabolites remaining toward the end of growth are used up; these metabolites can later be used for the synthesis of sporulation-specific components. In several organisms, spo0 mutants also fail to produce the antibiotic. In most cases this is due to a pleiotropic effect of the mutation which prevents most reactions that normally occur after the end of exponential growth. Certainly not all antibiotics are needed for sporulation because mutants of B. brevis unable to produce gramicidin S are still able to sporulate (12, 38). Nevertheless, some antibiotic may be needed for the initiation of sporulation. This possibility was advanced in particular for a strain of B. brevis which begins production of the tyrothricin complex (consisting of tyrocidine and the linear gramicidin D) at the end of growth (61, 64). The investigators observed that these lipophilic antibiotics not only affected the cell membrane but also inhibited transcription (tyrocidine by complexing with DNA), an effect which they regarded as essential for the initiation of sporulation. However, the decrease in the accumulation of RNA at the end of growth was much more drastic than the inhibition (less than 20%) which the small amount of tyrothricin present at that time could have produced; presumably, the decrease of RNA accumulation was not exclusively controlled by tyrothricin. Recently, Ristow et al. found that tyrothricin addition greatly increased the titer of spores produced by cells that had been transferred during exponential growth to a medium lacking all nitrogen sources. They argued that this effect may be due to the inhibition of transcription by tyrothricin: tyrocidine may be a nonspecific inhibitor of transcription, and gramicidin D, which reduces this inhibition, may specifically enable transcription of that part of the genome which is necessary for sporulation; they did show that tyrocidine or gramicidin D alone increased the sporulation frequency less effectively (60). A different interpretation of

their results is suggested by the experiments reported below, which demonstrate that cells must maintain an intermediate rate of macromolecular synthesis during the developmental period; if they are completely starved for nitrogen (or carbon) sources, only few cells sporulate. It is conceivable that tyrothricin reduces the rate at which the remaining intracellular nitrogen sources are consumed and thus reduces the rate of nucleotide synthesis and simultaneously enables the cell to synthesize RNA and other polymers for a longer time.

MUTATIONS AND INHIBITORS ENABLING SPORULATION IN THE PRESENCE OF EXCESS NUTRIENTS

If a signal compound of small molecular weight is needed to suppress sporulation, one should be able to identify the biochemical pathway in which it is made by demonstrating that specific inhibitors or mutants allow sporulation in a medium which contains an excess of nutrients so that the standard strain without added inhibition does not sporulate. Depending on the sporulation conditions, one can expect different results.

1. The medium is not changed. A high frequency of spores (per viable cell) may be observed 8 or more h after the end of growth because the mutant cells first grow to a high vegetative titer and then start massive sporulation, either because they have accumulated a compound which enables sporulation or because the concentration of a compound (different from the major carbon and nitrogen source) has decreased below a critical value (e.g., if a mutant enzyme or transport mechanism has a high K_m for a particular component of the medium). Alternatively, the mutant cells may continuously enter the sporulation development (asymmetric septation) at a probability which is much higher than for the standard strain. The frequency of spores will continuously increase because the sporulation process takes 6 to 8 h before heat-resistant cells can be detected; furthermore, the spore titer will continue to increase after the culture has stopped growth. These different possibilities have usually not been distinguished.

2. Sporulation is initiated by addition of an inhibitor or by transfer of auxotrophic cells from a growth medium to one containing growth-limiting amounts of the relevant substrate (the medium is so chosen that addition of an excess of this substrate prevents sporulation).

I will now enumerate the sporulation properties of various mutants and inhibitors used under these conditions.

Miscellaneous Observations

In a study on the regulation of extracellular protease in *B. cereus* T, Levisohn and Aronson isolated mutants able to sporulate in the presence of excess amino acids or glucose. They found auxotrophs for purines, pyrimidines, and lysine, as well as several mutants with no apparent nutritional requirement (45). Although these mutants also produced high activities of extracellular protease, Aronson et al. later found that other purine auxotrophs produced little extracellular protease but still sporulated (1).

Kunst et al. reported that *sacU*^{h} mutants of *B. subtilis*, which are hyperproducers of levansucrase and extracellular protease and which lack flagella, sporulated in the presence of excess glucose and high concentrations of several amino acids (43). Takahashi also isolated a number of *B. subtilis* mutants which sporulated in media containing glutamate and a carbohydrate at concentrations which prevented sporulation of the standard strain (77). It has not been reported whether the above mutants sporulate continuously or only after the end of growth. We have isolated several mutants of *B. subtilis* which sporulate continuously in medium containing excess glucose but not if the medium also contains carbon compounds that feed into the citrate cycle (aspartate, malate, glutamate, etc.) (Endo, Ishihara, and Freese, in preparation). The biochemical reason for the sporulation of any of the above mutants is unknown.

Sacks and Thompson (63) found that the sporulation frequency of the anaerobic *C. perfringens* in a complex medium containing dextrin (but not with glucose) can be greatly increased by addition of methylxanthines (caffeine, 3-isobutyl-1-methylxanthine or theophylline) or certain nucleosides (inosine). They mentioned the iron-binding properties of some methylxanthines as a possible cause. Because methylxanthines are also known to inhibit purine nucleotide synthesis, it seems more likely that the active compounds initiate sporulation by the same mechanism as that discussed below for *B. subtilis*. In fact, several inhibitors of purine synthesis (such as decoyinine), which were found to initiate sporulation of *B. subtilis* (see below), also initiated sporulation of *C. perfringens* (L. E. Sacks, Abstr. Am. Soc. Microbiol. 1980, I128, p. 105).

Pandey and Gollakota (51) observed that 100 μM amethopterin seems to induce microcycle sporulation of *B. brevis*. Since amethopterin mainly inhibited dTMP synthesis, they concluded that the inhibition of DNA synthesis was responsible for the sporulation induction. However, they did not attempt to counteract this induction by adding the different methylated compounds (including thymine and purines) whose synthesis is inhibited by amethopterin; therefore, deprivation of any of these compounds could be responsible for the sporulation.

Sporulation of Glutamine Auxotrophs

Elmerich and Aubert found that one particular glutamine (Gln) auxotroph of *B. megaterium* sporulated upon removal of glutamine in excess glucose and NH_4^+ (16), whereas other Gln auxotrophs, also deficient in glutamine synthetase, did not (54). Using this one Gln mutant as parent, they isolated many additional mutations that enabled sporulation in the presence of glutamine; these mutations were either blocked before 5'-phosphoribosyl-5-amidoimidazole (AIR) in the purine pathway or they were auxotrophic for some amino acid, in which case they sporulated less efficiently; mutants blocked beyond AIR did not sporulate after purine removal (16). Because only early blocked purine auxotrophs sporulated in the presence of glutamine, these investigators assumed that the repressive effect of glutamine was mediated by an effector made in one of the early steps of purine nucleotide synthesis, between phosphoribosyl-pyrophosphate and AIR, where glutamine intervenes twice as an amino donor. To explain why only one glutamine synthetase mutant sporulated after removal of glutamine whereas other Gln mutants sporulated badly even upon carbon deprivation in the presence of glutamine, Reysset et al. postulated that glutamine synthetase may play, in the control of sporulation, a regulatory role independent of its catalytic activity (55). Mutations of glutamine synthetase might impair the catalytic site, the regulatory site, or both sites. Therefore, glutamine synthetase might be the "receptor" of the "effector" made during purine deprivation in the early purine path, the combination of the two inducing sporulation. However, the authors could not explain why in a Gln Pur double mutant sporulation could be prevented only when both adenine and glutamine were present (17). The results obtained with *B. subtilis* (presented below) suggest that the relevant deficiency may have been the lack of guanine nucleotides; these compounds can be synthesized in a Gln Pur mutant only if both adenine and glutamine (which is required by the GMP synthetase reaction) are available.

Bott et al. (5) characterized two Gln auxotrophs of *B. subtilis*; one had apparently normal glutamine synthetase (reason for auxotrophy unknown) and sporulated normally, whereas the other had a reduced specific activity of the enzyme and sporulated badly in nutrient sporulation medium, regardless of the concentration of glutamine added. However, S. H. Fisher (personal communication) has shown that this mutant contained an additional sporulation mutation which could be genetically separated from the Gln⁻ phenotype. Dean et al. (11) and Fisher and Sonenshein (18) have found that all their Gln mutants of *B. subtilis* could sporulate in the presence of glutamine. Some of Fisher and Sonenshein's mutants failed to sporulate if the glutamine concentration in their resuspension medium was 0.005% or less. None of the results with Gln mutants of *B. subtilis* indicate a direct role of glutamine synthetase in the control of sporulation (they also do not disprove such a possibility). Recent studies by H. J. Schreier et al. (this volume) have shown that there is no correlation between sporulation and the concentrations of NH_4^+, glutamine, glutamate, or 2-oxoglutarate or the levels of the enzymes glutamate synthase, glutamine synthetase, or glutamate dehydrogenase.

Importance of a Partial Decrease in the Synthesis of Purine Nucleotides

For several years, we had tried in vain to initiate massive sporulation of *B. subtilis*, in the presence of glucose, NH_4^+ and HPO_4^-, by starving many auxotrophs and by adding many inhibitors. We succeeded only when we realized that it was essential to inhibit the synthesis of certain metabolites partially but not completely (27). The intracellular concentration of a compound required by an auxotroph usually cannot be controlled by the concentration of a nutrient in the medium because almost all natural nutrients are actively transported into bacteria with transport K_m values in the micromolar range. Cells grow on the nutrient at the maximal rate until its concentration decreases to the K_m range, whereupon the remaining amount is used up so rapidly that the cells, although their sporulation may no longer be suppressed, cannot sporulate because they are unable to synthesize the macromolecules required for differentiation. This problem, which arose as a result of the evolution of transport mechanisms, has been circumvented by nature in many ways as a result of the evolution of differentiation; the experimenter has to find his own tricks. The dilemma became apparent when we measured sporulation of purine auxotrophs 10 h after the cells had been transferred from a medium containing excess glucose, NH_4^+, HPO_4^-, and hypoxanthine to the same medium without hypoxanthine. Different mutants sporulated at different frequencies which were not correlated with the location of the mutational block in the purine pathway but which did depend on the leakiness of the mutation, as measured by the residual growth observed after purine removal (26, 27). We demonstrated the importance of incomplete inhibition of purine synthesis in a relatively nonleaky mutant blocked before 5'-phosphoribosyl-5-amino-4-imidazole-carboxamide (5'-phosphoribosyl-AICA). Like other purine mutants, the strain had a low K_m for the transport of purines and purine nucleosides, grew rapidly until these compounds were exhausted, and produced only few spores subsequently. In contrast, AICA was metabolized only slowly so that the mutant cells grew at different rates after transfer to media containing different AICA concentrations. They sporulated well at intermediate AICA concentrations (0.4 mM); at high AICA concentrations the cells simply grew without sporulating; without AICA sporulation was greatly reduced, presumably because the synthesis of compounds needed for sporulation was too much reduced (26). In similar experiments we showed that amethopterin induced sporulation of the standard strain when it caused partial purine deprivation but not when it caused thymine deprivation alone. Sporulation could also be induced by the addition of known inhibitors of purine synthesis or of purine analogs, whose detailed mode of action is still unknown. Again it was important to use the inhibitors at concentrations causing only partial growth inhibition.

Many inhibitors of pyrimidine nucleotide, DNA, RNA, and protein synthesis, used at different concentrations, failed to initiate sporulation (27, 35, 48; Zain, Lopez, and Freese, submitted for publication). (6-Azauracil was an exception [27]; however, it not only affected uracil nucleotides but also caused a decrease of GTP [46].) A uracil auxotroph also was unable to sporulate regardless of the rate of UMP synthesis, which could be controlled (as with AICA above) by the extracellular concentration of orotic acid, a slowly metabolizable precursor (this was true even when cytidine and thymidine derivatives were directly derived from extracellular cytidine and thymidine) (46; N. Vasantha and E. Freese, unpublished data). Amino acid auxotrophs are discussed in Initiation of Sporulation by the Stringent Response.

Because the synthesis of metabolites must be only partially inhibited, it is understandable that many investigators did not observe sporulation of auxotrophs that had been isolated to grow only in the presence of a supplement: they were probably too tightly blocked. In contrast, auxotrophs found among the mutants isolated as being able to sporulate (in a medium containing an excess of other nutrients) presumably were sufficiently leaky. This may explain why in *B. megaterium* only mutants blocked before AIR sporulated after purine removal (16); as the medium used for the isolation of these mutants did not contain thiamine, synthesis of which requires AIR, these early mutants had to be leaky to be able to grow with a purine but without additional thiamine.

Importance of the Deprivation of Guanine Nucleotides

To determine whether adenine or guanine nucleotides were responsible for the initiation of sporulation, we used different inhibitors and mutants of *B. subtilis* which specifically block only one of the two branches of purine nucleotide synthesis; we measured both sporulation and the changes in the intracellular concentration of nucleoside di- and triphosphates (26, 27, 46). Hadacidin, an inhibitor of adenylosuccinate synthetase, induced some sporulation at an intermediate concentration, and it caused a drastic decrease of ATP; but as the synthesis of all nucleotides depends on ATP, it also caused a decrease in all other nucleoside di- and triphosphates. In contrast, decoyinine, an inhibitor of GMP synthetase, induced 20 to 50% sporulation at some partially inhibitory concentration and caused a decrease of GTP but an increase of ATP. The "adenylate energy charge" did not change, as had also been observed under other sporulation conditions in *B. licheniformis* (15). This invalidates a proposal that the energy charge might control sporulation (37). Psicofuranine, another inhibitor of GMP synthetase (Zain, Lopez, and Freese, submitted for publication), and, less effectively, mycophenolic acid, an inhibitor of IMP dehydrogenase (29), also induced sporulation and caused a decrease of GTP but not ATP. Whereas adenine auxotrophs did not sporulate after adenine removal, guanine auxotrophs sporulated well and also showed the preferential decrease of GTP (26, 46). These results demonstrate that partial inhibition of the synthesis of guanine nucleotides suffices to initiate excellent sporulation.

Nucleotide changes were also measured under many conditions in which *B. subtilis* had previously been observed to sporulate. These included nutrient sporulation medium (22), transfer from Casamino Acids/glutamate to glutamate medium (75), transfer from glucose to lactate (53), nitrogen or phosphate limitation, etc. Under all these conditions, the concentrations of GDP and GTP decreased whereas those of the other nucleoside di- and triphosphates increased in some and decreased in other cases (45a, 46). When a GMP reductase mutant was used under sporulating conditions, GMP increased whereas GDP and GTP decreased (T. Endo and E. Freese, in preparation). These results indicated that the decrease of GDP and GTP occurred under all conditions in which sporulation was initiated by any nutritional deprivation.

Initiation of Sporulation by the Stringent Response

When amino acid auxotrophs were used for which the intracellular concentration of a particular amino acid could be regulated by the extracellular supply of an appropriate precursor, good (up to 10%) sporulation was observed at an intermediate precursor concentration (30; J. M. Lopez, K. Ochi, and E. Freese, this volume; Ochi, Kandala, and Freese, J. Biol. Chem., in press). The precursors had to be chosen so that they avoided the active transport mechanisms of the cell.

Because these as well as several of the sporulation conditions mentioned earlier involved the sudden decrease of one or more amino acids in the medium, they evoked the stringent response. This response includes the synthesis of (p)ppGpp and the inhibition of IMP dehydrogenase, resulting in a decrease of guanine nucleotides (45a). Sporulation observed under such conditions was produced by the stringent response because it was absent in mutants containing a relaxed (*relA*) mutation which abolishes the stringent response. In these *relA* mutants (p)ppGpp increased only little and GTP decreased only slowly to a level similar to that eventually reached in the stringent strain after it had adapted to the new medium (Lopez, Ochi, and Freese, this volume). This final level of GTP did not suffice to initiate sporulation; after adaptation, even cells of the stringent strain that had not yet entered sporulation continued to grow without further sporulation (30; Ochi et al., in press). Nevertheless, the relaxed strains could sporulate when the GTP decrease was further enhanced by decoyinine addition or by guanine starvation of a guanine auxotroph. One could argue that for sporulation resulting from the stringent response the increase of (p)ppGpp was more important than the decrease of GTP. This was invalidated by the isolation of mycophenolic acid-resistant mutants in which (p)ppGpp still increased while GTP decreased only as in the relaxed strain; this mutant no longer sporulated

TABLE 1. *Comparison of cellular responses under different sporulation conditions*

Condition	Sporulation initiated by:			References
	End of growth in nutrient medium	Stringent response	Direct deprivation of guanosine nucleotide synthesis	
Decrease of GTP	+[a]	+	+	45a, 46; Lopez et al., this volume; Ochi et al., in press
Sporulation of mutants blocked in citrate cycle or in gluconeogenesis	−	−	+	31, 83
Sporulation without Mn^{2+}	−	−	+	80
Induction of certain catabolite repressible enzymes	+		−	47
Development of transformability by DNA	+	+	−	30
Inhibition of sporulation by fusidic acid and kasugamycin	+	+	−	20, 46
Increase of intra- and extracellular protease	+	+	(+)	81, 82
Development of high protein turnover	+	+	+	8, 81
Decrease of RNA and protein synthesis	+	+	+	82; Vasantha and Freese, in preparation
Increase of citrate synthase and aconitase	+	+	+	83; Uratani-Wong, Lopez, and Freese, submitted for publication
Production of alkaline phosphatase	+	+	+	81, 82
Production of glucose dehydrogenase	+	+	+	81, 82

[a] +, The event occurs; −, the event does not occur; (+), comparatively little increase.

upon amino acid deprivation (Ochi, Kandala, and Freese, in preparation).

The stringent response could also be produced by the step-down from a rapidly metabolizable to a slowly metabolizable carbon source. Transfer from a medium containing glucose to one containing lactate as sole carbon source initiated sporulation (53) mainly via the stringent response, as shown by the greatly reduced sporulation of a *relA* mutant (45a).

These results demonstrate that changes in media leading to a sudden stringent response initiate sporulation of *B. subtilis* and maybe of other microorganisms. However, sporulation does not depend on this control mechanism. It can also be initiated by other starvation conditions that cause a decrease of GTP without an increase of (p)ppGpp.

Conclusions

(i) The specific and partial inhibition of the synthesis of guanine nucleotides initiates sporulation in the presence of excess NH_4^+, rapidly metabolizable carbon sources, and HPO_4^-. (ii) All nutritional conditions examined so far which cause the initiation of sporulation also result in a decrease of GTP (and GDP), whereas the other nucleotides increase in some and decrease in other cases. However, it is not known whether this GTP decrease is required in all these cases for the initiation of sporulation. (iii) GTP may suppress sporulation directly, in combination with a protein, or indirectly by enabling the synthesis of another metabolite.

COMPARISON OF CELLULAR CHANGES UNDER DIFFERENT CONDITIONS INITIATING SPORULATION

In the beginning of studies on endospore formation, cellular changes were determined only in certain sporulation media, such as the nutrient sporulation media used for *B. subtilis* (22, 66). It was assumed that the changes, observed at the end of growth, were correlated with or even necessary for sporulation. When certain mutations or inhibitors (antibiotics) prevented sporulation under a particular sporulation condition, it was again initially assumed that the molecules, affected by mutation or inhibitor, were always needed in normal form for sporulation. Although exceptions were occasionally reported, this rule remained acceptable because other sporulation conditions, such as the initiation of sporulation in *B. subtilis* by the transfer from Casamino Acids/glutamate to glutamate medium (75), produced essentially the same results (82). However, when we initiated sporulation by decreasing the synthesis of guanine nucleotides (using decoyinine or a guanine auxotroph) in the presence of a rapidly metabolizable carbon source (glucose), we no longer found some of the correlations usually observed. A comparison of different sporulation conditions can therefore help to identify cellular processes that are needed under certain but not all sporulation conditions. Such results are summarized in Table 1. For example, citrate cycle mutants which do not sporulate after a carbon step-down sporulate well after initiation by decoyinine in the presence of glucose (31). The same is true for sporulation in medium lacking manganese (80). In contrast, certain catabolite-repressible enzymes which can be induced after carbon step-down cannot be induced when cells sporulate in the presence of glucose (47, 77).

Certain inhibitors (antibiotics) inhibit sporulation more potently than growth. The previous explanation was that the macromolecule inhibited by an antibiotic can no longer perform its sporulation-specific function or change, so that certain sporulation genes cannot be properly transcribed or translated (7, 20, 42, 52, 73, 78, 79). However, it is also possible that the antibiotic action depended on the particular sporulation condition used. This was the case for the major effects of fusidic acid and kasugamycin when they were added to a glutamate replacement medium together with cells grown in a Casamino Acids/glutamate medium (Lopez, Ochi, and Freese, this volume). If decoyinine was added at the right intermediate concentration together with the antibiotic, the inhibitory effect on sporulation was almost abolished (Lopez, Ochi, and Freese, this volume). Apparently, these antibiotics decreased the stringent response [i.e., the increase of (p)ppGpp and the decrease of GTP] observed after the cell transfer and thereby prevented initiation of sporulation. Because the antibiotics bind to ribosomes, they may reduce the stringent response either by interfering with the function of the stringent factor or indirectly by decreasing the rate of protein synthesis so that less tRNA becomes uncharged.

A number of the reactions in Table 1 are observed under all presently known sporulation conditions. Some of them, such as the decrease in the accumulation of RNA and protein, are probably necessary for sporulation; other reactions, such as the increase in citrate synthase

and aconitase, are not necessary, as was shown by the sporulation of citrate cycle mutants.

LITERATURE CITED

1. **Aronson, A. I., N. Angelo, and S. C. Holt.** 1971. Regulation of extracellular protease production in *Bacillus cereus* T: characterization of mutants producing altered amounts of protease. J. Bacteriol. **106**:1016–1025.
2. **Aubert, J.-P., J. Millet, E. Pineau, and G. Milhaud.** 1961. L'acide N-succinyl-L(+)glutamique chez *Bacillus megaterium* en voie de sporulation. Biochim. Biophys. Acta **51**:529–537.
3. **Bernlohr, R. W., M. K. Haddox, and N. D. Goldberg.** 1974. Cyclic guanosine 3':5'-monophosphate in *Escherichia coli* and *Bacillus licheniformis*. J. Biol. Chem. **249**:4329–4331.
4. **Bohin, J. P., D. Rigomier, and P. Schaeffer.** 1976. Ethanol sensitivity of sporulation in *Bacillus subtilis*: a new tool for the analysis of the sporulation process. J. Bacteriol. **127**:934–940.
5. **Bott, K. F., G. Reysset, J. Gregoire, D. Islert, and J.-P. Aubert.** 1977. Characterization of glutamine requiring mutants of *Bacillus subtilis*. Biochem. Biophys. Res. Commun. **79**:996–1003.
6. **Burke, W. F., and J. Spizizen.** 1977. Selective inhibition of *Bacillus subtilis* sporulation by acridine orange and promethazine. J. Bacteriol. **129**:1215–1221.
7. **Cannon, J. G., and K. F. Bott.** 1979. Spectinomycin-resistant mutants of *Bacillus subtilis* with altered sporulation properties. Mol. Gen. Genet. **174**:149–162.
8. **Cheng, Y. E., and A. I. Aronson.** 1977. Alterations of spore coat processing and protein turnover in a *Bacillus cereus* mutant with a defective post exponential intracellular protease. Proc. Natl. Acad. Sci. U.S.A. **74**:1254–1258.
9. **Dawes, I. W., D. Kay, and J. Mandelstam.** 1971. Determining effect of growth medium on the shape and position of daughter chromosomes and on sporulation in *Bacillus subtilis*. Nature (London) **230**:567–569.
10. **Dawes, I. W., and J. Mandelstam.** 1970. Sporulation of *Bacillus subtilis* in continuous culture. J. Bacteriol. **103**:529–535.
11. **Dean, D. R., J. A. Hoch, and A. I. Aronson.** 1977. Alteration of the *Bacillus subtilis* glutamine synthetase results in overproduction of the enzyme. J. Bacteriol. **131**:981–987.
12. **Demain, A. L., J. M. Piret, T. E. Friebel, E. J. Vandamme, and C. C. Matteo.** 1976. Studies on *Bacillus brevis* directed towards the cell-free synthesis of gramicidin S, p. 437–443. *In* D. Schlessinger (ed.), Microbiology—1976. American Society for Microbiology, Washington, D.C.
13. **Doering, J. L., and K. F. Bott.** 1972. Differential amino acid requirements for sporulation in *Bacillus subtilis*. J. Bacteriol. **112**:345–355.
14. **Doi, R.** 1976. Changes in nucleic acids during sporulation, p. 125–166. *In* G. W. Gould and A. Hearst (ed.), The bacterial spore. Academic Press, Inc., New York.
15. **Donohue, T. J., and R. W. Bernlohr.** 1978. Effect of cultural conditions on the concentrations of metabolic intermediates during growth and sporulation of *Bacillus licheniformis*. J. Bacteriol. **135**:363–372.
16. **Elmerich, C., and J.-P. Aubert.** 1973. Involvement of the biosynthetic pathway of purine nucleotides in the repression of bacterial sporulation. Biochem. Biophys. Res. Commun. **55**:837–842.
17. **Elmerich, C., and J.-P. Aubert.** 1975. Involvement of glutamine synthetase and the purine nucleotide pathway in repression of bacterial sporulation, p. 385–390. *In* P. Gerhardt, R. N. Costilow, and H. L. Sadoff (ed.), Spores VI. American Society for Microbiology, Washington, D.C.

18. **Fisher, S. H., and A. L. Sonenshein.** 1977. Glutamine-requiring mutants of *Bacillus subtilis.* Biochem. Biophys. Res. Commun. **79:**987–995.

19. **Fitz-James, P. C.** 1965. Spore formation in wild and mutant strains of *B. cereus* and some effects of inhibitors, p. 529–544. *In* M. J. C. Senez (ed.), Regulations chez les microorganismes. Centre National de la Recherche Scientifique, Paris.

20. **Fortnagel, P., and R. Bergmann.** 1973. Alteration of the ribosomal fraction of *Bacillus subtilis* during sporulation. Biochim. Biophys. Acta **299:**136–141.

21. **Fortnagel, P., and E. Freese.** 1968. Inhibition of aconitase by chelation of transition metals causing inhibition of sporulation in *Bacillus subtilis.* J. Biol. Chem. **243:** 5289–5295.

22. **Fortnagel, P., and E. Freese.** 1968. Analysis of sporulation mutants. II. Mutants blocked in the citric acid cycle. J. Bacteriol. **95:**1431–1438.

23. **Freese, E.** 1976. Metabolic control of sporulation, p. 1–32. *In* A. N. Barker, G. W. Gould, and G. Wolf (ed.), Spore research 1976. Academic Press, London.

24. **Freese, E., P. Fortnagel, R. Schmitt, W. Klofat, E. Chappelle, and G. Picciolo.** 1969. Biochemical genetics of initial sporulation stages, p. 82–101. *In* L. L. Campbell (ed.), Spores IV. American Society for Microbiology, Bethesda, Md.

25. **Freese, E., and Y. Fujita.** 1976. Control of enzyme synthesis during growth and sporulation, p. 164–184. *In* D. Schlessinger (ed.), Microbiology—1976. American Society for Microbiology, Washington, D.C.

26. **Freese, E., J. E. Heinze, and E. M. Galliers.** 1979. Partial purine deprivation causes sporulation of *Bacillus subtilis* in the presence of excess ammonia, glucose and phosphate. J. Gen. Microbiol. **115:**193–205.

27. **Freese, E., J. Heinze, T. Mitani, and E. B. Freese.** 1978. Limitation of nucleotides induces sporulation, p. 277–285. *In* G. Chambliss and J. C. Vary (ed.), Spores VII. American Society for Microbiology, Washington, D.C.

28. **Freese, E., T. Ichikawa, Y. K. Oh, E. B. Freese, and C. Prasad.** 1974. Deficiencies or excesses of metabolites interfering with differentiation (vitamin deficiency, glycerol-phosphate dehydrogenase, glucose-6-phosphate/glycerol phosphate, etc.). Proc. Natl. Acad. Sci. U.S.A. **71:**4188–4193.

29. **Freese, E., J. M. Lopez, and E. B. Freese.** 1979. Initiation of bacterial and yeast sporulation by partial deprivation of guanine nucleotides, p. 127–143. *In* P. Richter and G. Koch (ed.), Regulation of macromolecular synthesis by low molecular weight mediators. Academic Press, Inc., New York.

30. **Freese, E., J. M. Lopez, and K. Ochi.** 1981. Role of guanine nucleotides and of the stringent response to amino acid deprivation in the initiation of bacterial sporulation, p. 11–16. *In* D. Schlessinger (ed.), Microbiology—1981. American Society for Microbiology, Washington, D.C.

31. **Freese, E. B., N. Vasantha, and E. Freese.** 1979. Induction of sporulation in developmental mutants of *Bacillus subtilis.* Mol. Gen. Genet. **170:**67–74.

32. **Fujita, Y., and E. Freese.** 1981. Isolation and properties of a *Bacillus subtilis* mutant unable to produce fructose-bisphosphatase. J. Bacteriol. **145:**760–767.

33. **Gallant, J. A.** 1979. Stringent control in *E. coli.* Annu. Rev. Genet. **13:**393–415.

34. **Haldenwang, W. G., and R. Losick.** 1979. A modified RNA polymerase transcribes a cloned gene under sporulation control in *Bacillus subtilis.* Nature (London) **282:**256–260.

35. **Heinze, J. E., T. Mitani, K. E. Rich, and E. Freese.** 1978. Induction of sporulation by inhibitory purines and related compounds. Biochim. Biophys. Acta **521:**16–26.

36. **Henner, D. J., and J. Hoch.** 1980. The *Bacillus subtilis* chromosome. Microbiol. Rev. **44:**57–82.

37. **Hutchison, K. W., and R. S. Hanson.** 1974. Adenine nucleotide changes associated with the initiation of sporulation in *Bacillus subtilis.* J. Bacteriol. **119:**70–75.

38. **Kambe, M., Y. Imae, and K. Kurahashi.** 1974. Biochemical studies on gramicidin S nonproducing mutants of *Bacillus brevis* ATCC 9999. J. Biochem. **75:**481–493.

39. **Katz, E., and A. L. Demain.** 1977. The peptide antibiotics of *Bacillus:* chemistry, biogenesis, and possible functions. Bacteriol. Rev. **41:**449–474.

40. **Keilman, G. R., K. Burtis, B. Tanimoto, and R. H. Doi.** 1976. Effect of netropsin on the derepression of enzymes during growth and sporulation of *Bacillus subtilis.* J. Bacteriol. **128:**80–85.

41. **Keynan, A., A. A. Berns, G. Dunn, M. Young, and J. Mandelstam.** 1976. Resporulation of outgrowing *Bacillus subtilis* spores. J. Bacteriol. **128:**8–14.

42. **Kobayashi, Y., H. Kobayashi, and H. Hirochika.** 1978. Role of elongation factor G in sporulation of *Bacillus subtilis,* p. 242–249. *In* G. Chambliss and J. C. Vary (ed.), Spores VII. American Society for Microbiology, Washington, D.C.

43. **Kunst, F., M. Pascal, J. Lepesant-Kejzlarova, J.-A. Lepesant, A. Billault, and R. Dedonder.** 1974. Pleiotropic mutations affecting sporulation conditions and the syntheses of extracellular enzymes in *Bacillus subtilis* 168. Biochimie **56:**1481–1489.

44. **Leighton, T.** 1977. New types of RNA polymerase mutations causing temperature-sensitive sporulation in *Bacillus subtilis.* J. Biol. Chem. **252:**268–272.

45. **Levisohn, S., and A. I. Aronson.** 1967. Regulation of extracellular protease production in *Bacillus cereus.* J. Bacteriol. **93:**1023–1030.

45a. **Lopez, J. M., A. Dromerick, and E. Freese.** 1981. Response of guanosine 5′-triphosphate concentration to nutritional changes and its significance for *Bacillus subtilis* sporulation. J. Bacteriol. **146:**605–613.

46. **Lopez, J. M., C. L. Marks, and E. Freese.** 1979. The decrease of guanine nucleotides initiates sporulation of *Bacillus subtilis.* Biochim. Biophys. Acta **587:**238–252.

47. **Lopez, J. M., B. Uratani-Wong, and E. Freese.** 1980. Catabolite repression of enzyme synthesis does not prevent sporulation. J. Bacteriol. **141:**1447–1449.

48. **Mitani, T., J. E. Heinze, and E. Freese.** 1977. Induction of sporulation in *Bacillus subtilis* by decoyinine or hadacidin. Biochem. Biophys. Res. Commun. **77:**1118–1125.

49. **Mychajlonka, M., A. M. Slee, and R. A. Slepecky.** 1975. Requirements for microcycle sporulation in outgrowing *Bacillus megaterium* cells, p. 434–440. *In* P. Gerhardt, R. N. Costilow, and H. L. Sadoff (ed.), Spores VI. American Society for Microbiology, Washington, D.C.

50. **Nishino, T., J. Gallant, P. Shalit, L. Palmer, and T. Wehr.** 1979. Regulatory nucleotides involved in the *rel* function of *Bacillus subtilis.* J. Bacteriol. **140:**671–679.

51. **Pandey, N. K., and K. G. Gollakota.** 1977. Induction of microcycle sporulation in *Bacillus brevis* spp. AG4 by amethopterin. Arch. Microbiol. **114:**189–191.

52. **Piggot, P., and J. Coote.** 1976. Genetic aspects of bacterial endospore formation. Bacteriol. Rev. **40:**908–962.

53. **Ramaley, R. F., and L. Burden.** 1970. Replacement sporulation of *Bacillus subtilis* 168 in a chemically defined medium. J. Bacteriol. **101:**1–8.

54. **Reysset, G., and J.-P. Aubert.** 1975. Relationship between sporulation and mutations impairing glutamine synthetase in *Bacillus megaterium.* Biochem. Biophys. Res. Commun. **65:**1237–1241.

55. **Reysset, G., K. F. Bott, and J.-P. Aubert.** 1978. Relationship between glutamine synthetase and sporulation, p. 271–276. *In* G. Chambliss and J. C. Vary (ed.), Spores

VII. American Society for Microbiology, Washington, D.C.

56. **Rhaese, H. J., and R. Groscurth.** 1976. Control of development: role of regulatory nucleotides synthesized by membranes of *Bacillus subtilis* in initiation of sporulation. Proc. Natl. Acad. Sci. U.S.A. **73**:331–335.

57. **Rhaese, H. J., R. Groscurth, and G. Rumpf.** 1978. Molecular mechanism of initiation of differentiation in *Bacillus subtilis*, p. 286–292. *In* G. Chambliss and J. C. Vary (ed.), Spores VII. American Society of Microbiology, Washington, D.C.

58. **Rhaese, H. J., R. Groscurth, R. Vetter, and H. Gilbert.** 1980. Regulation of sporulation by highly phosphorylated nucleotides in *Bacillus subtilis*, p. 145–159. *In* P. Richter and G. Koch (ed.), Regulation of macromolecular synthesis by low molecular weight mediators. Academic Press, Inc., New York.

59. **Rhaese, H. J., J. A. Hoch, and R. Groscurth.** 1977. Studies on the control of development: isolation of *Bacillus subtilis* mutants blocked early in sporulation and defective in synthesis of highly phosphorylated nucleotides. Proc. Natl. Acad. Sci. U.S.A. **74**:1125–1129.

60. **Ristow, H., W. Pschorn, J. Hansen, and U. Winkel.** 1979. Induction of sporulation in *Bacillus brevis* by peptide antibiotics. Nature (London) **280**:165–166.

61. **Ristow, H., B. Schazschneider, K. Bauer, and K. Kleinkauf.** 1975. Tyrocidin and the linear gramicidin: do these peptide antibiotics play an antagonistic regulative role in sporulation? Biochim. Biophys. Acta **390**:246–252.

62. **Rogolsky, M., and H. T. Nakamura.** 1974. Sensitivity of an early step in the sporulation of *Bacillus subtilis* to selective inhibition by ethidium bromide. J. Bacteriol. **119**:57–61.

63. **Sacks, L. E., and P. A. Thompson.** 1975. Influence of methylxanthines on sporulation of *Clostridium perfringens* cells, p. 341–345. *In* P. Gerhardt, R. N. Costilow, and H. L. Sadoff (ed.), Spores VI. American Society for Microbiology, Washington, D.C.

64. **Sarkar, N., and H. Paulus.** 1972. Function of peptide antibiotics in sporulation. Nature (London) New Biol. **239**:228–230.

65. **Schaeffer, P.** 1969. Sporulation and the production of antibiotics, exoenzymes, and exotoxins. Bacteriol. Rev. **33**:48–71.

66. **Schaeffer, P., H. Ionesco, A. Ryter, and G. Balassa.** 1965. La sporulation de *Bacillus subtilis*: etude genetique et physiologique. Colloq. Int. C.N.R.S. **124**:553–563.

67. **Schaeffer, P., J. Millet, and J. P. Aubert.** 1965. Catabolic repression of bacterial sporulation. Proc. Natl. Acad. Sci. U.S.A. **54**:704–711.

68. **Setlow, P.** 1973. Inability to detect cyclic AMP in vegetative or sporulating cells or dormant spores of *Bacillus megaterium*. Biochem. Biophys. Res. Commun. **52**:365–372.

69. **Slee, A. M., and R. A. Slepecky.** 1976. The formation in media affording different growth rates of spores of *Bacillus megaterium* containing varying amounts of deoxyribonucleic acid, p. 183–194. *In* A. N. Barker, L. J. Wolf, D. J. Ellar, G. J. Dring, and G. W. Gould (ed.), Spore research 1976. Academic Press, London.

70. **Slepecky, R.** 1969. Synchrony and the formation and germination of bacterial spores, p. 77–96. *In* G. M. Padilla, G. L. Whitson, and I. L. Cameron (ed.), The cell cycle, gene-enzyme interaction. Academic Press, Inc., New York.

71. **Slepecky, R. A., and J. W. Foster.** 1959. Alterations in metal content of spores of *Bacillus megaterium* and the effect of some spore properties. J. Bacteriol. **78**:117–123.

72. **Smith, I., P. Paress, K. Cabane, and E. Dubnau.** 1980. Genetics and physiology of the *rel* system of *Bacillus subtilis*. Mol. Gen. Genet. **178**:271–279.

73. **Sonenshein, A. L., and K. M. Campbell.** 1978. Control of gene expression during sporulation, p. 179–192. *In* G. Chambliss and J. C. Vary (ed.), Spores VII. American Society for Microbiology, Washington, D.C.

74. **Srinivasan, V. R., and H. O. Halvorson.** 1963. "Endogenous factor" in sporogenesis in bacteria. Nature (London) **197**:100–101.

75. **Sterlini, J. M., and J. Mandelstam.** 1969. Commitment to sporulation in *Bacillus subtilis* and its relationship to development of actinomycin resistance. Biochem. J. **113**:29–37.

76. **Sugae, K., and E. Freese.** 1970. Requirement for acetate and glycine (or serine) for sporulation without growth of *Bacillus subtilis*. J. Bacteriol. **104**:1074–1085.

77. **Takahashi, I.** 1979. Catabolite repression-resistant mutants of *Bacillus subtilis*. Can. J. Microbiol. **25**:1283–1287.

78. **Tipper, D. J., C. W. Johnson, C. L. Ginther, T. Leighton, and H. G. Wittmann.** 1977. Erythromycin-resistant mutations in *Bacillus subtilis* cause temperature sensitive sporulation. Mol. Gen. Genet. **150**:147–159.

79. **Tominaga, A., and Y. Kobayashi.** 1978. Kasugamycin-resistant mutants of *Bacillus subtilis*. J. Bacteriol. **135**:1149–1150.

80. **Vasantha, N., and E. Freese.** 1979. The role of manganese in growth and sporulation of *Bacillus subtilis*. J. Gen. Microbiol. **112**:29–37.

81. **Vasantha, N., and E. Freese.** 1980. Enzyme changes during sporulation of *Bacillus subtilis* caused by deprivation of guanine nucleotides. J. Bacteriol. **144**:1119–1125.

82. **Young, M., and J. Mandelstam.** 1979. Early events during bacterial endospore formation. Adv. Microb. Physiol. **20**:103–162.

83. **Youston, A. A., and R. S. Hanson.** 1972. Sporulation of tricarboxylic acid cycle mutants of *Bacillus subtilis*. J. Bacteriol. **109**:886–894.

Biochemistry of Bacterial Forespore Development and Spore Germination

PETER SETLOW

Department of Biochemistry, University of Connecticut Health Center, Farmington, Connecticut 06032

INTRODUCTION

The purpose of this article is to review bacterial forespore development and spore germination. It will be concerned primarily with biochemical events during these processes, but several recently reviewed aspects of this topic will not be covered in detail, including spore coat morphogenesis (4), spore resistance (110), activation (47), and the detailed biochemistry of the dormant spore (124). Because of space limitations, this review will emphasize only the current view(s) on topics such as metabolism during spore germination. I apologize in advance to those authors whose work has not been cited, but the details of early work can be found in previous reviews (26, 35, 37, 46, 89, 108). Because the majority of knowledge concerning the biochemistry of forespore development and germination has been obtained with *Bacillus* species, this review will focus on these organisms. However, some biochemical data are available for *Clostridium* species and will also be noted.

Two other things should be kept in mind by the reader of this review—one an assumption and one a definition. It is my assumption that the underlying mechanisms for fundamental spore properties and for forespore development and spore germination are similar for different species. Given this assumption, I will not cite all the different species in which fundamental observations have been made and will note partic-

ular species only when there is something unique or different about that species or when results from different species are fundamentally contradictory. I will define *spore germination* in this review as the time from mixing of spores with germinant to the completion of the first cell division. The first step(s) that begins or *triggers* spore germination (whatever it is or they are) will be referred to as the initiation of germination. The term *outgrowth* will not be used in this review.

FORESPORE DEVELOPMENT

Most work on the biochemistry of bacterial sporulation has either focused on the early events in sporulation or has examined the sporulating cell as a whole. However, the sporulating cell contains two distinct compartments, and it is in the developing forespore that many of the major biochemical changes take place. A major advance in the study of spore formation has been the development and refinement by Andreoli, Ellar, and co-workers of techniques for isolation of intact forespores early in their development (2, 27, 28). This has allowed analysis of various biochemical parameters of the developing forespore without interference from those of the physically larger mother-cell compartment. Of particular importance in some of the results discussed below has been the demonstration that even early in its development some of the biochemical properties of the forespore are distinctly different from those of the mother cell.

Energy and Small-Molecule Metabolism

At the earliest time at which forespores can be isolated (2 to 3 h prior to the accumulation of dipicolinic acid [DPA]), forespores contain a high ATP level and significant amounts of NADH (101). However, the metabolic pathway(s) generating high-energy compounds in the forespore is not known. Forespores do have enzymes of glycolysis and terminal electron transport, and their levels remain rather constant throughout forespore development (3, 101, 128). Tricarboxylic acid cycle enzymes are present at much lower levels in the forespore than in the surrounding mother cell, and in some reports they were not detected in forespores (3, 101). This appears to rule out the tricarboxylic acid cycle as having an important role in forespore energy metabolism. It appears probable that some of the forespore's nutrients must be supplied from the mother cell.

About 1 h *before* DPA accumulation, forespore ATP levels and the NADH/NAD ratio fall to the low values found in dormant spores. This is paralleled by an increase in 3-phosphoglyceric acid (3-PGA) levels in the forespore to those in the dormant spore. Despite these changes in the forespore, ATP levels and the NADH/NAD ratio in the mother cell remain constant, and no 3-PGA accumulates in this compartment (101). The forespore 3-PGA depot is stable despite the presence in forespores of phosphoglycerate mutase, enolase, and pyruvate kinase—enzymes that catalyze rapid 3-PGA catabolism in germination (89, 101). The enzyme which is regulated to allow 3-PGA accumulation in the forespore has been suggested to be phosphoglycerate mutase; the activity of this enzyme may be regulated in vivo by the level of *free* Mn^{2+} ions (65, 102).

After the fall in the forespore's NADH/NAD ratio, and at about the time of DPA accumulation, much of the coenzyme A (CoA) pool in the forespore becomes bound in disulfide linkage to protein (81). The proteins to which the CoA becomes linked and the significance of this process are unknown. The source of amino acids for the developing forespore is also not clear. A number of enzymes active in amino acid biosynthesis are absent or at low levels in forespores, suggesting that some amino acids must be provided by the mother cell (3, 101). In addition, protein turnover has been reported to be low in forespores—in particular after the time at which they can be isolated (27). However, it appears likely from recent work that some forespore proteins do turn over (A. J. Andreoli et al., this volume). Forespores of *Bacillus megaterium* accumulate a significant amount of arginine and glutamic acid in parallel with DPA. The source of these amino acids is not known, but they may be derived from the mother cell (101).

Dipicolinic Acid and Calcium Metabolism

Isolation of forespores late in sporulation and comparison of their DPA content with that of whole cells has shown that essentially all DPA is found in the forespore (26). However, the DPA may be synthesized in the mother cell, since DPA biosynthetic enzymes have been reported to be absent from the forespore but present in the mother cell (2). The mechanism by which DPA enters the forespore is not known.

Calcium accumulates in forespores at the time of DPA accumulation (40, 41). Isolated forespores neither take up Ca^{2+} by active transport nor concentrate it against a concentration gradient, but they do have a Ca^{2+}-specific facilitated diffusion system for Ca^{2+} uptake. In contrast, late in sporulation the mother cell contains a high-affinity active transport system which concentrates Ca^{2+} effectively. It has been proposed that active transport of Ca^{2+} by the mother cell results in elevated levels of mother-cell Ca^{2+}. The Ca^{2+} can now enter the forespore by facilitated diffusion, and there it complexes with DPA, thus lowering the effective free Ca^{2+} concentration in the forespore and thus allowing more uptake (26, 40, 41). In addition to Ca^{2+} uptake, there is also extensive uptake of other divalent metal ions (Mg^{2+}, Mn^{2+}) into the developing spore (102, 124). However, the mechanism for uptake of these other ions has not been studied with forespores.

Macromolecular Metabolism

Protein. As was indicated above, the protein complement of the forespore is different from that of the surrounding mother cell, and the protein complement changes throughout forespore development (Andreoli et al., this volume). Comparison of forespore and mother-cell proteins allows establishment of three classes of proteins: group I proteins are present in both compartments at similar levels, group II proteins are present in the mother cell but absent or at low (<12%) specific activity in the forespore, and group III proteins are those associated with the forespore but not the mother-cell compartment. Group I proteins include glycolytic enzymes (101, 125), some enzymes of terminal electron transport (26), enzymes of biosynthesis of vegetative cell walls (36, 115), and some enzymes of amino acid catabolism (101). In addition, the proteins of forespore ribosomes appear identical to those of mother-cell ribosomes (3).

Group II proteins include many enzymes of the tricarboxylic acid cycle (3, 101), alanine dehydrogenase (3), at least one enzyme unique to spore cortex biosynthesis (115), some enzymes of DPA biosynthesis (2), and the sporulation-associated nonspecific endoprotease(s) (101, 125). Several of these enzymes are synthesized late in sporulation—well after the mother-cell and forespore compartments have become separate. This would only require differential gene expression in the two compartments. However, some proteins made very early in sporulation (protease, tricarboxylic acid cycle enzymes) also do not appear in young forespores. If these enzymes are indeed synthesized before the septation dividing mother cell and forespore, then it is not clear why they do not appear in the forespore, since protein turnover in the forespore compartment has been reported to be low (27). However, more recent work indicates that some forespore proteins do turn over (Andreoli et al., this volume).

Group III proteins include glucose dehydrogenase (33), some enzymes for processing of spore cortex and for salvage of by-products of spore cortex synthesis (36), aspartase (3), the low-molecular-weight proteins degraded during germination and the protease that degrades them (91, 101), and the spore coat proteins (4). Some of these proteins (glucose dehydrogenase, the low-molecular-weight proteins and their protease, spore coat proteins) are synthsized at defined times in sporulation, and the low-molecular-weight proteins, the protease which initiates their degradation, and the coat proteins appear to be sporulation-specific gene products.

In the case of the spore-specific low-molecular-weight proteins, it has been shown that the newly synthesized proteins are found only in the forespore; consequently, these species may be synthesized within the forespore (22). Intact young *isolated* forespores can synthesize protein, but it is not known whether the proteins made under these semi-in vitro conditions are the same as those made in vivo (2). However, it seems reasonable to suppose that young forespores do synthesize protein and that consequently there is differential gene expression in the two compartments of the sporulating cell. In contrast to the low-molecular-weight spore proteins which are probably synthesized in the forespore, the spore coat proteins are synthesized in

precursor form in the mother cell and deposited on the developing spore (4, 5, 14, 60).

RNA. There is incorporation of radioactive precursors into forespore RNA in vivo, but the nature of this RNA has not been determined. The results on protein accumulation within the forespore (see above) certainly suggest that transcription within the forespore takes place and is different from that in the mother cell. However, there has as yet been no definitive comparison of the mother-cell and the forespore transcriptional apparatus, i.e., RNA polymerase. If this enzyme is different in the two compartments, this could explain some of the heterogeneity in the RNA polymerase extracted from whole sporulating cells (forespore plus mother cell).

Membrane and cell wall. As was expected on the basis of morphological studies, Ellar and co-workers have shown by enzyme assays of intact forespores that the outer forespore membrane has an orientation opposite to that of the inner membrane, with the normal cytoplasmic side (ATPase side) of the outer forespore membrane facing the mother-cell cytoplasm (126, 127). The presence of these forespore membranes in opposite orientation raises unanswered questions about mechanisms of transport into the forespore during its development. It is known that the inner and outer membranes of the forespore have different lipid and protein compositions in the mature spore (19, 26, 28). However, the time in forespore maturation when these changes are effected is not clear.

Spore cortex is deposited between the inner and outer forespore membranes. It appears likely that much of this synthesis, in particular that of various precursors, is catalyzed by mother-cell enzymes (36, 115).

GERMINATION—EARLY EVENTS

Spore germination, the process whereby a dormant spore is converted into a vegetative cell, completes the procaryotic differentiation cycle initiated by sporulation. Although spores can and do remain dormant for long periods of time, the initiation of the germination process leads within minutes to the loss of most of the unique spore properties and constituents. Germination of spores of any given strain can be initiated or triggered by a specific group of low-molecular-weight compounds, but the specific germination requirements for this same strain can vary significantly depending on how the spores were prepared and treated subsequently (35, 46, 120). Even more striking are the differences in requirements for specific germinants between spores of different species or strains. The agents able to initiate spore germination of at least one strain include: sugars (glucose, fructose), amino acids (alanine, leucine, proline), purines (adenosine), salts (calcium-DPA, KBr), detergents (dodecylamine), and high pressure (35, 46). Whereas spores of most strains respond to glucose and alanine, only a few germinate readily with salts such as KBr. Similarly, spores of many strains require mixtures of these germinants for rapid and complete initiation of germination. Mutation can also abolish the response of a strain to one germinant while not affecting the response to others, or it can result in new requirements for initiation of germination (see below).

Sequence of Events

The overall germination process (from addition of germinant to vegetative cell) can take from 40 to 100 min depending on the growth medium. However, the upper limit can be much higher in practice, since spores of some strains have very long lag periods between addition of germinant and the first measurable steps in germination. This long lag period, and the fact that individual spores in these populations may have very different lag times, often make studies of spore germination in these strains difficult. Unfortunately, the reasons for the strain differences in lag period and its heterogeneity are not understood (35). The sequence of specific events in germination (35, 46, 54, 72, 89, 108) is as follows:

1. **Commitment.** The first step, commitment, is defined as some change in the spore such. that the germinant can now be removed without aborting the remainder of the germination process (46). The biochemical change(s) during commitment is not understood. Attempts to detect specific binding or metabolism (or both) of a germinant during commitment have given negative results (46, 78).

2. **Loss of heat resistance and excretion of calcium-DPA.** The next step is loss of heat resistance and excretion of calcium-DPA. Again, the biochemistry of these changes is not understood, but it appears likely that Ca and DPA are excreted in a 1:1 ratio (77). Most studies have found calcium-DPA excretion to be slightly later than loss in heat resistance.

3. **Loss of spore cortex, spore refractility, and UV resistance.** It has been known for some time that an early event in spore germination is the degradation of much of the spore cortex and excretion of the fragments (35, 108, 124). This loss of spore cortex may result in the loss in spore refractility (42), and several lytic

(lysozyme-like) enzymes have been identified in and in some cases purified from spores (12, 35, 124). These enzymes will (as will lysozyme) often cause the germination of spores whose coats have been removed to make the cortex accessible. Although these enzymes are prime candidates for involvement in this early step of germination, confirmation of their role by isolation and analysis of an appropriate mutant has yet to be achieved. UV resistance of spores is also lost at about the time of spore cortex degradation; however, the possible relationship between the spore cortex and UV resistance is unclear. Steps 1–3 may take only 5 min for a spore population, and almost certainly much less time for individual spores.

4. **Initiation of lipid, protein, and RNA hydrolysis.** Degradation of macromolecules other than peptidoglycan also begins in the first minutes of germination; this includes degradation of spore lipid, RNA, and protein, and these processes continue throughout germination (26, 89). The degradation of spore lipid early in spore germination may be particularly important, since it begins well before lipid synthesis and could result in significant reorganization of spore membranes. However, the magnitude of this lipid turnover and its substrates have not yet been reported in detail (26). The major function of the RNA breakdown early in germination is to supply nucleotides for RNA synthesis (see below) (89). The massive proteolysis (~20% of total spore protein) which begins early in germination (see below) also generates building blocks for macromolecular synthesis. It has been shown that a mutation in the enzyme catalyzing this proteolysis has no effect on spore germination even though proteolysis itself is slowed significantly (67).

5. **Metabolism.** Many workers have shown that active metabolism resumes early in spore germination with initiation of oxygen consumption and accumulation of high-energy compounds such as ATP, NADH, and NADPH (46, 80, 89); the major metabolic pathways early in germination are discussed below. Clearly, metabolism will be necessary for the overall germination process, but a major question is whether metabolism is necessary for the initial steps in germination—in particular for initiation or completion (or both) of steps 1–3 above. One reason for the initial belief that energy metabolism was necessary for these early steps was that most common germinants can be readily metabolized to yield ATP and NADH. However, detailed studies in several laboratories have shown that: (i) detectable ATP and NADH accumulations begin only *after* initiation of the early

steps of germination (74, 78); (ii) there is no detectable catabolism of metabolizable germinants when steps 1–3 are beginning (69, 73); (iii) non-metabolizable analogs will often initiate germination as well as their metabolite analogs (73, 100); (iv) mutants blocked in catabolism of a particular metabolite will still germinate with that metabolite (46); and (v) anaerobiosis or inhibition of energy metabolism by use of fluoride or electron transport chain inhibitors often has no effect on the early events (steps 1–4) of spore germination (23, 26, 69). Data of this type have been interpreted to mean that metabolism of exogenous germinants is not necessary for spores to proceed through steps 1–3 and probably 4 of germination. Indeed, the finding that spores of several strains can readily initiate germination with salts alone would indicate a priori that *metabolism of exogenous germinants is not necessary for the initial steps in the germination process.*

However, spores have abundant endogenous energy reserves (see below), and it is possible that their catabolism could be essential for early steps in germination. This is much more difficult to assess (in particular to disprove), but several of the experiments noted above also suggest that, if metabolism of endogenous compounds is essential early in germination, its magnitude is very small. In particular, since significant ATP and NADH accumulations occur only after steps 1 and 2, and inhibitors of metabolism do not block early events in germination, metabolism of endogenous compounds apparently is not necessary early in germination. However, measurements of ATP and NADH are only steady-state measurements (see below), and metabolic inhibitors might not be effective in the first seconds of germination as a result of permeability problems.

Results of several other experiments have also indicated little or no role for metabolism of endogenous compounds early in germination. Ellar and his co-workers found no evidence for fixation of ^{3}H from ^{3}H$_2$O during steps 1 and 2 (26). However, metabolism involving transfer of protons to an exchangeable position, such as oxidation with concomitant reduction of NAD and with subsequent reduction of a disulfide, would not have been detected by this procedure. In my experiments with spores of *B. megaterium* germinating in KBr alone, inhibition of ATP production had no effect on steps 1–4 of germination (89); however, NADH production was not affected (80). Rapid synthesis of ATP and NADH could be balanced by equally rapid utilization during this period, although it has been pointed out that this would be unlikely (78).

In contrast to these results, several types of experiments support a role for endogenous metabolism in initiating germination. Thus, some workers have found that inhibitors of energy metabolism *will* inhibit the initiation of the germination process (24). Although these results have been questioned, and in some cases could not be reproduced (69), as noted above, spore permeability barriers may make interpretation of the results of these experiments ambiguous— in particular if they are negative.

Two other results which suggest a role for endogenous metabolism in initiating germination come from analysis of mutants. In one study a mutant of *B. megaterium* was isolated which required γ-aminobutyric acid (GABA) for initiation of germination (30), and it was suggested that GABA might be a germinant normally generated endogenously from decarboxylation of glutamic acid (31). Indeed, GABA formation from endogenous glutamic acid was detected early in germination, and studies of the GABA-requiring mutant spores suggested that there was some defect in their glutamate decarboxylase (29). Unfortunately, the mutant spores required not only GABA for initiation of germination, but also KI. Similarly, the work on the enzyme levels in the mutant spores by no means proved that there was any specific enzyme defect. In addition, in work with another strain of *B. megaterium*, GABA formation during germination was not detected (78). Unfortunately, there has been no work on this interesting mutant for ~7 years. Possibly it should be reexamined, especially in light of recent suggestions that GABA formation from glutamic acid may be an early event in fungal spore germination (76).

Whereas most mutants in glycolytic or oxidative enzymes exhibited no germination defects, two glycolytic mutants of *B. subtilis* did (68). Spores of 3-phosphoglycerate kinase mutants initiated germination on alanine much more poorly than did wild-type spores. Similarly, spores of a fructose-1-phosphate kinase mutant would no longer initiate germination on glucose, fructose, and asparagine, but would on glucose, mannose, and asparagine. The analysis of these mutants led to the suggestion that the early steps (1–3) in germination required the generation of NADH, an amino donor, and some metabolite in the upper metabolic subdivision of the glycolytic pathway. Although accumulation of NADH and such a metabolite has not been detected during steps 1–2, as noted above, it is difficult to assess negative results. The isolation and analysis of similar mutants in a strain whose spores will respond to salts alone could prove very informative.

6. **Macromolecular synthesis.** Since dormant spores contain extremely low levels of ribonucleoside triphosphates and aminoacyl-tRNA's, metabolism is clearly a prerequisite for macromolecular biosynthesis during germination. Indeed, macromolecular biosynthesis does begin early in germination, starting with RNA synthesis followed a few minutes later by protein synthesis (51, 89). After a lag of 30 to 60 min, cell wall synthesis and DNA replication begin (89, 108). Since inhibition of ATP accumulation has no effect on steps 1–4 of germination, it is generally held that macromolecular biosynthesis is not involved in these early steps.

Germination Mutants

Mutants have already proved useful in ruling out a number of metabolic enzymes, including glycolytic enzymes and ATPase (68) as well as a specific protease, as participants early in germination, while suggesting a role for other reactions (see above). Mutants defective in the early steps or the initiation of spore germination have been isolated in several laboratories (57, 104, 121); in general, these have been conditional mutants of some type. In *B. subtilis* these mutants have been assigned to at least seven different groups on the basis of analysis of 29 mutants plus the phosphoglycerate kinase mutant noted above (57). It is somewhat discouraging that a major conclusion from analysis of these mutants is that mutations often do not affect the response to all germinants equally, but rather abolish germination with one compound and have little or no effect on the response to others (57, 119, 121). This leads to the suspicion either that there are multiple independent pathways for initiation of germination or that there may be a series of steps unique to a given germinant, these independent steps eventually triggering or activating a common pathway. For none of the initiation-of-germination mutants isolated (with the exception of the phosphoglycerate kinase mutant) has the biochemical lesion been identified.

Another type of mutant has a lesion in spore coats. *B. cereus* strains carrying lesions presumed to reside either in the protease which processes the spore coat precursor or in some other aspect of spore coat deposition produce spores which can only germinate readily upon addition of lysozyme (15, 107), but it is not clear how coat alterations affect spore germination. Indeed, in *B. megaterium* removal of the great majority of spore coat protein by extraction has little effect on the germination of the treated spores (120), although this is not true with spores of some other species.

A second group of germination mutants which have blocks at points *after* the initiation of germination but have normal vegetative growth have also been isolated (1, 34). These mutations also map in a number of different loci, but the biochemical lesions involved have not been identified.

Nature of the Trigger

Much of the work on the early events in spore germination has been an attempt to discern the molecular nature of the earliest, or triggering, event in this process. As is undoubtedly evident from the discussion above, the nature of this event(s)—in particular the first event—is not clear, although the initial event is apparently not metabolism of the exogenous germinant. Since the germinant need be present only briefly to elicit its effect, it has been suggested that binding of a germinant to some receptor molecule (possibly in the inner forespore membrane) might trigger some biophysical change which is or results in the early events in germination. One such event may possibly be early initiation of endogenous metabolism (43, 46). Although neither the existence of such a receptor nor the identification of such a biophysical change has been proved, there are several lines of evidence which either suggest such a model or make it readily testable. (i) The existence of specific receptors for specific germinants would fit with the findings that germination mutants lose response to one germinant without losing their response to others. (ii) This model would not require generation of any metabolites for the initiation of germination. (iii) The receptor molecules should be able to be isolated. Unfortunately, attempts at isolation of such receptors have not yet been successful (74; F. M. Racine, Fed. Proc. **37**:1613, 1978). Clearly, our understanding of the mechanism(s) for initiating or triggering spore germination is far from complete.

GERMINATION—LATER EVENTS

Although there is still very little understanding of the biochemistry of the earliest events (steps 1–3) in spore germination, much more is known about subsequent events (steps 4–6). In many cases the biochemistry in steps 4–6 is similar to that in growing cells, but there are a number of unique processes in germination.

Energy Metabolism

As noted above, dormant spores carry out no detectable metabolism, and they have very low levels of ATP and reduced pyridine nucleotides (80, 89). Other common high-energy compounds such as CTP, GTP, UTP, deoxynucleotide triphosphates, acyl-CoA's, and aminoacyl-tRNA's are also low or absent (81, 88, 89). However, the low-energy forms of many of these compounds (AMP, CMP, GMP, CoA, etc.) are present in spores at levels similar to those of the high-energy forms in growing cells. Thus, spores of *B. cereus, B. megaterium*, and *Clostridium bifermentans* contain *total* free adenine nucleotide levels similar to those in growing cells, but in spores 80% is AMP whereas in cells 80% is ATP (38, 89). Since metabolism begins in the first minutes of spore germination, one would expect production of high-energy compounds at this time. Indeed, production of ATP, CTP, GTP, UTP, deoxynucleotide triphosphates, NADH, NADPH, sugar phosphates, and acyl-CoA has been found to occur early in spore germination (38, 63, 80, 81, 89).

Production of much of the high-energy compounds needed early in spore germination can be driven by metabolism of energy reserves stored in the dormant spore. This can be shown most readily with spores of *B. megaterium* QM B1551, which can be readily induced to germinate by salts alone. During the first 10 to 15 min of germination in salt, spores of this strain accumulate as much ATP, NADH, NADPH, and acyl-CoA as are accumulated by spores germinating in a complete nutrient medium (80, 81, 89). These compounds are then capable of supporting rapid protein and RNA synthesis during this period. However, subsequent rapid production of high-energy compounds (as well as macromolecular synthesis) requires exogenous metabolites (89).

The ATP generated from endogenous spore reserves is generated by both substrate level and oxidative phosphorylation, the latter process using NADH also generated from endogenous reserves (80). A major energy reserve stored in the dormant spore is 3-PGA, which represents up to 5% of spore dry weight. 3-PGA has been found in spores of a number of *Bacillus* and *Clostridium* species, but has been reported to be absent from spores of *C. tyrobutyricum* (7, 38, 89). As would be expected, 3-PGA is catabolized in the first 10 to 15 min of germination. Under some conditions, 3-PGA catabolism is known to proceed via the action of phosphoglycerate mutase, enolase, pyruvate kinase, and pyruvate dehydrogenase, with production of acetate (84, 89). In spores of *B. megaterium* germinating in salt, addition of fluoride (a potent enolase inhibitor) blocks both ATP production and 3-PGA utilization. If exogenous metabolites are present during germination, 3-PGA is still utilized (78, 89). However, under these conditions fluoride does not block 3-PGA utilization, which may proceed by reversal of glycolysis to the triose phosphate level and then feed into the hexose monophosphate (HMP) shunt (89).

A second major endogenous energy reserve is the free amino acids (in particular glutamate) stored in the dormant spore, as well as the amino acids which are generated by the degradation of 15 to 20% of dormant spore protein during the first 20 min of germination (97). In *B. megaterim* ~50% of these amino acids are catabolized, predominantly via oxidative reactions, and the energy generated in this process is 10 to 20 times that obtained from 3-PGA catabolism (84, 97). Utilization of this reserve continues two to four times longer than for 3-PGA. Other possible energy reserves in spores include α-glycerophosphate, small amounts of which have been found in *B. megaterium* spores, and polysaccharides, which have been found in *C. tyrobutyricum* (possibly replacing 3-PGA?) (7, 62). In contrast to the situation in fungal spores, levels of free mono- and disaccharides are extremely low in dormant bacterial spores (B. Setlow and P. Setlow, unpublished data). Levels of phosphoenolpyruvate, pyruvate, lactate, malate, and α-ketoglutarate are also low or undetectable (26, 84).

In addition to substantial energy reserves, most enzymes necessary for metabolism are also present in the spore. These include enzymes of glycolysis, the HMP shunt, oxidative phosphorylation, and sugar transport (46, 52, 100, 101), but possibly not the Entner-Douderoff pathway (100). Completion of germination requires exogenous metabolites as well as exogenous sources of nitrogen, sulfur, and phosphorus (35, 46, 89, 108).

On the basis of measurements of $^{14}CO_2$ released from specifically labeled glucose, levels of intermediates such as fructose-1,6-diphosphate and gluconate, and enzyme levels in spores, it appears likely that glucose can be catabolized during germination by both the glycolytic and the HMP pathways (46, 63, 89). There is also evidence that some glucose may feed into the HMP pathway by first being oxidized to gluconate (56). It appears likely that the HMP pathway alone may suffice for ATP generation during this period, since fluoride does not affect germination of *B. megaterium* in a nutrient medium (89). The major end product of glucose catabolism during germination is acetate. This is not metabolized further, since a number of enzymes of the tricarboxylic acid cycle are absent (52, 84, 101). Similarly, fluorocitrate, a potent inhibitor of the tricarboxylic acid cycle, does not inhibit ATP production throughout spore germination (89). Spores do contain a system for terminal electron transport; however, this system may be inactive in the dormant spore and activated only upon germination (26, 128, 129). Throughout germination of spores of *Bacillus* species, terminal electron transport is sensitive to cyanide and azide (80, 89).

Metabolism of Other Small Molecules

Amino acids. Dormant spores of *B. cereus, B. megaterium,* and *B. subtilis* contain high levels (1 to 2% of their dry weight) of free glutamic acid with significant levels of arginine or lysine, or both, but levels of other free amino acids are very low (61). Spores of at least one *Clostridium* species have low levels of all free amino acids (38). However, free amino acids are produced by proteolysis of ~20% of the dormant spore's protein during the first 30 min of germination (97). All of the common amino acids are generated in this process, and a significant amount (up to 50%) is reutilized directly in protein synthesis and can support all protein synthesis through 30 to 45 min of germination (97). Generation of amino acids in this fashion explains the observation that there is extensive protein synthesis early in the germination of spores of amino acid-auxotrophic strains, despite the lack of the required free amino acid in the dormant spore (45, 117). Indeed, germinating spores of even *prototrophic* strains appear to be functional multiple amino acid auxotrophs as a result of the absence of key enzymes of amino acid biosynthesis. Absent enzymes include those for biosynthesis of arginine (ornithine transcarbamylase), histidine (imidazolylacetolphosphate transaminase), lysine (aspartokinase), methionine (aspartokinase), threonine (threonine deaminase), and tryptophan (tryptophan synthetase) (97). These enzymes are lost during sporulation (52) and are only resynthesized at defined times during germination—in some cases not until 60 min after its initiation (97). Only after these enzymes are resynthesized does de novo amino acid biosynthesis begin (97).

In contrast to the lack of amino acid biosynthesis early in spore germination, there is extensive amino acid catabolism. Enzymes for catabolism of alanine, arginine, glutamate, isoleucine, leucine, and valine have been found in spores, and up to one-half of the amino acids generated

by proteolysis are catabolized to generate energy (3, 52, 97, 101).

Nucleotides. Throughout the first 30 min of germination, the supply of ribonucleotides for RNA synthesis is provided by RNA breakdown, as nucleotide biosynthesis does not take place (89). Again, key biosynthetic enzymes are lost during sporulation and are only resynthesized at defined times in germination (52, 89). Only then does de novo nucleotide biosynthesis begin. Despite the absence of nucleotide synthesis early in germination, the spore does have the capacity for purine and pyrimidine salvage and consequently must be able to generate phosphoribosylpyrophosphate. The spore also is able to carry out the interconversion of various ribonucleotides (89).

In contrast to ribonucleotides, deoxyribonucleotides have not been detected in spores. However, all four common deoxynucleotide triphosphates appear in the first minutes of germination and comprise ~4% of the total nucleotide triphosphate pool (85). Ribonucleotide reductase, deoxynucleoside kinases, and thymidylate kinase have all been found in spores (85, 89, 111). In *B. megaterium* the deoxynucleotide triphosphate pool remains rather constant during germination until just prior to the onset of DNA replication. At this time there is a sharp increase in ribonucleotide reductase activity due to new protein synthesis, followed by a two- to threefold increase in the deoxynucleotide triphosphate pool size (85).

Fatty acids and phospholipids. Fatty acid synthesis begins in the first minutes of spore germination, using enzymes present in the dormant spore (64). Since total fatty acid levels in the spore remain constant through 60 min of germination, the amount of fatty acid synthesis during this period is either small or is balanced by an equivalent amount of fatty acid catabolism (75). There are quantitative differences in the type of fatty acids synthesized at different times in germination, eventually resulting in conversion of the fatty acid complement from that typical of spores into that typical of vegetative cells (64, 75).

Phospholipid synthesis begins early in spore germination, using preexisting spore enzymes (20, 26); lipid turnover also begins extremely early in germination (26). However, the relative magnitudes of these processes are unclear. Phospholipid synthesis is extremely rapid at the times of chromosome replication and cell division (20).

Coenzymes, sulfhydryls, and disulfides. Pyridine nucleotide levels in spores are ~5% of those in growing cells (80). In spores all pyridine nucleotides are in the oxidized form, but the reduced forms are generated in the first minutes of germination, through the use of endogenous reserves if necessary (80). During germination, significant biosynthesis of NAD does not begin until the 30th min of germination. However, there may be some NADP biosynthesis prior to this time, presumably by direct phosphorylation of preexisting NAD (80).

CoA also is present predominantly in an oxidized form in dormant spores, with <1% as acyl-CoA, ~25% as a low-molecular-weight CoA-disulfide, 50% as CoA in disulfide linkage to spore core protein, and the remainder as CoA-SH (81). The disulfide forms (comprising 1% of the spore's disulfides) are reduced in the first minutes of spore germination (possibly even before much NADH or NADPH accumulation) through the use of endogenous reserves of reducing power; the majority of the CoA is then converted to acyl-CoA (81). The enzyme involved in cleaving the CoA-protein disulfides during germination has not been identified. However, several pyridine nucleotide-dependent disulfide reductases have been identified in spores (9, 109). The CoA level remains constant for 60 min after the start of germination and is only two- to threefold below the level of growing cells (81).

Glutathione, a second candidate for formation of disulfides with protein, has not been found in bacterial spores (81). However, non-peptide-bound cysteine has been reported in disulfide linkage to coat proteins in B. cereus (4). The fate of this latter type of disulfide during germination is not known. Spore coat proteins are known to contain high levels of cystine, and an early report had suggested that much of this disulfide was cleaved early in germination (10). However, this is now known to be incorrect (93).

Potential regulatory nucleotides. Where tested, the potential regulatory nucleotides cyclic AMP, cyclic GMP, and HPN (high phosphorylated nucleotide) I to HPN IV are absent from dormant and germinating spores (46, 71, 82; P. Setlow, unpublished data). ppGpp and pppGpp, nucleotides which mediate the stringent response in bacteria, are absent from dormant spores but appear early in spore germination (88). However, their levels are similar to the steady-state levels in growing cells.

Polyamines. Spermidine is the only polyamine found in spores of Bacillus species although it can be replaced by spermine with no ill effects (86). Spores contain two to four times as much spermidine per unit of RNA as do vegetative cells. During germination, the spermidine content remains constant until the sper-

midine/RNA ratio reaches that of the growing cell, at which time spermidine synthesis begins (87). Spermidine is wholly synthesized from arginine via agmatine, and may involve a nondissociable enzyme-putrescine intermediate. At least one enzyme of spermidine synthesis must be synthesized during germination (87).

Cations. Most of the high level of divalent cations stored in the dormant spores is excreted early in germination—the great majority presumably in conjunction with DPA (35, 77). The mechanism for this excretion is not known. Although dormant and activated spores exhibit no cation transport, K^+ and Mg^{2+} transport begin in the first minutes of germination of B. subtilis spores through the use of preexisting transport systems (25). This cation transport requires addition of glucose, indicating that endogenous energy reserves may be unable to supply the appropriate high-energy compound to drive cation transport.

Macromolecular Metabolism

mRNA synthesis. Dormant bacterial spores do not appear to contain any functional mRNA, since there is no protein synthesis in spores germinated with actinomycin D (51, 89). mRNA-like species have been detected in spores, but their functional significance is questionable (21, 44); possibly this RNA is some remnant from sporulation. mRNA is synthesized in the first minutes of germination, and its synthesis precedes that of protein by several minutes (37, 51, 103). Early reports had indicated that the percentage of the total RNA made during germination which was mRNA varied depending on the time in germination (37). However, a more recent report indicated that the percentage of total RNA synthesis represented by mRNA remained approximately constant throughout germination (103). Addition of inhibitors of RNA synthesis at any time during germination shuts off protein synthesis rapidly and has indicated (i) the absence of significant stable mRNA and (ii) that mRNA's in germinating spores have half-lives of ~2 min (37, 106).

Studies using spores of several species have indicated that mRNA synthesis (and thus almost certainly protein synthesis) during germination is a highly ordered process (37, 103). There are variations in the types of mRNA synthesized at different times during germination, and these differ in part from mRNA's of vegetative cells. Similar conclusions can be drawn from analysis of the proteins synthesized at various times during germination (106, 117). Since germinating spore mRNA is not stable, the type of proteins made and the time of their synthesis during germination should be a direct reflection of the type and time of synthesis of mRNA species. If this is indeed the case, then knowledge of the time of synthesis of a specific protein will allow a strong inference about the time of synthesis of its mRNA. Consequently, in the discussion below concerning synthesis of specific proteins during germination, I will refer to synthesis of their mRNA's, even though experimental proof for this is not yet available.

Throughout germination, protein synthesis (and by inference mRNA synthesis) is a highly ordered process. Individual proteins are made at defined times in germination, and often only for a discrete time period (37, 45, 106). In some cases, the time in which a given gene is available for transcription during germination appears preprogrammed. Thus, the enzymes histidase and α-glucosidase can only be induced during specific times in germination (106). Unfortunately, the mechanism for controlling this sequential ordered transcription during germination is not clear. Early work in *B. subtilis* had suggested that the order in which genes were transcribed during germination paralleled their map order on the chromosome (45). However, the elegant experiments of Yeh and Steinberg have disproved this idea (129).

Given current knowledge concerning the regulation of transcription during sporulation, it is possible that modifications in RNA polymerase could bring about the ordered transcriptional pattern seen in germination. There have been a number of studies on the RNA polymerase of dormant and germinating spores. Unfortunately, as yet there is no clear agreement on even the subunit composition of the dormant-spore enzyme from different species—let alone possible changes in germination (6, 48, 49, 55). Changes in the in vitro transcriptional specificity of RNA polymerase extracted from different stages in germination have been noted, and synthesis of a specific transcriptional "factor" early in germination has been suggested (6, 13, 18). However, the molecular nature of these changes or factors and their possible significance in vivo have not yet been established.

rRNA and tRNA synthesis. Dormant spores contain both 30S and 50S ribosomal subunits, and although some workers have reported the spore ribosomes to be "defective," the rRNA species from these ribosomes appear identical to those in vegetative cells (21, 50). Synthesis of rRNA begins early in and continues throughout germination; a recent report indicates that rRNA makes up a constant percentage of the total RNA made during germination (37, 103).

The amount of tRNA in spores (relative to that of rRNA) is similar to that in growing cells. However, significant quantitative differences have been noted between tRNA's in dormant spores and growing cells. Thus, spores and cells of *B. subtilis* contain different amounts of tRNA's for leucine, histidine, and glycine and different amounts of isoaccepting tRNA species for a number of amino acids (122a). At present, neither the significance nor the reason for these differences is completely understood.

A major difference between spore and cell tRNA's is that 20 to 33% of total spore tRNA lacks the 3'-terminal AMP residue, with smaller amounts of the penultimate CMP absent (98, 122). In *B. megaterium* spores, 90% of the tRNA's for tyrosine lack the terminal AMP residue. The missing CMP and AMP residues are added back in the first minutes of spore germination, even under conditions of very low steady-state levels of ATP (98). tRNA biosynthesis also begins early in spore germination, and there is little detectable change in the tRNA species made at different times in germination (39).

RNA breakdown. As was discussed above, mRNA made throughout germination is labile, turning over with a half-life of ~2 min. In addition, with spores of *B. megaterium* ~5% of the RNA stored in the dormant spore is degraded in the first 20 to 30 min of germination. The identity of the RNA degraded is unclear, although it is not tRNA (98). One possible candidate for the RNA degraded is the nonfunctional mRNA identified in spores of several species.

RNases have not been studied in detail in spores or during germination, although bulk RNase activity on a variety of substrates is similar in germinating spores and vegetative cells (G. Primus and P. Setlow, unpublished data). Spores of *B. subtilis* contain a phosphodiesterase which generates 5'-mononucleotides, but not the 3'-mononucleotide–generating phosphodiesterase which is found in cells (79). 3'-Nucleotidase and 5'-nucleotidase are also present in spores (79).

Protein synthesis. Dormant spores contain all the components necessary for protein synthesis, including aminoacyl-tRNA synthetases, methionyl-tRNA formylase, tRNA's, ribosomes, and initiation and elongation factors (8, 21, 122a). Some workers have found that the dormant spore's protein synthetic apparatus (in particular that from *B. cereus* spores) is defective, and these defects have been localized in the ribosomes (50). However, other workers, using either *B. subtilis* or *B. cereus*, have found active ribosomes in spores (8, 48). If there is a defect in dormant-spore ribosomes, this defect is corrected in the first minutes of germination without need of RNA or protein synthesis. Protein synthesis generally begins a few minutes after the initiation of germination, slightly after the initiation of RNA synthesis. Only a few proteins are made in the first minutes of germination, but their function is unknown; subsequently, many different proteins are made (106, 117).

Protein degradation. In addition to protein synthesis, there is also a rapid degradation of 15 to 20% of dormant-spore protein early in germination (97). In *B. megaterium* the degradation products are free amino acids, and this proteolysis requires neither ATP nor RNA or protein synthesis. The dormant spore proteins degraded are a group of at least seven acid-soluble, low-molecular-weight (7,000 to 15,000) proteins located in the spore core. Three of these species generally comprise the majority of protein degraded during germination, and these proteins from different spore species are similar (97, 99). The major proteins have been purified from spores of *B. megaterium, B. cereus* (K. Yuan and P. Setlow, unpublished data), and *B. subtilis* (97; C. Johnson and D. J. Tipper, unpublished data), and the primary sequences of the *B. megaterium* proteins have been determined (94–96). These proteins are synthesized late in sporulation and are found only in the developing forespore and dormant spore, where they comprise ~50% of the protein in the spore core (22, 97). The proteins bind to DNA in vitro, and the majority are associated with DNA in vivo as well (83). It has been suggested that they may be necessary (but not sufficient) for UV resistance of spores.

The degradation of the proteins described above is initiated during germination by an endoprotease active only on these species (91). This enzyme also appears late in sporulation only within the developing spore; it disappears rapidly during germination (91; C. A. Loshon and P. Setlow, unpublished data). The protease is an amino acid sequence-specific serine endoprotease which cleaves its substrates into large oligopeptides which are then degraded to amino acids by the active peptidase(s) present in the spore (91, 92). Mutants with decreased levels of this protease have been isolated. These show decreased proteolysis of dormant-spore protein during germination, but no other physiological defect (67).

There is also a second type of proteolytic system known to operate during germination—in general on higher-molecular-weight proteins, especially those synthesized early in germination (97). Thus, with spores of both *B. cereus* and *B. megaterium*, the proteins made in the first min-

utes of germination showed a high rate of turnover (25 to 80%/h), with proteins synthesized at the end of germination being degraded more slowly (3 to 4%/h) (70, 97). This proteolysis requires metabolic energy and produces amino acids. A protease-negative mutant of *B. cereus* did not show this proteolysis, but was otherwise normal (70). The mutants defective in degradation of the low-molecular-weight spore proteins showed normal turnover of proteins newly synthesized during germination (67). The energy-dependent proteolysis may also degrade the sequence-specific spore protease and glucose-6-phosphate dehydrogenase during spore germination (66, 91).

In addition to the protease active on the low-molecular-weight proteins, spores also contain protease activity on a wide variety of other substrates as noted above. In general, the level of this latter activity is similar to that in vegetative cells (8, 90). However, in at least one case a supposed spore protease was shown to be absorbed on the spore surface (114). Spores also contain peptidase activity toward a variety of substrates, again at levels similar to those in growing cells (90).

DNA synthesis and repair. Rapid DNA synthesis (i.e., chromosome replication) during spore germination does not begin until 30 to 90 min after the start of germination (53, 70, 85, 130). The same is true when permeabilized germinated spores are assayed (32). The initiation of this rapid DNA synthesis requires synthesis of one or more proteins (70, 85); however, their identity has not been established. One of these proteins could be DNA polymerase III, which has been reported to be absent from *B. subtilis* spores (17). However, other workers found this enzyme in the dormant spore (112). Both DNA polymerases I and II as well as DNA ligase are present in the dormant spore and throughout germination (17, 85, 112). In addition to chromosome replication late in germination, a slow rate of DNA synthesis, beginning soon after initiation of germination, has sometimes been observed (53, 70, 85, 130). However, it has not been observed during germination of all spores tested (123), and there is no agreement on whether it is replicative or repair synthesis; its significance is at present unclear.

DNA repair mechanisms can also operate during spore germination. These include mechanisms for repair of damage caused by both UV and ionizing radiation (59, 113). There appear to be two mechanisms for repair of UV damage: one to excise and repair thymine dimers, the normal lesion caused by UV light, and a second spore-specific system which attacks and repairs the thymidyl-thymidine adduct (initially called spore protoproduct) which results when dormant spores are irradiated with UV light (58, 105, 113, 118). The nature of this spore-specific repair system is not clear, but mutants defective in this repair process have been isolated and produce very UV-sensitive spores (58).

Cell wall. As noted earlier, much of the peptidoglycan in the spore cortex is degraded early in germination through the use of preexisting spore enzymes. Estimates of the magnitude of this peptidoglycan hydrolysis range from 30 to 80% of the total amount in the spore (35, 124). It appears likely that the material not hydrolyzed has a different structure from the material which is digested (124). The products of this hydrolysis have molecular weights of 10,000 to 20,000 and are excreted from the spore. Several studies have shown that the digestion fragments, in particular the DPA present in the fragments, are not reutilized (124). A number of lytic enzymes capable of depolymerization of spore cortex peptidoglycan have been isolated from dormant spores (12, 35, 124). Although several of these enzymes have been purified to some degree, their roles in peptidoglycan digestion during germination are unclear.

In addition to peptidoglycan hydrolases, spores of at least one species, *B. sphaericus*, contain a number of the enzymes needed for synthesis of new peptidoglycan during germination (36, 115), and some of these enzymes are synthesized rapidly during germination (116). Spores of *B. subtilis* also contain the enzymes for synthesis of nucleotide sugars (16). However, peptidoglycan synthesis during germination has not been studied extensively, although synthesis of these cell wall polymers is coordinated with the synthesis of other cell wall polymers such as teichoic acids (16).

Dormant spores of several species contain no teichoic acids, although there have been reports of cell wall-bound phosphate in several types of spores (62, 124). In *B. subtilis* and *B. licheniformis* rapid synthesis of teichoic acid begins only 20 to 40 min after initiation of spore germination, long before septum formation (11, 16). There is coordinate synthesis of the different classes of teichoic acid as well as coordinate synthesis of teichoic acid and cell wall peptidoglycan. In *B. subtilis*, enzymes for synthesis of some teichoic acid species are present in the dormant spore whereas enzymes for synthesis of other teichoic acids probably must be synthesized during germination (11). The latter is also true for the enzymes of teichoic acid synthesis in *B. licheniformis* (11).

CONCLUSIONS

As is evident from the preceding sections, the biochemistry of some aspects of forespore development and spore germination is understood—in particular some (but not all) aspects of energy and small-molecule metabolism as well as some aspects of macromolecular degradation. Similarly, work in recent years now strongly suggests that neither metabolism of exogenous germinants nor accumulation of significant levels of high-energy compounds is necessary for initiating germination. However, it is also clear that the biochemistry of a number of fundamental processes in forespore development and spore germination is not understood. In particular, questions remain concerning (i) the functioning of the double membrane of the forespore; (ii) the mechanisms controlling the sequential regulation of gene expression during both forespore development and spore germination; (iii) the nature of the reactions whereby spore enzymes become "turned off" during forespore maturation, thus resulting in spore dormancy; and (iv) the nature of the initial step(s) in spore germination and the possible role for metabolism of endogenous reserves in this step(s). Answers to several of these questions, in particular the first two, seem unlikely to come quickly. Similarly, the answers to the second two questions are at present unclear. However, a major reason for the latter is our basic lack of understanding at present of those physical and chemical parameters of *dormant* spores which cause their properties of dormancy and resistance. If these properties could be understood at the biochemical level, then the elucidation of the biochemistry of their acquisition (forespore development) or their loss (spore germination) would be much much simpler.

ACKNOWLEDGMENTS

The work carried out in my laboratory was supported by Public Health Service grant GM-19698 from the National Institutes of Health and by a grant from the Army Research Office.

LITERATURE CITED

1. **Albertini, A. M., M. L. Baldi, E. Ferrari, E. Isnenghi, M. T. Zambelli, and A. Galizzi.** 1979. Mutants of *Bacillus subtilis* affected in spore outgrowth. J. Gen. Microbiol. **110:**351–363.
2. **Andreoli, A. J., J. Saranto, P. A. Baecker, S. Suehiro, E. Escamilla, and A. Steiner.** 1975. Biochemical properties of forespores isolated from *Bacillus cereus*, p. 418–424. *In* P. Gerhardt, R. N. Costilow, and H. L. Sadoff (ed.), Spores VI. American Society for Microbiology, Washington, D.C.
3. **Andreoli, A. J., J. Saranto, N. Caliri, E. Escamilla,**

and E. Piña. 1978. Comparative study of proteins from forespore and mother-cell compartments of *Bacillus cereus*, p. 260–264. *In* G. Chambliss and J. C. Vary (ed.), Spores VII. American Society for Microbiology, Washington, D.C.

4. **Aronson, A. I., and P. C. Fitz-James.** 1976. Structure and morphogenesis of the bacterial spore coat. Bacteriol. Rev. **40**:360–402.

5. **Aronson, A. I., and N. K. Pandey.** 1978. Comparative structural and functional aspects of spore coats, p. 54–61. *In* G. Chambliss and J. C. Vary (ed.), Spores VII. American Society for Microbiology, Washington, D.C.

6. **Ben-Ze'ev, T. H., J. Hattori, Z. Silberstein, C. Tesone, and A. Torriani.** 1975. Ribonucleic acid polymerase from dormant and germinating spores of *Bacillus cereus* T, p. 472–477. *In* P. Gerhardt, R. N. Costilow, and H. L. Sadoff (ed.), Spores VI. American Society for Microbiology, Washington, D.C.

7. **Bergère, J.-L., C. Zevaco, C. Cherrier, and H. Petit-demange.** 1975. La germination de la spore de *Clostridium tyrobutyricum*. III. Hypothèse sur le méchanisme de la phase initiale. Ann. Microbiol. (Paris) **126A**:421–434.

8. **Bishop, H. L., L. K. Migita, and R. H. Doi.** 1969. Peptide synthesis by extracts from *Bacillus subtilis* spores. J. Bacteriol. **99**:771–778.

9. **Blankenship, L. C., and J. R. Mencher.** 1971. A disulfide reductase in spores of *Bacillus cereus* T. Can. J. Microbiol. **17**:1273–1277.

10. **Blankenship, L. C., and M. J. Pallansch.** 1966. Differential analysis of sulfhydryl and disulfide groups of intact spores. J. Bacteriol. **92**:1615–1617.

11. **Boylen, C. W., and J. C. Ensign.** 1968. Ratio of teichoic acid and peptidoglycan in cell walls of *Bacillus subtilis* following spore germination and during vegetative growth. J. Bacteriol. **96**:421–427.

12. **Brown, W. C., and R. C. Cuhel.** 1975. Surface-localized cortex-lytic enzyme in spores of *Bacillus cereus* T. J. Gen. Microbiol. **91**:429–432.

13. **Buu, A., and A. L. Sonenshein.** 1975. Nucleic acid synthesis and ribonucleic acid polymerase specificity in germinating and outgrowing spores of *Bacillus subtilis*. J. Bacteriol. **124**:190–200.

14. **Cheng, Y. S. E., and A. I. Aronson.** 1977. Alterations of spore coat processing and protein turnover in a *Bacillus cereus* mutant with a defective postexponential intracellular protease. Proc. Natl. Acad. Sci. U.S.A. **74**:1254–1258.

15. **Cheng, Y. E., P. C. Fitz-James, and A. I. Aronson.** 1978. Characterization of a *Bacillus cereus* protease mutant defective in an early stage of spore germination. J. Bacteriol. **133**:336–344.

16. **Chin, T., J. Younger, and L. Glaser.** 1968. Teichoic acids. VII. Synthesis of teichoic acids during spore germination. J. Bacteriol. **95**:2044–2050.

17. **Ciarrocchi, G., C. Attolini, F. Cobianchi, S. Riva, and A. Falaschi.** 1977. Modulation of deoxyribonucleic acid polymerase III level during the life cycle of *Bacillus subtilis*. J. Bacteriol. **131**:776–783.

18. **Cohen, A., H. Ben-Ze'ev, and Z. Silberstein.** 1975. Control of temporal gene expression during outgrowth of *Bacillus cereus* spores, p. 478–482. *In* P. Gerhardt, R. N. Costilow, and H. L. Sadoff (ed.), Spores VI. American Society for Microbiology, Washington, D.C.

19. **Crafts-Lighty, A., and D. J. Ellar.** 1980. The structure and function of the outer membrane in dormant and germinated spores of *Bacillus megaterium*. J. Appl. Bacteriol. **48**:135–145.

20. **Dawes, I. W., and H. O. Halvorson.** 1972. Membrane synthesis during outgrowth of bacterial spores, p. 449–455. *In* H. O. Halvorson, R. Hanson, and L. L. Campbell (ed.), Spores V. American Society for Microbiology, Washington, D.C.

21. **Deutscher, M. P., P. Chambon, and A. Kornberg.** 1968. Biochemical studies of bacterial sporulation and germination. XI. Protein-synthesizing systems from vegetative cells and spores of *Bacillus megaterium*. J. Biol. Chem. **243**:5117–5125.

22. **Dignam, S. S., and P. Setlow.** 1980. *In vivo* and *in vitro* synthesis of the spore specific proteins A and C of *Bacillus megaterium*. J. Biol. Chem. **255**:8417–8423.

23. **Dills, S. S., and J. C. Vary.** 1978. An evaluation of respiration chain-associated functions during initiation of germination of *Bacillus megaterium* spores. Biochim. Biophys. Acta **541**:301–311.

24. **Dring, G. J., and G. W. Gould.** 1975. Electron transport-linked metabolism during germination of *Bacillus cereus* spores, p. 488–494. *In* P. Gerhardt, R. N. Costilow, and H. L. Sadoff (ed.), Spores VI. American Society for Microbiology, Washington, D.C.

25. **Eisenstadt, E., and S. Silver.** 1972. Restoration of cation transport during germination, p. 443–448. *In* H. O. Halvorson, R. Hanson, and L. L. Campbell (ed.), Spores V. American Society for Microbiology, Washington, D.C.

26. **Ellar, D. J.** 1978. Spore specific structures and their function. Symp. Soc. Gen. Microbiol. **28**:295–324.

27. **Ellar, D. J., M. W. Eaton, C. Hogarth, B. J. Wilkinson, J. Deans, and J. La Nauze.** 1975. Comparative biochemistry and function of forespore and mother-cell compartments during sporulation of *Bacillus megaterium* cells, p. 425–433. *In* P. Gerhardt, R. N. Costilow, and H. L. Sadoff (ed.), Spores VI. American Society for Microbiology, Washington, D.C.

28. **Ellar, D. J., and J. A. Postgate.** 1974. Characterization of forespores isolated from *Bacillus megaterium* at different stages of development into mature spores, p. 21–40. *In* A. N. Barker, G. W. Gould, and J. Wolf (ed.), Spore Research 1973. Academic Press, London.

29. **Foerster, C. W., and H. F. Foerster.** 1973. Glutamic acid decarboxylase in spores of *Bacillus megaterium* and its possible involvement in spore germination. J. Bacteriol. **114**:1090–1098.

30. **Foerster, H. F.** 1971. γ-Aminobutyric acid as a required germinant for mutant spores of *Bacillus megaterium*. J. Bacteriol. **108**:817–823.

31. **Foerster, H. F.** 1972. Spore pool glutamic acid as a metabolite in germination. J. Bacteriol. **111**:437–442.

32. **Fujita, Y., T. Komano, and H. Tanooka.** 1973. Increasing activity of germinating *Bacillus subtilis* spores to incorporate thymidine triphosphate into deoxyribonucleic acid after detergent treatment. J. Bacteriol. **113**:558–564.

33. **Fujita, Y., R. Ramaley, and E. Freese.** 1977. Location and properties of glucose dehydrogenase in sporulating cells and spores of *Bacillus subtilis*. J. Bacteriol. **132**:282–293.

34. **Galizzi, A., A. M. Albertini, M. L. Baldi, E. Ferrari, E. Isnenghi, and M. T. Zambelli.** 1978. Genetic studies of spore germination and outgrowth in *Bacillus subtilis*, p. 150–157. *In* G. Chambliss and J. C. Vary (ed.), Spores VII. American Society for Microbiology, Washington, D.C.

35. **Gould, G. W.** 1969. Germination, p. 397–444. *In* G. W. Gould and A. Hurst (ed.), The bacterial spore. Academic Press, London.

36. **Guinand, M., M. J. Vacheron, G. Michel, and D. J. Tipper.** 1979. Location of peptidoglycan lytic enzymes in *Bacillus sphaericus*. J. Bacteriol. **138**:126–132.

37. **Hansen, J. N., G. Spiegelman, and H. O. Halvorson.** 1970. Bacterial spore outgrowth: its regulation. Science **168**:1291–1298.

38. **Hausenbauer, J. M., W. M. Waites, and P. Setlow.** 1977. Biochemical properties of *Clostridium bifermentans* spores. J. Bacteriol. **129:**1148–1150.

39. **Henner, D. J., and W. Steinberg.** 1979. Transfer ribonucleic acid synthesis during sporulation and spore outgrowth in *Bacillus subtilis* studied by two-dimensional polyacrylamide gel electrophoresis. J. Bacteriol. **140:**555–566.

40. **Hogarth, C., and D. J. Ellar.** 1978. Calcium accumulation during sporulation of *Bacillus megaterium* KM. Biochem. J. **176:**197–203.

41. **Hogarth, C., and D. J. Ellar.** 1979. Energy dependence of calcium accumulation during sporulation of *Bacillus megaterium* KM. Biochem. J. **178:**627–632.

42. **Imae, Y., and J. L. Strominger.** 1976. Conditional spore cortex-less mutants of *Bacillus sphaericus* 9602. J. Biol. Chem. **251:**1493–1499.

43. **Janoff, A. S., R. T. Coughlin, F. M. Racine, E. J. McGroarty, and J. C. Vary.** 1979. Use of electron spin resonance to study *Bacillus megaterium* spore membranes. Biochem. Biophys. Res. Commun. **89:**569–570.

44. **Jeng, Y., and R. H. Doi.** 1974. Messenger ribonucleic acid of dormant spores of *Bacillus subtilis*. J. Bacteriol. **119:**514–521.

45. **Kennett, R. H., and N. Sueoka.** 1971. Gene expression during outgrowth of *Bacillus subtilis* spores. The relationship between gene order on the chromosome and temporal sequence of enzyme synthesis. J. Mol. Biol. **60:**31–44.

46. **Keynan, A.** 1978. Spore structure and its relations to resistance, dormancy, and germination, p. 43–53. *In* G. Chambliss and J. C. Vary (ed.), Spores VII. American Society for Microbiology, Washington, D.C.

47. **Keynan, A., and Z. Evenchik.** 1969. Activation, p. 359–396. *In* G. W. Gould and A. Hurst (ed.), The bacterial spore. Academic Press, London.

48. **Kieras, R. M., R. A. Preston, and H. A. Douthit.** 1978. Isolation of stable ribosomal subunits from spores of *Bacillus cereus*. J. Bacteriol. **136:**209–218.

49. **Klier, A., and M.-M. Lecadet.** 1974. Evidence in favor of the modification *in vivo* of RNA polymerase subunits during sporogenesis in *Bacillus thuringiensis*. Eur. J. Biochem. **47:**111–119.

50. **Kobayashi, Y., and H. O. Halvorson.** 1968. Evidence for a defective protein synthesizing system in dormant spores of *Bacillus cereus*. Arch. Biochem. Biophys. **123:**622–663.

51. **Kobayashi, Y., W. Steinberg, A. Higa, H. O. Halvorson, and C. Levinthal.** 1965. Sequential synthesis of macromolecules during outgrowth of bacterial spores, p. 200–212. *In* L. L. Campbell and H. O. Halvorson (ed.), Spores III. American Society for Microbiology, Ann Arbor, Mich.

52. **Kornberg, A., J. A. Spudich, D. L. Nelson, and M. P. Deutscher.** 1968. Origin of proteins in sporulation. Annu. Rev. Biochem. **37:**51–78.

53. **Lammi, C. J., and J. C. Vary.** 1972. Deoxyribonucleic acid synthesis during outgrowth of *Bacillus megaterium* QM B1551 spores, p. 277–282. In H. O. Halvorson, R. Hanson, and L. L. Campbell (ed.), Spores V. American Society for Microbiology, Washington, D.C.

54. **Levinson, H. S., and M. T. Hyatt.** 1966. Sequence of events during *Bacillus megaterium* spore germination. J. Bacteriol. **91:**1811–1818.

55. **Maia, J. C. C., P. Kerjan, and J. Szulmajster.** 1971. DNA-dependent RNA polymerase from vegetative cells and from spores of *Bacillus subtilis*. FEBS Lett. **13:**269–274.

56. **Maruyama, T., M. Otani, K. Sano, and C. Umezawa.** 1980. Glucose catabolism during germination of *Bacillus megaterium* spores. J. Bacteriol. **141:**1443–1446.

57. **Moir, A., E. Lafferty, and D. A. Smith.** 1979. Genetic analysis of spore germination mutants of *Bacillus subtilis* 168: the correlation of phenotype with map location. J. Gen. Microbiol. **111:**165–180.

58. **Munakata, N., and C. Rupert.** 1972. Genetically controlled removal of "spore photoproduct" from deoxyribonucleic acid of ultraviolet-irradiated *Bacillus subtilis* spores. J. Bacteriol. **111:**192–198.

59. **Munakata, N., and C. Rupert.** 1974. Dark repair of DNA containing "spore photoproduct" in *Bacillus subtilis*. Mol. Gen. Genet. **130:**239–250.

60. **Munoz, L., Y. Sadaie, and R. Doi.** 1978. Spore coat protein of *Bacillus subtilis*: structure and precursor synthesis. J. Biol. Chem. **253:**6694–6701.

61. **Nelson, D. L., and A. Kornberg.** 1970. Biochemical studies of bacterial sporulation and germination. XVIII. Free amino acids in spores. J. Biol. Chem. **245:**1101–1107.

62. **Nelson, D. L., and A. Kornberg.** 1970. Biochemical studies of bacterial sporulation and germination. XIX. Phosphate metabolism during sporulation. J. Biol. Chem. **245:**1137–1145.

63. **Nelson, D. L., and A. Kornberg.** 1970. Biochemical studies of bacterial sporulation and germination. XX. Phosphate metabolism during germination. J. Biol. Chem. **245:**1146–1155.

64. **Nickerson, K. W., J. DePinto, and L. A. Bulla.** 1975. Lipid metabolism during bacterial growth, sporulation, and germination: kinetics of fatty acid and macromolecular synthesis during spore germination and outgrowth of *Bacillus thuringiensis*. J. Bacteriol. **121:**227–233.

65. **Oh, Y. K., and E. Freese.** 1976. Manganese requirement of phosphoglycerate phosphomutase and its consequences for growth and sporulation of *Bacillus subtilis*. J. Bacteriol. **127:**739–746.

66. **Orlowski, M., and M. Goldman.** 1975. Inactivation of glucose 6-phosphate dehydrogenase during germination and outgrowth of *Bacillus cereus* T endospores. Biochem. J. **148:**259–268.

67. **Postemsky, C. J., S. S. Dignam, and P. Setlow.** 1978. Isolation and characterization of mutants of *Bacillus megaterium* having decreased levels of spore protease. J. Bacteriol. **135:**841–850.

68. **Prasad, C., M. Diesterhaft, and E. Freese.** 1972. Initiation of spore germination in glycolytic mutants of *Bacillus subtilis*. J. Bacteriol. **110:**321–328.

69. **Racine, F. M., S. S. Dills, and J. C. Vary.** 1979. Glucose-triggered germination of *Bacillus megaterium* spores. J. Bacteriol. **138:**442–445.

70. **Rana, R. S., and H. O. Halvorson.** 1972. Nature of deoxyribonucleic acid synthesis and its relationship to protein synthesis during outgrowth of *Bacillus cereus* T. J. Bacteriol. **109:**606–615.

71. **Rhaese, H. J., H. Dichtelmüller, R. Grade, and R. Groscurth.** 1975. High phosphorylated nucleotides involved in regulation of sporulation in *Bacillus subtilis*, p. 335–340. *In* P. Gerhardt, R. N. Costilow, and H. L. Sadoff (ed.), Spores VI. American Society for Microbiology, Washington, D.C.

72. **Rossignol, D. P., and J. C. Vary.** 1977. A unique method for studying the initiation of *Bacillus megaterium* spore germination. Biochem. Biophys. Res. Commun. **79:**1098–1103.

73. **Rossignol, D. P., and J. C. Vary.** 1979. Biochemistry of L-proline-triggered germination of *Bacillus megaterium* spores. J. Bacteriol. **138:**431–441.

74. **Rossignol, D. P., and J. C. Vary.** 1979. L-Proline site for triggering *Bacillus megaterium* spore germination. Biochem. Biophys. Res. Commun. **89:**547–551.

75. **Scandella, C. J., and A. Kornberg.** 1969. Biochemical studies of bacterial sporulation and germination. XV.

Fatty acids in growth, sporulation, and germination of *Bacillus megaterium*. J. Bacteriol. **98**:82–86.

76. **Schmit, J. C., and S. Brody.** 1975. *Neurospora crassa* conidial germination: role of endogenous amino acid pools. J. Bacteriol. **124**:232–242.

77. **Scott, I. R., and D. J. Ellar.** 1978. Study of calcium dipicolinate release during bacterial spore germination by using a new, sensitive assay for dipicolinate. J. Bacteriol. **135**:133–137.

78. **Scott, I. R., G. S. A. B. Stewart, M. A. Koncewicz, D. J. Ellar, and A. Crafts-Lighty.** 1978. Sequence of biochemical events during germination of *Bacillus megaterium* spores, p. 95–103. *In* G. Chambliss and J. C. Vary (ed.), Spores VII. American Society for Microbiology, Washington, D.C.

79. **Senesi, S., R. A. Felicioli, P. L. Ipata, and G. Falcone.** 1975. Regulation of polyribonucleotide turnover in vegetative cells and spores of *Bacillus subtilis*, p. 265–270. *In* P. Gerhardt, R. N. Costilow, and H. L. Sadoff (ed.), Spores VI. American Society for Microbiology, Washington, D.C.

80. **Setlow, B., and P. Setlow.** 1977. Levels of oxidized and reduced pyridine nucleotides in dormant spores and during growth, sporulation, and spore germination of *Bacillus megaterium*. J. Bacteriol. **129**:857–865.

81. **Setlow, B., and P. Setlow.** 1977. Levels of acetyl coenzyme A, reduced and oxidized coenzyme A, and coenzyme A in disulfide linkage to protein in dormant and germinating spores and sporulating cells of *Bacillus megaterium*. J. Bacteriol. **132**:444–452.

82. **Setlow, B., and P. Setlow.** 1978. Levels of cyclic GMP in dormant, germinated, and outgrowing spores and growing and sporulating cells of *Bacillus megaterium*. J. Bacteriol. **136**:433–436.

83. **Setlow, B., and P. Setlow.** 1980. Localization of low-molecular-weight basic proteins in *Bacillus megaterium* spores by cross-linking with ultraviolet light. J. Bacteriol. **139**:486–494.

84. **Setlow, B., L. K. Shay, J. C. Vary, and P. Setlow.** 1977. Production of large amounts of acetate during germination of *Bacillus megaterium* spores in the absence of exogenous carbon sources. J. Bacteriol. **132**:744–746.

85. **Setlow, P.** 1973. Deoxyribonucleic acid synthesis and deoxynucleotide metabolism during bacterial spore germination. J. Bacteriol. **114**:1099–1107.

86. **Setlow, P.** 1974. Polyamine levels during growth, sporulation, and spore germination of *Bacillus megaterium*. J. Bacteriol. **117**:1171–1177.

87. **Setlow, P.** 1974. Spermidine biosynthesis during germination and subsequent vegetative growth of *Bacillus megaterium* spores. J. Bacteriol. **120**:311–315.

88. **Setlow, P.** 1974. Percent charging of transfer ribonucleic acid and levels of ppGpp and pppGpp in dormant and germinated spores of *Bacillus megaterium*. J. Bacteriol. **118**:1067–1074.

89. **Setlow, P.** 1975. Energy and small-molecule metabolism during germination of *Bacillus* spores, p. 443–450. *In* P. Gerhardt, R. N. Costilow, and H. L. Sadoff (ed.), Spores VI. American Society for Microbiology, Washington, D.C.

90. **Setlow, P.** 1975. Protease and peptidase activities in growing and sporulating cells and dormant spores of *Bacillus megaterium*. J. Bacteriol. **122**:642–649.

91. **Setlow, P.** 1976. Purification of a highly specific protease from spores of *Bacillus megaterium*. J. Biol. Chem. **251**:7853–7862.

92. **Setlow, P., C. Gerard, and J. Ozols.** 1980. The amino acid sequence specificity of a protease from spores of *Bacillus megaterium*. J. Biol. Chem. **255**:3624–3628.

93. **Setlow, P., and A. Kornberg.** 1969. Biochemical studies of bacterial sporulation and germination. XVII.

Sulfhydryl and disulfide levels in dormancy and germination. J. Bacteriol. **100**:1155–1160.

94. **Setlow, P., and J. Ozols.** 1979. Covalent structure of protein A: a low molecular weight protein degraded during germination of *Bacillus megaterium* spores. J. Biol. Chem. **254**:11938–11942.

95. **Setlow, P., and J. Ozols.** 1980. Covalent structure of protein C. A second major low molecular weight protein degraded during *Bacillus megaterium* spore germination. J. Biol. Chem. **255**:8413–8416.

96. **Setlow, P., and J. Ozols.** 1980. The complete covalent structure of protein B: the third major protein degraded during germination of *Bacillus megaterium* spores. J. Biol. Chem. **255**:10445–10450.

97. **Setlow, P., and G. Primus.** 1975. Protein degradation and amino acid metabolism during germination of *Bacillus megaterium* spores, p. 451–457. *In* P. Gerhardt, R. N. Costilow, and H. L. Sadoff (ed.), Spores VI. American Society for Microbiology, Washington, D.C.

98. **Setlow, P., G. Primus, and M. P. Deutscher.** 1974. Absence of 3′-terminal residues from transfer ribonucleic acid of dormant spores of *Bacillus megaterium*. J. Bacteriol. **117**:126–132.

99. **Setlow, P., and W. M. Waites.** 1976. Identification of several unique low-molecular-weight basic proteins in dormant spores of *Clostridium bifermentans* and their degradation during spore germination. J. Bacteriol. **127**:1015–1017.

100. **Shay, L. K., and J. C. Vary.** 1978. Biochemical studies on glucose initiated germination in *Bacillus megaterium*. Biochim. Biophys. Acta **538**:284–292.

101. **Singh, R. P., B. Setlow, and P. Setlow.** 1977. Levels of small molecules and enzymes in the mother cell compartment and the forespore of sporulating *Bacillus megaterium*. J. Bacteriol. **130**:1130–1138.

102. **Singh, R. P., and P. Setlow.** 1979. Regulation of phosphoglycerate phosphomutase in developing forespores and germinated spores of *Bacillus megaterium* by the level of free manganous ion. J. Bacteriol. **139**:889–898.

103. **Sloma, A., and I. Smith.** 1980. RNA synthesis during spore germination in *Bacillus subtilis*. Mol. Gen. Genet. **175**:113–120.

104. **Smith, D. A., A. Moir, and R. Sammons.** 1978. Progress in genetics of spore germination in *Bacillus subtilis*, p. 158–163. *In* G. Chambliss and J. C. Vary (ed.), Spores VII. American Society for Microbiology, Washington, D.C.

105. **Stafford, R. S., and J. E. Donnellan, Jr.** 1968. Photochemical evidence for conformation changes in DNA during germination of bacterial spores. Proc. Natl. Acad. Sci. U.S.A. **59**:822–829.

106. **Steinberg, W., and H. O. Halvorson.** 1968. Timing of enzyme synthesis during outgrowth of *Bacillus cereus*. I. Ordered enzyme synthesis. J. Bacteriol. **95**:469–478.

107. **Stelma, G. N., A. I. Aronson, and P. C. Fitz-James.** 1979. A *B. cereus* mutant defective in spore coat deposition. J. Gen. Microbiol. **116**:173–185.

108. **Strange, R. E., and J. R. Hunter.** 1969. Outgrowth and the synthesis of macromolecules, p. 445–484. *In* G. W. Gould and A. Hurst (ed.), The bacterial spore. Academic Press, London.

109. **Swerdlow, R. D., C. L. Green, B. Setlow, and P. Setlow.** 1979. Identification of an NADH-linked disulfide reductase from *Bacillus megaterium* specific for disulfides containing pantethine-4′,4″-diphosphate moieties. J. Biol. Chem. **254**:6835–6837.

110. **Tallentire, A.** 1970. Radiation resistance of spores. J. Appl. Bacteriol. **33**:141–146.

111. **Tanooka, H., H. Terano, and H. Otsuka.** 1971. Increase of thymidine, thymidylate and deoxycytidine

kinase activities during germination of bacterial spores. Biochim. Biophys. Acta **228**:26–37.

112. **Terano, H., Y. Fujita, S. Hiroshi, H. Kadota, and T. Komano.** 1975. Three kinds of deoxyribonucleic acid polymerases in spores of *Bacillus subtilis*. Agric. Biol. Chem. **39**:2057–2063.

113. **Terano, H., H. Tanooka, and H. Kadota.** 1971. Repair of radiation damage to deoxyribonucleic acid in germinating spores of *Bacillus subtilis*. J. Bacteriol. **106**:925–930.

114. **Tesone, C., and A. Torriani.** 1975. Protease associated with spores of *Bacillus cereus*. J. Bacteriol. **124**:593–594.

115. **Tipper, D. J., and P. E. Linnett.** 1976. Distribution of peptidoglycan synthetase activities between sporangia and forespores in sporulating cells of *Bacillus sphaericus*. J. Bacteriol. **126**:213–221.

116. **Tipper, D. J., I. Pratt, M. Guinand, S. C. Holt, and P. E. Linnett.** 1977. Control of peptidoglycan synthesis during sporulation in *Bacillus sphaericus*. p. 50–68. *In* D. Schlessinger (ed.), Microbiology—1977. American Society for Microbiology, Washington, D.C.

117. **Torriani, A., and C. Levinthal.** 1967. Ordered synthesis of proteins during outgrowth of spores of *Bacillus cereus*. J. Bacteriol. **94**:176–183.

118. **Varghese, A. J.** 1970. Photochemistry of thymidine in ice. Biochemistry **9**:4781–4787.

119. **Vary, J. C.** 1972. Spore germination of *Bacillus megaterium* QM B1551 mutants. J. Bacteriol. **112**:640–642.

120. **Vary, J. C.** 1973. Germination of *Bacillus megaterium* spores after various extraction procedures. J. Bacteriol. **116**:797–802.

121. **Vary, J. C., and A. Kornberg.** 1970. Biochemical studies of bacterial sporulation and germination. XXI. Temperature-sensitive mutants for initiation of germination. J. Bacteriol. **101**:327–329.

122. **Vold, B.** 1974. Degree of completion of 3′-terminus of transfer ribonucleic acids of *Bacillus subtilis* 168 at various development stages and asporogenous mutants. J. Bacteriol. **117**:1361–1362.

122a.**Vold, B. S., and S. Minatogawa.** 1972. Characterization of changes in transfer ribonucleic acids during sporulation in *Bacillus subtilis*, p. 254–263. *In* H. O. Halvorson, R. Hanson, and L. L. Campbell (ed.), Spores V. American Society for Microbiology, Washington, D.C.

123. **Wake, R. G.** 1967. A study of the possible extent of synthesis of repair DNA during germination of *Bacillus subtilis* spores. J. Mol. Biol. **25**:217–234.

124. **Warth, A. D.** 1978. Molecular structure of the bacterial spore. Adv. Microb. Physiol. **17**:1–47.

125. **Watabe, K., S. Iida, R. Wada, T. Ichikawa, and M. Kondo.** 1980. Characterization of forespores isolated from *Bacillus subtilis* at each stage of sporulation. Microbiol. Immunol. **24**:79–82.

126. **Wilkinson, B. J., J. A. Deans, and D. J. Ellar.** 1975. Biochemical evidence for the reversed polarity of the outer membrane of the bacterial forespore. Biochem. J. **152**:561–569.

127. **Wilkinson, B. J., and D. J. Ellar.** 1975. Morphogenesis of the membrane-bound electron-transport system in sporulating *Bacillus megaterium* KM. Eur. J. Biochem. **55**:131–139.

128. **Wilkinson, B. J., D. J. Ellar, I. R. Scott, and M. A. Konceivicz.** 1977. Rapid chloramphenicol-resistant activation of membrane electron transport on germination of *Bacillus megaterium* spores. Nature (London) **226**:174–176.

129. **Yeh, E. C., and W. Steinberg.** 1978. The effect of gene position, gene dosage and a regulatory mutation on the temporal sequence of enzyme synthesis accompanying outgrowth of *Bacillus subtilis* spores. Mol. Gen. Genet. **158**:287–296.

130. **Yoshikawa, H.** 1965. DNA synthesis during germination of *Bacillus subtilis* spores. Proc. Natl. Acad. Sci. U.S.A. **53**:1476–1483.

Advances in the Genetics of *Bacillus subtilis* Differentiation

P. J. PIGGOT, A. MOIR, AND D. A. SMITH

*National Institute for Medical Research, Mill Hill, London, NW7 1AA, and Department of Genetics,
University of Birmingham, Birmingham, B15 2TT, England*

INTRODUCTION

Much current effort is being devoted to the analysis of genes associated with differentiation in *Bacillus subtilis*, and much of the progress achieved has depended on the existence of a body of information about the relevant genetic loci. We review here the advances during the past few years in the identification and characterization of such loci, and we consider their implications for the nature and control of the differentiation cycle. We emphasize particularly the links between the different parts of the cycle, namely, spore formation, germination, and outgrowth, illustrating them with specifically genetic examples. These links have implications for the analysis of regulation at the molecular level during the differentiation cycle. We also discuss genetic techniques that are new to *B. subtilis* and that show promise for the future. We have not attempted a comprehensive review; in particular, we have not considered in any detail several related areas of considerable interest covered elsewhere in this volume in the reviews on biochemistry, molecular biology, and cloning strategies.

GENETIC LOCI

Many loci have been identified by the genetic mapping of relevant mutations causing blocks at different places in the differentiation cycle (Fig. 1, Table 1). Among the mutations recognized, most of those that block sporulation (*spo*) impose an absolute block, whereas those that block germination (*ger*) and outgrowth (*out*) are con-ditional. This is a consequence of the difficulty of isolating and recognizing absolute *ger* or *out* mutants. In addition to those that block differentiation, some mutations alter differentiation without blocking it; so far, these are only available for spore formation.

Sporulation Loci

Most genetic studies of the differentiation cycle have concentrated on spore formation, and the majority of loci identified have been *spo* loci, of which some 35 have been described. Piggot and Coote provided an extensive review in 1976 (56), and we will limit our discussion to material published since then. In addition, fuller descriptions of the phenotypes associated with *spo0* loci and *spoII* loci were given by Hoch et al. (36) and Young and Mandelstam (79), respectively.

The *spoVF* (or *dpa*) locus deserves detailed consideration. There has been considerable debate about the role of dipicolinic acid (DPA) in the maintenance of heat resistance, influenced in part by the existence of mutants which apparently form DPA-less but heat-resistant spores. Strains CB22 and DG47 with *spoVF* mutations form DPA-less, heat-sensitive spores (9, 13). Balassa et al. (7) showed that addition of very small amounts (about 2 µg/ml) of DPA to sporulating cells of CB22 allowed them to form heat-resistant spores; this is also true for DG47 (P. Piggot, unpublished data). The amount of DPA incorporated into the spores was increased by this addition but only to a level that was very low compared with that present in wild-type spores and barely distinguishable from the level

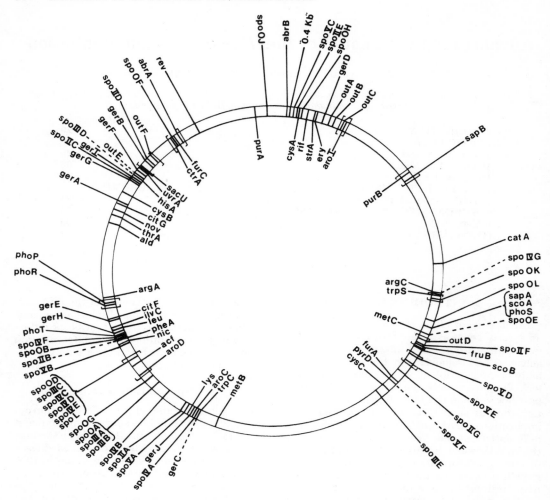

FIG. 1. *Genetic map of* B. subtilis *168 showing differentiation-associated loci. The distribution of genetic markers is based on that of Henner and Hoch (33). The relative positions of the loci are estimates and are based on the data in reference 56 and Table 1. Groups of loci whose orientation relative to each other has not been determined are shown as mapping at a single site. Where no orientation is known and linkage is uncertain, this is indicated by a dotted line. Map positions placed in parentheses have not been ordered relative to outside markers and may not have been ordered relative to each other.*

in untreated mutant cells. This work suggests a role for DPA in determining heat resistance; the mutants offer the possibility of manipulating the level of DPA, or its analogs, within the spore, so as to explore that role further. In this context, addition of exogenous DPA restored heat resistance in the presence of chloramphenicol but not KCN (7).

Spore formation involves two cell types, the mother cell and the forespore; transcription and translation take place in both (19, 63). Lencastre and Piggot described a method for identifying which *spo* loci are expressed in each cell type

(46). Asporogenous mutants were transformed with *spo⁺* DNA so as to be able to form spores with a minimum of DNA replication between transformation and spore formation. Under the conditions used, it was likely that only one of the two copies of the chromosome in the sporangium was transformed to *spo⁺*. If the *spo⁺* chromosome segregated into the mother cell, and expression there was sufficient for spore formation, then the mutant chromosome would segregate into the forespore, and the resulting heat-resistant spores would be genetically asporogenous. Alternatively, if the *spo⁺* chromosome seg-

TABLE 1. *Description of sporulation, germination and outgrowth loci*[a]

Locus	Comments
Sporulation	
spo0A	Spo⁻ now established as recessive to Spo⁺ (73). Map order, *spo0C spo0A strC spoIIIA* (36).
spo0C	Relationship to *spo0A* uncertain (36, 73).
spo0F	Associated with synthesis of highly phosphorylated nucleotides (61).
spo0J	Map order *purA spo0J novA nalA gua-1* (74). Possibly involved in RNA polymerase modification (10).
spo0L	Near *spo0K* (J. A. Hoch, unpublished data; listed by Henner and Hoch [33]).
spoIIIA	Map order *spo0A strC spoIIIA* (36).
spoIVC	33% cotransformed with *aroD* (P. J. Piggot and K. F. Chak, unpublished data).
spoVC	Map order *gua abrB tms-26 spoVC lysS* (53).
spoVF	Also called *dpa* (7); mutations *spoVF224* (DG 47) and *dpa-1* (CB22) are probably in the same locus. 50% cotransduced with *pyrD1* (7; P. J. Piggot, unpublished data). Mutants form octanol- and chloroform-resistant spores (9, 13). They form heat-resistant spores in the presence of DPA (7; Piggot, unpublished data).
spoL	17% cotransformed with *aroD*; mutant shows "decadent" sporulation (8).
"0.4-kb gene"	The first mutation in *B. subtilis* generated by use of restriction enzymes was formed by deleting part of the "0.4-kb gene" (R. Losick, this volume). During spore formation, the mutant acquires toluene resistance normally, but is delayed in attaining heat and lysozyme resistance; only 5 to 10% of the population ever display these later resistances. Germination characteristics not yet determined. Maps between *abrB* and *tms-26* (31).
abrA	Antibiotic resistant. Mutation suppresses some of the pleiotropic effects of *spo0* mutations. 50% cotransformed with *ctrA*, 80% with *furC* (72).
abrB	Map order *gua recG abrB pac tms-26 lysS* (74). Same locus as *cpsX* (30) and probably as *absA*, *absB* (40) and *tolA* (39). Ribosomes from mutants show altered proteins on polyacrylamide gels.
abrC	Maps close to, or in, *spo0A* locus. A *spo0A12 abrC1* double mutant produced spores at a low frequency (72).
absA, B	See *abrB*.
catA	Overproduces extracellular proteases; can sporulate in certain circumstances in presence of glucose (41). Identical to *scoC* (51) and possibly *hpr* (34).
cpsX	See *abrB*.
dpa	See *spoVF*.
phoS	Mutation leads to constitutive production of alkaline phosphatase; uncertain relationship to sporulation. Mutations 78% cotransduced with *metC3*, 37% with *argC4* (58). Deletions of this region also lead to constitutive enzyme production, suggesting that the gene product exerts negative control on alkaline phosphatase synthesis (R. S. Buxton and P. J. Piggot, unpublished data).
phoT	Mutation leads to constitutive production of alkaline phosphatase; uncertain relationship to sporulation. Map order *phe phoT leu phoR argA*; 65% cotransduced with *phe-12*; 80% with *leuA8* (Piggot, unpublished data).
sapA	Mutations overcome sporulation-phosphatase-negative phenotype of early-blocked *spo* mutants. 74% cotransduced with *metC3*; 23% with *argC4*; maps very close to, and not definitely separate from, *phoS* (58) and *scoA* (42).
sapB	88% cotransduced with *purB6*; 32%, with *tre-12* (58).
scoA	Mutation causes protease and phosphatase overproduction. Delayed spore formation; 68% cotransduced with *metC3*; 45%, with *argC4* (51).
scoB	*scoD* is probably the same locus as *scoB*. Protease and phosphatase overproduction; delayed spore formation. 70% cotransduced with *pyrA*, 50% with *metC*; map order *scoC argC scoA metC scoB pyrA* (18, 42).
scoC	See *catA*.
Germination	
gerA	Defective in germination response to alanine and related amino acids (52, 66, 76). 70 to 90% linked to *citG4* by SPP1 transduction (66). Map order *cysA gerA citG*; *gsp-10* (2) may map in this locus.
gerB	Defective in germination response to the combination of glucose, fructose, asparagine, and KCl. 30% cotransduced with *hisA1*, <5% with *cysB* by PBS1. Map order *cysB hisA gerB* (52).
gerC	Temperature-sensitive germination in alanine (76). Has not been separated from other linked germination and growth-associated mutations in the original isolate. Maps near *trpC* (52).

TABLE 1. *Continued*

Locus	Comments
gerD	Defective in germination in a range of germinants (52). Very slow response is improved by the addition of monovalent cations (R. Warburg, personal communication). 77 to 88% cotransduced with *cysA* by PBS1. Order *cysA ery gerD* (52).
gerE	Defective in a range of germinants. Blocked at a late stage in germination (after loss of spore heat resistance). Spores are coat defective (A. Moir, in preparation). 90% cotransduced with *citF2* by SPP1. Order *citF gerE ilvB*. Present on an SPβ defective transducing phage (A. Moir, unpublished data).
gerF	Defective in germination in a range of germinants. Cotransduced with *hisA* at 89 to 93% by PBS1 (52), at 58% by SPP1 (Warburg, personal communication). Probable order is *cysB hisA gerF* (52).
gerG (pgk)	Mutant lacks phosphoglycerate kinase activity (20). Sporulates poorly, and spores are reported to germinate poorly in alanine (59). 58% cotransduced with *cysB* and 38% with *hisA* by PBS1. Order *cysB pgk hisA* (52).
gerH	Defective in germination in a range of germinants. 7% cotransduced with *citF* by SPP1, 85% with *hemA* by PBS1. Probably distinct from *gerE*. Germination phenotype of transductants has not been tested (S. Taylor and D. A. Smith, unpublished data).
gerI	Defective in germination in a range of germinants. 83% cotransduced with *hisA* by PBS1 (Taylor and Smith, unpublished data). May be identical to *gerF*.
gerJ	Defective in germination in a range of germinants. Blocked at a late stage in germination, after loss of spore heat resistance. Development of resistance to chemicals during sporulation is delayed and incomplete (R. J. Warburg and A. Moir, in preparation; R. J. Warburg, this volume).
fruB	Defective in fructose-1-phosphate kinase. Mutant is also defective in response to fructose as a component of germinant mixtures (59). The mutation responsible for enzyme deficiency is 32% cotransduced by PBS1 with *pyrD* (26), but the germination phenotype of transductants has not been examined. Map order is *pyrD fruB pts metC*.
Outgrowth	
outA (gspIV)	Temperature-sensitive in outgrowth. RNA, protein, and DNA synthesis resumes, but cell division is impaired, and large, spherical cells are formed. Mutants are 44 to 54% cotransduced with *cysA* by PBS1 (2, 25). Map order *cysA ery outA*.
outB (gsp-81)	Temperature sensitive. Blocked in outgrowth before the synthesis of most RNA and protein (1, 23). 72% cotransduced with *aroI* by PBS1. Map order *narB aroI outB* (1).
outC (gsp-25)	Temperature sensitive. Blocked in outgrowth after RNA and protein synthesis but before the synthesis of DNA. 85% cotransformed with *aroI* (2); order unknown.
outD (gsp-1)	Temperature sensitive. Blocked in outgrowth. Protein and DNA synthesis reduced. 17% cotransduced with *pyrA* by PBS1. Map order *outD fur pyrA* (25).
outE (gsp-42)	Temperature sensitive. Blocked in outgrowth. Extent of RNA synthesis is normal, protein synthesis is reduced, and DNA synthesis is prevented. 10% cotransduced with *cysB* by PBS1; 23% cotransformed with *hisA*. Map order uncertain (2).
outF (gsp-4)	Temperature sensitive. Blocked in outgrowth. Very little RNA or protein synthesis; no DNA synthesis. 30% cotransduced with *hisA* by PBS1. Map order *hisA outF glyC* (2).

a The loci listed are not all-inclusive, but only update the descriptions and map locations given previously in references 36 and 56.

regated into the forespore and expression there was sufficient, the resulting spore would be genetically *spo*$^+$. It was found that one locus, *spoVA*, had to be expressed in the forespore, whereas for eight other loci expression in the mother cell was sufficient for spore formation; for one of the latter, *spoVB*, Lencastre and Piggot (46) went on to establish that correct expression in the mother cell was necessary as well as sufficient. The interpretations offered depend upon the assumption that the various loci were expressed after stage II, and this has still to be established unequivocally. However, the results suggest that the forespore chromosome is relatively inactive, and this could be a consequence

of a change of the microenvironment of the DNA in the forespore as it becomes dormant (4). This transformation/sporulation system has also been adapted to provide a rapid method for constructing multiply marked strains (57).

In a novel but related approach, B. N. Dancer and J. Mandelstam (this volume) were able to induce late-blocked *spo* mutants to form spores by fusing protoplasts of mutant and *spo*$^+$ strains at stage III of sporulation. Many of the spores formed retained the *spo* genotype of the mutant (and could also be distinguished by other genetic traits from spores formed by the *spo*$^+$ strains). This suggested that transient correction had occurred in the mother cell. Fusions between *spoIVC* mutants revealed two complementing groups in this locus. The approach has considerable promise for studies of compartmentalization of gene expression and for more extensive complementation analysis.

Various loci (Table 1, Fig. 1) have been identified by mutations that modify sporulation without blocking it. All are potential control mutations. Balassa and co-workers identified a collection of *sco* (or *lad*) mutants by their overproduction of protease (17, 18, 42, 51). All the mutants sporulated, although *scoA* and *scoB* mutants were slightly delayed in the production of heat-resistant spores; electron microscopic studies suggested that sporulation was also delayed in *scoC* mutants (69). The original mutants containing the *scoA* and *scoB* mutations overproduced spore coat, often laying down extra layers (69), but it is not clear that this is a consequence of *sco* mutations.

Several groups have described partial-suppressor, or bypass, mutations which overcome some of the pleiotropic effects of *spo0* or *spoII* mutations. These were isolated by selecting or screening for reversion to such *spo*+ characters as resistance to polymyxin (*cpsX* [29, 30]) or the antibiotic produced by *B. subtilis* (*cpsX* [29, 30], *abs* [40], *abr* [72]), resistance to bacteriophage (*tol* [39], *abr* [72]), formation of extracellular protease (29), or formation of alkaline phosphatase during sporulation (*sap* [58]). These were not true revertants and, with the exception of *abrC* (72), none had regained the ability to sporulate. These mutants can be studied in their own right or used as tools to work out dependent sequences of events (55, 58). Studies of the *sap* mutants indicated the existence of sporulation-specific control quite separate from that effected by the known *spo* loci (58). All the studies confirm the complexity of the early stages of sporulation. Mutations in the most complex of the loci, *abrB* (or *cpsX* or *abs*) (29, 30, 40, 74, 75), were thought to cause substantial changes in ribosomal proteins (75). Trowsdale et al. made use of *abrB* mutations to establish that *spo0A* mutations were recessive to the wild type in dominance tests (73); previous studies had given anomalous results.

Mutations to drug resistance that also affect sporulation have been widely used. Most work has concerned mutations affecting ribosomes or RNA polymerase. Perhaps the most intriguing cases are erythromycin-resistant mutants causing a block at stage IV (27) and different rifampin-resistant (Rfm^r) mutants causing a block at stages 0, II, III, IV, and V of sporulation (71). The interpretation of these and other results with drug-resistant mutants is difficult. As an indication of the complexities, at least some of the *rfm spo* mutants can have their *spo* defect corrected by amino acids (60). Other Rfm^r mutants are affected (apparently as a consequence of the *rfm* mutation) in dihydrofolate reductase (43) or glutamate synthase (64, 65), or have a Spo0C phenotype modified to that of Spo0A (38); indeed, some are able to suppress the sporulation defect resulting from a mutation to fusidic acid resistance (35). Moreover, mutation in the *rev* locus can suppress the sporulation defect in *rfm* and *ery* mutants without affecting drug resistance (T. Leighton, personal communication), and phage PMB12 can correct the sporulation defects resulting from both *rfm* and *str* mutations (10; M. Bramucci and D. Kinney, personal communication). One cannot with confidence deduce a specific role for RNA polymerase in spore formation from all these observations, although very powerful evidence for such a role is found elsewhere (see R. Losick, this volume). There has been little work directed toward finding mutants resistant to drugs that block germination and outgrowth (28).

Germination Loci

A variety of Ger$^-$ phenotypes have been recognized, and the mutations responsible have been mapped (see Table 1). The mapping of most of the *ger* mutants has been greatly simplified by a test of the capacity of spore-containing colonies to reduce a tetrazolium redox dye (45, 52, 76). Colonies of most *ger* mutants, unlike the wild type, remain unstained in this test. Other *ger* mutants have been mapped by replica plating of spore-containing colonies (52). Most of the mutants have been characterized in a range of potential germinants including alanine, the combination of glucose, fructose, asparagine, and KCl, and a nutrient broth (52, 76).

The defect in some mutants is conditional on the germinant employed (52, 59, 66, 68, 76). Mutants of the *gerA* group, for example, are altered in their response to alanine and its analogs, but germinate normally in the glucose, fructose, asparagine, and KCl mixture. This group includes the vast majority of *ger* mutants isolated (at least 50, compared with 1 to 6 mutants representing each of the other groups). Some *gerA* mutants will germinate in alanine at concentrations higher than that required by the wild type and characteristic for the particular mutant. This locus (which maps near *cysB* and may comprise one or more genes) may represent the region encoding a germination receptor for alanine and its analogs, enabling them to act as sole germinants. No defect has been found in vegetative cells of *gerA* mutants, either in their use of alanine as a carbon or nitrogen source or in their chemotactic response to alanine (66). A mutation called *gsp-10* (2) that causes defective germination in nutrient broth plus glucose has not yet been characterized in terms of its response to amino acid germinants. Its map position is also near *cysB* (2). Transformation crosses and germination studies of *gsp-10* strains are needed to establish whether this mutation is a member of the *gerA* group or represents a separate locus.

The one known *gerB* mutant has a conditional phenotype different from that of *gerA* mutants, germinating normally in alanine but blocked in some aspects of the specific response to glucose, fructose, KCl, and asparagine (52). A *fruB* (fructose-1-phosphate kinase) mutant also fails to respond to this combination. Unlike *gerB* strains, however, it germinates well when mannose is substituted for fructose (59); *gerC* and *gerG* mutants may also have a defect in only one of the possible pathways of germination.

Mutants carrying *gerD*, *gerE*, *gerF*, *gerH*, *gerI*, or *gerJ* mutations are defective in their response to a wide range of germinants and may thus be defective in some component of the germination system common to many germinants. In contrast to *gerD* and *gerF* mutants, however, *gerE* and *gerJ* mutants respond by losing heat resistance at a normal rate, whereas some other germination-associated events, including loss of absorbance and phase darkening, are incomplete (R. J. Warburg and A. Moir, in preparation). These mutations also have pleiotropic effects on spore formation and spore properties which are described later in this review and in Table 1 (A. Moir, in preparation; R. J. Warburg, this volume). The *gerH* and *gerI* mutations map fairly close to *gerE* and *gerF* loci, respectively; their exact location is not yet established, and their phenotypes have not been compared in detail with those of the *E* and *F* groups (D. A. Smith and S. Taylor, unpublished data).

Outgrowth Loci

Many temperature-sensitive spore outgrowth mutants have been isolated; Galizzi and co-workers (1, 2, 24, 25; summarized in 22) have mapped mutations that exhibit a range of phenotypes, scoring the outgrowth defect by replica plating at the nonpermissive temperature, and have identified a number of loci (Table 1). (After discussion with A. Galizzi the *gsp* terminology previously used has been replaced in this review by *out* designations. The difference between the properties of the mutants, as defined by the stages reached in successive RNA, protein, and DNA synthesis, is not reflected by the subdivision of germination and outgrowth into morphological stages as defined by Young and Wilson [78].) Most *out* loci are represented by a single mutation; the exception, *outA*, includes at least six mutations. The phenotypic blocks in the mutants matched the temporal sequence of biosynthesis of RNA, protein, and DNA during outgrowth and suggest the existence of a dependent sequence of events. The order of phenotypic blocks was (*outF*, *outB*) → *outE* → *outD* → *outC* → *outA*; this apparent order was supported by experiments on the time of temperature sensitivity. The periods at the nonpermissive temperature required to block outgrowth were progressively later for *outF*, *outE*, and *outC*, in that order (22). All these outgrowth mutants grew normally in nutrient broth plus glucose at the nonpermissive temperature. However, for some mutants, the efficiency of plating of vegetative cells at the nonpermissive temperature was low (8), suggesting that in some cases there may be an alteration in their vegetative growth. The phenotypes of several classes of *out* mutants are repaired in the presence of 2% NaCl, 20% sucrose, or sublethal concentrations of distamycin A (1, 24).

Previous studies on *B. subtilis* 160 resulted in the isolation of temperature-sensitive outgrowth mutants classified by morphological criteria of defects in swelling, elongation, and septation (54). These were also temperature sensitive during vegetative growth. Mapping by marker frequency analysis suggested that a number of loci along the chromosome were represented. One of the mutants blocked before swelling, *ts-115*, was 11% cotransduced with *purB* by phage PBS1; there are insufficient mapping data to establish whether or not this has the same location as an *out* mutation. Dawes and Halvorson (14) examined both the biosynthetic capacity and the morphological block in a number of outgrowth mutants, all of which were temperature sensitive for vegetative growth. These are interesting collections of mutants which deserve further study. In the absence of transduction, transformation, or reversion analysis, it is difficult to determine the relationship between outgrowth and vegetative functions in the mutants, or to relate them to the *out* loci described above. Enrichment or screening procedures might unconsciously have selected for defects in both processes, resulting from either a single pleiotropic mutation or two separate mutations.

LINKS BETWEEN VEGETATIVE GROWTH, SPORULATION, GERMINATION, AND OUTGROWTH

Studies of spore formation, germination, and outgrowth are generally conducted independently, although the processes are all part of the same differentiation cycle. The time is ripe to begin to integrate genetic studies of the different aspects of *B. subtilis* differentiation.

Since the differentiation cycle is naturally interrupted at the stage of the mature spore, it might seem that the distinction between *spo* and *ger* or *out* loci would be clear. However, germination takes place without the synthesis of measurable amounts of RNA, protein, or DNA, so that gene products "specific to," or rather required specifically for, germination are almost certainly synthesized during sporulation or vegetative growth and incorporated into the spore. The mutations blocking outgrowth before the resumption of RNA and protein synthesis should also represent genes whose products, present in the spore, were elaborated during preceding stages of the differentiation cycle.

The link between spore formation and the subsequent germination of the spore is illustrated by the successive acquisition of germination requirements, as assayed by intrasporangial germination, during the course of spore formation (16). Specifically genetic illustrations can be taken from the properties of *ger* mutants. The germination phenotype of several *gerA* mutants, for example, depends on the temperature of spore formation (45). More dramatically, the *gerE* and *gerJ* mutations are pleiotropic, affecting both spore formation and late stages of germination: *gerE36* strains form heat-resistant, lysozyme-sensitive, coat-defective spores at 26°C, and some, but not all, strains carrying *gerE36* are temperature sensitive for sporulation (A. Moir, in preparation; see Table 1); *gerJ50* strains are altered in the progress of sporulation (R. J. Warburg, this volume), and the spores formed are more sensitive to heat than are normal spores. The link between spore structure and germination is also emphasized by the altered germination properties of coat mutants of *B. cereus* (3, 12).

The distinction between late phases of germination and early phases of outgrowth is also not an absolute one; mutants blocked after the initial response to germinant but before macromolecular synthesis resumes might be ascribed to either *ger* or *out* mutations.

We should also acknowledge that even the sharp division between "differentiation-specific" and other functions is likely to prove blurred.

For instance, Freese and co-workers (see E. Freese, this volume) have established a very strong correlation between the levels of GTP and the onset of sporulation, and there are already several indications that *spo0* gene products are present during exponential growth (although not necessarily activated until the onset of sporulation).

PROMISING NEW GENETIC TECHNIQUES

The cloning of genes involved in differentiation onto plasmids or phages, followed by the analysis of their organization and regulation and the identification of mRNA and protein gene products, is being actively pursued and is reviewed elsewhere in this volume. Consequently, although it is very important and of direct relevance to a consideration of the genetics of differentiation, it is not discussed here. Genetic analysis, also, can reveal aspects of the organization and regulation of genes through fine structure mapping and dominance and complementation tests.

Methods are now available for protoplast fusion (67) and for the uptake of DNA by protoplasts (11, 47). These are proving useful both in the introduction of intact plasmid DNA into *Bacillus* (11) and in facilitating the construction of diploid strains (48). However, in many diploids formed by protoplast fusion one chromosome is not expressed (37), and this may complicate the interpretation of complementation experiments, as may also the occurrence of multiple fusions (21).

An in vivo method of transferring chromosomal genes onto a bacteriophage has been described. The temperate phage SPβ will transduce chromosomal markers on either side of its site of insertion into the *B. subtilis* genome (62, 80). Specialized transducing phages for other regions can be generated by forcing the SPβ prophage to integrate into a chromosome lacking the normal *att* site. *ilv-leu* transducing phages have already been used for complementation and dominance studies of the *ilv-leu* operon of *B. subtilis* (S. A. Zahler and R. Z. Korman, this volume), and the methods adopted should be generally applicable, e.g., to differentiation genes. Transducing phages also provide a means of isolating the transduced genes away from the bulk of the chromosome, and this might prove useful for their molecular analysis or as a source of DNA for subcloning. SPβ transducing phages carrying nonsense suppressor genes (Zahler and Korman, this volume) permit a quick and efficient identification of nonsense mutations; such mutations could be of use, for example, in the identification of the polypeptide product of a particular cloned gene.

A refinement of the SPβ technique uses a derivative of SPβ into which the *Streptococcus faecalis* transposon Tn*917* has been inserted (P. Youngman and S. Zahler, personal communication). This provides a selection for insertion and a convenient marker for locating the inserted prophage. If this element proves to act as a transposon in *B. subtilis*, it would make available a wide range of techniques exploiting transposons (44) for the study of genes concerned with differentiation, providing, for example, a selection for the *mutant* genotype where this has not previously been available.

DISCUSSION

Among the large numbers of *spo* loci identified (Table 1), there are some regions where many loci are concentrated. It can be noted also that *spo* mutations that cause blocks at very different stages can map close together (e.g., *spo0B* and *spoIVF*). The *ger* and *out* loci map at a variety of positions round the chromosome. There is (as yet) no evidence of clustering either with each other or with the major clusters of identified sporulation genes, although the absence of fine structure genetic analysis and complementation studies make this comment a rather tentative one. It may be that *gerG*, for example, is close to *spoIIC*, *gerB* is close to *spoIID*, and *gerJ* is close to *spoIVA*, although whether such juxtapositions are more than coincidental remains to be established. In addition, it must be borne in mind that the range of *spo*, *ger*, and *out* mutants currently available reflects the selection proce-

dures employed and does not represent a complete (or random) collection of the loci involved in differentiation. It is probably also worth pointing out that the time of expression of a gene may not correspond to either the moment its product functions or the stage at which a mutation in the gene would block sporulation; spore coat protein mutants, for example, might be a case in point. Notwithstanding the paucity of relevant data, it is still worth considering how information on map locations might be integrated into models of the regulation of differentiation which integrate sporulation, germination, and outgrowth genetics.

There is much evidence favoring sequentially dependent induction of gene transcription during spore formation (6, 32, 49, 50, 55, 56). For example, *spo* mutations cause sporulation-associated events to be lost in a characteristic and ordered manner (5, 77), and new sporulation-specific mRNA species continue to be synthesized during much of spore formation (15, 70). We shall consider a model focusing on a short sequence of operon inductions. However, the actual system is likely to be more complex. For example, there may be several sequences with converging and diverging paths, and with different paths in mother cell and forespore. Regulation may involve interactions at transcriptional, translational, and post-translational levels and may include negative as well as positive regulation. Several regulatory factors may control any given operon.

The integration of sporulation, germination, and some outgrowth processes discussed above predicts strongly that some *ger* and early *out* genes would be coordinately controlled with *spo* genes. Such genes could map separately or could

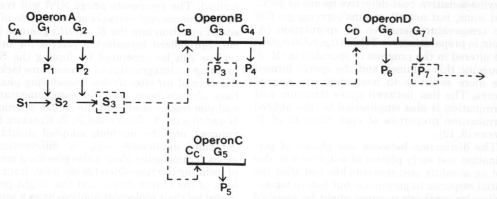

FIG. 2. *Schematic representation of a sequence of operon induction during sporulation. There are four operons, A, B, C, and D. Each contains structural genes G_1, G_2, G_3, etc., coding for proteins P_1, P_2, P_3, etc. Where the proteins are enzymes, they act on a substrate S: P_1 on S_1, P_2 on S_2. Each operon also contains a control element (C_A in operon A, C_B in operon B, etc.) controlling its transcription. Compound S_3 acts on C_B and C_C to switch on operons B and C, respectively; protein P_3 acts on C_D to switch on operon D.*

be part of a single operon. We focus here on a simplistic model of regulation of sporulation genes, incorporating both types of organization (Fig. 2), and we consider some of the consequences of such a scheme for the genetics of sporulation.

In this model, the switching on of successive operons via elements C_A, C_B, C_C, etc., could be mediated by low-molecular-weight products of enzyme action (as S_3 on C_B and C_C) or by protein (P_3 on C_D). Different consequences of point mutations that inactivate the corresponding gene product might be anticipated. In operon A, mutation in G_1 or G_2 would prevent expression of operon B and consequently subsequent operons; the resulting phenotypes would be difficult to distinguish unless, by chance, any of P_1, P_2, S_1, S_2, or S_3 was known. In operon B, mutation in G_3 would prevent expression of subsequent operons. Since operons A and B together represent only a very small part of the overall sporulation sequence, the phenotype resulting from a G_3 mutation might be very difficult to distinguish from that resulting from a mutation in G_1 or G_2. Thus, mutations producing very similar or identical phenotypes may map quite separately. In contrast to a G_3 mutation, mutation in G_4 of operon B would not prevent the expression of subsequent operons. Mutation in G_3 would be a typical pleiotropic *spo* mutation, whereas, if P_4 had no regulatory role and functioned solely in the final assembly of a complete spore, then mutation in G_4 would give an incomplete spore—causing, for example, incomplete spore resistance or altered spore germination, or both—and could be represented by a late *spo* or a *ger* mutation. Mutation in the controlling element C_B could lead to constitutive functioning of operon B and of other operons whose activity depended solely on operon B; such mutations might be partial suppressor mutations or bypass mutations as described earlier. Thus, mutations mapping close together, in C_B, G_3, and G_4, could give rise to very different phenotypes. Operon C is subject to the same regulation as operon B, but maps in a separate location. If its products have no regulatory role in spore formation, then a mutation in G_5 would generate a phenotype similar to one in G_4 and could, for example, give rise to a Ger⁻ phenotype. Thus, genes mapping far apart on the chromosome and represented by mutations generating very different phenotypes (Spo⁻ for G_3 and Ger⁻ for G_5) could still be subject to coordinate regulation.

The potential for progress in the analysis of differentiation lies not only with in vitro isolation and characterization of genes but also with in vivo methods; the combination of these new methods could lead to considerable progress in our understanding of the differentiation genes of *B. subtilis* and their regulation. An exciting prospect is that the validity of models such as that presented above will soon be testable.

ACKNOWLEDGMENTS

Some of the work reported was supported by the Science Research Council in the form of a research fellowship (A.M.) and research grants (D.A.S.).

LITERATURE CITED

1. **Albertini, A. M., M. L. Baldi, E. Ferrari, E. Isnenghi, M. T. Zambelli, and A. Galizzi.** 1979. Mutants of *Bacillus subtilis* affected in spore outgrowth. J. Gen. Microbiol. **110**:351–363.
2. **Albertini, A. M., and A. Galizzi.** 1975. Mutant of *Bacillus subtilis* with a temperature-sensitive lesion in ribonucleic acid synthesis during germination. J. Bacteriol. **124**:14–25.
3. **Aronson, A. I., and P. C. Fitz-James.** 1975. Properties of *Bacillus cereus* spore coat mutants. J. Bacteriol. **123**:354–365.
4. **Baillie, E., G. R. Germaine, W. G. Murrell, and D. F. Ohye.** 1974. Photoreactivation, photoproduct formation, and deoxyribonucleic acid state in ultraviolet-irradiated sporulating cultures of *Bacillus cereus*. J. Bacteriol. **120**:516–523.
5. **Balassa, G.** 1969. Biochemical genetics of bacterial sporulation. I. Unidirectional pleiotropic interactions among genes controlling sporulation in *Bacillus subtilis*. Mol. Gen. Genet. **104**:73–103.
6. **Balassa, G.** 1971. The genetic control of spore formation in Bacilli. Curr. Top. Microbiol. Immunol. **56**:99–192.
7. **Balassa, G., P. Milhaud, E. Raulet, M. T. Silva, and J. C. F. Sousa.** 1979. A *Bacillus subtilis* mutant requiring dipicolinic acid for the development of heat resistant spores. J. Gen. Microbiol. **110**:365–379.
8. **Balassa, G., P. Milhaud, J. C. F. Sousa, and M. T. Silva.** 1979. Decadent sporulation mutants of *Bacillus subtilis*. J. Gen. Microbiol. **110**:381–392.
9. **Balassa, G., and T. Yamamoto.** 1970. A *Bacillus subtilis* mutant with heat-sensitive but chloroform resistant spores. Microbios **5**:73–76.
10. **Bramucci, M. G., K. M. Keggins, and P. S. Lovett.** 1977. Bacteriophage PMB12 conversion of the sporulation defect in RNA polymerase mutants of *Bacillus subtilis*. J. Virol. **24**:194–200.
11. **Chang, S., and S. N. Cohen.** 1979. High frequency transformation of *Bacillus subtilis* protoplasts by plasmid DNA. Mol. Gen. Genet. **168**:111–115.
12. **Cheng, Y. S. E., P. Fitz-James, and A. I. Aronson.** 1978. Characterization of a *Bacillus cereus* protease mutant defective in an early stage of spore germination. J. Bacteriol. **133**:336–344.
13. **Coote, J. G.** 1972. Sporulation in *Bacillus subtilis*. Characterisation of oligosporogenous mutants and comparison of their phenotypes with those of asporogenous mutants. J. Gen. Microbiol. **71**:1–15.
14. **Dawes, I. W., and H. O. Halvorson.** 1974. Temperature-sensitive mutants of *Bacillus subtilis* defective in spore outgrowth. Mol. Gen. Genet. **131**:147–157.
15. **DiCioccio, R. A., and N. Strauss.** 1973. Patterns of transcription in *Bacillus subtilis* during sporulation. J. Mol. Biol. **77**:325–336.
16. **Dion, P., and J. Mandelstam.** 1980. Germination properties as marker events characterizing later stages of

Bacillus subtilis spore formation. J. Bacteriol. **141**:786–792.

17. **Dod, B., and G. Balassa.** 1978. Spore control (*Sco*) mutations in *Bacillus subtilis*. III. Regulation of extracellular protease synthesis in the spore control mutation *ScoC*. Mol. Gen. Genet. **163**:57–63.

18. **Dod, B., G. Balassa, E. Raulet, and V. Jeannoda.** 1978. Spore control (*Sco*) mutations in *Bacillus subtilis*. II. Sporulation and the production of extracellular proteases and α-amylase by *Sco* mutants. Mol. Gen. Genet. **163**:45–56.

19. **Eaton, M. W., and D. J. Ellar.** 1974. Protein synthesis and breakdown in the mother cell and forespore compartments during spore morphogenesis in *Bacillus megaterium*. Biochem. J. **144**:327–337.

20. **Freese, E., Y. K. Oh, E. B. Freese, M. D. Diesterhaft, and C. Prasad.** 1972. Suppression of sporulation of *Bacillus subtilis*, p. 212–221. *In* H. O. Halvorson, R. Hanson, and L. L. Campbell (ed.), Spores V. American Society for Microbiology, Washington, D.C.

21. **Frehel, C., A. M. Lheritier, C. Sanchez-Rivas, and P. Schaeffer.** 1979. Electron microscopic study of *Bacillus subtilis* protoplast fusion. J. Bacteriol. **137**:1354–1361.

22. **Galizzi, A., A. M. Albertini, M. L. Baldi, E. Ferarri, E. Isnenghi, and M. T. Zambelli.** 1978. Genetic studies of spore germination and outgrowth in *Bacillus subtilis*, p. 150–157. *In* G. Chambliss and J. C. Vary (ed.), Spores VII. American Society for Microbiology, Washington, D.C.

23. **Galizzi, A., A. M. Albertini, P. Plevani, and G. Cassani.** 1976. Synthesis of RNA and protein in a mutant of *Bacillus subtilis* temperature sensitive during spore germination. Mol. Gen. Genet. **148**:159–164.

24. **Galizzi, A., F. Gorrini, A. Rollier, and M. Polsinelli.** 1973. Mutants of *Bacillus subtilis* temperature sensitive in the outgrowth phase of spore germination. J. Bacteriol. **113**:1482–1490.

25. **Galizzi, A., A. G. Siccardi, A. M. Albertini, A. R. Amileni, G. Meneguzzi, and M. Polsinelli.** 1975. Properties of *Bacillus subtilis* mutants temperature sensitive in germination. J. Bacteriol. **121**:450–454.

26. **Gay, P., and A. Delobbe.** 1977. Fructose transport in *Bacillus subtilis*. Eur. J. Biochem. **79**:363–373.

27. **Goldman, R. C., and D. J. Tipper.** 1979. Morphology and patterns of protein synthesis during sporulation of *Bacillus subtilis* Ery^r Spo(Ts) mutants. J. Bacteriol. **138**:625–637.

28. **Gould, G. W.** 1964. Effect of food preservatives on the growth of bacteria from spores, p. 17–24. *In* N. Molin and A. Ericksen (ed.), 4th International Symposium on Food Microbiology. Almquist and Wiksell, Stockholm.

29. **Guespin-Michel, J. F.** 1971. Phenotypic reversion in some early blocked sporulation mutants of *Bacillus subtilis*: isolation and phenotype identification of partial revertants. J. Bacteriol. **108**:241–247.

30. **Guespin-Michel, J. F.** 1971. Phenotypic reversion in some early blocked sporulation mutants of *Bacillus subtilis*. Genetic study of polymyxin resistant partial revertants. Mol. Gen. Genet. **112**:243–254.

31. **Haldenwang, W. G., C. D. B. Banner, J. F. Ollington, R. Losick, J. A. Hoch, M. B. O'Connor, and A. L. Sonenshein.** 1980. Mapping of a cloned gene under sporulation control by insertion of a drug resistance marker into the *Bacillus subtilis* chromosome. J. Bacteriol. **142**:90–98.

32. **Halvorson, H. O.** 1965. Sequential expression of biochemical events during intracellular differentiation. Symp. Soc. Gen. Microbiol. **15**:343–368.

33. **Henner, D. J., and J. A. Hoch.** 1980. The *Bacillus subtilis* chromosome. Microbiol. Rev. **44**:57–82.

34. **Higerd, T. B., J. A. Hoch, and J. Spizizen.** 1972.

Hyperprotease-producing mutants of *Bacillus subtilis*. J. Bacteriol. **112**:1026–1028.

35. **Hirochika, H., and Y. Kobayashi.** 1978. Suppression of temperature-sensitive sporulation of a *Bacillus subtilis* elongation factor G mutant by RNA polymerase mutations. J. Bacteriol. **136**:983–993.

36. **Hoch, J. A., M. A. Shiflett, J. Trowsdale, and S. M. H. Chen.** 1978. Stage 0 genes and their products, p. 127–130. *In* G. Chambliss and J. C. Vary (ed.), Spores VII. American Society for Microbiology, Washington, D.C.

37. **Hotchkiss, R. D., and M. H. Gabor.** 1980. Biparental products of bacterial protoplast fusion showing unequal parental chromosome expression. Proc. Natl. Acad. Sci. U.S.A. **77**:3553–3557.

38. **Ikeuchi, T., K. Babasaki, and K. Kurahashi.** 1974. Genetic evidence for possible interaction between a ribonucleic acid polymerase subunit and the *spo0C* gene product of *Bacillus subtilis*. J. Bacteriol. **139**:327–332.

39. **Ito, J.** 1973. Pleiotropic nature of bacteriophage tolerant mutants obtained in early-blocked asporogeneous mutants of *Bacillus subtilis* 168. Mol. Gen. Genet. **124**:97–106.

40. **Ito, J., G. Mildner, and J. Spizizen.** 1971. Early blocked asporogeneous mutants of *Bacillus subtilis* 168. I. Isolation and characterisation of mutants resistant to antibiotic(s) produced by sporulating *Bacillus subtilis* 168. Mol. Gen. Genet. **112**:104–109.

41. **Ito, J., and J. Spizizen.** 1973. Genetic studies of catabolite repression insensitive sporulation mutants of *Bacillus subtilis*. Colloq. Int. C.N.R.S. **227**:81–82.

42. **Jeannoda, V., and G. Balassa.** 1978. Spore control (*Sco*) mutations in *Bacillus subtilis*. IV. Synthesis of alkaline phosphatase during sporulation of *Sco* mutants. Mol. Gen. Genet. **163**:65–73.

43. **Kane, J. F., V. J. Wainscott, and M. A. Hurt.** 1979. Increased levels of dihydrofolate reductase in rifampin-resistant mutants of *Bacillus subtilis*. J. Bacteriol. **137**:1028–1030.

44. **Kleckner, N., J. Roth, and D. Botstein.** 1977. Genetic engineering *in vivo* using translocatable drug-resistance elements. J. Mol. Biol. **116**:125–159.

45. **Lafferty, E., and A. Moir.** 1977. Further studies on conditional germination mutants of *Bacillus subtilis* 168, p. 87–105. *In* A. N. Barker, J. Wolf, D. J. Ellar, G. J. Dring, and G. W. Gould (ed), Spore research 1976. Academic Press, London.

46. **Lencastre, H. de, and P. J. Piggot.** 1979. Identification of different sites of expression for *spo* loci by transformation of *Bacillus subtilis*. J. Gen. Microbiol. **114**:377–389.

47. **Lévi-Meyrueis, C., K. Fodor, and P. Schaeffer.** 1980. Polyethylene glycol-induced transformation of *Bacillus subtilis* protoplasts by bacterial chromosomal DNA. Mol. Gen. Genet. **179**:589–594.

48. **Lévi-Meyrueis, C., C. Sanchez-Rivas, and P. Schaeffer.** 1980. Formation de bactéries diploides stables par fusion de protoplasts de *Bacillus subtilis* et effet de mutations *rec⁻* sur les produits de fusion formés. C.R. Acad. Sci. Ser. D **291**:67–70.

49. **Mandelstam, J.** 1969. Regulation of bacterial spore formation. Symp. Soc. Gen. Microbiol. **19**:377–402.

50. **Mandelstam, J.** 1976. Bacterial sporulation: a problem in the biochemistry and genetics of a primitive developmental system. Proc. R. Soc. London Ser. B **193**:89–106.

51. **Milhaud, P., G. Balassa, and J. Zucca.** 1978. Spore control (*Sco*) mutations in *Bacillus subtilis*. I. Selection and genetic mapping of *Sco* mutants. Mol. Gen. Genet. **163**:35–44.

52. **Moir, A., E. Lafferty, and D. A. Smith.** 1979. Genetic analysis of spore germination mutants of *Bacillus sub-*

tilis 168: the correlation of phenotype with map location. J. Gen. Microbiol. **111:**165–180.

53. **Moran, C. P., R. Losick, and A. L. Sonenshein.** 1980. Identification of a sporulation locus in cloned *Bacillus subtilis* deoxyribonucleic acid. J. Bacteriol. **142:**331–334.

54. **Nukushina, J.-I., and Y. Ikeda.** 1969. Genetic analysis of the developmental processes during germination and outgrowth of *Bacillus subtilis* spores with temperature-sensitive mutants. Genetics **63:**63–74.

55. **Piggot, P. J.** 1979. Genetic strategies for studying bacterial differentiation. Biol. Rev. **54:**347–367.

56. **Piggot, P. J., and J. G. Coote.** 1976. Genetic aspects of bacterial endospore formation. Bacteriol. Rev. **40:**908–962.

57. **Piggot, P. J., and H. de Lencastre.** 1978. A rapid method for constructing multiply marked strains of *Bacillus subtilis.* J. Gen. Microbiol. **106:**191–194.

58. **Piggot, P. J., and S. Y. Taylor.** 1977. New types of mutation affecting formation of alkaline phosphatase by *Bacillus subtilis* in sporulation conditions. J. Gen. Microbiol. **102:**69–80.

59. **Prasad, C., M. Diesterhaft, and E. Freese.** 1972. Initiation of spore germination in glycolytic mutants of *Bacillus subtilis.* J. Bacteriol. **110:**321–328.

60. **Pun, P. P. T., C. D. Murray, and N. Strauss.** 1975. Characterization of a rifampin-resistant, conditional asporogenous mutant of *Bacillus subtilis.* J. Bacteriol. **123:**346–353.

61. **Rhaese, H.-J., J. A. Hoch, and R. Groscurth.** 1977. Studies on the control of development: isolation of *Bacillus subtilis* mutants blocked early in sporulation and defective in synthesis of highly phosphorylated nucleotides. Proc. Natl. Acad. Sci. U.S.A. **74:**1125–1129.

62. **Rosenthal, R., P. A. Toye, R. Z. Korman, and S. A. Zahler.** 1979. The prophage of SPβc2dcitK$_1$, a defective specialised transducing phage of *Bacillus subtilis.* Genetics **92:**721–739.

63. **Ryter, A., B. Bloom, and J. P. Aubert.** 1966. Localisation intracellulaire des acides ribonucléiques synthétisés pendant la sporulation chez *Bacillus subtilis.* C.R. Acad. Sci. **262:**1305–1307.

64. **Ryu, J.-I.** 1978. Pleiotropic effect of a rifampin-resistant mutation in *Bacillus subtilis.* J. Bacteriol. **135:**408–414.

65. **Ryu, J.-I., and S. Takanayagi.** 1979. Ribonucleic acid polymerase mutation affecting glutamate synthase activity in and sporulation of *Bacillus subtilis.* J. Bacteriol. **139:**652–656.

66. **Sammons, R. L., A. Moir, and D. A. Smith.** 1981. Isolation and properties of spore germination mutants of *Bacillus subtilis* 168 deficient in the initiation of germination. J. Gen. Microbiol., in press.

67. **Schaeffer, P., B. Cami, and R. D. Hotchkiss.** 1976. Fusion of bacterial protoplasts. Proc. Natl. Acad. Sci. U.S.A. **73:**2151–2155.

68. **Smith, D. A., A. Moir, and R. Sammons.** 1978. Progress in genetics of spore germination in *Bacillus subtilis,* p. 158–163. *In* G. Chambliss and J. C. Vary (ed.), Spores VII. American Society for Microbiology, Washington D.C.

69. **Sousa, J. C. F., M. T. Silva, and G. Balassa.** 1978. Spore control (*Sco*) mutations in *Bacillus subtilis.* V. Electron microscope studies of the kinetics of morphogenesis in wild type and *Sco* strains. Mol. Gen. Genet. **163:**285–291.

70. **Sumida-Yasumoto, C., and R. H. Doi.** 1974. Transcription from the complementary deoxyribonucleic acid strands of *Bacillus subtilis* during various stages of sporulation. J. Bacteriol. **117:**775–782.

71. **Sumida-Yasumoto, C., and R. H. Doi.** 1977. *Bacillus subtilis* ribonucleic acid polymerase mutants conditionally temperature sensitive at various stages of sporulation. J. Bacteriol. **129:**433–444.

72. **Trowsdale, J., S. M. H. Chen, and J. A. Hoch.** 1978. Genetic analysis of phenotypic revertants of *spo0A* mutants in *Bacillus subtilis:* a new cluster of ribosomal genes. p. 131–135. *In* G. Chambliss and J. C. Vary (ed.), Spores VII. American Society for Microbiology, Washington, D.C.

73. **Trowsdale, J., S. M. H. Chen, and J. A. Hoch.** 1978. Evidence that *spo0A* mutations are recessive in *spo0A⁻/spo0A⁺* merodiploid strains of *Bacillus subtilis.* J. Bacteriol. **135:**99–113.

74. **Trowsdale, J., S. M. H. Chen, and J. A. Hoch.** 1979. Genetic analysis of a class of polymyxin resistant partial revertants of stage 0 sporulation mutants of *Bacillus subtilis:* a map of the chromosome region near the origin of replication. Mol. Gen. Genet. **173:**61–70.

75. **Trowsdale, J., M. Shiflett, and J. A. Hoch.** 1978. New cluster of ribosomal genes in *Bacillus subtilis* with regulatory role in sporulation. Nature (London) **272:**179–181.

76. **Trowsdale, J., and D. A. Smith.** 1975. Isolation, characterization, and mapping of *Bacillus subtilis* 168 germination mutants. J. Bacteriol. **123:**83–95.

77. **Waites, W. M., D. Kay, I. W. Dawes, D. A. Wood, S. C. Warren, and J. Mandelstam.** 1970. Sporulation in *Bacillus subtilis.* Correlation of biochemical events with morphological changes in asporogenous mutants. Biochem. J. **118:**667–676.

78. **Young, F. E., and G. A. Wilson.** 1972. Genetics of *Bacillus subtilis* and other gram-positive sporulating bacilli, p. 77–106. *In* H. O. Halvorson, R. Hanson, and L. L. Campbell (ed.), Spores V. American Society for Microbiology, Washington, D.C.

79. **Young, M., and J. Mandelstam.** 1979. Early events during bacterial endospore formation. Adv. Microb. Physiol. **20:**103–162.

80. **Zahler, S. A., R. Z. Korman, R. Rosenthal, and H. E. Hemphill.** 1977. *Bacillus subtilis* bacteriophage SPβ: localization of the prophage attachment site and specialized transduction. J. Bacteriol. **129:**556–558.

Cloning Strategies in *Bacillus subtilis*

PAUL S. LOVETT

Department of Biological Sciences, University of Maryland Baltimore County, Catonsville, Maryland 21228

INTRODUCTION

Among procaryotes that undergo developmental changes, *Bacillus subtilis* possesses the most extensively studied genetic system. Generalized transduction and DNA-mediated transformation have been used to establish a circular genetic map for the organism, on which a variety of genetic loci, including those involved in bacterial sporulation, are positioned (21). In recent years, genetic studies in *B. subtilis* identified a system of specialized transduction (40) and a system of conjugal gene transfer (O. E. Landman, S. Kim, B. A. Koehler, and C. W. Finn, Jr., Abstr. Annu. Meet. Am. Soc. Microbiol. 1980, H35, p. 114). Studies of plasmids in *B. subtilis* and the concomitant development of molecular cloning have added an additional dimension to genetic analysis in this organism. Gene cloning technology in *B. subtilis* is in the early stages of development. In this brief review several approaches to gene cloning in *B. subtilis* are described.

The principles of molecular cloning, established in *Escherichia coli* (6, 24), rely on the ability to join, in vitro, small segments of non-self-replicating DNA to small replicons such as plasmids or phage genomes. The method most commonly used for this purpose involves cleavage of vector (e.g., plasmid) and target (e.g., chromosome) DNA with an endonuclease that generates single-stranded cohesive termini. The fragments of target and vector DNA join by annealing of their cohesive termini, and nicks are sealed with DNA ligase. A second method of in vitro joining of unrelated DNA fragments requires the synthetic generation of single-stranded termini by use of terminal transferase

(37). The synthesis of polyadenylate tails on vector DNA and polythymidylate tails on target DNA permits the joining of the two classes of molecules when combined. Regardless of the method of joining, the resulting composite molecules are then introduced into recipient cells by transformation and in one case by transduction (6–8).

PLASMIDS THAT REPLICATE AUTONOMOUSLY IN *B. SUBTILIS*

Plasmids occur naturally in strains of *B. subtilis* and related bacteria (29, 31, 32, 48). Those plasmids identified in *B. subtilis* strains either are cryptic or specify host traits not amenable to direct selection. As a result, plasmids currently most useful as cloning vectors for *B. subtilis* originated in other gram-positive bacteria. Ehrlich demonstrated that several small drug resistance plasmids identified in *Staphylococcus aureus* could be transformed into *B. subtilis* where the plasmids were stably maintained and the appropriate drug resistance trait was expressed (12). This is in contrast to the observation that several drug resistance plasmids isolated from *E. coli* are not stably maintained in *B. subtilis* nor are the plasmid-specified drug resistance traits expressed in *B. subtilis* even when the enteric plasmids are joined to replicons capable of autonomous replication in *Bacillus*. Subsequent to Ehrlich's key observation, other drug resistance plasmids from gram-positive bacteria were found to be capable of both autonomous replication in *B. subtilis* and expression of their drug resistance traits (Table 1; references 1, 12, 13, 16, 30).

Two fundamentally distinct systems exist for

TABLE 1. *Some drug resistance plasmids in B. subtilis*

Plasmid	Marker	Mol wt	Reference
Native			
pBC16	Tet^r	2.8×10^6	1
pT127	Tet^r	2.9×10^6	12
pCM194	Cm^r	1.8×10^6	12
pC221	Cm^r	3.0×10^6	12
pC223	Cm^r	3.0×10^6	12
pUB112	Cm^r	3.0×10^6	12
pUB110	Kan^r/Neo^r	2.9×10^6	16, 41
pSA2100	Cm^r Sm^r	4.7×10^6	16
pS194	Cm^r Sm^r	3.0×10^6	30
pC194	Sm^r	2.0×10^6	30
Constructed			
pBD6	Kan^r Sm^r	5.8×10^6	19
pBD8	Kan^r Sm^r Cm^r	6.8×10^6	19
pBD9	Kan^r Em^r	5.4×10^6	19
pBD10	Kan^r Cm^r Em^r	4.4×10^6	19
pBD11	Kan^r Em^r	4.4×10^6	19
pBD12	Kan^r Cm^r	4.5×10^6	19
pBD64	Kan^r Cm^r	3.2×10^6	19
pTL12	Leu$^+$ Tmp^r	6.4×10^6	47

rendering *B. subtilis* transformable by plasmid DNA. The classical Spizizen technique (9, 46), which is routinely used to render *B. subtilis* competent for transformation by chromosome or bacteriophage DNA, also renders cells competent for the uptake of plasmid DNA (12, 18, 27, 33). However, the transformation frequency of small plasmids (2 to 3 megadaltons) is low, on the order of 10^3 and 10^5 transformants per microgram of plasmid DNA, when cells are brought to competence by the Spizizen procedure. Recent studies have shown that partially purified monomeric pCM194 transforms *B. subtilis* poorly, whereas multimeric forms of the plasmid appear to be much more efficient in generating stable transformants (3, 38). The available evidence suggests that most plasmid transformants of *B. subtilis* cells made competent by the Spizizen technique result from uptake of multimeric covalently closed circular forms of plasmids (3). Since plasmid transformants harbor predominantly the monomeric form of introduced plasmids, it must be inferred that recombination events occur within the transformed multimer.

Chang and Cohen (4) developed a second method for transforming plasmids into *B. subtilis* that relies on inducing the cells to form protoplasts. This method gives much higher transformation efficiencies with plasmid DNA than those achieved with cells made competent by the Spizizen technique. Moreover, it has been reported that the protoplasts incorporate both monomeric and multimeric plasmid forms at high efficiency (17). The protoplast system is of great value in that its use extends the range of *Bacillus* species that can be made competent for plasmid transformation. Thus, although *B. pumilus* and *B. megaterium* cannot be made competent by the Spizizen technique, protoplasts of both species can be transformed with plasmids by a slight variation of the Chang and Cohen procedure (2; P. S. Lovett, unpublished data).

Certain small plasmids introduced into mutants of *B. subtilis* that are temperature sensitive for DNA replication replicate in cells shifted to the temperature that is nonpermissive for chromosome replication (44). As a consequence, the copy number of the plasmids per cell selectively increases. This type of plasmid amplification does not require inhibition of protein synthesis as in the amplification of ColE1 by chloramphenicol (5). Therefore, it has the potential of being used to detect in cells increased levels of plasmid-encoded gene products. At present, segregation of plasmids into minicells of *B. subtilis* has permitted the determination of several proteins specified by pUB110 and pCM194 (45).

MOLECULAR CLONING IN *B. SUBTILIS* BY USE OF PLASMID VECTORS

Genetically active DNA segments cloned on plasmids and maintained extrachromosomally in *B. subtilis* can be placed, perhaps artificially, into two broad categories depending on the frequency with which the DNA segments are encountered in *Bacillus* strains prior to cloning. Fragments of DNA that naturally occur in *Bacillus* at high frequency seem, in general, to be amenable to plasmid cloning in *Bacillus*. The initial studies demonstrating this point involved in vitro joining of endonuclease-generated fragments derived from pairs of plasmids, each of which is maintained in *Bacillus* at high copy number (13, 18, 19). The application of this in vitro manipulation of plasmid fragments has generated a variety of derivative plasmids, some of which contain two or three drug resistance loci (Table 1). Such derivative plasmids offer great potential for further gene cloning in *Bacillus*.

A second example of DNA sequences that occur at high frequency in *Bacillus* are phage genomes, and plasmid cloning of the endonuclease-generated fragments of the genomes of two well-characterized *B. subtilis* temperate phages, ϕ105 and SPO2, has presented no difficulties (35). Although no direct selection appeared to exist for the phage DNA fragments, it was found that insertion of nearly any SPO2 and ϕ105 sequence into a vector plasmid, pCM194 or pUB110, rendered the chimeric derivative transducible by the phage whose DNA was represented in the chimera (Table 2). As a result, cloned fragments spanning a large portion of the SPO2 and ϕ105 genomes were readily identified.

Cloning and expression of DNA segments that are normally represented in *Bacillus* at low copy number, such as segments of the bacterial chromosome, have been successful in a few laboratories. However, this aspect of *Bacillus* cloning has clearly presented difficulties to many investigators interested in obtaining a specific *Bacillus* DNA segment. Two approaches have resulted in the successful cloning of certain genetically active *Bacillus* chromosome fragments on plasmids in *B. subtilis*. Keggins et al. (26, 27) cloned *Eco*RI-generated segments of the chromosomes of *B. subtilis, B. pumilus,* and *B. licheniformis* that complemented the *trpC2* mutation in *B. subtilis*. The approach used involved in vitro ligation of *Eco*RI-cleaved pUB110 with similarly digested chromosome DNA and transformation into *B. subtilis* (made competent by the Spizizen system) with selection in liquid for the neomycin resistance trait specified by pUB110, followed by selection for TrpC⁺ cells (34). By this approach many segments were cloned from various *Bacillus* species that complemented two to four contiguous *trp* genes (27; Lovett, unpublished data). Subsequent analysis of the levels of the tryptophan biosynthetic enzymes in *B. subtilis* harboring pUB110 with cloned *trp* inserts demonstrated an elevation in those enzymes specified by the *trp* genes represented on each cloned segment (28). The *trp* segments cloned from *B. subtilis* 168, *B. pumilus* NRS576, and *B. licheniformis* 9945A each contained a single *Hin*dIII-sensitive site (28). Interestingly, insertions into this *Hin*dIII site specifically inactivated *trpC* complementing activity and eliminated detectable levels of the plasmid-specified *trpC* gene product (28). Using a similar approach, we have recently successfully cloned chloramphenicol acetyltransferase structural genes from nine different strains of *B. pumilus*. However, this direct cloning approach has not been successful for cloning a variety of other chromosome segments.

A second approach successfully used to clone segments of the *B. licheniformis* chromosome on plasmids in *B. subtilis* was reported by Gryczan et al. (17). If, as suggested by Canosi et al. (3), monomeric plasmid forms inefficiently transform *B. subtilis* (made competent by the Spizizen technique), then most plasmid derivatives cut with an endonuclease and ligated to similarly cut foreign DNA would be monomeric and thereby transform poorly. To circumvent this possibility and to avoid the protoplast system which necessitates the use of a nutritionally complex growth medium, Gryczan et al. attempted to clone foreign DNA segments (*B. licheniformis* DNA) into a vector plasmid (pBD64) and to recombine these cloned segments into a recipient plasmid (pUB110) which is largely homologous with the vector plasmid and is resident in the transformation recipient. *B. licheniformis* DNA was cleaved with *Bgl*II or *Bcl*I, and a chloramphenicol resistance (Cmʳ), neomycin resistance (Neoʳ) derivative of pUB110, designated pBD64, was cleaved with *Bgl*II. The two DNA species were combined, ligated, and used to transform *B. subtilis* harboring pUB110. By applying double selection for Cmʳ and a nutritional requirement carried by the recipient, segments of *B. licheniformis* DNA were cloned that complemented *trpD* → *hisH*, *trpC* → *trpA,* as well as a segment complementing *aroB* and *aroF.* It is thought that after entry in *B. subtilis* (pUB110) of monomeric pBD64 or monomeric pBD64 containing a DNA insert, recombination events occur to insert portions of the incoming plasmid into the homologous resident plasmid. Thus, the need for a multimer to successfully cause transformation is eliminated.

GENE CLONING IN BACTERIOPHAGES

In the *E. coli* system, several derivatives of λ have been developed that permit the phage to

TABLE 2. *Partial list of endonuclease fragments of ϕ105 and SPO2 isolated by the transductional cloning method*[a]

Plasmid designation	Phage DNA/vector plasmid	Insert fragment size(s) (megadaltons)	Transduction/ PFU by homologous phage
pPL1000	ϕ105/pUB110	0.25	3×10^{-7}
pPL1001	ϕ105/pUB110	0.70	2×10^{-6}
pPL1002	ϕ105/pUB110	3.48	1×10^{-5}
pPL1003	ϕ105/pCM194	1.22	1×10^{-5}
pPL1004	ϕ105/pCM194	1.62	3×10^{-4}
pPL1005	ϕ105/pCM194	0.72	7×10^{-5}
pPL1006	ϕ105/pCM194	3.44, 1.91	2×10^{-4}
pPL1007	ϕ105/pCM194	2.71	6×10^{-4}
pPL1008	ϕ105/pCM194	2.71, 1.91, 1.22	3×10^{-6}
pPL1009	ϕ105/pCM194	0.98	8×10^{-5}
pPL1010	SPO2/pUB110	1.67	1×10^{-2}
pPL1011	SPO2/pPLFHL	0.29	1×10^{-5}
pPL1012	SPO2/pPLFHL	1.57	3×10^{-5}
pPL1013	SPO2/pPLFHL	1.06, 0.98, 0.33	2×10^{-5}
pPL1014	SPO2/pPLFHL	0.60	1×10^{-6}

[a] Data from reference 35.

be used as a cloning vector (see 37). No equivalents to the λ cloning vectors are currently available in *B. subtilis*. However, a novel approach to gene cloning in *B. subtilis* involving the generation of specialized transducing phages was developed by Kawamura, Saito, and Ikeda (25). The DNA from the temperate phage ρ11 and *B. subtilis* chromosome DNA (isolated from PBSX particles) are digested in vitro with a chosen endonuclease that generates cohesive termini. The DNA is then ligated and used to transform *B. subtilis* that is lysogenic for ρ11. Selection is applied to detect transformants for specific chromosome markers (e.g., His$^+$, Lys$^+$). Transformant clones are induced with mitomycin C, and the resulting phage lysates are used to transduce *B. subtilis* for those traits selected in the initial cloning. By this approach several defective and nondefective phages were obtained that transduced *hisA* and *lys* loci in *B. subtilis*. Figure 1 shows the mechanism proposed to explain the origin of such phages. It is suggested that adjacent endonuclease fragments of phage DNA flank a restriction fragment of *B. subtilis* DNA harboring the trait to be selected (e.g., *hisA*). The ligated DNA (linears or circles?) is transformed in *B. subtilis* (ρ11), and a recombination event occurs with the resident prophage. The transducing phage is presumably defective or nondefective depending on the location within the phage genome of the *B. subtilis* DNA insert. This, in turn, must be determined by the ρ11 fragments that flank the *B. subtilis* DNA insert during ligation. The mechanism described appears similar to that suggested by Gryczan et al. (17) to explain their system of recombinational cloning into plasmids.

Iljima et al. (23) have demonstrated that the above cloning approach can also be used to generate *metB* and *lys* specialized transducing derivatives of the temperate phage φ105. The

defective phage φ105 *metB*, characterized in some detail, was found to consist of an *Eco*RI-generated fragment of *B. subtilis* DNA (*Eco*RI-M; 2.7 megadaltons) inserted in place of two normally adjacent *Eco*RI fragments of φ105 DNA, fragments G and E (42). Compelling transformation evidence was presented indicating that the *metB* gene likely resides on the *Eco*RI-M fragment (23). These results tend to support the model proposed by Kawamura, Saito, and Ikeda (25) to explain the origin of the transducing phages. Insertion of the *Eco*RI M fragment into the prophage with concomitant elimination of the G and E fragments could be achieved if the M fragment was flanked in vitro by the φ105 I and B fragments (see 42).

Yoneda, Graham, and Young (50) used the phage cloning system to generate a specialized, apparently nondefective, transducing derivative of φ3T carrying the structural gene for α-amylase, a commercially important enzyme. The source of the α-amylase gene was *B. amyloliquefaciens*. Thus, the phage cloning system has been successfully used to clone homologous and heterologous genes in *B. subtilis*.

In addition to the obvious differences between the phage cloning system and the plasmid cloning system described earlier, a possibly less obvious difference exists. Chromosome genes cloned in *B. subtilis* on available plasmid vectors increase the copy number of the gene 5- to 10-fold. Although in the case of *trpE → A* loci the *hisH* locus and *aroB* and *aroF* genes in this increased copy number seem to have no deleterious effects on the host cell, it is possible to envision other genes or regulatory sites that will cause cell death when present in high copy number. The phage cloning system allows genes to be maintained in *B. subtilis* in an isolatable form and in low copy number.

INSERTION OF FOREIGN GENES INTO THE *B. SUBTILIS* CHROMOSOMES

Small segments of DNA that are nonhomologous with *B. subtilis* DNA have been inserted into the *B. subtilis* chromosome by in vitro joining of a nonhomologous DNA sequence to a fragment of DNA that shares homology with the *B. subtilis* chromosome, generating a chimera that appears to integrate into the *B. subtilis* chromosome if appropriate selection can be applied. Duncan et al. (11) first demonstrated that a pMB9 derivative containing a *Bacillus* thymidylate synthetase gene could successfully transform *B. subtilis* to Thy$^+$. Hybridization analysis of the DNA from the Thy$^+$ transformants demonstrated the integration of pMB9 se-

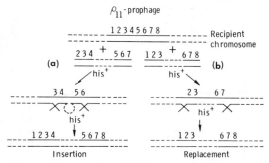

FIG. 1. *Suggested mechanism to explain the recombination of an endonuclease-generated fragment of* B. subtilis *DNA flanked by ρ11 DNA fragments with a ρ11 prophage. Redrawn from reference 25.*

quences into the *B. subtilis* chromosome (11). Haldenwang et al. (20) used this principle to allow genetic mapping of a cloned *B. subtilis* gene whose transcription was known to be under sporulation control. pCM194 was joined in vitro with a pMB9 derivative (p213–1) harboring a presumptive *B. subtilis* sporulation gene. The joining was made through the *Hpa*II site on pCM194, which renders pCM194 incapable of replicating in *B. subtilis* although the expression of the pCM194 Cm^r trait remains unaffected (4). The resulting plasmid, p1949-2, could replicate extrachromosomally in *E. coli* but not in *B. subtilis*. Transformation of p1949-2 into *B. subtilis* did generate Cm^r transformants, which were shown to have resulted from the integration of the Cm^r segment into the *purA cysA* region of the chromosome. p1949-2 transformation of *B. subtilis* is dependent on the *recE4* function since no Cm^r transformants were detected when a *B. subtilis recE4* recipient was used (20). Similarly, removal of the *B. subtilis* sequence from p1949-2 by in vitro digestion and ligation rendered the derivative incapable of transforming *B. subtilis* to Cm^r. Figure 2 shows a plausible mechanism

to explain the insertion of p1949-2 into the *B. subtilis* chromosome.

M. S. Rosenkrantz and A. L. Sonenshein (Abstr. Annu. Meet. Am. Soc. Microbiol. 1980, I95, p. 100) suggested an extension of the above technique that may be useful for cloning specific segments of the *B. subtilis* chromosome in *E. coli*. An *E. coli* plasmid such as pMB9 is joined to a gene that expresses in *B. subtilis* but is not homologous with the *B. subtilis* chromosome, e.g., the chloramphenicol acetyltransferase gene of a nonreplicating derivative of pCM194. The chimera is used to clone random fragments of the *B. subtilis* chromosome. The resulting plasmid molecules are transformed into *B. subtilis* with selection for Cm^r. The molecules will presumably integrate at locations on the chromosome determined by the cloned segment of *B. subtilis* DNA. The location of the chloramphenicol acetyltransferase gene on the chromosome is used as an indication of the location of the integrated pMB9. If the integrated pMB9 can be recovered in *E. coli*, either by direct transformation or after cleavage of the *B. subtilis* DNA with an appropriate endonuclease and ligation,

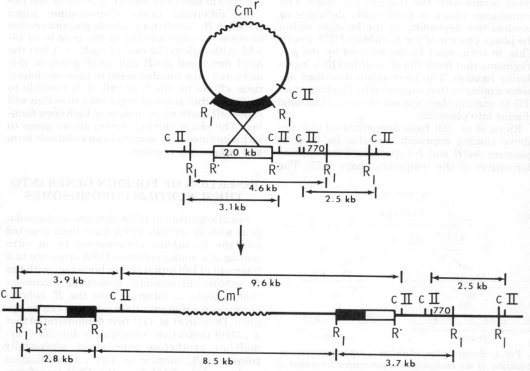

FIG. 2. *Proposed mechanism for the insertion of p1949-2 into the* B. subtilis *chromosome. Redrawn from reference 20.*

it may carry the segment of the *B. subtilis* chromosome flanking the insertion site.

GENE CLONING DIFFICULTIES IN *B. SUBTILIS*

Many segments of *B. subtilis* DNA have been cloned into *E. coli* on phage or plasmid vectors to create several gene banks (22, 39; J. A. Hoch, personal communication). Specific segments cloned in *E. coli* have been identified in one of three ways. First, if the cloned *Bacillus* segment is appropriately expressed in *E. coli*, a specific genetic function can be identified. A thymidylate synthetase encoding fragment cloned in *E. coli* from the DNA of the *B. subtilis* phage φ3T was identified by its expression (14). Similarly, segments spanning the *leu* region of the *B. subtilis* chromosome and the *trp* region of *B. licheniformis* DNA were identified on the basis of expression in *E. coli* (49; W. Brammar, personal communication). Moreover, drug resistance plasmids such as pCM194 or pUB110 which replicate in *B. subtilis* can be joined to *E. coli* plasmids, e.g., pBR322, and the resulting chimeras express in *E. coli* those drug resistance traits specified by the plasmids from *B. subtilis* (e.g., 10). A second approach to identifying *Bacillus* segments cloned in *E. coli* involves direct transformation of appropriate auxotrophic and sporulation mutants of *B. subtilis* to screen for *B. subtilis* genes represented on the cloned fragment (e.g., 36, 39). Lastly, *E. coli* colonies containing cloned *Bacillus* DNA fragments can be screened by colony hybridization (using a DNA or RNA probe; e.g., 15, 43).

Certain *B. subtilis* DNA fragments cloned initially in *E. coli* have been excised from the *E. coli* vector, joined to a *Bacillus* plasmid, and inserted into *B. subtilis*. One of the first such experiments of this type was done by Tanaka and Sakaguchi (49), who cloned a *B. subtilis leu* marker in *E. coli* and subsequently used this cloned segment to mark a cryptic *Bacillus* plasmid that is maintained in *B. subtilis* at high copy number. The cloned segment complemented mutations in *leuA, leuB,* and *leuC*. In later studies Tanaka and Kawano (47) transferred a segment of *B. subtilis* DNA specifying a trimethoprim-resistant dihydrofolate reductase from an *E. coli* plasmid vector to a *Bacillus* plasmid with subsequent transformation into *B. subtilis*, where the plasmid was maintained extrachromosomally and expressed trimethoprim resistance. A 0.4-kilobase segment of *B. subtilis* DNA cloned in *E. coli* by Segall and Losick (43) contains a region whose transcription is activated during an early stage of sporulation. C. D.

B. Banner has reported the joining of a subclone of this segment containing a possible sporulation gene promoter to the high-copy plasmid pUB110 (Abstr. Annu. Meet. Am. Soc. Microbiol. 1980, I97, p. 100). Introduction of the chimera into *B. subtilis* caused the cells to become sporulation negative. One interpretation is that the high copy of the sporulation promoter may titrate proteins essential to sporulation.

The "gene bank" developed by Rapaport et al. (39) was constructed in *E. coli* using a plasmid vector that is capable of autonomous replication in both *B. subtilis* and *E. coli*. These investigators have identified clones that carry the following *B. subtilis* loci: *thr, leuA, hisA glyB,* and *purB*. It appears that when the chimeras were transformed into recombination-proficient strains of *B. subtilis* recombination occurred between the plasmid and the chromosome. Use of a recipient containing the *recE4* mutation prevented recombination.

Currently, it seems evident that far more *Bacillus* DNA had been successfully cloned in *E. coli* than in *B. subtilis*. Possible explanations for the general lack of success for large-scale cloning of *Bacillus* genes in *B. subtilis* are many, but the following seem most likely. The majority of cloning studies in *B. subtilis* have used the high-copy drug resistance plasmids originally isolated from *S. aureus*. Cloning *Bacillus* genes on such plasmids increases the copy number of cloned DNA well above that normally encountered in *Bacillus*. Many genes or genetic regulatory elements in high copy number may titrate essential regulatory proteins, thereby causing death of cells carrying the cloned segment. Similarly, high copy number of a cloned gene may result in overproduction of gene products. Alternatively, joining a chromosome segment to a plasmid may permit unregulated gene expression, causing, for example, expression of a spore gene at an incorrect point in the growth cycle. If plasmid copy number is a difficulty, then the phage cloning system (23, 25) may ultimately prove useful for general cloning in *B. subtilis*. Similarly, the suggestion by Gryczan et al. (17) that plasmids into which DNA segments have been cloned are selected against during transformation led them to develop a novel method for gene cloning involving recombination between "donor" and "recipient" plasmids. This approach requires the use of a recombination-proficient transformation recipient, but it may prove valuable for the isolation of *B. subtilis* DNA segments if *B. licheniformis* can be used as cloning recipient (17).

LITERATURE CITED

1. **Bernhard, K., H. Schrempf, and W. Goebel.** 1978. Bacteriocin and antibiotic resistance plasmids in *Bacillus cereus* and *Bacillus subtilis*. J. Bacteriol. **133**:897–903.

2. **Brown, B. J., and B. C. Carlton.** 1980. Plasmid-mediated transformation in *Bacillus megaterium*. J. Bacteriol. **142**:508–512.

3. **Canosi, U., G. Morelli, and J. A. Trantner.** 1978. The relationship between molecular structure and transformation efficiency of some *Staphylococcus aureus* plasmids isolated from *Bacillus subtilis*. Mol. Gen. Genet. **166**:259–267.

4. **Chang, S., and S. N. Cohen.** 1979. High frequency transformation of *Bacillus subtilis* protoplasts by plasmid DNA. Mol. Gen. Genet. **168**:111–115.

5. **Clewell, D. B.** 1972. Nature of ColE1 plasmid replication in *Escherichia coli* in the presence of chloramphenicol. J. Bacteriol. **110**:667–676.

6. **Cohen, S. N., A. C. Y. Chang, H. W. Boyer, and R. B. Helling.** 1973. Construction of biologically functional plasmids in vitro. Proc. Natl. Acad. Sci. U.S.A. **72**:3240–3244.

7. **Collins, J., and H. J. Bruning.** 1978. Plasmids usable as gene-cloning vectors in an *in vitro* packaging by coliphage λ: "cosmids." Gene **4**:85–107.

8. **Collins, J., and B. Hohn.** 1978. Cosmids: a type of plasmid gene-cloning vector that is packageable in vitro in λ levels. Proc. Natl. Acad. Sci. U.S.A. **75**:4242–4246.

9. **Contente, S., and D. Dubnau.** 1979. Characterization of plasmid transformation in *Bacillus subtilis*: kinetic properties and the affect of DNA conformation. Mol. Gen. Genet. **167**:251–258.

10. **Courvalin, P., and M. Fiandt.** 1980. Aminoglycoside-modifying enzyme of *Staphylococcus aureus*: expression in *Escherichia coli*. Gene **9**:247–269.

11. **Duncan, C. H., G. A. Wilson, and F. E. Young.** 1978. Mechanism of integrating foreign DNA during transformation of *Bacillus subtilis*. Proc. Natl. Acad. Sci. U.S.A. **75**:3664–3668.

12. **Ehrlich, S. D.** 1977. Replication and expression of plasmids from *Staphylococcus aureus* in *Bacillus subtilis*. Proc. Natl. Acad. Sci. U.S.A. **74**:1620–1682.

13. **Ehrlich, S. D.** 1978. DNA cloning in *Bacillus subtilis*. Proc. Natl. Acad. Sci. U.S.A. **75**:1433–1436.

14. **Ehrlich, S. D., H. Bursetyn-Pettegrew, I. Stroynowski, and J. Lederberg.** 1976. Expression of the thymidylate synthetase gene of the *Bacillus subtilis* bacteriophage Phi-3-T in *Escherichia coli*. Proc. Natl. Acad. Sci. U.S.A. **73**:4145–4149.

15. **Grunstein, M., and D. S. Hogness.** 1975. Colony hybridization: a method for the isolation of cloned DNA's that contain a specific gene. Proc. Natl. Acad. Sci. U.S.A. **72**:3961–3965.

16. **Gryczan, T. J., S. Contente, and D. Dubnau.** 1978. Characterization of *Staphylococcus aureus* plasmids introduced by transformation into *Bacillus subtilis*. J. Bacteriol. **134**:318–323.

17. **Gryczan, T., S. Contente, and D. Dubnau.** 1980. Molecular cloning of heterologous chromosomal DNA by recombination between a plasmid vector and a homologous resident plasmid in *Bacillus subtilis*. Mol. Gen. Genet. **177**:459–467.

18. **Gryczan, T. J., and D. Dubnau.** 1978. Construction and properties of chimeric plasmids in *Bacillus subtilis*. Proc. Natl. Acad. Sci. U.S.A. **75**:1428–1432.

19. **Gryczan, T., A. G. Shivakumar, and D. Dubnau.** 1980. Characterization of chimeric plasmid cloning vehicles in *Bacillus subtilis*. J. Bacteriol. **141**:246–253.

20. **Haldenwang, W. G., C. D. B. Banner, J. F. Ollington, R. Losick, J. A. Hoch, M. B. O'Connor, and A. L. Sonnenshein.** 1980. Mapping a cloned gene under sporulation control by insertion of a drug resistance marker into the *Bacillus subtilis* chromosome. J. Bacteriol. **142**:92–98.

21. **Henner, D. J., and J. A. Hoch.** 1980. The *Bacillus subtilis* chromosome. Microbiol. Rev. **44**:57–82.

22. **Hutchison, K. W., and H. O. Halvorson.** 1980. Cloning of randomly cleaved DNA fragments from a φ105 lysogen of *Bacillus subtilis*: identification of prophage containing clones. Gene **8**:267–278.

23. **Iljima, T., F. Kawamura, H. Saito, and Y. Ikeda.** 1980. A specialized transducing phage constructed from *Bacillus subtilis* phage φ105. Gene **9**:115–126.

24. **Jackson, D. A., R. H. Symons, and P. Berg.** 1972. Biochemical method for inserting new genetic information into DNA of simian virus 40: SV40 DNA molecules containing lambda phage genes and the galactose operon of *Escherichia coli*. Proc. Natl. Acad. Sci. U.S.A. **69**:2904–2909.

25. **Kawamura, F., H. Saito, and Y. Ikeda.** 1979. A method for construction of specialized transducing phage p11 of *Bacillus subtilis*. Gene **5**:87–91.

26. **Keggins, K. M., E. J. Duvall, and P. S. Lovett.** 1978. Recombination between compatible plasmids containing homologous segments requires the *Bacillus subtilis* recE gene product. J. Bacteriol. **134**:514–520.

27. **Keggins, K. M., P. S. Lovett, and E. J. Duvall.** 1978. Molecular cloning of genetically active fragments of *Bacillus* DNA in *Bacillus subtilis* and properties of the vector plasmid pUB110. Proc. Natl. Acad. Sci. U.S.A. **75**:1423–1427.

28. **Keggins, K. M., P. S. Lovett, R. Marrero, and S. O. Hoch.** 1979. Insertional inactivation of *trpC* in cloned *Bacillus trp* segments: evidence for a polar effect on *trpF*. J. Bacteriol. **139**:1001–1006.

29. **LeHagaret, J.-C., and C. Anagnostopoulos.** 1977. Detection and characterization of naturally occurring plasmids in *Bacillus subtilis*. Mol. Gen. Genet. **157**:167–174.

30. **Lofdahl, S., J.-E. Sjostrom, and L. Philipson.** 1978. A vector for recombinant DNA in *Staphylococcus aureus*. Gene **3**:161–172.

31. **Lovett, P. S.** 1973. Plasmid in *Bacillus pumilus* and the enhanced sporulation of plasmid-negative variants. J. Bacteriol. **115**:291–298.

32. **Lovett, P. S., and M. G. Bramucci.** 1975. Plasmid deoxyribonucleic acid in *Bacillus subtilis* and *Bacillus pumilus*. J. Bacteriol. **124**:484–490.

33. **Lovett, P. S., E. J. Duvall, and K. M. Keggins.** 1976. *Bacillus pumilus* plasmid pPL10: properties and insertion into *Bacillus subtilis* by transformation. J. Bacteriol. **127**:817–828.

34. **Lovett, P. S., and K. M. Keggins.** 1979. *Bacillus subtilis* as a host for molecular cloning. Methods Enzymol. **68**:342–357.

35. **Marrero, R., and P. S. Lovett.** 1980. Transductional selection of cloned bacteriophage φ105 and SP02 deoxyribonucleic acids in *Bacillus subtilis*. J. Bacteriol. **143**:879–886.

36. **Moran, C. P., Jr., R. Losick, and A. L. Sonenshein.** 1980. Identification of a sporulation locus in cloned *Bacillus subtilis* deoxyribonucleic acid. J. Bacteriol. **142**:331–334.

37. **Morrow, J. F.** 1979. Recombinant DNA techniques. Methods Enzymol. **68**:3–24.

38. **Mottes, M., G. Grandi, V. Sgaramella, U. Canosi, G. Morelli, and T. A. Trautner.** 1979. Different specific activities of the monomeric and oligomeric forms of plasmid DNA in transformation of *B. subtilis* and *E. coli*. Mol. Gen. Genet. **174**:281–286.

39. **Rapaport, G., A. Klier, A. Billault, F. Fargette, and R. Dedonder.** 1979. Construction of a colony bank of *E. coli* containing hybrid plasmids representative of the

Bacillus subtilis 168 genome. Expression of functions harbored by the recombinant plasmids in *B. subtilis*. Mol. Gen. Genet. **176**:239–245.

40. **Rosenthal, R., P. A. Joye, R. Z. Korman, and S. A. Zahler.** 1979. The prophage of SP*Bc2dcitk,* a defective specialized transducing phage of *Bacillus subtilis*. Genetics **92**:721–739.

41. **Sadaie, Y., K. C. Burtis, and R. H. Doi.** 1980. Purification and characterization of a kanamycin nucleotidyltransferase from plasmid pUB110 carrying cells of *Bacillus subtilis*. J. Bacteriol. **141**:1178–1182.

42. **Scher, B. M., M. F. Law, and A. J. Garro.** 1978. Correlated genetic and *EcoRI* cleavage map of *Bacillus subtilis* bacteriophage φ105 DNA. J. Virol. **28**:395–402.

43. **Segall, J., and R. Losick.** 1977. Cloned *Bacillus subtilis* DNA containing a gene that is activated early during sporulation. Cell **11**:751–761.

44. **Shivakumar, A. G., and D. Dubnau.** 1978. Plasmid replication in *dna*Ts mutants of *Bacillus subtilis*. Plasmid **1**:405–416.

45. **Shivakumar, A. G., J. Hahn, and D. Dubnau.** 1979.

Studies on the synthesis of plasmid-coded proteins and their control in *Bacillus subtilis*. Plasmid **2**:279–289.

46. **Spizizen, J.** 1958. Transformation of biochemically deficient strains of *Bacillus subtilis* by deoxyribonucleate. Proc. Natl. Acad. Sci. U.S.A. **44**:1072–1078.

47. **Tanaka, T., and N. Kawano.** 1980. Cloning vehicles for the homologous *Bacillus subtilis* host-vector system. Gene **10**:131–136.

48. **Tanaka, T., M. Kuroda, and K. Sakaguchi.** 1977. Isolation and characterization of four plasmids from *Bacillus subtilis*. J. Bacteriol. **129**:1487–1494.

49. **Tanaka, T., and K. Sakaguchi.** 1978. Construction of a recombinant plasmid composed of *B. subtilis* leucine genes and a *B. subtilis* (*natto*) plasmid: its use as a cloning vehicle in *B. subtilis* 168. Mol. Gen. Genet. **165**: 269–276.

50. **Yoneda, Y., S. Graham, and F. E. Young.** 1979. Cloning of a foreign gene coding for α-amylase in *Bacillus subtilis*. Biochem. Biophys. Res. Commun. **91**:1556–1564.

Sigma Factors, Stage 0 Genes, and Sporulation

RICHARD LOSICK

The Biological Laboratories, Harvard University, Cambridge, Massachusetts 02138

INTRODUCTION

The process of endospore formation in *Bacillus subtilis* is triggered by depriving growing cells of a source of carbon, nitrogen, or phosphorus. In the first stages of sporulation, nutrient-deprived bacteria cease normal cell division and form instead polar septa that partition the sporulating cells into mother-cell and forespore compartments. At least 10 genetic loci are involved in the initiation of this developmental process (10, 11, 17). These loci, which are known as the stage 0 or *spo0* genes, are located at widely scattered sites on the *B. subtilis* chromosome (Fig. 1). A mutation in any one of these genes blocks spore formation prior to completion of any of the morphological changes characteristic of early spore development. The *spo0* genes are thought, therefore, to be regulatory cistrons whose products are directly involved in the initiation stage of the sporulation process. Although several *spo0* cistrons have now been isolated by gene cloning methods (J. A. Hoch, personal communication; E. Dubnau et al., this volume; Y. Kobayashi et al., this volume), little is known about the nature and function of their products. Here I will briefly review the genetic and physiological properties of the *spo0* genes and consider in detail a model system for studying their involvement in the control of sporulation RNA synthesis and in the modification of RNA polymerase.

spo0 GENE PRODUCTS

At Least Certain *spo0* Gene Products Are Vegetative Proteins

The most striking property of mutations in *spo0* genes is their pleiotropic character (for reviews, see 10, 17). These mutations not only block spore formation at its earliest stage but also alter several of the properties of vegetative cells. The degree of pleiotropy varies with the genetic locus. Mutations in *spo0A*, the most highly pleiotropic locus, prevent sporulation-associated antibiotic and protease production, interfere with the acquisition of genetic competence, and confer sensitivity during growth to surface-active antibiotics and certain phages such as $\phi2$ and $\phi15$, which are otherwise unable to grow in Spo$^+$ bacteria; mutations in other *spo0* loci exhibit a subset of these effects. For example, *spo0F* is less permissive for $\phi2$ and $\phi15$ than is *spo0A* but resembles *spo0A* in other properties. These pleiotropic effects are exerted, at least in part, at the level of gene expression; mutations in the *spo0A, spo0B, spo0E, spo0F,* and *spo0H* genes cause global changes in the pattern of protein synthesis in growing cells (3, 12). I conclude, therefore, that some, if not all, of the *spo0* genes are expressed in growing cells and that their products probably interact, directly or indirectly, with the transcriptional or

48

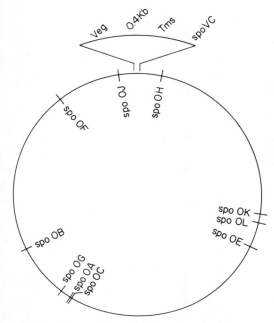

FIG. 1. *Genetic map of* spo0 *genes on the* B. subtilis *chromosome. See P. J. Piggot, A. Moir, and D. A. Smith (this volume) for a detailed map of* spo *loci. The figure also shows the chromosomal location of the cloned gene cluster of Fig. 2.*

translational machinery, or both, of vegetative cells. Since nonsense mutations have been identified in both the *spo0A* and the *spo0B* loci (3; Hoch, personal communication), these two stage 0 cistrons must encode vegetative proteins.

Although some, if not all, *spo0* genes encode vegetative proteins, in no case has a *spo0* gene product been definitively identified. Rhaese and his co-workers (18) have proposed that the *spo0F* gene encodes a synthetase for adenosine-bis-triphosphate (p_3Ap_3), a novel nucleotide suggested to be involved in the initiation of sporulation. The synthetase protein has not, however, been isolated, and the evidence that it is encoded by the *spo0F* gene is preliminary. Clear-cut evidence that the *spo0F* gene codes for the synthetase would support the notion that synthesis of that highly phosphorylated nucleotide is required in sporulation. Bramucci et al. (2) discovered that phage PMB12 (a relative of PBS2) suppresses the sporulation defect in Rfm^r Spo$^-$ RNA polymerase mutants as well as in a mutant with a lesion at the *spo0J* locus. This finding suggested that *spo0J*, like the *rfm* locus, encodes a subunit of RNA polymerase. Although this interesting possibility is not excluded, Bramucci

(personal communication) has now been able to show that PMB12 also suppresses sporulation-defective ribosomal mutations, a finding which indicates that the cellular target(s) of PMB12 conversion is not limited to (or necessarily inclusive of) the host-cell RNA polymerase.

Kobayashi et al. (this volume) have constructed a specialized transducing phage of $\rho11$ that carries the *spo0B* locus. This phage induces in UV-irradiated cells the synthesis of a 39,000-dalton polypeptide that is absent from cells infected with the parent phage. The 39,000-dalton polypeptide is therefore a strong candidate for the product of the *spo0B* gene, although this remains to be established directly.

spo0 Gene Products Are Essential for Sporulation

Mutations at the *abrB* locus, which maps near the origin of chromosome replication, suppress many but not all of the effects of *spo0A* and *spo0B* mutations (10, 25). Thus, *abrB* mutations restore an Spo0A mutant to polymyxin resistance, genetic competence, antibiotic and protease production, and phage resistance. *abrB spo0A* double mutants do not, however, sporulate. Indeed, no suppressor locus has as yet been identified that converts Spo0A, Spo0B, or any other Spo mutant to Spo$^+$. This could mean that some or all of the *spo0* gene products play an indispensable role in the initiation of sporulation.

Certain drug resistance mutations that affect the *B. subtilis* transcriptional and translational machinery are also blocked in sporulation at stage 0. Such mutants frequently do not exhibit altered growth properties (reviewed by Sonenshein and Campbell [22]) and can be distinguished from Spo0 mutants by the following criteria. (i) The drug resistance mutants are altered in known and essential components of the vegetative cell; at least certain *spo0* genes are not essential for growth, as null mutations exist in both the *spo0A* and *spo0B* loci. (ii) Some, if not all, sporulation-defective ribosomal and RNA polymerase mutants are suppressed by the PMB12 converting phage; of the *spo0* loci, PMB12 converts only a *spo0J* mutant to Spo$^+$ (M. G. Bramucci, personal communication). In addition, transcriptional-translation mutations that are defective in sporulation can be partially suppressed by agents that affect the cell membrane or by mutations at a locus termed *rev* (T. Leighton, personal communication). It will be interesting to know whether or not *rev* and membrane-active agents also suppress *spo0* mutations.

I conclude that the initiation stage of sporulation is controlled by at least two pathways. One of these pathways involves components of the *B. subtilis* transcriptional and translational machinery, which are also involved in normal vegetative growth. The other pathway includes the *spo0* gene products, some of which are apparently indispensable for sporulation, but play no critical role in normal growth.

A CLONED GENE WHOSE TRANSCRIPTION IS UNDER *spo0* CONTROL

A Cloned Cluster of Genes That Are Actively Transcribed During Sporulation

Figure 2 is an endonuclease restriction map of a cloned cluster of genes that are actively transcribed during sporulation. As will be discussed below, the transcription of at least one gene in this cluster requires the function of five of the *spo0* loci. The cloned cluster is approximately 8.5 kilobases (kb) in length and maps within the *purA-cysA* region (see Fig. 1) of the *B. subtilis* chromosome (6). (As oriented in Fig. 1 and 2, the right side of the segment is proximal to the *cysA* locus.) Five transcription units, of which three correspond to genetically defined loci, have so far been mapped in this chromosomal segment. These define the following genes: *spoVC*, *tms*, *0.4 kb*, *veg*, and *L*. I briefly summarize their properties below.

spoVC. The rightmost transcription unit in the cluster overlaps with the site of a mutation, *spoVC285* (26) (which maps within the right-hand *Hin*dIII segment of the physical map of Fig. 1 [14, 16]), that blocks spore formation at stage IV to V. I will assume here that this transcription unit corresponds to the *spoVC* locus. Although the position of its right-hand terminus is uncertain, this sporulation gene is at least 365 base pairs in length (J. F. Ollington et al., manuscript in preparation). The *spoVC* gene is actively transcribed at both early and late times in sporulation but is only expressed at a low rate in vegetative cells (less than 10% of the sporulation rate).

tms. The vegetative gene *tms* is defined by a mutation (*tms-26*; reference 5) causing temperature-sensitive growth. Although *B. subtilis* RNA polymerase containing σ^{55} (see below)

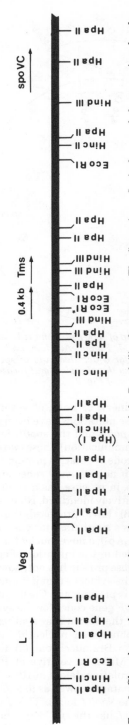

FIG. 2. *Cloned sporulation genes from the purA-cysA region of the B. subtilis chromosome. The figure is a physical map of cleavage sites for the indicated endonuclease restriction enzymes. The arrows indicate the location and direction of transcription of the L, veg, 0.4 kb, tms, and spoVC genes. The right-hand endpoints of the L, tms, and spoVC genes are uncertain; minimum estimates of these gene lengths are shown. The data in the figure are summarized from references 6, 7, 14, and 16 and Ollington et al., in preparation.*

readily transcribes the *tms-26*–containing region of cloned DNA in vitro (9), this transcript has not been detected in vivo (presumably because the rate of *tms* RNA synthesis is too low to be detected by Southern hybridizations). I will assume here that this in vitro transcription unit corresponds to the *tms* locus.

0.4 kb. The sporulation gene *0.4 kb* is about 400 base pairs in length, as determined both by the size of its transcript (19; Ollington et al., in preparation) and by the nuclease S1 mapping (N. Lang, unpublished data) procedure of Berk and Sharp (1). Like the *spoVC* gene, the *0.4 kb* gene is actively transcribed at both early and late times in sporulation but at only a low rate in growing cells. No previously known sporulation locus has been assigned to the *0.4 kb* gene. However, A. Rosenbluh, C. D. B. Banner, and R. Losick (unpublished data) have constructed by recombinant DNA techniques a 200-base pair deletion within the *0.4 kb* gene and inserted this mutation into the *B. subtilis* chromosome by genetic transformation. The mutation has no detectable effect on growth but interferes with spore formation at a late stage, a finding which defines the *0.4 kb* gene as an *spo* locus.

veg. The "vegetative" gene is about 340 base pairs in length and is actively transcribed in both growing cells and sporulating bacteria (Ollington et al., in preparation). No known genetic locus has been found to correspond to the *veg* transcription unit.

L. The newly discovered transcription unit *L* is defined by its recognition in vitro by a sporulation-specific form of RNA polymerase (7; see below). If the *L* sequence is expressed in vivo, its level of transcription is too low to be detected by Southern hybridizations. No genetic locus is known that could correspond to the *L* transcription unit.

Regulation of the *veg*, *0.4 kb*, and *spoVC* Genes

The effect of *spo0* mutations upon expression of the three most actively transcribed cloned genes (*veg*, *0.4 kb*, and *spoVC*) is summarized in Table 1. The *veg* gene is actively transcribed in wild-type sporulating cells and in stationary-phase cells of all seven of the Spo0 mutants that were examined. Similarly, *spoVC* is actively transcribed in stationary-phase cells of the stage 0-blocked mutants. In contrast, *0.4 kb* RNA synthesis was severely restricted in five Spo0 mutants (Spo0A, Spo0B, Spo0E, Spo0F, and

Spo0H). Two other stage 0 mutations, *spo0C* and *spo0J*, had no detectable effect on *0.4 kb* gene transcription. I conclude that activation of the *0.4 kb* gene is dependent upon the products of at least five *spo0* loci; these gene products, which are likely to be present in vegetative cells (see above), must somehow be involved in controlling the transcription of a gene turned on at the onset of sporulation.

The experiments of Table 1 suggest that the *spoVC* gene is not under *spo0* control; we may at least conclude that *0.4 kb* and *spoVC* differ in their mode of regulation. Nevertheless, it is conceivable that there is more than one *spo0*-dependent pathway for activating the *spoVC* gene and that a mutation in any one *spo0* gene is not sufficient to block *spoVC* RNA synthesis. This could be tested by examining *spoVC* expression in various combinations of *spo0* gene double mutations (e.g., an *spo0A spo0B* double mutant).

It is interesting to note that both the *0.4 kb* and the *spoVC* genes are activated early in sporulation, even though mutations at both these loci block spore formation at a late stage. This is consistent with the observation of Young (26) that the period of temperature sensitivity in the conditional sporulation mutant SpoVC285 precedes significantly the stage of morphological block (stage IV–V). These findings probably indicate that the *0.4 kb* and *spoVC* gene products are accumulated in sporulating cells well before their time of involvement in spore morphogenesis.

MODIFICATION OF RNA POLYMERASE

Multiple Sigma Factors in *B. subtilis*

B. subtilis contains at least three species of sigma factor, termed σ^{55}, σ^{37}, and σ^{29} (7, 8, 21; W. G. Haldenwang, N. Lang, and R. Losick, submitted for publication). Doi and co-workers refer to these polypeptides as σ, ϵ, and δ^{1}, respectively (see R. H. Doi, T. Kudo, and C. Dickel, this volume). Each of these factors confers on core RNA polymerase, which is composed of the subunits β', β, and α, distinct promoter recognition specificities. σ^{55} is the predominant sigma factor in vegetative cells and is believed to direct (along with other regulatory factors) most, if not all, RNA synthesis in growing cells. σ^{37} is a minor species of sigma factor which directs the transcription of two sporulation genes in vitro and may be involved, as proposed below, in the initiation of sporulation. σ^{29} is a sporulation-specific sigma factor and may control RNA synthesis at

TABLE 1. *Regulation of the* veg, 0.4 kb, *and* spoVC *genes*[a]

Gene	Growth	Sporulation Spo$^+$	Stationary phase						
			Spo0A	Spo0B	Spo0C	Spo0E	Spo0F	Spo0H	Spo0J
veg	+	+	+	+	+	+	+	+	+
0.4 kb	−	+	−	−	+	−	−	+	+
spoVC	−	+	+	+	+	+	+	+	+

[a] Expression of the *veg, 0.4 kb,* and *spoVC* genes was measured by hybridization of pulse-labeled RNA to gene-specific endonuclease restriction fragments. These results are summarized from Ollington et al. (in preparation). A minus sign represents less than 10% of the level of transcription indicated by a plus sign.

an early to intermediate stage of sporulation. Purified *B. subtilis* RNA polymerases containing σ^{55}, σ^{37}, and σ^{29} are displayed in the sodium dodecyl sulfate-polyacrylamide gel of Fig. 3.

An experiment to illustrate the time of appearance of these factors in the *B. subtilis* life cycle is shown in Fig. 4 and summarized in Table 2. RNA polymerase was partially purified and subjected to sodium dodecyl sulfate-polyacrylamide gel electrophoresis. Although the enzyme was impure, gel electrophoresis readily resolved the known subunits of RNA polymerase. Both σ^{55} and σ^{37} were present in enzyme from vegetative cells (track 1) but disappeared from polymerase during the first hours of sporulation (tracks 2 and 3). As reported previously (e.g., 4, 23) and confirmed here (Fig. 4, tracks 4 and 5), the loss of these sigma factors is a sporulation-specific event, as enzyme from an Spo0A mutant retained high levels of σ^{55} and σ^{37} during stationary phase.

The pattern of appearance of σ^{29} was reciprocal to that observed for σ^{55} and σ^{37}. σ^{29}-containing RNA polymerase accumulated during the first 2 h of sporulation (Fig. 4, tracks 2 and 3); little or no σ^{29}-containing enzyme could be detected in vegetative cells (track 1). Like the loss of σ^{55} and σ^{37}, the induction of σ^{29}-polymerase is a sporulation-specific event, as this modified form of polymerase was absent in stationary-phase cells of the Spo0A mutant (Fig. 5, tracks 4 and 5).

Transcription of Cloned Genes

Table 2 summarizes the transcriptional specificities of the three forms of *B. subtilis* RNA polymerase on the cloned template shown in Fig. 2. Each sigma factor confers on core RNA polymerase distinct but overlapping transcriptional specificities. Thus, σ^{55} enzyme transcribes only the *veg* and *tms* genes. σ^{37} polymerase, in contrast, transcribes the *veg, 0.4 kb,* and *spoVC* genes, but not the *tms* gene. Finally, σ^{29} polymerase exhibits a third transcription pattern.

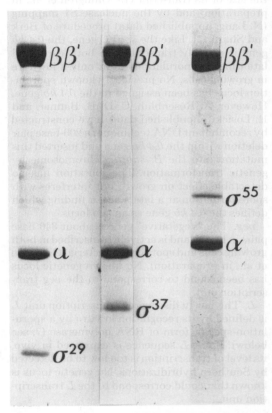

FIG. 3. *Sodium dodecyl sulfate-polyacrylamide gel electrophoresis of* B. subtilis *RNA polymerases containing* σ^{29}, σ^{37}, *or* σ^{55}. *Reproduced with permission from reference 7.*

Like σ^{37} polymerase, it transcribes the *veg, 0.4 kb,* and *spoVC* genes. In addition, however, σ^{29} polymerase transcribes a sequence called *L* which is not recognized by either σ^{55} or σ^{37} polymerase.

An example of the kind of transcription experiment that distinguishes the specificities of these sigma factors is shown in Fig. 5. In this RNA "runoff" experiment, which was performed

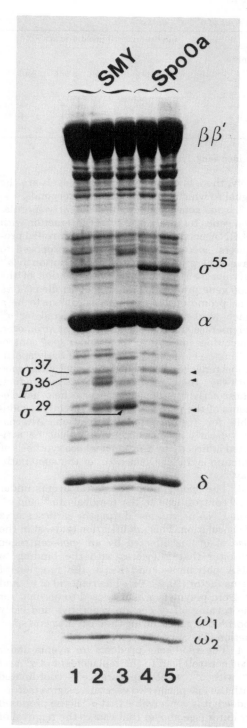

FIG. 4. Association of σ^{29}, σ^{37}, and σ^{55} with RNA polymerase during growth and sporulation in B. sub-

by W. G. Haldenwang (personal communication), hybrid plasmid DNA containing the *tms* and *spoVC* genes was cleaved with the restriction endonucleases HindIII and HpaII and was used as a template for in vitro RNA synthesis by core RNA polymerase alone or by core RNA polymerase reconstituted by the addition of σ^{55} or σ^{37}. (HindIII cuts within the *tms* gene at a site 85 base pairs to the right of its promoter, whereas HpaII cleaves within the *spoVC* gene at a site 365 base pairs to the right of its starting point [Fig. 2].) The products of transcription were then displayed by polyacrylamide gel electrophoresis. Core RNA polymerase alone did not generate a detectable transcript, but σ^{55} directed the synthesis of an 85-base RNA which was diagnostic of *tms* transcription. σ^{37} elicited the synthesis of a 365-base RNA which was diagnostic of *spoVC* transcription.

These experiments provide, therefore, direct evidence for the proposal, first advanced by Losick and Sonenshein (15), that sporulation genes are controlled, at least in part, by sigma factors of novel promoter recognition specificities.

A MODEL FOR THE INVOLVEMENT OF THE STAGE 0 GENE PRODUCTS IN RNA POLYMERASE MODIFICATION

The relationship between *spo0* gene products and RNA polymerase sigma factors is summarized in the following proposed scheme for the regulation of sporulation RNA synthesis. In this RNA polymerase modification model, the *spo0* gene products are proposed to be components of a pathway that senses nutrient deprivation and translates this environmental signal into the transcriptional activation of sporulation genes by causing, directly or indirectly, modifications of RNA polymerase. The principal features of this model are as follows.

tilis. *Growth and sporulation were in Schaeffer nutrient broth medium (4). Cells were harvested during growth or at 1 h (T_1) or 2 h (T_2) after the end of exponential growth. RNA polymerase was partially purified by phase partitioning, gel filtration, and stepwise elution from DNA-cellulose and was displayed by electrophoresis through a sodium dodecyl sulfate-polyacrylamide slab gel. The figure displays partially purified RNA polymerase from: track 1, mid-exponential-phase cells of strain SMY; track 2, T_1 cells of strain SMY; track 3, T_2 cells of strain SMY; track 4, T_1 cells of strain Spo0A-22NA; track 5, T_2 cells of strain Spo0A-22NA. Reproduced with permission from reference 7.*

TABLE 2. *Summary of the properties of σ^{55}, σ^{37}, and σ^{29}*

Sigma factor	Time of appearance[a]					In vitro transcriptional specificities[b]				
	Growth mid-log	Sporulation (spo^+)		Stationary phase (*spo0A*)		*L*	*veg*	*0.4 kb*	*tms*	*spoVC*
		T_1	T_2	T_1	T_2					
σ^{55}	++	+	±	++	++	−	+	−	+	−
σ^{37}	++	+	±	++	++	−	+	+	−	+
σ^{29}	−	+	++	−	−	+	+	+	−	+

[a] Summarized from Haldenwang, Lang, and Losick (7).
[b] Summarized from Haldenwang and Losick (8, 9) and Haldenwang, Lang, and Losick (7).

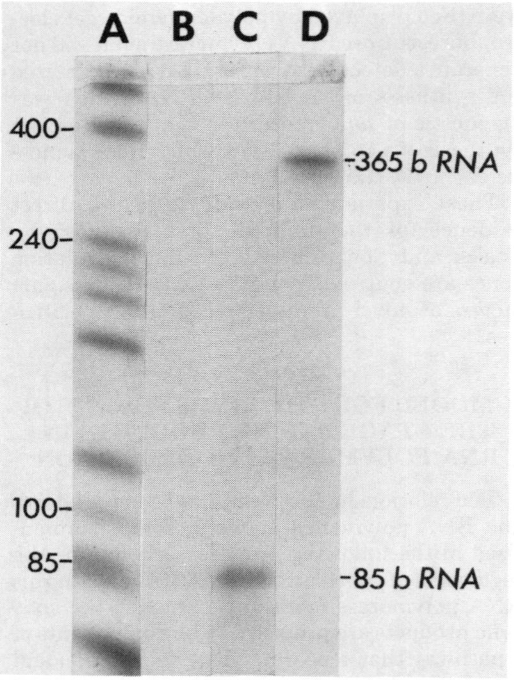

FIG. 5. *"Runoff" transcription of the* tms *and* spoVC *genes. ^{32}P-labeled RNA was transcribed in vitro from 2.5 μg of hybrid plasmid DNA (p63 of reference 9) that had been cleaved with* HindIII *and* HpaII *by core RNA polymerase alone (track B), core RNA polymerase plus 40 ng of σ^{55} (track C) or core RNA polymerase plus 8 ng of σ^{37}. The conditions for in vitro RNA synthesis were as described previously (8). The in vitro-synthesized RNAs were subjected to electrophoresis through a 6% polyacrylamide gel containing 7 M urea. Track A displays as molecular weight markers denatured fragments of ^{32}P end-labeled,* HaeIII*-cut pMB9 DNA. Radioactive RNAs were visualized by autoradiography.*

1. The *spo0* gene products are vegetative proteins, one or more of which respond to nutrient deprivation. As proposed by E. Freese (this volume), nutrient limitation leading to low guanine nucleotide levels may trigger sporulation. Possibly, then, low purine nucleotide levels are the signal to which the *spo0* pathway responds.

2. *spo0* gene products turn on early sporulation genes by direct or indirect interaction with σ^{37} RNA polymerase or with σ^{37}-controlled promoters. *spo0*-controlled genes are proposed to have weak promoters whose transcription by σ^{37} polymerase is stimulated by the action of the *spo0* gene products. The *spo0*-controlled *0.4 kb* gene promoter, for example, is known to be a much weaker in vitro initiation signal for σ^{37} polymerase than is the *spoVC* gene promoter, which does not appear to be under *spo0* control (Ollington et al., in preparation). If early sporulation genes are controlled directly by the *spo0* pathway, then only one *spo0* gene product, whose activity is dependent upon the other elements of the pathway, need be directly responsible for transcriptional activation. Alternatively, early sporulation genes might be activated indirectly by a product whose synthesis or function is under the control of the *spo0* pathway.

3. The *spo0* gene products (or events under *spo0* control) lead to the removal of σ^{55} and σ^{37} from RNA polymerase during the earliest stages of sporulation. This modification (at least in the case of σ^{55}) is effected by an *spo0*-controlled inhibitor that interferes with the binding to RNA polymerase and hence the function of sigma factor (20, 23, 24). The removal of σ^{55} and σ^{37} from polymerase is proposed to account for the turning off of certain vegetative and early sporulation genes during the early stages of sporulation (13).

4. The *spo0* gene products (or events under *spo0* control) lead to the replacement of σ^{55} and σ^{37} with σ^{29}, whose appearance is coordinated with the loss of the two vegetative sigma factors. Indeed, it is conceivable that σ^{29} is the proposed inhibitor (see above) that causes the removal of σ^{55} and σ^{37} from polymerase. Two regulatory functions are proposed for σ^{29}. First, based on its overlapping transcriptional specificities with σ^{55}

and σ^{37}, σ^{29} could continue the transcription of certain vegetative genes (e.g., the *veg* gene) and certain early sporulation genes (e.g., the *0.4 kb* gene and the *spoVC* genes) whose transcription persists through several developmental stages. Second, based on its unique transcriptional specificities, σ^{29} could turn on at stage II to III sporulation genes (with *L*-like promoters) whose transcription had not been previously activated by σ^{37}. Thus, σ^{29} would be responsible for the temporally delayed turning on of certain sporulation genes at an early to intermediate stage of sporulation.

This model undoubtedly oversimplifies the complicated events of early sporogenesis. Two important reservations should be emphasized. First, σ^{37} and σ^{29} are not likely to be the only sigma-like regulatory proteins that control sporulation RNA synthesis. This laboratory (see references 7, 8, and 13, for example) and that of Doi (Doi et al., this volume) have reported on several sporulation-associated components of RNA polymerase that could also play a regulatory role in spore formation. One of these, a polymerase subunit of 36,000 daltons (P^{36}) that appears transiently early in sporulation (7, 8) is displayed in the experiment of Fig. 4. Second, although a variety of RNA polymerase modifications appear to be under *spo0* control, it is not known whether *spo0* gene products effect these changes by interacting *directly* with RNA polymerase or by as yet undefined indirect mechanisms. Nevertheless, RNA polymerase modification is proposed to be an important, though not exclusive, consequence of *spo0* gene product action. A test of the RNA polymerase modification model now requires the identification of the genetic loci that specify σ^{55}, σ^{37}, and σ^{29} and an investigation of the mode of action of the *spo0* gene products.

ACKNOWLEDGMENTS

I thank P. Youngman, A. L. Sonenshein, and D. Tipper for critical suggestions.

This work was supported by Public Health Service grant GM 18568 from the National Institutes of Health.

LITERATURE CITED

1. **Berk, A., and P. Sharp.** 1976. Sizing and mapping of early adenovirus mRNAs by gel electrophoresis of S_1 endonuclease-digested hybrids. Cell **12**:721–732.
2. **Bramucci, M. G., K. M. Keggins, and P. S. Lovett.** 1977. Bacteriophage conversion of spore-negative mutants to spore-positive in *Bacillus pumilus*. J. Virol. **22**: 194–202.
3. **Brehm, S. P., F. Le Hegarat, and J. A. Hoch.** 1975. Deoxyribonucleic acid-binding proteins in vegetative *Bacillus subtilis*: alterations caused by stage 0 sporulation mutations. J. Bacteriol. **124**:977–984.
4. **Brevet, J., and A. L. Sonenshein.** 1972. Template specificity of ribonucleic acid polymerase in asporogenous mutants of *Bacillus subtilis*. J. Bacteriol. **112**:1270–1274.
5. **Copeland, J. C., and J. Marmur.** 1968. Identification of conserved genetic functions in *Bacillus* by use of temperature-sensitive mutants. Bacteriol. Rev. **32**:302–312.
6. **Haldenwang, W. G., C. D. B. Banner, J. F. Ollington, R. Losick, J. A. Hoch, M. B. O'Connor, and A. L. Sonenshein.** 1980. Mapping a cloned gene under sporulation control by insertion of a drug resistance marker into the *Bacillus subtilis* chromosome. J. Bacteriol. **142**:90–98.
7. **Haldenwang, W. G., N. Lang, and R. Losick.** 1981. A sporulation-induced sigma-like regulatory protein from *Bacillus subtilis*. Cell **23**:615–624.
8. **Haldenwang, W. G., and R. Losick.** 1979. A modified RNA polymerase transcribes a cloned gene under sporulation control in *Bacillus subtilis*. Nature (London) **282**:256–260.
9. **Haldenwang, W. G., and R. Losick.** 1980. A novel RNA polymerase sigma factor from *Bacillus subtilis*. Proc. Natl. Acad. Sci. U.S.A. **77**:7000–7004.
10. **Hoch, J. A.** 1976. Genetics of bacterial sporulation. Adv. Genet. **18**:67–98.
11. **Hoch, J. A., M. A. Shiflett, J. Trowsdale, and S. M. H. Chen.** 1978. Stage 0 genes and their products, p. 127–130. *In* G. Chambliss and J. C. Vary (ed.), Spores VIII. American Society for Microbiology, Washington, D.C.
12. **Linn, T., A. L. Greenleaf, and R. Losick.** 1975. RNA polymerase from sporulating *Bacillus subtilis*: purification and properties of a modified form of the enzyme containing two sporulation polypeptides. J. Biol. Chem. **250**:9256–9261.
13. **Linn, T., and R. Losick.** 1976. The program of protein synthesis during sporulation in Bacillus subtilis. Cell **8**: 103–114.
14. **Losick, R.** 1981. Sporulation genes and their regulation. *In* D. A. Dubnau (ed.), Molecular biology of the bacilli, vol. 1: *Bacillus subtilis*. Academic Press, Inc., New York. In press.
15. **Losick, R., and A. L. Sonenshein.** 1969. Change in the template specificity of RNA polymerase during sporulation of *Bacillus subtilis*. Nature (London) **224**:35–37.
16. **Moran, C. P., Jr., R. Losick, and A. L. Sonenshein.** 1980. Identification of a sporulation locus in cloned *Bacillus subtilis* deoxyribonucleic acid. J. Bacteriol. **142**:331–334.
17. **Piggot, P. J., and J. G. Coote.** 1976. Genetic aspects of bacterial endospore formation. Bacteriol. Rev. **40**:908–962.
18. **Rhaese, H. J., R. Groscurth, and G. Rumpf.** 1978. Molecular mechanism of initiation of differentiation in *Bacillus subtilis*, p. 286–292. *In* G. Chambliss and J. C. Vary (ed.), Spores VII. American Society for Microbiology, Washington, D.C.
19. **Segall, J., and R. Losick.** 1977. Cloned *Bacillus subtilis* DNA containing a gene that is activated early during sporulation. Cell **11**:751–761.
20. **Segall, J., R. Tjian, J. Pero, and R. Losick.** 1974. Chloramphenicol restores sigma factor activity to sporulating *Bacillus subtilis*. Proc. Natl. Acad. Sci. U.S.A. **71**:4860–4863.
21. **Shorenstein, R., and R. Losick.** 1973. Comparative size and properties of the sigma subunits of ribonucleic acid

polymerase from *Bacillus subtilis* and *Escherichia coli.* J. Biol. Chem. **248**:6170–6173.

22. **Sonenshein, A. L., and K. M. Campbell.** 1978. Control of gene expression during sporulation, p. 179–192. *In* G. Chambliss and J. C. Vary (ed.), Spores VII. American Society for Microbiology, Washington, D.C.

23. **Tjian, R., and R. Losick.** 1974. An immunological assay for the sigma subunit of RNA polymerase in extracts of vegetative and sporulating *Bacillus subtilis.* Proc. Natl. Acad. Sci. U.S.A. **71**:2872–2876.

24. **Tjian, R., D. Stinchcomb, and R. Losick.** 1975. Antibody directed against *Bacillus subtilis* σ factor purified by sodium dodecyl sulfate slab gel electrophoresis: ef-

fect on transcription by RNA polymerase in crude extracts of vegetative and sporulation cells. J. Biol. Chem. **250**:8824–8828.

25. **Trowsdale, J., S. M. H. Chen, and J. A. Hoch.** 1978. Genetic analysis of phenotype revertants of *spo0A* mutants in *Bacillus subtilis*: a new cluster of ribosomal genes, p. 131–135. *In* G. Chambliss and J. C. Vary (ed.), Spores VII. American Society for Microbiology, Washington, D.C.

26. **Young, M.** 1976. Use of temperature-sensitive mutants to study gene expression during sporulation in *Bacillus subtilis.* J. Bacteriol. **126**:928–936.

Toxins of Sporeforming Bacteria[1]

ROBERT E. ANDREWS, JR., AND LEE A. BULLA, JR.

Division of Biology, Section of Microbiology and Immunology, Kansas State University, and U.S. Grain Marketing Research Laboratory, Science and Education Administration, U.S. Department of Agriculture, Manhattan, Kansas 66502

INTRODUCTION

There are several gram-positive, sporeforming bacteria that produce proteinaceous toxins of economic, commercial, and biomedical significance. Probably the best known organisms belong to the genera *Bacillus* and *Clostridium*, and of these, *B. thuringiensis, C. perfringens, C. botulinum*, and *C. tetani* have been investigated most intensely. *B. thuringiensis* synthesizes a potent entomocidal toxin which is the basis of commercial insecticides designed to control economically important insect pests (10), *C. perfringens* produces a number of toxins that are involved in human and animal pathogenesis (30), and *C. botulinum* and *C. tetani* form neurotoxins of very similar structure and action (42).

A number of reviews have been written about toxins of various gram-positive, sporeforming bacteria (2–4, 25, 27, 30, 31, 42, 46). Therefore, it is not our intent to review the literature on this subject but, rather, to address special features of selected toxins formed by particular sporeformers. We have chosen to focus on *B. thuringiensis, C. perfringens*, and *C. botulinum*. These organisms were chosen because *B. thuringiensis* and *C. botulinum* produce protoxic or progenitor molecules from which smaller toxic molecules are derived, spore coats of *B. thuringiensis* and *C. perfringens* contain their respective toxic proteins, and the temporal relationship of toxin synthesis to growth and sporulation is similar in all three organisms. In addition to their insect or mammalian pathogenesis, *B. thuringiensis* and *C. botulinum* toxins may be useful as therapeutic agents. *B. thuringiensis* parasporal crystalline toxic protein has been reported to cause regression of tumors in vivo (33, 34) and to enhance the overall immune response in rats (35). *C. botulinum* protein neurotoxin may be potentially useful for correcting strabismus in lieu of surgery (42).

BACILLUS THURINGIENSIS

B. thuringiensis is a gram-positive, aerobic sporeformer. There are a number of subspecies of *B. thuringiensis*, all of which synthesize insecticidal toxins with some variation in host specificity (9). Generally, all of the strains are toxic to lepidopteran insects (moths and butterflies), but recently isolated strains called subspecies *israelensis* and *kyushuensis* kill mosquitoes and black flies (16, 32, 44, 45). For the purpose of discussion, we will primarily address subspecies *kurstaki*, which is representative of the moth-killing strains and which currently is the basis for commercial microbial insecticides (10).

The insecticidal toxin of *B. thuringiensis* subsp. *kurstaki* is contained primarily in a parasporal crystalline inclusion that is synthesized during stages III to V of sporulation (1). The major component of the native crystal is a glycoprotein and represents 20 to 30% of the cell dry weight. This glycoprotein is a protoxic subunit of the crystal that is activated after ingestion by an insect susceptible to the toxic product (8). The subunit, which has an apparent molecular weight of 1.34×10^5 (7), can be solubilized in denaturing solvents or by mild alkali titration. Carbohydrate constitutes about 5% of the crystal, and the remainder is protein. The sugar

[1] Contribution No. 81-364-j, Division of Biology, Kansas Agricultural Experiment Station, Manhattan.

portion is composed of neutral hexoses (glucose and mannose, 3.8% and 1.8%, respectively, of the crystal dry weight), but the structure of the individual carbohydrate side chains and their specific site(s) of attachment to the protein are not known. It has been shown, however, that the carbohydrate is released via β-elimination in alkali, and this fact, together with the formation of lysinoalanine under alkaline conditions, indicates that the sugars are attached to serine and threonine (15).

Crystal solubilization is dependent upon hydroxide ion concentration and is accompanied by a release of protons. Consequently, it is mandatory that a solvent of relatively high buffering capacity or titration with hydroxide ion be used during solubilization to maintain constant pH. Soluble components are generated from the glycoprotein by increasing the pH of the solvent (7). Radiolabeled peptides (molecular weight $\leq$ 1×10^4) appear in solution after radioactive protoxin is incubated for 30 min at pH values greater than 9; none appear at lower pH values.

The solubilized protoxin apparently contains thiol groups (four cystine residues per subunit), possibly some proteolytic activity, and degradation products. Dissolution and activation are brought about, partially, by a mechanism involving a protease(s) associated with the crystal that becomes activated under alkaline conditions. The action of this protease(s) induces a gradual degradation of the 1.34×10^5-dalton protoxin, with formation of lower-molecular-weight products. Proteolytic activity occurs at a pH of 9.0 and higher. By using synthetic peptide substrates and specific inhibitors, it has been shown that proteinases contained in the crystal are sulfhydryl, serine, and metalloproteases (8, 11). Some of these can be removed from the crystal either by acid treatment followed by repeated water washings or by anion-exchange chromatography of solubilized material. If the contaminating protease(s) is removed, the solubilized protoxin has chemical and biological properties similar to those of the native crystal. The best solubilization procedure involves titration of Renografin-purified crystals (7) with 1 N NaOH until solvation is obtained. Approximately 4 μmol of alkali is required to dissolve 1 mg of crystal, or 400 mol of hydroxide ion for each mole of subunit. Only the highly charged anionic form of the crystal subunit is completely soluble, consistent with an isoelectric point near physiological pH. The pI, determined by isoelectric focusing on polyacrylamide, is 7.2 ± 0.1.

The insecticidal toxin of B. thuringiensis subsp. kurstaki is a glycoprotein (apparent molecular weight, 6.8×10^4) that is generated from the protoxin upon prolonged incubation (several days) at slightly alkaline pH (6), or it can be generated more rapidly when the solubilized crystal is incubated with insect gut juice or with trypsin (26). Such conversion may be similar to what occurs within the midgut of susceptible lepidopteran larvae. The midgut pH of such insects ranges from approximately 9.5 to 11.0. Interestingly, the toxin is stable at room temperature and at neutral pH and, like the protoxin, it is two to three times more insecticidal than the native crystal (6).

Apparently, synthesis of the crystalline protein protoxin is a sporulation-specific event. Not only do the times of appearance of the spore and crystal overlap, but also they are formed in close proximity. Of special importance is the presence of the crystal protoxic protein in the spore coat. The identity of the crystal protein has been established in the coat (L. A. Bulla, Jr., K. J. Kramer, and L. I. Davidson, Abstr. Annu. Meet. Am. Soc. Microbiol. 1977, I62, p. 165), and its toxicity to insect larvae has been determined to be identical to that of protoxin purified from crystals (38, 43).

Several hypotheses have been proposed (4) on the toxicological events related to parasporal crystal toxin action. None has been proven. These include: (i) separation of gut cells from each other and their detachment from the basement membrane; (ii) enhancement of secretory activity of gut epithelial cells; (iii) increase in permeability of the gut wall to sodium ions, with a slower rate of glucose uptake into the hemolymph; and (iv) gut paralysis and sometimes general paralysis of the body. Generally, the gross pathology of a susceptible larva includes necrosis of the midgut epithelial cells, dehydration, and subsequent death. Once the crystal is ingested, it dissolves and interacts with the midgut epithelial cells and transforms them into abnormal cells; physiological functions are impaired so that death occurs. In short, the alkalinity of the lepidopteran midgut (pH ≅ 10) along with a mixture of digestive enzymes, including proteases that are active in extremely alkaline conditions, provides an environment suitable for facilitating protoxin activation and subsequent toxin action. Because the toxic molecule recently has been purified from crystals of B. thuringiensis (6, 26), its characterization now can be accomplished and the mode of action at the molecular and cellular levels can be defined accurately.

CLOSTRIDIUM PERFRINGENS

C. perfringens is a gram-positive, sporeforming obligate anaerobe. A variety of toxins are synthesized by the different types of *C. perfrin-gens.* At least 12 toxins have been identified and are thought to be involved in mammalian pathogenesis (30). These toxins are soluble antigens and have been utilized in grouping the different strains of *C. perfringens* into toxigenic types.

Of the various toxins produced by *C. perfrin-gens,* the proteinaceous enterotoxin is the only one that has been demonstrated to be a sporulation-specific product (18, 19). Apparently, it is synthesized during stages III to V of sporulation, as is the insecticidal toxin of *B. thuringiensis.* The enterotoxin is responsible for clostridial food poisoning and, like *B. thuringiensis* toxin, affects the intestinal tract of susceptible hosts (22). Although there are discrepancies in the literature regarding the molecular size of enterotoxin, most evidence indicates that it is a protein with an apparent molecular weight of 3.5 $\times$ 10^4 (30). J. L. McDonel and W. P. Smith (this volume) have reported an in vitro system able to synthesize enterotoxin, and from such a synthesis they found three proteinaceous components with molecular weights of approximately 17,000, 35,000, and 52,000. These investigators hypothesized that the 3.5 $\times$ 10^4-dalton polypeptide is a precursor of a smaller spore coat component (apparent molecular weight, 1.7 $\times$ 10^4).

Unlike the entomocidal toxin of *B. thuringiensis,* *C. perfringens* enterotoxin apparently is not a glycoprotein nor does it exhibit associated proteolytic activity (30). Chemical analysis of the enterotoxin indicates that it is an acidic protein whose major amino acids are aspartic acid, serine, leucine, glutamic acid, isoleucine, glycine, and threonione (21). Its isoelectric point is 4.3. No amino sugars or carbohydrate moieties have been reported. The enterotoxin is heat labile; it loses biological activity, but not serological activity, upon heating at 60°C for 5 min. Serological activity generally is destroyed by heating at pH values lower than 6.0.

Enterotoxin-like protein occurs in spores of *C. perfringens* (20). In fact, Duncan et al. (17) have observed parasporal crystalline inclusions in sporulating cells which presumably are aggregates of enterotoxin protein. As is true for parasporal crystal protein of *B. thuringiensis,* the enterotoxin probably is a structural component of the spore coat. Frieben and Duncan (20) identified three enterotoxin-like proteins in the coat of *C. perfringens* with apparent molecular weights of 14,500, 23,000, and 36,500. All three proteins had biological and serological activities similar to those of pure enterotoxin. As mentioned above, McDonel and Smith (this volume) observed that the three proteins synthesized in vitro were precipitated by anti-enterotoxin se-

rum but had somewhat different molecular weights than those proteins described by Frieben and Duncan (20). Nevertheless, all of the proteins described are sporulation-specific products because they are spore coat components and probably contribute in some way to the structural integrity of the spore.

Reports are numerous of the pathological and physiological effects of *C. perfringens* enterotoxin on the mammalian gut. Apparently, the enterotoxin acts directly on the gut epithelial cell membrane (28) to interfere with energy production and to destroy cell structure and function. The initial site of interaction probably is the brush border membrane of villus tip epithelial cells (29), which is the same as the site on midgut columnar epithelial cells where the primary action of *B. thuringiensis* toxin occurs. Further, toxicological effects and symptoms resulting from enterotoxin action are similar to those occasioned by *B. thuringiensis* toxin in insects. These include fluid and electrolyte accumulation and inhibition of glucose uptake, energy production, and macromolecular synthesis (30).

CLOSTRIDIUM BOTULINUM

C. botulinum, another sporeforming anaerobe, synthesizes a proteinaceous toxin that causes, in mammals, a profound neurological disorder commonly referred to as botulism. Although several toxin types are primarily identified by serological properties, all have a similar mode of action which involves the peripheral nervous system. The recognized toxin types are A, B, C_1, C_2, D, E, F, and G. Serotypes A, B, E, and F are highly toxic to humans, whereas types C and D chiefly affect animals. The average molecular weight of these neurotoxic molecules is about 1.5 $\times$ 10^5, although there have been reports of toxic components of lower molecular weight (42). The specific activities of these smaller components, however, are less than those reported for the 1.5 $\times$ 10^5-dalton proteins.

Generally, neurotoxin synthesis begins during middle to late exponential growth and becomes maximal during the stationary phase. Correlation has been made between type C_2 toxin production and sporulation (30), but types A and E are synthesized by asporogenic L-form cultures, thus placing in question the significance of sporulation to neurotoxin synthesis and vice versa. In this respect, the neurotoxin of *C. botulinum* differs from the toxins of *C. perfringens* and *B. thuringiensis.*

An interesting feature shared by both *C. botulinum* and *B. thuringiensis* is that the protein toxins are derived from parent molecules that exist as protoxins or progenitor toxins. Furthermore, the protoxin of *C. botulinum* is enzyme activated (14), just as the protoxin of *B. thuringiensis* may be (8). One major difference in the two parent molecules, however, is that the protoxin complex of *C. botulinum* contains molecules other than the neurotoxic protein, whereas the protoxic crystal of *B. thuringiensis* is composed of a single repeating subunit. The toxic molecules of *C. botulinum* are associated in a complex with other nontoxic botulinum proteins that are present naturally in the culture medium. These complexes, with molecular weights of 5×10^5 to 9×10^5, can be recovered easily from culture fluid, facilitating efficient biochemical and biophysical characterization of the toxin.

Activation of the neurotoxin from various *C. botulinum* serotypes apparently involves a sulfhydryl protease (12, 13). Activation of *B. thuringiensis* protoxin also may involve a sulfhydryl protease that renders the protoxin insecticidal (8). Whether the protease is essential to activation of the *B. thuringiensis* protoxin is an unresolved issue, but, certainly, it may expedite activation. Indeed, reagents that complex or modify sulfhydryl groups inhibit solubilization and activation of whole parasporal crystalline protoxin; furthermore, these reagents prevent toxicity of crystals that are already solubilized.

As we have pointed out already, the toxic proteins of *B. thuringiensis* and *C. perfringens* are spore coat components. Neurotoxin of *C. botulinum* is produced mainly by vegetative cells but is sometimes incorporated within the spore as it is formed. The amount of toxin is never as great in spores as in vegetative cells. The mode and site of action of *C. botulinum* toxin also are different from those of either *B. thuringiensis* or *C. perfringens*. Botulinum toxin acts on the peripheral nervous system, the result being flaccid paralysis. Some of the eventual symptoms are impairment of cranial nerve functions and respiratory failure which ultimately causes death (42). The only symptom that is similar to those caused by *B. thuringiensis* and *C. perfringens* toxins is diarrhea, which sometimes precedes neuromuscular paralysis.

OTHER ORGANISMS

B. popilliae and *B. lentimorbus* are gram-variable, facultatively anaerobic bacteria that are insecticidal to various scarabaeid beetles. *B. popilliae* forms a parasporal crystal adjacent to the endospore in much the same manner as *B. thuringiensis*; *B. lentimorbus* does not form a crystal. *B. popilliae* causes a so-called type A milky disease, and *B. lentimorbus* causes a type B disease (5). Whether parasporal crystals of *B. popilliae* kill beetle larvae has not been established. The crystals are proteinaceous and contain glycoprotein components. No toxic compounds have been found in either *B. popilliae* or *B. lentimorbus*. The two bacteria are fastidious and do not sporulate efficiently outside their insect hosts. All reported attempts to determine the factor(s) that regulates spore formation have failed. The best means currently available for producing spores commercially is by artificially infecting individual larvae.

B. sphaericus is another gram-variable facultative anaerobe. Its insecticidal activity is against mosquitoes. As yet, no single toxic molecule has been elucidated, but most evidence, to date, indicates that the toxin is associated with the cell wall of vegetative cells, although spores of this organism also are lethal to mosquito larvae. Research on this particular organism has not progressed at a very rapid rate, but it is hoped that future endeavors will help to define the toxic compound(s) synthesized by this bacterium.

B. anthracis synthesizes a toxic mixture or complex composed of three factors or components: edema factor, protective antigen, and lethal factor (27). Although each component alone has some biological activity, full toxicity occurs only when the three components are combined. Evidently, all three components are proteinaceous, the protective antigen functioning as a coenzyme and the edema factor and lethal factor as apoenzymes. The effects that anthrax toxin has on susceptible animals are complex and have not been fully delineated. However, it is believed that a major effect is on the central nervous system. Much more work needs to be done on determining the toxic mode of action and on gaining a better biophysical and biochemical characterization of the toxic components.

B. cereus produces two protein toxins associated with food poisoning. They are synthesized during exponential growth and sporulation, with maximal synthesis occurring during the transition between vegetative proliferation and sporulation. Apparently, sporulation is not dependent upon toxin synthesis because sporulation occurs in media that prohibit toxin production (2). In addition to these two toxins, *B. cereus* synthesizes phospholipase C and hemolysin molecules that have toxic activity. Interestingly, *C. perfringens* also produces phospholipase C (α toxin) that has some similarity in biological and biochemical properties to that produced by *B. cereus*.

C. tetani produces a neurotoxic protein that is

similar to that of *C. botulinum* (46). The apparent molecular weight of the toxic protein of *C. tetani* is approximately 150,000 and it is activated by culture enzymes and by trypsin, just as is the *C. botulinum* toxic protein. The activated toxin consists of disulfide-linked α and β polypeptides which cause muscular paralysis by blocking transmitter release and primarily affect the spinal cord, whereas *C. botulinum* neurotoxin exhibits its interference at cholinergic junctions. The initial site of infection by *C. tetani* is normally in wounds, whereas *C. botulinum* toxin generally must be ingested, although there are rare cases of wound botulism.

GENERAL CONSIDERATIONS

Significant questions regarding toxins of sporeformers concern their origin and function. Why are they formed? What controls their synthesis? Why do the parent organisms need them? Just as has been proposed for peptide antibiotics of the various *Bacillus* species (24) there are several possible answers: (i) they are evolutionary vestiges, (ii) they represent an intermediary or end product of metabolism, (iii) they serve as reserve food, (iv) they are the result of specific syntheses which provide the bacteria some selective advantage, (v) they represent loosely regulated protein synthesis, (vi) they represent detoxification mechanisms, (vii) they afford a method of escaping death due to unbalanced growth, and (viii) they inhibit spore germination.

As we have already mentioned, there is some correlation between toxin synthesis and sporulation in a few of the organisms described. Whether the toxins are important compounds in cellular differentiation is not known, but we believe that such a possibility exists because in the case of *B. thuringiensis* and *C. perfringens* not only does the time of protein toxin synthesis and sporulation overlap but also the proteins are contained within the spore coats. In addition, it has been shown for *B. thuringiensis* subsp. *kurstaki* that the parasporal crystal protein may have some function in spore germination and, consequently, play a protective role (41). Tyrell et al. (43) have shown that spore coats of several subspecies of *B. thuringiensis* all contain their respective crystal proteins as major components. Interestingly, the mosquito-killing subspecies *israelensis* has a unique spore coat protein profile unlike those of moth-toxic *B. thuringiensis* strains such as subspecies *kurstaki*. Stahly et al. (41) have demonstrated that spores of acrystalliferous mutants of *B. thuringiensis* subsp. *kurstaki* germinate more rapidly than do spores of

B. cereus. Curiously, the rapidly germinating spores lacked the 134,000-dalton crystal protein in their coats and the spores were no longer toxic to moth larvae, as they normally are. Besides insecticidal activity, then, survival in nature may be dependent upon spore germinability which, in turn, may be determined by certain coat proteins such as the proteinaceous toxins. Certainly, the differences in coat proteins of subspecies *kurstaki* and *israelensis*, as reflected by differences in their crystal proteins and other coat components, could affect the stability of their spores in certain environments. If so, the unique spore coat protein profile of *B. thuringiensis* subsp. *israelensis* may afford that organism a distinct survival advantage over other bacilli in mosquito breeding sites.

At the present time, little is known about the genetic basis of toxin production, although botulinum toxins C_1 and D have been observed to be associated with a bacteriophage (23) and *B. thuringiensis* toxin has been reported to be associated with a plasmid (41). Plasmids have been identified in *C. perfringens* (36, 37, 39, 40), but plasmid control of toxins, particularly enterotoxin, has not been established. Also, the mode of action of most protein toxins from sporeformers is not understood at the molecular level. Perhaps the best studied system is *C. botulinum*. A more detailed knowledge of toxicological action of the various toxins would be significant for fundamental and applied research. For example, a thorough understanding of the mode of action of *B. thuringiensis* toxin may lead to the development of better and more efficacious insecticides. A better perception of the pharmacological properties of botulinum toxin may prove useful in attaining greater appreciation of the mechanism of neuromuscular transmission (42) as well as provide practical methods for preventing food contamination.

ACKNOWLEDGMENTS

We are grateful to H. Sugiyama, University of Wisconsin, J. L. McDonel, Pennsylvania State University, and J. J. Iandolo, Kansas State University, for their critical review of this paper.

LITERATURE CITED

1. **Bechtel, D. B., and L. A. Bulla, Jr.** 1976. Electron microscopic study of sporulation and parasporal crystal formation in *Bacillus thuringiensis*. J. Bacteriol. **127**: 1472–1481.
2. **Bonventre, P. T., and C. E. Johnson.** 1970. *Bacillus cereus* toxin, p. 415–435. *In* T. C. Montie, S. Kadis, and S. J. Ajl (ed.), Microbial toxins. Academic Press, Inc., New York.
3. **Boroff, D. A., and B. R. Das Gupta.** 1971. Botulinum

toxin, p. 1–68. In S. Kadis, T. C. Montie, and S. J. Ajl (ed.), Microbial toxins. Academic Press, Inc., New York.

4. **Bulla, L. A., Jr., D. B. Bechtel, K. J. Kramer, Y. Shethna, A. I. Aronson, and P. C. Fitz-James.** 1980. Ultrastructure, physiology, and biochemistry of Bacillus thuringiensis. Crit. Rev. Microbiol. 8:147–204.

5. **Bulla, L. A., Jr., R. N. Costilow, and W. S. Sharpe.** 1978. Biology of Bacillus popilliae. Adv. Appl. Microbiol. 23:1–18.

6. **Bulla, L. A., Jr., L. I. Davidson, K. J. Kramer, and B. L. Jones.** 1979. Purification of the insecticidal toxin from the parasporal crystal of Bacillus thuringiensis subsp. kurstaki. Biochem. Biophys. Res. Commun. 91: 1123–1130.

7. **Bulla, L. A., Jr., K. J. Kramer, D. J. Cox, B. L. Jones, L. I. Davidson, and G. L. Lookhart.** 1981. Purification and characterization of the entomocidal protoxin of Bacillus thuringiensis. J. Biol. Chem. 256:3000–3004.

8. **Bulla, L. A., Jr., K. J. Kramer, and L. I. Davidson.** 1977. Characterization of the entomocidal parasporal crystal of Bacillus thuringiensis. J. Bacteriol. 130:375–383.

9. **Bulla, L. A., Jr., R. A. Rhodes, and G. St. Julian.** 1975. Bacteria as insect pathogens. Annu. Rev. Microbiol. 29: 163–190.

10. **Bulla, L. A., Jr., and A. A. Yousten.** 1979. Bacterial insecticides, p. 91–114. In A. H. Rose (ed.), Economic microbiology, vol. 4. Academic Press, Inc., New York.

11. **Chestukhina, G. G., I. A. Zalunin, L. I. Kostina, T. S. Kotova, S. P. Katrukha, L. A. Lyublinskaya, and V. M. Stepanov.** 1978. Proteinases bound to crystals of Bacillus thuringiensis. Biokhimiya 43:857–864.

12. **DasGupta, B. R.** 1971. Activation of Clostridium botulinum type B toxin by an endogenous enzyme. J. Bacteriol. 108:1051–1057.

13. **DasGupta, R. B., and H. Sugiyama.** 1972. Isolation and characterization of a protease from Clostridium botulinum type B. Biochim. Biophys. Acta 268:719–729.

14. **Das Gupta, B. R., and H. Sugiyama.** 1972. Role of a protease in natural activation of Clostridium botulinum neurotoxin. Infect. Immun. 6:587–590.

15. **Dastidar, P. G., and K. W. Nickerson.** 1978. Lysinoalanine in alkali-solubilized protein crystal toxin from Bacillus thuringiensis. FEMS Microbiol. Lett. 4:331–333.

16. **deBarjac, H.** 1978. Toxicité de Bacillus thuringiensis var. israelensis pour les larves d'Aedes aegypti et d'Anopheles stephensi. C. R. Acad. Sci. Ser. D 286: 797–800.

17. **Duncan, C. L., G. J. King, and W. R. Frieben.** 1973. A paracrystalline inclusion formed during sporulation of enterotoxin-producing strains of Clostridium perfringens type A. J. Bacteriol. 114:845–859.

18. **Duncan, C. L., and D. H. Strong.** 1969. Ileal loop fluid accumulation and production of diarrhea in rabbits by cell-free products of Clostridium perfringens. J. Bacteriol. 100:86–94.

19. **Duncan, C. L., D. H. Strong, and M. Sebald.** 1972. Sporulation and enterotoxin production by mutants of Clostridium perfringens. J. Bacteriol. 110:378–391.

20. **Frieben, W. R., and C. L. Duncan.** 1975. Heterogeneity of enterotoxin-like protein extracted from spores of Clostridium perfringens type A. Eur. J. Biochem. 55: 455–463.

21. **Hauschild, A. H., R. Hilsheimer, and W. G. Martin.** 1973. Improved purification and further characterization of Clostridium perfringens type A. Can. J. Microbiol. 19:1379–1382.

22. **Hobbs, B. C.** 1969. Clostridium perfringens and Bacillus cereus infections, p. 131–173. In H. Rieman (ed.), Food-borne infections and intoxications. Academic Press, Inc., New York.

23. **Inoue, K., and H. Iida.** 1970. Conversion of toxigenicity in Clostridium botulinum type C. Jpn. J. Microbiol. 14: 87–89.

24. **Katz, E., and A. L. Demain.** 1977. The peptide antibiotics of Bacillus: chemistry, biogenesis, and possible function. Bacteriol. Rev. 41:449–474.

25. **Lecadet, M. M.** 1970. Bacillus thuringiensis toxins—the proteinaceous crystal, p. 437–471. In T. C. Montie, S. Kadis, and S. J. Ajl (ed.), Microbial toxins. Academic Press, Inc., New York.

26. **Lilley, M., R. N. Ruffell, and H. J. Somerville.** 1980. Purification of the insecticidal toxin in crystals of Bacillus thuringiensis. J. Gen. Microbiol. 118:1–11.

27. **Lincoln, R. E., and D. C. Fish.** 1978. Anthrax toxin, p. 362–414. In T. C. Montie, S. Kadis, and S. J. Ajl (ed.), Microbial toxins. Academic Press, Inc., New York.

28. **McDonel, J. L.** 1979. The molecular mode of action of Clostridium perfringens enterotoxin. Am. J. Clin. Nutr. 32:210–218.

29. **McDonel, J. L.** 1980. Mechanism of action of Clostridium perfringens enterotoxin. J. Food Technol. 34:91–95.

30. **McDonel, J. L.** 1980. Clostridium perfringens toxins (Type A, B, C, D, E). Pharmacol. Ther. 10:617–655.

31. **Nickerson, K. W.** 1980. Structure and function of the Bacillus thuringiensis protein crystal. Biotechnol. Bioeng. 22:1305–1333.

32. **Ohba, M., and K. Aizawa.** 1979. A new subspecies of Bacillus thuringiensis possessing 11a:11c flagellar antigenic structure: Bacillus thuringiensis subsp. kyushuensis. J. Invertebr. Pathol. 33:387–388.

33. **Prasad, S. S. S. V., H. Lalithakumari, and Y. I. Shethna.** 1973. Inhibitory activity of the proteinaceous crystal of Bacillus thuringiensis. Curr. Sci. 42:568–570.

34. **Prasad, S. S. S. V., and Y. I. Shethna.** 1974. Purification, crystallization and partial characterization of the antitumour and insecticidal subunit from the δ-endotoxin of Bacillus thuringiensis. Biochim. Biophys. Acta 363:558–566.

35. **Prasad, S. S. S. V., and Y. I. Shethna.** 1975. Enhancement of immune response by the proteinaceous crystal of Bacillus thuringiensis. Biochem. Biophys. Res. Commun. 62:517–523.

36. **Rood, J. I., E. M. Maher, E. B. Somes, E. Campos, and C. L. Duncan.** 1978. Isolation and characterization of multiple-antibiotic-resistant Clostridium perfringens strains from porcine feces. Antimicrob. Agents Chemother. 13:871–880.

37. **Rood, J. I., V. N. Scott, and C. L. Duncan.** 1978. Identification of a transferable tetracycline resistance plasmid (pCW3) from Clostridium perfringens. Plasmid 1:563–570.

38. **Schesser, J. H., and L. A. Bulla, Jr.** 1978. Toxicity of Bacillus thuringiensis spores to the tobacco hornworm, Manduca sexta. Appl. Environ. Microbiol. 35:121–123.

39. **Sebald, M., D. Bouanchaud, and G. Bieth.** 1975. Nature plasmidique de la resistance a plusiers antibiotiques chez C. perfringens type A, souche 659. C. R. Acad. Sci. 280:2401–2404.

40. **Sebald, M., and G. Brefort.** 1975. Transfer du plasmide tetracycline-chloramphenicol chez Clostridium perfringens. C. R. Acad. Sci. 281:317–319.

41. **Stahly, D. P., D. W. Dingman, L. A. Bulla, and A. I. Aronson.** 1978. Possible origin and function of the parasporal crystals in Bacillus thuringiensis. Biochem. Biophys. Res. Commun. 84:581–588.

42. **Sugiyama, H.** 1980. Clostridium botulinum neurotoxin. Microbiol. Rev. 44:419–448.

43. **Tyrell, D. J., L. A. Bulla, Jr., R. E. Andrews, Jr., K. J. Kramer, L. I. Davidson, and P. Nordin.** 1981. Comparative biochemistry of entomocidal parasporal

crystals of selected strains of *Bacillus thuringiensis*. J. Bacteriol. **145**:1052–1062.

44. **Tyrell, D. J., L. I. Davidson, L. A. Bulla, Jr., and W. A. Ramoska.** 1979. Toxicity of parasporal crystals of *Bacillus thuringiensis* subsp. *israelensis* to mosquitoes. Appl. Environ. Microbiol. **38**:656–658.

45. **Undeen, A. H., and W. L. Nagel.** 1978. The effect of *Bacillus thuringiensis* ONR-60A strain (Goldberg) on *Simulium* larvae in the laboratory. Mosq. News **38**:524–527.

46. **vanHeyningen, W. E., and J. Mellanby.** 1971. Tetanus toxin, p. 69–108. *In* S. Kadis, T. C. Montie, and S. J. Ajl (ed.), Microbial toxins. Academic Press, Inc., New York.

Biophysical Studies on the Molecular Mechanisms of Spore Heat Resistance and Dormancy

W. G. MURRELL

*Commonwealth Scientific and Industrial Research Organisation, Division of Food Research, North Ryde,
New South Wales, 2113, Australia*

INTRODUCTION

The basic mechanisms of heat resistance of bacterial spores are still not satisfactorily explained (12). The various theories proposed fail to define fully the physicochemical basis for stabilization of essential structures and macromolecules and the physiological mechanism for attainment of this stabilization (12, 27). In my review of the various theories at the last International Spore Conference, it was concluded "that the factors involved in the development and maintenance of the heat-resistant state are complex and, although there may be one mechanism more important in achieving this state, all the mechanisms such as enzyme stabilization, chelation, stable gel formation, contraction or expansion of the cortical polymer, and osmoregulation may well be involved to some extent. To prove conclusively, experimentally, which mechanism is the most important may be a very difficult task. Certainly, however, no theory to date is fully acceptable as a basic resistance mechanism" (quoted in 27).

A key to resolution of the problem of understanding the basic resistance mechanisms of spores appears to lie with the unique in situ properties which are lost when the spore is analyzed by the usual fractionation techniques. Instead, these properties apparently must be characterized by analytical probes that are nondestructive to the cellular and molecular configurations conferring resistance.

In a program designed to define some of the molecular mechanisms involved in heat resistance and dormancy and the biophysical state of the protoplast (in this paper "protoplast" is correctly and specifically restricted to the inner forespore and spore membrane and its enclosed cytoplasm), several nondestructive biophysical techniques and probes have been used by the Australian collaborators of the U.S.-Australia Cooperative Science Program on spores. A variety of spores and spore preparations were examined, and several in vitro model systems were studied. The results of these studies are described and interpreted.

BIOPHYSICAL STUDIES

Volume Change During Germination

Following a report of large values for the volume change of activation of *Bacillus cereus* T spores (4, 42) germinated by hydrostatic pressure and suggestions of a decrease in molecular volume during germination (5, 33), M. Izard and P. A. Willis used a packed cell volume technique to study volume changes during the initial stages of germination (Fig. 1).

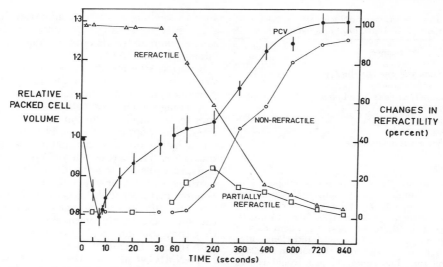

FIG. 1. *Changes in packed cell volume and refractility of spores of* B. cereus T *during germination at 35°C (Izard and Wills, unpublished data). Heat-activated spores (0.2 ml, 10^{10}/ml) were mixed rapidly with 0.1 ml of germinant containing adenosine (5 mg/ml), L-alanine (25 mg/ml), and glucose (50 mg/ml) at pH 8.0. Germination was stopped at the times shown by addition of mercuric chloride (0.05 ml, 0.05 mol/liter). Packed cell volume (ratio of length of pellet to pellet length + length of supernatant) was measured in microhematocrit tubes after centrifuging for 1 min at 2,700 × g. Symbols: △, percentage fully refractile; □, percentage partially refractile; ○, percentage nonrefractile; ●, packed cell volume relative to value at t_0 (bars show standard error of mean). The spores were previously heat activated in water at 60°C for 30 min.*

At 35°C the packed cell volume fell immediately, reaching a minimal value that was about 84% of the initial level at 10 s, before refractility and absorbance changed. Between 10 and 30 s the packed cell volume recovered to the initial level, and a small proportion (1.5%) of the spores partially lost refractility; however, the absorbance remained constant. During the next 15 min, packed cell volumes and the percentage of non-refractile spores continued to rise, with a corresponding decrease in absorbance until initiation of germination was complete. The final packed cell volume was 30% higher than that for dormant spores. Similar, but less extensive, changes occurred at 20°C.

Assuming that changes in packed cell volumes reflect changes in spore volumes, Izard and Wills interpreted the results as supporting the expanded cortex concept; i.e., triggering leads to cortex collapse, and the volume of the spore decreases. As germination proceeds, the core rehydrates and expands, and the spore volume increases.

The above decrease in volume would occur well before the loss of calcium, dipicolinic acid (DPA), and peptidoglycan fragments (7, 24). A possible biophysical explanation of this reduction in volume is mentioned later.

Density of Intact Spores and Components

The densities of intact spores and integuments were determined in the dry state by I. C. Watt (Table 1). For *B. stearothermophilus*, the coat material had a dry density of 1.29 g/cm³, similar to that of keratinous materials. The dry density of the coat plus cortex preparation was 1.32 g/cm³, and that of the intact spores was 1.395 g/cm³. The results confirm that the material of highest density is in the interior and that the densities of the cortex and core are >1.4 g/cm³.

Water State in Spores

Water sorption. Water sorption data give information about some very fundamental properties of spores and their components. Using a very sensitive quartz spring balance, Watt (this volume) determined their sorption properties more precisely than had been done previously (19, 34, 63). Since previous studies were at 20 to 30°C, data were also obtained at lethal temperatures. Watt's results show that the sorption properties were generally similar at the higher temperature and indicate that it is valid to apply similar sorption theories to water sorption in the lethal range.

TABLE 1. *Dry densities of bacterial spores and components (I. C. Watt, unpublished data)*

Sample	Density[a] (g/cm³)
B. megaterium vegetative cells	1.46
B. cereus T intact spores	1.405
Clostridium perfringens intact spores	1.40
B. stearothermophilus	
Intact spores	1.395
Spore coat material	1.29
Spore coat + cortex material	1.32
Cortex	1.40[b]
Protoplast	>1.40[b]

[a] The spore samples (intact spores or spore fractions) were vacuum dried at room temperature and then immersed in a 1:5 *t*-butyl alcohol–perchlorethylene mixture. The samples were transferred by pipette to the top of a calibrated density gradient column charged with *t*-butyl alcohol–perchlorethylene with a density range from 1.50 g/cm³ at the bottom to 1.25 g/cm³ at the top. The spore samples would remain essentially dry during the transfer process, and in the column the large volume of dry solvent would act as a dehydrating agent of zero a_w. The samples were allowed to attain their equilibrium position and their density was calculated.

[b] Deduced densities.

At low humidities (0 to 0.1 water activity [a_w]), the isolated protoplast fraction absorbs almost twice the amount of water adsorbed by the intact spores, and the order of most sorption is isolated protoplast contents > intact spores > isolated coat plus cortex > isolated coat (Watt, this volume). This order is most significant since in the a_w region of 0.5 the isolated fractions and spores have similar water contents. The greater variability at low humidities, where Langmuir adsorption occurs, is directly related to a variation between the fractions in the number of strongly hydrophilic sites (specific sites with highest energy of binding). Since the water content of intact spores lies between the extremes of water content for the fractions, this suggests that the water content of intact spores is directly attributable to the sorptivity of the fractions in the low a_w region.

The order of sorption at high humidities (near saturation) is isolated protoplast contents > isolated coat plus cortex > intact spores > isolated coat material. Multilayer adsorption, characteristic of this region, contributes its greatest effect at high humidities. Although the physical structure of the substrate does not affect the amount of water associated with specific sites, multilayer formation is limited by constraints imposed by the substrate. The greater the amount of water which condenses in the multilayer, the closer it becomes to liquid water in its properties. Resistance to swelling of the coat material therefore apparently restricts the amount of water that can be taken up by the other components in the intact spores.

Sorption kinetics are largely determined by the molecular structure of the absorbent. The spore samples exhibit the characteristic kinetics of swelling systems, indicating that sorption occurs by a coupled diffusion-relaxation mechanism. The approach to equilibrium is determined by the rate of relaxation processes within the spores. The permeability of spores to water is a prerequisite for adsorption. The diffusion coefficient for water vapor uptake by all components increases with higher equilibrium relative humidity, indicating that all components of the spore are highly permeable to water and that they therefore will have water associated with them in the intact spore.

¹H-NMR study of the water state in spores. The line width ($\Delta \nu_{1/2}$) of the nuclear magnetic resonance (NMR) water signal (64) was measured by J. H. Bradbury and colleagues in spores, isolated coat, and coat plus cortex samples equilibrated at a particular a_w for five different species of spores. The transverse relaxation time, T_2, and the transverse relaxation rate, $1/T_2$ (s⁻¹), were calculated by the equation $1/T_2 = \pi \Delta \nu_{1/2}$ = function (constants, τ_c, r), in which $1/T_2$ is related by a complex equation to the correlation time, τ_c (the time for translation or rotation of the water molecule), and r (the distance from the observed proton to the nearest relaxing center). As $1/T_2$ increases, τ_c also increases, and the water molecule becomes less mobile. Thus, water in spores has a much greater correlation time and hence much lower mobility than liquid water.

Paramagnetic ions increase the relaxation rate of water, and since these occur in spores, it is necessary to account for this. Wool keratin was used as a model system, and known amounts of Ca(II) (diamagnetic), Mn(II), Ni(II), and Fe(III) (all paramagnetic) were introduced into wool fibers. Diamagnetic Ca(II) does not alter the relaxation rate, and the results for the paramagnetic ions are shown in Fig. 2. Mn(II) has a dominant effect on increasing $1/T_{2T}$ ($1/T_2$ total), and since Mn(II) occurs to a greater extent in spores than do the other paramagnetic ions, J. H. Bradbury et al. (Biochim. Biophys. Acta, in press) interpreted the results for spores and spore fractions in terms of the concentration of Mn(II) present (Fig. 3).

The relaxation rate at a particular Mn(II) concentration is consistently less for spores than for isolated coat or coat plus cortex preparations (Fig. 3). Thus, the correlation time for intact

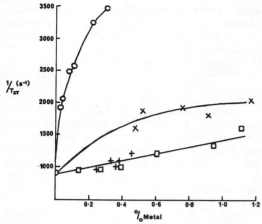

FIG. 2. *Graph of the relaxation rate* $1/T_{2T}$ (s^{-1}) *of the water resonance in wool keratin equilibrated at* $a_w = 0.98$ *against the amount of paramagnetic metal ions as follows:* ○, *Mn(II);* ×, *Cu(II);* □, *Ni(II);* +, *Fe(III) (Bradbury et al., in press). Samples (0.2 g) of cleaned Merino wool were soaked in aqueous solutions of the salts* $MnSO_4$, $Fe(SO_4)_3$, $Ni(NO_3)_2$, $CuSO_4$, $Gd(NO_3)_3$, *and* $Ca(NO_3)_2$ *at a range of concentrations. The samples were washed four times with 20 ml of water to remove surface loosely bound metal ions and air dried. The samples were equilibrated in vacuo, and then the ^1H-NMR spectrum was measured with a Varian CFT-20 pulsed NMR spectrometer operating at 80.0 MHz and 37° C for protons.*

spores is less than for the outer integuments, and the mobility of the water in the intact spore (and hence in the protoplast of the spore) is greater than that in the outer integuments (Fig. 3). This finding, together with the results of Watt (this volume), which show that the protoplast contains at least as much water as the coat and cortex, indicates that the protoplast cannot be more dehydrated than the coat and cortex.

There are only low concentrations of free low-molecular-weight solutes other than DPA, sulfolactate, and glutamate in the spore (for reviews, see 36, 37, 43, 65), and these three solutes may not be fully osmotically active; consequently, it is possible that the effective a_w is not very low.

Element Distribution in Spores

An important feature of bacterial spores is the high content of DPA, calcium, and other cations, and any theory based on heat resistance or biophysical state must take into account their presence and distribution. Techniques are now available to determine accurately elemental distribution in cells. Previous studies on elemental

distribution were handicapped by less precise methods or instruments (30, 50).

Stewart, using cryosectioned *B. cereus* spores and high-resolution scanning electron probe X-ray microanalysis, studied the distribution of a number of key elements in *B. cereus* T wild-type spores, spores of *B. cereus* T DPA⁻ mutants with and without DPA, and wild-type spores grown with low Ca (59).

The results indicated that almost all of the Ca, Mg, and Mn and most of the P were located in the protoplast region, with an unexpectedly high concentration of Si in the cortex/coat layer. Granules containing high concentrations of Ca, Mn, and P occurred in the spores with reduced DPA levels. Spot mode analyses, in which a fine stationary beam was located over the region of interest in the spore cryosection, confirmed the scanning mode results and provided an accurate quantitation of the elemental concentrations on

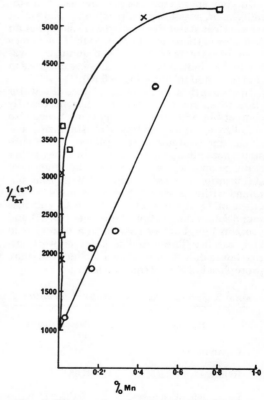

FIG. 3. *Water relaxation rate* ($1/T_{2T}$) *for spores and spore components equilibrated at* $a_w = 0.98$ *against the indicated percent Mn (Bradbury et al., in press). Symbols:* ○, *spores;* □, *isolated coat;* ×, *isolated coat + cortex. See legend to Fig. 2 for experimental details.*

a dry weight basis (59). Similar elemental distributions for *B. megaterium* have recently been obtained (26).

Enzymes

Heat stability of spore enzymes at various a_w's. Most spore enzymes are heat labile when extracted (48). Considerable stabilization is commonly achieved by reduction in water content. It is important to know the water content or a_w needed to stabilize the enzymes to the same extent that they are stabilized in the spore in order to assess the significance of partial dehydration theories. A. D. Warth therefore investigated the relationship between the amount of stabilization of spore enzymes and a_w and water content (this volume) and defined these values. The inactivation temperatures increased steadily as a_w was reduced, reaching about 136°C for three enzymes at 0.1 a_w.

In vivo, spore enzymes also increased in stability as a_w was lowered, reaching a plateau at 0.3 a_w. Heat stabilities in vitro were similar at an a_w 0.3 lower than those in vivo. At 85°C, enzymes in spore extracts at 0.73 a_w, containing 20% water, had stabilities similar to those of enzymes in spores heated in water ($a_w > 0.99$).

Stabilization of enzymes. Since the water state in spores (see above) is not necessarily compatible with the a_w conditions needed to stabilize enzymes, methods of stabilizing enzymes and proteins were explored. Many substances stabilize proteins to heat, but they either stabilize only a few proteins or the amount of stabilization is small compared with that obtained within a spore. Previous studies on the effect of DPA on heat resistance of enzymes found some stabilization of enzymes (20, 21) and proteins (35), but not enough to account for in vivo stability. These studies, however, were carried out in dilute aqueous media. Since the spore protoplast is a highly concentrated gel, attempts were made to simulate the in vivo spore protoplast state, in which opportunity for molecular interactions such as hydrogen bonding and noncovalent bonding between components exists. Preliminary attempts were therefore made to "cocrystallize" calcium, DPA, and calcium dipicolinate (CaDPA) with enzymes and to study their resistance under reduced a_w conditions. Various enzymes were repeatedly freeze-dried in the presence of CaDPA and equilibrated to various a_w's, after which the heat stability was determined (J. H. Bradbury et al., unpublished data). Significant increases in stability of lactic dehydrogenase and RNase, but not lysozyme, were obtained (Table 2). Although the experiments were preliminary, the degree of "cocrystallization" was unknown, and the a_w conditions during heating were not reliably controlled, it is significant that greater stabilization than the control could be demonstrated.

Warth (this volume), however, observed no appreciable decrease in the heat stability of glucose-6-phosphate dehydrogenase, heated at controlled a_w in crude spore extracts, on the removal of CaDPA and other low-molecular-weight solutes, demonstrating that, under conditions approximating the concentrated protoplasm of the spore core, DPA and other low-molecular-weight substances collectively did not play a stabilizing role for this enzyme. Further purification of the enzyme did, however, reduce its stability. These observations are apparently contradictory to the results above, but Bradbury et al. used pure enzyme systems whereas Warth was using crude extracts. The cooperative effect of enzymes/proteins in stabilizing other enzymes is well known. The data are therefore not incompatible and may indicate a stabilizing role for enzyme/protein interactions in the spore.

DNA State and Stabilization

Photoproduct studies. An important insight

TABLE 2. *Stabilization of enzymes "cocrystallized" with CaDPA (J. H. Bradbury et al., unpublished data)*

Enzyme[a]	Heat treatment	CaDPA	Activity remaining (%) in expt:				
			1	2	3	4	5
Lysozyme	130°C, 21 h	−	5.8	9	2	2	0.35
		+	12	27	4	18	3.2
RNase A	130° C, 21 h	−	40	37			
		+	82	62			
Lactic dehydrogenase	60°C, 20.5 h	−	0				
		+	47				

[a] A 1:3 enzyme-CaDPA solution by weight was made by preparing a solution of CaDPA (4 mg/ml) in warm water and adding the enzyme to it. The solution was lyophilized, the complex was dissolved in a minimum of warm water to get a clear solution, and the solution was again lyophilized. The complex was equilibrated in sealed tubes in vacuo over saturated $MgNO_3 \cdot 6H_2O$ ($a_w = 0.52$) for at least 1 week and then heated in an oven at $130 \pm 2°C$.

into the state of the spore protoplast and the most vital component, the DNA, is made possible by a study of UV photoproducts. Previous studies (1, 6, 13, 14, 45, 57) indicated that during sporulation the photoproducts change from the typical thymine dimer type to the spore type. These results suggest that the DPA is closely associated with the DNA and that the DNA may be in the more tightly coiled A state. J. A. Lindsay (unpublished data) studied the photoproducts from several *Bacillus* species with a range in growth temperatures and found that new photoproducts were produced from species with altered genomes (temperature-sensitive mutants or strains with genomes composed of minicircles). Further, *B. cereus* T DPA$^-$ mutants exhibited photoproduct profiles different from that of their wild-type counterparts. On the addition of exogenous DPA during sporulation, these mutants produced a "pseudophotoproduct" different from their normal type (Fig. 4). The tentative conclusion from the mass spectrum degradation data is that DPA has interacted with the radioactively labeled thymine to form a DPA-thymine adduct with a molecular weight of 293, which during hydrolysis or photolysis is modified to a decarboxylated thymine-DPA adduct with a molecular weight of 233.

The essential deduction from these results supports the idea that there is a close association of DPA with DNA and that DPA has a profound effect on the state of the DNA, and hence the type of photoproduct formed.

DPA stabilization of DNA. Whereas the DPA content of species varying widely in heat resistance is not markedly different (41), reduction in DPA content within a species has reduced heat resistance greatly (3, 39).

DPA$^-$ mutants of *B. cereus* have reduced heat resistance, but, on the other hand, *B. subtilis* DPA$^-$ mutants were reported as not having lower resistance than the wild type (22, 70). Recent results have not confirmed this (2). Several lines of evidence suggest that DPA is located in the protoplast (9, 14, 31, 59).

Therefore, J. A. Lindsay investigated the effect of DPA on the stability of DNA by measuring its effect on the melting profile of DNA and on RNA synthesis. DPA or CaDPA in a 1:1 molar ratio with DNA increased the T_m by more than 20°C, but glutamate, sulfolactate, Ca, and Mn, or combinations of these, had very little effect on T_m (Table 3). Further, DPA and CaDPA inhibited in vitro RNA synthesis whereas the other components tested had no effect (Table 4).

These effects and the chemical structure of DPA suggest that the DPA has interacted with

the DNA and that an appreciable fraction of DPA may be bound on the outside of the DNA molecule. Such an interaction would lead to an increase in nucleic acid stability, prevent uncoiling, and hence increase heat resistance within the spore.

DPA/DNA interactions. The interaction of

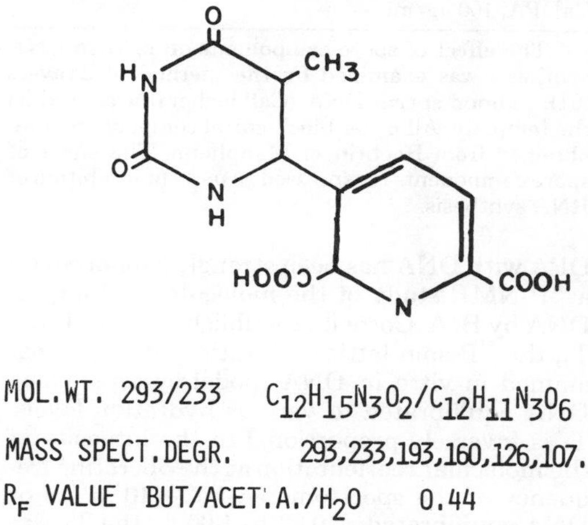

MOL. WT. 293/233 $C_{12}H_{15}N_3O_2/C_{12}H_{11}N_3O_6$

MASS SPECT. DEGR.   293,233,193,160,126,107.

R_F VALUE   BUT/ACET.A./H_2O 0.44

FIG. 4. *Spore photoproduct T/DPA.A tentatively as a thymine-DPA adduct, produced by 254-nm UV irradiation of a B. cereus HW$_1$ DPA$^-$ mutant grown with exogenous DPA (100 µg/ml) during sporulation (Lindsay and Murrell, unpublished data). The photoproduct was isolated and purified.*

TABLE 3. *Effect of spore components on ΔT_m of salmon DNA^a (J. A. Lindsay and W. G. Murrell, unpublished data)*

Component	ΔT_m (°C)
Control	—
DPA (20 µg/ml)	+4
CaDPA (20 µg/ml)	+7
DPA (100 µg/ml)	>20
CaDPA (100 µg/ml)	>20
Glutamic acid (50 µg/ml)	−2
Glutamic acid (200 µg/ml)	−5
Sulfolactic acid (50 µg/ml)	−1.5
Sulfolactic acid (200 µg/ml)	−1.5
Calcium (1 mg/ml)	−2
Glutamic acid (200 µg/ml) + calcium (1 mg/ml)	−1
Sulfolactic acid (200 µg/ml) + calcium (1 mg/ml)	−3

^{a} Melting profile behavior of salmon sperm DNA in 0.01 × SSC (SSC = 0.15 M NaCl plus 0.015 M sodium citrate), pH 7.0, in the presence of various sporal components, examined on a Pye Unicam SP2000 spectrophotometer. Calculation of T_m was made by the method of Marmur and Doty (32) with a temperature increase of 1°C/min. ΔT_m is defined as the T_m (experimental) − T_m (control).

TABLE 4. *Effect of DPA/CaDPA on in vitro RNA synthesis[a] (Lindsay and Murrell, unpublished data)*

Condition	% Inhibition
Control	Nil
DPA, 20 µg/ml	80.4
CaDPA, 20 µg/ml	88.04
DPA, 100 µg/ml	98.8
CaDPA, 100 µg/ml	98.8

[a] The effect of spore components on in vitro RNA synthesis was examined by the method of Travers (61). Salmon sperm DNA (Calbiochem) was used as the template. All other biochemical components were obtained from Boehringer Mannheim. The effect of spore components is expressed as percent inhibition of RNA synthesis.

DPA with DNA has been strongly supported by a ^{31}P-NMR study of the molecular mobility of DNA by B. A. Cornell (unpublished data). First, T_1, the ^{31}P spin lattice relaxation time, was examined in vitro in DNA model systems using DNA equilibrated to various hydration levels. T_1 is inversely proportional to the intensity of the molecular reorientation at the operating frequency of the spectrometer, i.e., ~10^8 s^{-1}. For DNA equilibrated at 0.63 to 1.00 a_w the T_1 was constant. At lower a_w, the DNA underwent a phase transition, stabilizing again with constant T_1 at about 0.38 a_w and below (Fig. 5).

These data are consistent with a DNA transition from the B to the A state. T_1 analysis of fully hydrated DNA in the presence of DPA and CaDPA showed a longer T_1, equivalent in motion to that of DNA equilibrated to 0.40 to 0.45 a_w and in the A state. DPA- or CaDPA-DNA rotary evaporated to an a_w of 0.97 showed no molecular motion (infinite T_1).

Preliminary proton-enhanced ^{13}C-NMR analyses of *B. cereus* T and *B. cereus* T DPA$^-$ mutant spores labeled with [^{13}C]thymine suport the above findings (B. A. Cornell, J. A. Lindsay, and W. G. Murrell, unpublished data).

CONCLUSIONS FROM THE BIOPHYSICAL STUDIES

Although these studies are incomplete, they indicate several phenomena that probably play a significant role in heat resistance and are involved in determining the biophysical state of the spore protoplast. The conclusions may be summarized as follows:

1. On the triggering of germination, the volume of the spore decreases briefly (Izard and Wills, unpublished data). This may be the result of changes associated with electrorestrictive forces accompanying solvation (4) or the breaking of hydrophobic interactions associated with initiation of germination, specifically with hy-

dration of the core (33) or changes in the protoplast membrane in response to the trigger action.

2. The isolated protoplast fraction binds more water than the isolated coats and cortex, and these relative sorptive properties persist at lethal temperatures. The coat is the layer that restricts greater hydration of the protoplast (Watt, this volume). The proton NMR results on water state support the water sorption observation that there is more water in the protoplast than in the coat and cortical macromolecular structures and that this water is equally as mobile as or more mobile than the water absorbed in the coat and cortex layers. Stewart et al. (58) interpreted their NMR results to indicate that spores contain more nonexchangeable water than vegetative cells, adsorbed to macromolecules in a relatively anhydrous environment.

3. The presence of most of the major ions, Ca, Mg, Mn, and probably DPA, in the protoplast largely rules out an osmoregulatory role of the cortex (16–18) in heat resistance and points to a more general type of biophysical interaction of low-molecular-weight components and macromolecules in the stabilization of the protoplast.

4. The removal of water markedly increases enzyme stability, and the a_w levels required to

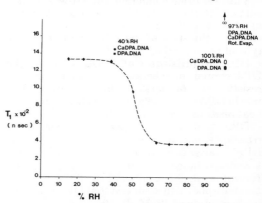

FIG. 5. *Effect of DPA and CaDPA on the ^{31}P T_1 (molecular reorientation) motion of salmon sperm DNA at various equilibrium relative humidities (Cornell, Lindsay, and Murrell, unpublished data). A 300-mg amount of salmon sperm DNA (Calbiochem) was hydrated to the desired a_w by equilibration in a sealed evacuated desiccator using H_2SO_4/H_2O solutions (40). All solutions were degassed to remove molecular oxygen. Equilibration time was minimally 10 days. All a_w's were checked after analysis by drying the samples and correlating water content to a_w by a water sorption curve. The error in a_w was less than 2%. The symbols (●, ■, ◆, ○) represent experimental points for the observed T_1 of DNA in the presence of DPA or CaDPA at the indicated equilibrium relative humidities.*

stabilize enzymes to the extent that they are in vivo has been established (Warth, this volume).

5. "Cocrystallization" of enzymes with CaDPA increased their heat stability. This possibly simulates conditions in the protoplast at stage IV and thereafter of sporulation. This is suggestive evidence that DPA plays a role in the stabilization of labile macromolecules.

6. DPA/CaDPA interacts with nucleic acids, possibly by intercalation and polymeric aggregation, stabilizing these molecules to high temperatures and at the same time preventing transcription, indicated by the inhibition of RNA synthesis. Whether there is a sequential interaction starting from one point on the chromosome or at points in a selective sequence of DNA regions in this interaction of DPA/CaDPA with DNA, thus exerting some control of cell differentiation during sporulation, as suggested earlier (1), is not known.

7. There is a general molecular immobilization of macromolecules such as DNA and enzymes and constituents such as DPA as indicated by the ^{31}P-, [^{13}C]DNA-, and [^{19}F]DPA-NMR studies. This reduced macromolecular mobility has been observed also by Stewart et al. (58).

Thus, a picture emerges of the biophysical state of the protoplast—a high-density matrix of macromolecules interacting with DPA/CaDPA and ions, with much hydrogen, electrostatic, and noncovalent bonding, a great reduction in intermolecular space and in molecular mobility, a region still with water of hydration and possibly even a very small amount of free water which can readily permeate in and out of the spore (34, 66; Watt, this volume), and a region that is not necessarily dehydrated in the absolute sense but has a relatively low water content ($\sim$30%, wt/wt, on a dry weight basis [34]) and yet has an a_w that is not essentially markedly reduced.

The mechanisms of stabilization involve specific and nonspecific molecular interactions, a solid support system involving CaDPA, constraint of molecular motion (thus preventing uncoiling and denaturation) associated with a restricted protoplast volume within the surrounding integuments, and an envelope which does not allow inhibition of water and solution of potentially soluble components. Differential permeability of the protoplast membrane must also operate to prevent loss of low-molecular-weight solutes such as Ca^{2+} and DPA.

ATTAINMENT OF THE RESISTANT STATE OF THE DORMANT SPORE PROTOPLAST

With an insight into the biophysical state of the protoplast, it is essential to describe the physiological dynamics of how this state is attained. There have been very few attempts to do this, mainly because of the lack of the necessary basic information.

Sporulation is a long, slow process. In this presentation the interest is mainly in stage IV and its ensuing development. This usually takes 2 to 4 h in most species under near optimal conditions. Determination of the sequence of events and duration of each is complicated by the degree of asynchrony present in sporulation studies (36). The events that appear to be most relevant to the changing biophysical state of the protoplast are DNA changes, basic protein formation, Ca uptake, DPA formation, protoplast volume reduction, cortex formation, and development of refractility. The relation of these in *B. cereus* T is summarized in Fig. 6 and Table 5. The first change is apparently in the DNA. Thymine dimer formation on UV irradiation decreases, and the ability to photoreactivate the dimers is lost, spore-type photoproducts appear, and the cytological appearance of the DNA changes from the vegetative to the spore type (Fig. 6 and reference 1). The protoplast volume begins to decrease at almost the same time that these changes occur; about half the reduction has occurred by the beginning of DPA synthesis. Spore-type photoproduct formation precedes the cytological changes in DNA by about 40 min (Table 5). Calcium uptake tends to occur about 30 min ahead of DPA formation and development of refractility.

Earlier results with *B. cereus* T and its DPA$^-$ mutant (Fig. 4 and 5 in reference 39) showed that synthesis of DPA and development of refractility and cortex (labeled diaminopimelate uptake) are closely associated with Ca^{2+} uptake. This and the data in Fig. 6 suggest that the protoplast volume reduction precedes the major uptake of Ca^{2+} and that CaDPA is not involved in the reduction. It must be recognized, however, that any consistent error in estimation of these parameters (either over or under as a result of analytical procedures, extraction errors, or loss from cell samples) will displace each of these events forward or backward on the time scale (see footnote a, Table 5).

In *B. cereus* T DPA$^-$ mutants without an exogenous supply of DPA, only about 30% of the Ca^{2+} uptake occurred (39); i.e., DPA is required to fix Ca^{2+} in the spore. The reduction in protoplast volume, although not as great, also occurred in the mutant without DPA and with only 30% of the normal Ca^{2+} (Table 6). This also suggests that DPA and probably Ca^{2+} are not involved in the volume reduction.

The reduction in protoplast volume in two

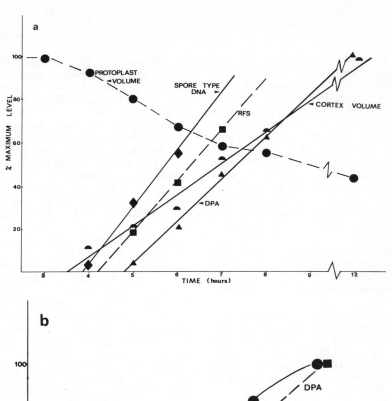

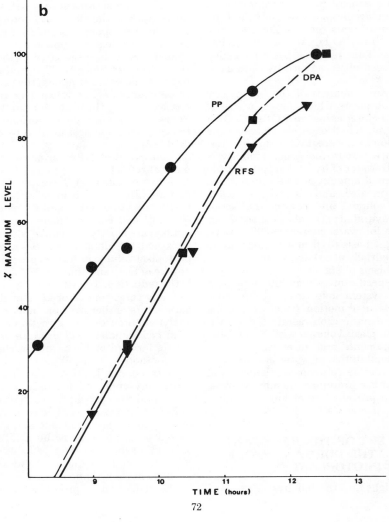

other species is shown in Table 6.

These changes, however, precede cortex completion (Fig. 6a) and encirclement and completion of the initial coat layer (Fig. 6d in reference 1). That is, many of the biochemical and cytological changes in the protoplast occur before cortex formation is complete and well before the end of coat formation, hence before an osmoregulatory cortical or expanded cortex dehydration mechanism could operate.

Since all these events are contingent on time-dependent uptake and synthesis (both biochemical and cellular development), it is reasonable to assume that the timing and duration of the cultural changes reflect temporal changes in individual cells. Thus, initial biochemical events which bring about the reduction in protoplast volume (34, 38) are possibly uptake of cations other than Ca^{2+} or the synthesis of polyamines and basic proteins (53, 54). The basic proteins A and B are formed about 30 min before development of refractility in *B. megaterium* (Fig. 4 and 5 in reference 54). Interaction of DPA/CaDPA with DNA and immobilization of the DNA appear to be associated in time with calcium uptake; this is deduced from the cytological appearance of spore-type DNA. Enzymes and other macromolecules possibly undergo similar stabilization changes. These changes would result in a general metabolic shutdown in the protoplast. The reported (in the literature) metabolic activities such as protein synthesis and mRNA activity during late sporulation stages can probably all be explained as the result of tardy sporulation in some cells in asynchronous cultures or residual activities of the mother cell.

The macromolecular rearrangements involved in the development of a solid support system, the polymerized matrix of CaDPA-nucleic acid and RNA-protein macromolecules, are considered to occur during the reduction of protoplast volume with loss of water by syneresis. As indicated above, this occurs before completion of the cortex and coats.

Kinetic studies on the development of heat resistance (Murrell and Lindsay, unpublished data) show that the changes during the reduction in protoplast volume only result in an initial step-up in heat resistance from that of the vegetative cell, not full heat resistance. The latter occurs during further development and completion of the cortex and coats. As previously postulated (34, 38), it is believed that the cortex is built in a contracted state about the protoplast—in the contracted state because the existing pH and ionic environment in the sporulating cell

TABLE 5. *Relationship of sporulation events in* B. cereus *T*

Event	Time[a] relative to development of refractility (h)		
	Mean	Expt	Range
Spore photoproduct	−1.23	2	−1.12 to −1.34
[³H]DPA[b] incorporation	−0.6	5	−1.6 to −0.7
Spore-type DNA[c]	−0.55	1	—
⁴⁵Ca uptake	−0.5	6	−3.6 to +2.7
DPA content	−0.5	19	−3.15 to +1.9
Refractility development	0	—	—
Cortex formation[d]	+0.4	1	—
Heat stability[e]	+0.5	3	−0.25 to +1.2

[a] The times are measured at the midpoint in the linear phase of the curve depicting development of refractile forespores; hence, differences in rate of each event from that of development of refractile forespores will accentuate their displacement on the time scale.
[b] Indicator of cortical peptidoglycan synthesis. Hot (10 min at 90°C) trichloroacetate (5%) insoluble.
[c] See Baillie et al. (1).
[d] See Fig. 6 legend.
[e] Resistant to 12 min at 80°C in 0.05 M phosphate buffer.

FIG. 6. *(a) Reduction in protoplast volume in relation to other sporulation events in* B. cereus *T. Based on the data of Baillie et al. (1) and unpublished studies of Ohye, Warth, and Murrell. Refractile forespore counts (RFS), DPA, stages of sporulation, and photoproducts were determined as described previously (1, 14). The protoplast and cortex volumes of developing spores in the culture were obtained by summing the products of the proportion of sporulation stages in each sample (from electron micrographs) times the average protoplast or cortex volume of each stage. For example, at 6 h:*

	Avg forespore protoplast vol (μm³)	Relative protoplast vol in culture (μm³)
10.8% cells with vegetative DNA	0.497	5.4
33.8% cells with transition DNA	0.402	13.6
55.4% stage V	0.255	14.1
100.0		33.1

Electron micrographs of medial longitudinal sections only were used for measurements. At least 60 suitable cell sections were examined in each sample. Cortex volume was the difference between the volumes enclosed within the inner and outer forespore membranes. Volumes were calculated by using the formula for a prolate spheroid. (b) Relation of sporetype photoproducts (PP) to DPA formation and development of refractile forespores (RFS).

TABLE 6. *Forespore protoplast volume of four* Bacillus *species at different stages of maturity as indicated by DNA state and cytological structure (D. F. Ohye, A. D. Warth, and W. G. Murrell, unpublished data)*

Forespore development stage	Avg protoplast vol (μm³)	df[a]	SE[b] of mean	Per cent
B. cereus T				
Stage IV				
Vegetative DNA ...	0.497	30	0.017	100
Transition DNA ...	0.440	23	0.016	88.5
Stage V	0.255	34	0.010	51.3
Mature spore	0.209	28		42.0
B. coagulans				
Stage IV				
Vegetative DNA ...	0.219	26	0.015	100
Transition DNA ...	0.158	9	0.019	72.1
Stage V	0.113	24	0.005	51.6
Mature	0.0875			39.9
B. stearothermophilus				
Stage IV				
Vegetative DNA ...	0.461	18	0.0172	100
Transition DNA ...	0.356	7	0.0198	77.2
Stage V	0.242	21	0.0087	52.5
Mature	0.177			38.4
B. cereus T HW 1 DPA⁻ mutant				
No DPA				
Stage IV vegetative DNA	0.605	17	0.0215	100
Stage V	0.391	18	0.0235	64.6
Exogenous DPA[c]				
Stage IV vegetative DNA	0.509	12	0.0182	100
Stage V	0.321	9	0.0195	63.1

[a] Degrees of freedom.
[b] Standard error.
[c] Concentration: 100 μg/ml of culture.

would necessitate this (38). Even though the contraction and reordering of macromolecules (syneresis), during stage IV, is not responsible for full heat resistance, it is considered essential to the final resistant state. In fact, it may be that the cortex and coat are required to maintain this state only during heating by preventing swelling and inhibition of water (38). It is not possible to test the potential heat resistance of the protoplast in its reduced-volume, reordered state, as it rapidly becomes disordered during heating. Postulated contraction or expansion of the cortex may not be necessary (38), but such additional changes after completion of the cortex would certainly aid in increasing resistance (65) and together with the inextensible coats would be important in maintaining the fully resistant state.

DORMANCY

Dormancy will be maintained while conditions for initiation of germination are unfavorable. However, as soon as a suitable trigger, physical treatment with pressure (4) or a surface-active substance (44) starts germination, there is a rapid reversal of the biophysical state of the protoplast of the mature resting spore. Ca^{2+} and DPA are lost rapidly from the cell, and the cortex begins to be degraded. Many metabolic activities are regained rapidly, but it is some time before protein, RNA, and DNA synthesis occurs (67).

If it is assumed that all the calcium and DPA are in the protoplast, one of the first events in this reversal must be a change in the permeability of the protoplast membrane to allow loss of Ca and DPA. This seems to be an essential deduction from the biophysical studies on the state of the resting spore protoplast. It is also the view that has currently gained predominance from germination studies (27, 47, 52, 62).

In this presentation, the trigger events in relation to the release of the spore from its resistant resting biophysical state are our major concern. The physicochemical reactions involved in the initial trigger event have not been defined precisely (27). There appears to be no metabolism associated with this event (46, 52, 62). It possibly involves hydrolytic, degradative, or biophysical mechanisms (52). Vary (62) has suggested a receptor protein in the inner spore membrane as the possible primary target of the trigger. This may undergo a conformational change which alters the permeability of the membrane and results in loss of inorganic ions and organic acids in exchange for water and counterions. Attempts are being made to isolate and identify this specific receptor site for the trigger L-proline (47).

The inhibitory action of alcohols on initiation of germination is believed to occur by an interaction on a hydrophobic region near the L-alanine trigger site (69). The effect of inert gases and CO_2 points to a reaction in the lipid bilayer of the membrane, resulting in a shift in the lipid-phase equilibrium toward a more fluid configuration which would disturb the activity of a membrane-bound enzyme essential for initiation of germination (10). This change in fluidity of membranes on germination is paralleled by the rapid release of Ca^{2+} from the spore, although it was not possible to relate Ca^{2+} or CaDPA to the change to the polycrystalline state of the membranes that occurs after stage VI of sporulation (58).

Whether the extensive degradative changes in the protoplast membrane that have been observed in the first 10 min of germination (56) are part of the initial trigger reaction is difficult to decide. Since the population of spores undergoing germination does not germinate perfectly synchronously, it is possible that some proteolysis of the membrane is a very early event in the triggering of germination and possibly is responsible for the permeability changes.

It would seem that the biophysical state of the spore as presented earlier must provide explanations of three phenomena or requirements of physiological germination—a mechanism for activation, a trigger site and mechanism, and release of macromolecular immobility. Activation must allow the trigger compound to reach the site of its actions more rapidly or increase the rate of the trigger action. This probably requires permeability changes in the integuments (29), a temporary increase (because activation is reversible [28]) in the mobility of macromolecules or macromolecular side chains in the protoplast membrane, in nucleic acids, and in proteins, and an increase in intermolecular space so that metabolic reactions can take place. An increase in free SH groups may be associated with these changes (28, 58).

Since the coats are discarded largely unchanged in structure and composition during germination (49) except perhaps for the loss of some alkali-soluble protein (68), and since cortical degradation follows loss of heat stability, Ca^{2+}, and DPA (7, 25), the initial changes brought about by the trigger reaction would seem to have to occur in the protoplast membrane along the lines indicated earlier. That is, the trigger reaction must directly or indirectly (almost simultaneously) result in a change in the permeability of the protoplast membrane. Perhaps the trigger reaction occurs in the membrane itself, activating enzymes or bringing about ionic changes that create protein channels or reverse the directionality of protein channels (11) or the movement of solutes ("carrier mechanism" [8]), permitting rapid efflux of Ca, DPA, or CaDPA. Alternatively, it may create a more general permeability change.

The mobility of the macromolecules in the protoplast must be regained rapidly by increased hydration, loss of noncovalently bound solutes such as calcium and DPA, and proteolysis of proteins with the release of low-molecular-weight basic proteins (54). Ca and DPA are lost very rapidly and are associated with a very early event in the triggering of germination (51). As soon as calcium and DPA are lost, the macromolecules have more space and become molecularly mobile and active metabolically in a now fully hydrated system.

DNA increases in stability to UV light during the first few minutes of germination (57, 60) because DPA is lost and there is no transfer of photochemical energy from DPA (14, 15); yet the DNA has not assumed its fully expanded hydrated state. This also suggests that the DNA state is not dependent solely on a presence of DPA/CaDPA but possibly also depends on the presence of basic proteins which have not been fully degraded at this time (55).

Cortical lysis rapidly follows the loss of Ca and DPA (7, 25). On cytological evidence, lysis appears to occur generally throughout the cortex (23), and this suggests that the lytic enzymes may be located throughout the cortical region. With major breakdown of the cortex, the protoplast hydrates fully, expanding into the vacated spaces and becoming metabolically active, perhaps reaching full activity before the coats are disrupted or absorbed.

CONCLUDING REMARKS

The results presented here are consistent with a rather well-designed model by which the spore can readily attain dormancy and considerable heat resistance. This may be by reduction of the effective protoplast a_w to about 0.7 by one of several mechanisms (12, 65) or by the interaction of CaDPA with labile macromolecules to immobilize the macromolecules within a high-density, reduced-space system which simulates a solid support system. In the latter model, partial dehydration may be less important than previously believed.

Further work is needed to confirm some of the ideas presented and to define which mechanisms are most important in determining heat resistance, the DNA state (A, B, or otherwise), and the arrangement of CaDPA and polynucleotides. Intercalation of DPA into DNA appears feasible, and the DPA-DNA photoadducts indicate that such an interaction does occur in the spore. Further work is also required to define membrane sites for germination enzymes, trigger mechanisms, and the biophysical basis of the permeability changes that occur during germination.

ACKNOWLEDGMENTS

I greatly appreciate the help and cooperation of my colleagues Pamela A. Wills, J. H. Bradbury, B. A. Cornell, J. A. Lindsay, M. Stewart, I. C. Watt, and A. D. Warth in the preparation of this paper.

LITERATURE CITED

1. **Baillie, E. C., G. R. Germaine, W. G. Murrell, and D. F. Ohye.** 1974. Photoreactivation, photoproduct for-

mation, and deoxyribonucleic acid state in ultraviolet-irradiated sporulating cultures of *Bacillus cereus*. J. Bacteriol. **120**:516–523.

2. **Balassa, G., P. Milhaud, E. Raulet, M. T. Silva, and J. C. F. Sousa.** 1979. A *Bacillus subtilis* mutant requiring dipicolinic acid for the development of heat-resistant spores. J. Gen. Microbiol. **110**:365–379.

3. **Church, B. D., and H. O. Halvorson.** 1959. Dependence of the heat resistance of bacterial endospores on their dipicolinic acid content. Nature (London) **183**:124–125.

4. **Clouston, J. G., and P. A. Wills.** 1969. Initiation of germination and inactivation of *Bacillus pumilus* spores by hydrostatic pressure. J. Bacteriol. **97**:684–690.

5. **Clouston, J. G., and P. A. Wills.** 1970. Kinetics of initiation of germination of *Bacillus pumilus* spores by hydrostatic pressure. J. Bacteriol. **103**:140–143.

6. **Donnellan, J. E., and R. Setlow.** 1965. Thymine photoproducts but not thymine dimers found in ultraviolet irradiated bacterial spores. Science **149**:308–310.

7. **Dring, G., and G. W. Gould.** 1971. Sequence of events during rapid germination of spores of *Bacillus cereus*. J. Gen. Microbiol. **65**:101–104.

8. **Eisenman, G., G. Szabo, S. Ciani, S. McLaughlin, and S. Krasne.** 1973. Ion binding and ion transport produced by neutral lipid soluble molecules. Recent Prog. Surf. Membrane Sci. **6**:139–241.

9. **Ellar, D. J., and J. A. Posgate.** 1974. Characterization of forespores isolated from *Bacillus megaterium* at different stages of development into mature spores, p. 21–40. *In* A. N. Barker, G. W. Gould, and J. Wolf (ed.), Spore Research 1973. Academic Press, London.

10. **Enfors, S. O., and G. Molin.** 1978. Mechanism of the inhibition of spore germination by inert gases and carbon dioxide, p. 80–84. *In* G. Chambliss and J. C. Vary (ed.), Spores VII. American Society for Microbiology, Washington, D.C.

11. **Finkelstein, A., and R. Holtz.** 1973. Aqueous pores created in thin lipid membranes by the polyene antibiotics nystatin and amphotericin B, p. 377–408. *In* G. Eisenman (ed.), Membranes—a series of advances, vol. 2. Marcel Dekker, New York.

12. **Gerhardt, P., and W. G. Murrell.** 1978. Basis and mechanism of spore resistance: a brief preview, p. 18–20. *In* G. Chambliss and J. C. Vary (ed.), Spores VII. American Society for Microbiology, Washington, D.C.

13. **Germaine, G. R., E. Coggiola, and W. G. Murrell.** 1973. Development of ultraviolet resistance in sporulating *Bacillus cereus*. J. Bacteriol. **116**:823–831.

14. **Germaine, G. R., and W. G. Murrell.** 1973. Effect of dipicolinic acid on the ultraviolet radiation resistance of *Bacillus cereus* spores. Photochem. Photobiol. **17**:145–154.

15. **Germaine, G. R., and W. G. Murrell.** 1974. Use of ultraviolet radiation to locate dipicolinic acid in *Bacillus cereus* spores. J. Bacteriol. **118**:202–208.

16. **Gould, G. W.** 1978. Practical implications of compartmentalization and osmotic control of water distribution in spores, p. 21–26. *In* G. Chambliss and J. C. Vary (ed.), Spores VII. American Society for Microbiology, Washington, D.C.

17. **Gould, G. W., and G. J. Dring.** 1975. Heat resistance of bacterial endospores and concept of an expanded osmoregulatory cortex. Nature (London) **258**:402–405.

18. **Gould, G. W., and G. J. Dring.** 1975. Role of an expanded cortex in resistance of bacterial endospores, p. 541–546. *In* P. Gerhardt, R. N. Costilow, and H. L. Sadoff (ed.), Spores VI. American Society for Microbiology, Washington, D.C.

19. **Grecz, N., R. F. Smith, and C. C. Hoffmann.** 1970. Sorption of water by spores, heat-killed spores, and vegetative cells. Can. J. Microbiol. **16**:573–579.

20. **Hachisuka, Y., K. Tochikubo, Y. Yokoi, and T. Murachi.** 1967. The action of dipicolinic acid and its chemical analogues on the heat stability of glucose dehydrogenase of *Bacillus subtilis* spores. J. Biochem. (Tokyo) **61**:659–661.

21. **Halvorson, H. O., and C. Howitt.** 1961. The role of DPA in bacterial spores, p. 149–164. *In* H. O. Halvorson (ed.), Spores II. Burgess Publishing Co., Minneapolis, Minn.

22. **Hanson, R. S., M. V. Curry, J. V. Garner, and H. O. Halvorson.** 1972. Mutants of *Bacillus cereus* strain T that produce thermoresistant spores lacking dipicolinate and have low levels of calcium. Can. J. Microbiol. **18**:1139–1143.

23. **Hashimoto, T., and S. F. Conti.** 1971. Ultrastructural changes associated with activation and germination of *Bacillus cereus* T spores. J. Bacteriol. **105**:361–368.

24. **Hashimoto, T., W. R. Frieben, and S. F. Conti.** 1969. Microgermination of *Bacillus cereus* spores. J. Bacteriol. **100**:1385–1392.

25. **Hsieh, L. K., and J. C. Vary.** 1975. Peptidoglycan hydrolysis during initiation of spore germination in *Bacillus megaterium*, p. 465–471. *In* P. Gerhardt, R. N. Costilow, and H. L. Sadoff (ed.), Spores VI. American Society for Microbiology, Washington, D.C.

26. **Johnstone, K., D. J. Ellar, and T. C. Appleton.** 1980. Location of metal ions in *Bacillus megaterium* spores by high-resolution electron probe X-ray microanalysis. FEMS Microbiol. Lett. **7**:97–101.

27. **Keynan, A.** 1978. Spore structure and its relation to resistance, dormancy, and germination, p. 43–53. *In* G. Chambliss and J. C. Vary (ed.), Spores VII. American Society for Microbiology, Washington, D.C.

28. **Keynan, A., Z. Evenchik, H. O. Halvorson, and J. W. Hastings.** 1964. Activation of bacterial endospores. J. Bacteriol. **88**:313–318.

29. **Keynan, A., and H. Halvorson.** 1965. Transformation of a dormant spore into a vegetative cell, p. 174–179. *In* L. L. Campbell and H. O. Halvorson (ed.), Spores III. American Society for Microbiology, Ann Arbor, Mich.

30. **Knaysi, G.** 1965. Further observations on the spodogram of *Bacillus cereus* endospore. J. Bacteriol. **90**:453–455.

31. **Leanz, G., and C. Gilvarg.** 1972. Dipicolinic acid location in intact spores of *Bacillus megaterium*. J. Bacteriol. **114**:455–456.

32. **Marmur, J., and P. Doty.** 1962. Determination of the base composition of deoxyribonucleic acid from its thermal denaturation temperature. J. Mol. Biol. **5**:109–118.

33. **Marquis, R. E.** 1978. An electrochemical appraisal of bacterial endospores. Spore Newsl. **6**(Special Issue):29–31.

34. **Marshall, B. J., and W. G. Murrell.** 1970. Biophysical analysis of the spore. J. Appl. Bacteriol. **33**:103–129.

35. **Mishiro, Y., and M. Ochi.** 1966. Effect of dipicolinate on the heat denaturation of proteins. Nature (London) **211**:1190.

36. **Murrell, W. G.** 1967. The biochemistry of the bacterial endospore. Adv. Microb. Physiol. **1**:133–251.

37. **Murrell, W. G.** 1969. Chemical composition of spores and spore structures, p. 215–273. *In* G. W. Gould and A. Hurst (ed.), The bacterial spore. Academic Press, Inc., New York.

38. **Murrell, W. G.** 1978. Development of the resistant state in bacterial spores. Spore Newsl. **6**(Special Issue):27–28.

39. **Murrell, W. G., D. F. Ohye, and R. A. Gordon.** 1969. Cytological and chemical structure of the spore, p. 1–19. *In* L. L. Campbell (ed.), Spores IV. American Society for Microbiology, Bethesda, Md.

40. **Murrell, W. G., and W. J. Scott.** 1966. The heat resistance of bacterial spores at various water activities. J. Gen. Microbiol. **43**:411–425.

41. **Murrell, W. G., and A. D. Warth.** 1965. Composition and heat resistance of bacterial spores, p. 1–24. *In* L. L. Campbell and H. O. Halvorson (ed.), Spores III. American Society for Microbiology, Ann Arbor, Mich.

42. **Murrell, W. G., and P. A. Wills.** 1977. Initiation of *Bacillus* spore germination by hydrostatic pressure: effect of temperature. J. Bacteriol. **129:**1272–1280.

43. **Nelson, D. L., J. A. Spudich, P. P. M. Bonsen, L. L. Bertsch, and A. Kornberg.** 1969. Biochemical studies of bacterial sporulation and germination, p. 59–71. *In* L. L. Campbell (ed.), Spores IV. American Society for Microbiology, Bethesda, Md.

44. **Rode, L. J., and J. W. Foster.** 1960. The action of surfactants on bacterial spores. Arch. Mikrobiol. **36:**67–94.

45. **Romig, W. R., and O. Wyss.** 1957. Some effects of ultraviolet radiation on sporulating cultures of *Bacillus cereus*. J. Bacteriol. **74:**386–391.

46. **Rossignol, D. P., and J. C. Vary.** 1978. L-Proline-initiated germination in *Bacillus megaterium* spores, p. 90–94. *In* G. Chambliss and J. C. Vary (ed.), Spores VII. American Society for Microbiology, Washington, D.C.

47. **Rossignol, D. P., and J. C. Vary.** 1979. L-Proline site for triggering *Bacillus megaterium* spore germination. Biochem. Biophys. Res. Commun. **89:**547–551.

48. **Sadoff, H. L.** 1970. Heat resistance of spore enzymes. J. Appl. Bacteriol. **33:**130–140.

49. **Santo, L. Y., and R. H. Doi.** 1974. Ultrastructural analysis during germination and outgrowth of *Bacillus subtilis* spores. J. Bacteriol. **120:**475–481.

50. **Scherrer, R., and P. Gerhardt.** 1972. Location of calcium within *Bacillus* spores by electron probe X-ray microanalysis. J. Bacteriol. **112:**559–568.

51. **Scott, I. R., and D. J. Ellar.** 1978. Study of calcium dipicolinate release during bacterial spore germination by using a new, sensitive essay for dipicolinate. J. Bacteriol. **135:**133–137.

52. **Scott, I. R., G. S. A. B. Stewart, M. A. Koncewicz, D. J. Ellar, and A. Crafts-Lighty.** 1978. Sequence of biochemical events during germination of *Bacillus megaterium* spores, p. 95–103. *In* G. Chambliss and J. C. Vary (ed.), Spores VII. American Society for Microbiology, Washington, D.C.

53. **Setlow, P.** 1974. Polyamine levels during growth, sporulation, and spore germination of *Bacillus megaterium*. J. Bacteriol. **117:**1171–1177.

54. **Setlow, P.** 1975. Identification and localization of the major proteins degraded during germination of *Bacillus megaterium* spores. J. Biol. Chem. **250:**8159–8167.

55. **Setlow, P., and G. Primus.** 1975. Protein metabolism during germination of *Bacillus megaterium* spores. J. Biol. Chem. **250:**623–630.

56. **Seto-Young, D., and D. J. Ellar.** 1979. Membrane changes during germination of *Bacillus megaterium* K M spores. Microbios **26:**7–15.

57. **Stafford, R. S., and J. E. Donnellan.** 1968. Photochemical evidence for conformation changes in DNA during germination of bacterial spores. Proc. Natl. Acad. Sci. U.S.A. **59:**822–828.

58. **Stewart, G. S. A. B., M. W. Eaton, K. Johnstone, M.** D. Barrett, and D. J. Ellar. 1980. An investigation of membrane fluidity changes during sporulation and germination of *Bacillus megaterium* K M measured by electron spin and nuclear magnetic resonance spectroscopy. Biochim. Biophys. Acta **600:**270–290.

59. **Stewart, M., A. P. Somlyo, A. V. Somlyo, H. Shuman, J. A. Lindsay, and W. G. Murrell.** 1980. Distribution of calcium and other elements in cryosectioned *Bacillus cereus* T spores, determined by high-resolution scanning electron probe X-ray microanalysis. J. Bacteriol. **143:**481–491.

60. **Stuy, J. H.** 1956. Studies on the mechanism of radiation inactivation of microorganisms. III. Inactivation of germinating spores of *Bacillus cereus*. Biochim. Biophys. Acta **22:**241–246.

61. **Travers, A.** 1974. On the nature of DNA promoter conformations. The effects of glycerol and dimethylsulphoxide. Eur. J. Biochem. **47:**435–441.

62. **Vary, J. C.** 1978. Glucose-initiated germination in *Bacillus megaterium* spores, p. 104–108. *In* G. Chambliss and J. C. Vary (ed.), Spores VII. American Society for Microbiology, Washington, D.C.

63. **Waldham, D. G., and H. O. Halvorson.** 1954. The relationship between equilibrium vapor pressure and moisture content of bacterial endospores. Appl. Microbiol. **2:**333–338.

64. **Walter, J. A., and A. B. Hope.** 1971. Nuclear magnetic resonance and the state of water in cells. Prog. Biophys. Mol. Biol. **23:**1–20.

65. **Warth, A. D.** 1978. Molecular structure of the bacterial spore. Adv. Microb. Physiol. **17:**1–45.

66. **Watt, I. C.** 1978. Water vapor sorption by intact spores and isolated fractions. Spore Newsl. **6**(Special Issue): 37–38.

67. **Woese, C. R., and M. Bleyman.** 1969. Characterization of the ribonucleic acid formed during germination of *Bacillus subtilis* spores, p. 223–234. *In* L. L. Campbell (ed.), Spores IV. American Society for Microbiology, Bethesda, Md.

68. **Wyatt, L. R., and W. M. Waites.** 1971. Spores of *Clostridium bifermentans*. Comparison of germination mutants, p. 123–131. *In* A. N. Barker, G. W. Gould, and J. Wolf (ed.), Spore Research 1971. Academic Press, London.

69. **Yasuda-Yasaki, Y., S. Namiki-Kanie, and Y. Hachisuka.** 1978. Inhibition of germination of *Bacillus subtilis* spores by alcohols, p. 113–116. *In* G. Chambliss and J. C. Vary (ed.), Spores VII. American Society for Microbiology, Washington, D.C.

70. **Zytkovicz, T. H., and H. O. Halvorson.** 1972. Some characteristics of dipicolinic acid-less mutant spores of *Bacillus cereus*, *Bacillus megaterium*, and *Bacillus subtilis*, p. 49–52. *In* H. O. Halvorson, R. Hanson, and L. L. Campbell (ed.), Spores V. American Society for Microbiology, Washington, D.C.

A. Genetics of Sporeforming Bacteria and Cloning of Sporulation Genes

Genetic Map of *Bacillus licheniformis*

FREDERICK J. PERLAK[1] AND CURTIS B. THORNE

Department of Microbiology, University of Massachusetts, Amherst, Massachusetts 01003

Previous mapping data for *Bacillus licheniformis* strains ATCC 9945A and ATCC 11946 were combined with new data obtained primarily by cotransduction with bacteriophage SP-15 to construct partial chromosomal maps. The organization of the *B. licheniformis* chromosome appeared to be similar to that of *B. subtilis*.

Genetic analysis of *Bacillus* species has focused primarily on *B. subtilis*. With the accumulation of a large inventory of mutants and the use of transformation, generalized transduction, and marker frequency analysis techniques, the *B. subtilis* chromosome has been linked in a circular map (6).

B. licheniformis has been reported to be similar to *B. subtilis*. The two organisms have similar guanine plus cytosine ratios, susceptibility to infection by the same bacteriophages, and similar defective phages (2, 16). Transformation, generalized transduction, and gene frequency analysis have been utilized by researchers working on *B. licheniformis*, but unlike *B. subtilis*, mapping has been restricted to small sections of the chromosome (8, 15). This article is a review of the mapping studies of two *B. licheniformis* strains, 9945A and 11946. We have been successful in mapping new mutations of strain 11946 and demonstrating linkage between previously unlinked regions of the chromosome. The linkages of strain 11946 reported in this paper are primarily the result of bacteriophage SP-15–mediated generalized transduction. SP-15 is similar to phage PBS-1 in size and morphology and is capable of transducing large segments of the chromosome (15). The organization of the mutations of *B. licheniformis* is also compared with the *B. subtilis* chromosome.

RESULTS AND DISCUSSION

Each mapped segment will be discussed and compared with analogous segments of the *B.*

subtilis chromosome. The transductions reported in this paper were done by previously reported procedures (8). The genetic markers and their phenotypes are listed in Table 1.

ade-1 to str-2. McCuen and Thorne (8) had established in strain 9945A linkage between a mutation with a strict adenine requirement, *ade-2*, a streptomycin resistance mutation, *str-1*, and a glutamyl polypeptide mutation, *pepB7*. *B. licheniformis* produces a capsule composed of glutamyl polypeptide. Colonial variants of 9945A auxotrophic mutants which were deficient in the production of glutamyl polypeptide (Pep⁻) could be transformed at higher frequencies than the parental type (5, 8, 13). By screening a number of Pep⁻ mutants by SP-15–mediated transduction, at least three groups of *pep* mutations could be distinguished. Group 1 mutations mapped near a glycine mutation (*gly-42*) and a histidine marker (*his-6*), group II mutations mapped near *ade-2* and *str-1*, and group III mutations, which contained a majority of the Pep⁻ mutations, could not be linked to either *gly-42* or *ade-2*.

N. D. Rapoport (M.S. thesis, University of Massachusetts, Amherst, 1974) mapped the positions of mutations of 11946 analogous to the mutations of 9945A described above. She was able to link a mutation conferring a strict adenine requirement to two Str^r mutations, one of high-level resistance (*str-2*) and the other of low-level resistance (*str-1*). In addition, a *pep* mutation, *pep-3*, was mapped in this region of the chromosome. Heterologous transduction between 9945A and 11946 indicated that the mutations mapped by Rapoport were analogous to

[1] Present address: Department of Microbiology, Ohio State University, Columbus, OH 43210.

TABLE 1. *Genetic markers of* B. licheniformis *in Fig. 1–4*

Marker	Phenotype, enzyme deficiency, or other characteristic	Marker	Phenotype, enzyme deficiency, or other characteristic
Strain 9945A			mutant in acetohydroxyacid synthase
ade-2	Adenine requirement		
arg-3	Arginine requirement	*ilvC1*	Isoleucine and valine requirement; mutant in acetohydroxyacid isomerase
chlE[a]	Chlorate resistance		
gly-42	Glycine requirement		
his-6	Histidine requirement	*ilvD1*	Isoleucine and valine requirement; mutant in dihydroxyacid dehydratase
ilvA[b]	Isoleucine requirement; mutant in threonine deaminase		
ilvD[b]	Isoleucine and valine requirement; mutant in dihydroxyacid dehydratase	*leu-2*	Leucine requirement
		liv-5	Leucine, isoleucine, and valine requirement; could be a regulatory site affecting expression of aminotransferase
leu-2	Leucine requirement		
lys-3	Lysine requirement		
met-1	Methionine requirement	*lys-1*	Lysine requirement
pepA1	Glutamyl polypeptide negative	*metA1*	Methionine, homocysteine, or cystathionine requirement
pepB7	Glutamyl polypeptide negative		
ser-1[c]	Serine requirement	*metB1*	Methionine or homocysteine requirement
spoA[c]	Sporulation mutant		
spoB[c]	Sporulation mutant	*metC1*	Methionine requirement
spoLR5[c]	Sporulation mutant	*metD1*	Methionine requirement
spoLR6[c]	Sporulation mutant	*pep-1*	Mutant in glutamyl polypeptide production
spoLR52[c]	Sporulation mutant		
str-1	Streptomycin resistance	*pep-2*	Mutant in glutamyl polypeptide production
trp-1	Tryptophan requirement		
tyr-1	Tyrosine requirement	*pep-3*	Mutant in glutamyl polypeptide production
ura-1	Uracil requirement		
Strain 11946		*pha-1*	Resistance to bacteriophage LP-52
ade-1	Adenine requirement	*phe-1*	Phenylalanine requirement
argA1	Arginine requirement	*smo-1*	Smooth colony morphology
argC1	Arginine or citrulline requirement	*smo-2*	Smooth colony morphology
argO1	Arginine, citrulline, or ornithine requirement	*str-1*	Resistant to 400 µg of streptomycin sulfate per ml on enriched media
cysC1	Methionine or cysteine requirement	*str-2*	Resistant to 1 mg of streptomycin sulfate per ml
gly-1	Glycine requirement		
his-1	Histidine requirement	*thi-1*	Thiamine requirement
hom-1	Homoserine or methionine and threonine requirement	*thi-2*	Thiamine requirement
		thr-1	Threonine requirement
ilvA1	Isoleucine requirement; mutant in threonine deaminase	*trp-3*	Tryptophan requirement
		tyr-1	Tyrosine requirement
ilvB1	Isoleucine and valine requirement;	*ura-2*	Uracil requirement

[a] Reference 10.
[b] Reference 11.
[c] Reference 9.

the mutations of strain 9945A mapped by McCuen and Thorne (8).

This region in *B. licheniformis* may be analogous to the ribosomal protein region of the *B. subtilis* chromosome. Interspecies transformation between *B. subtilis* and *B. licheniformis* indicated that this region of the chromosome may be conserved between the two species (1). Transformation between the two species was unsuccessful for markers in other areas of the chromosome.

***metD1* to *smo-1*.** We have linked several markers in the *metD1*-to-*smo-1* section of the chromosome of 11946 (Fig. 1). A group of auxo-

trophic markers were ordered with respect to each other by SP-15–mediated transduction. A mutation in strain 11946 resulting in a smooth colony morphology (*smo-1*) is linked to *cysC1* and *ura-2*. The *smo-1* mutant was isolated as a spontaneous mutant with a smooth, glistening morphology from a well-marked strain. The order of the markers, *metC1 ura-2 cysC1 smo-1,* in *B. licheniformis* 11946 is the result of several factors. First, the cotransduction was higher between *cysC1* and *smo-1* than between *ura-2* and *smo-1*. Second, there is a lack of linkage between *metC1* and *cysC1* or *smo-1*. Finally, the order of mutations implied by three-point cross-trans-

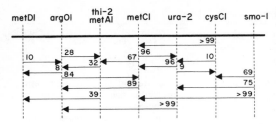

FIG. 1 *Summary of linkage data for the region between* metD1 *and* smo-1 *in strain 11946. Distances are expressed as 100 minus percent cotransduction with SP-15. The arrows point from donor markers to recipient markers.*

duction data (data not shown) agrees with the order of mutations from the transduction studies described above.

It appears that the mutations mapped in this section of 11946 are analogous to those in a segment of linked mutations in *B. subtilis*. The strong linkage observed between *metA1* and *thi-2* has been reported with analogous markers in *B. subtilis*, as well as linkage between markers phenotypically similar to *metD1, argO1,* and *metC1*. However, Henner and Hoch (3) pointed out that linkages in this area of the *B. subtilis* map vary greatly in different reports. They attributed the discrepancy of the reported values to multiple arginine mutations with the same phenotypic requirements and a possible role for PBS-X, the defective prophage of *B. subtilis* 168, which maps in this region of the chromosome.

An arginine locus, *argC*, maps in this area of the *B. subtilis* chromosome. We have isolated arginine mutants of *B. licheniformis* which phenotypically correspond to *argC* mutants of *B. subtilis*, but the mutations do not cotransduce with other markers of this area of the chromosome. Thus far, we have been unsuccessful in linking the outside markers of this segment, *metD1* and *smo-1*, to other mutations of strain 11946.

thi-1 to phe-1. The sequential replication data obtained by Tyeryar et al. (14) with strain 9945A suggest that a segment of linked markers including a glycine mutation (*gly-3*), a glutamyl polypeptide mutation (*pep-1*), and a histidine mutation (*his-9*) is located near the origin of replication. McCuen and Thorne (8) were able to link three analogous mutations, *gly-42, pepA1,* and *his-6*, by cotransduction.

In mapping studies in *B. licheniformis* 10716, a strain which appears identical to 11946, J. B. Kowalski (4, 12; Ph.D. thesis, University of Massachusetts, Amherst, 1975) was able to link a mutation (*pha-1*) conferring resistance to bac-

teriophage LP-52 to other mutations in strain 10716, and he established the order *thi-1 gly-1 pha-1 his-1*. These mutations were linked by SP-15–mediated interstrain transduction to the *gly-42 pepA1 his-6* segment of 9945A.

Studies in *B. subtilis* (17) showed a segment of the *B. subtilis* chromosome analogous to the regions of 9945A and 11946 described above. This segment contains a *hisA* mutation, several phage resistance mutations, and a glycine mutation. The *hisA* locus of *B. subtilis* has been linked to a threonine mutation (*thrA*) and a homoserine mutation (*hom-1*) by cotransduction with a mutation conferring a smooth colony morphology (*smo*). This mutation is distinct from the phage resistance mutation *gtaA*.

The *smo-2* mutation in 11946, which confers smooth colony morphology, was cotransduced with *his-1, hom-1,* and *thr-1* (Fig. 2). The established order, *his-1 smo-2 hom-1 thr-1*, is analogous to the order of similar mutations in *B. subtilis*.

Additional markers are linked to the segment described above. Rapoport (M.S. thesis, University of Massachusetts, Amherst, 1974), working with strain 11946, established linkage (37%) be-

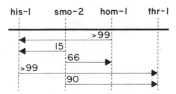

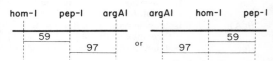

FIG. 2. *Summary of linkage data for the region between* his-1 *and* argA1 *in strain 11946. See legend to Fig. 1 for explanation of distances and arrows. In the crosses represented by lines without arrows, both markers were in the recipient.*

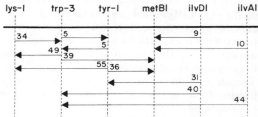

FIG. 3. *Summary of linkage data for the region between* lys-1 *and* ilvA1 *in strain 11946. See legend to Fig. 1 for explanation of distances and arrows.*

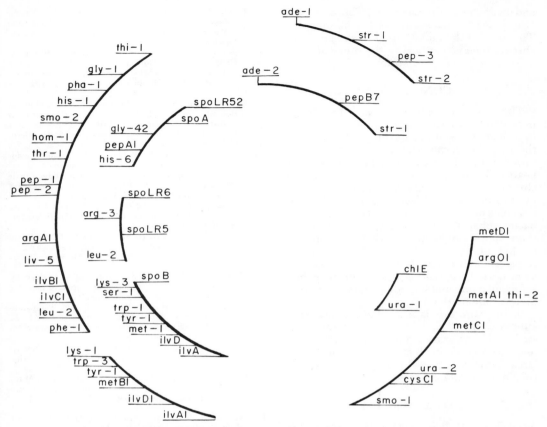

FIG. 4. *Chromosomal maps of* B. licheniformis *strains 11946 and 9945A. The outer segments represent mapped sections of strain 11946, and the inner segments represent mapped sections of strain 9945A. See Table 1 for explanations of the markers. The maps are depicted as circular by analogy to* B. subtilis.

tween *hom-1* and a polypeptide mutation, *pep-2*. She was also able to demonstrate by transformation that two polypeptide mutations of 11946 (*pep-1* and *pep-2*) were identical or overlapping deletions. In addition, *arg-1* was linked to *pep-1* by cotransduction in 11946, although strain 9945A was the donor. The linkage between *pep-1* and *arg-1* should be considered tentative because the cotransduction value is low, and a heterologous strain, 9945A, was the donor.

The *argA1* mutation of 11946 is linked to markers in well-characterized mutants of *B. licheniformis*. Lessie and Thorne (7) reported the isolation of mutants of *B. licheniformis* 11946 with mutations in branched-chain amino acid synthesis. The mutants were characterized for enzyme deficiencies, and the mutations were ordered on the chromosome by three-factor crosses with phages SP-15 and SP-10. Except for the inclusion of an unusual mutation (*liv-5*), this linkage group appears to correspond to one

situated between *argA* and *pheA* in *B. subtilis*.

lys-1 to *ilvA1*. Sequential replication experiments with 9945A (14) placed a number of auxotrophic markers (*tyr, trp, lys, met*) near the terminus of replication. Subsequent mapping of *B. licheniformis* 9945A by SP-10– and SP-15– mediated transduction resulted in a genetically linked segment of several of these auxotrophic markers (15). Rogolsky (9) added a sporulation mutation to this segment, and Sherratt and Collins (11) added, by transformation, two branched-chain amino acid mutations, *ilvD* and *ilvA*, and two penicillinase markers. However, the penicillinase markers were in strain 749.

We have conducted mapping studies with mutations of 11946 analogous to those described above for strain 9945A (Fig. 3). The order, *lys-1 trp-3 tyr-1*, was confirmed by SP-15–mediated three-point cross-transductions (data not shown). The *ilvD1* and *ilvA1* markers isolated and characterized by Lessie and Thorne (7) were also linked to this section of the chromosome. The order of the mutations of this region of the chromosome is identical to the order of analogous mutations in *B. subtilis*.

CONCLUSION

The new mapping data presented in this paper, along with those from previous mapping experiments, result in large mapped segments of the *B. licheniformis* chromosome. The organization of the chromosome of 11946 (Fig. 4) appears to be analogous to that of the *B. subtilis* chromosome. Although *B. subtilis* and *B. licheniformis* share similar characteristics (guanine plus cytosine ratio, susceptibility to several bacteriophages, morphology, and physiology), there appears to be little DNA-DNA homology by hybridization studies (2, 16). Interspecies transformation of auxotrophic markers has been unsuccessful, suggesting that there may be a sufficient number of heterologous regions to prohibit recombination.

B. licheniformis is a good candidate as an alternative to *B. subtilis* for recombinant DNA studies. Transformation and transduction systems are well defined, and the mapping of the large segments reported here should serve as a good foundation for a research system.

LITERATURE CITED

1. **Goldberg, I. D., D. D. Gwinn, and C. B. Thorne.** 1966. Interspecies transformation between *Bacillus subtilis* and *Bacillus licheniformis*. Biochem. Biophy. Res. Commun. **23**:543–548.
2. **Hemphill, H. E., and H. R. Whiteley.** 1975. Bacteriophages of *Bacillus subtilis*. Bacteriol. Rev. **39**:257–315.
3. **Henner, D. J., and J. A. Hoch.** 1980. The *Bacillus subtilis* chromosome. Microbiol. Rev. **44**:57–82.
4. **Kowalski, J. B., and C. B. Thorne.** 1979. Genetic mapping of a bacteriophage resistance mutation, *pha-1*, and the *thi-1* mutation in *Bacillus licheniformis*. J. Bacteriol. **137**:689–691.

5. **Leonard, C. G., and M. J. Mattheis.** 1965. Different transforming characteristics of colonial variants from auxotrophic mutants of *Bacillus licheniformis*. J. Bacteriol. **90**:558–559.
6. **Lepesant-Kejzlarova, J., J.-A. Lepesant, J. Walle, A. Billault, and R. Dedonder.** 1975. Revision of the linkage map of *Bacillus subtilis* 168: indications for circularity of the chromosome. J. Bacteriol. **121**:823–834.
7. **Lessie, T. G., and C. B. Thorne.** 1976. Unusual mutations affecting branched-chain amino acid biosynthesis in *Bacillus licheniformis*, p. 91–100. *In* D. Schlessinger (ed.), Microbiology, Washington, D.C. American Society for Microbiology, Washington, D.C.
8. **McCuen, R. W., and C. B. Thorne.** 1971. Genetic mapping of genes concerned with glutamyl polypeptide production by *Bacillus licheniformis* and a study of their relationship to the development of competence for transformation. J. Bacteriol. **107**:636–645.
9. **Rogolsky, M.** 1970. The mapping of genes for spore formation on the chromosome of *Bacillus licheniformis*. Can. J. Microbiol. **16**:595–600.
10. **Schulp, J. A., and A. H. Stouthamer.** 1972. Isolation and characterization of mutants resistant against chlorate of *Bacillus licheniformis*. J. Gen. Microbiol. **73**:95–112.
11. **Sherratt, D. J., and J. F. Collins.** 1973. Analysis by transformation of the penicillinase system in *Bacillus licheniformis*. J. Gen. Microbiol. **76**:217–230.
12. **Thorne, C. B., and J. B. Kowalski.** 1976. Temperate bacteriophage for *Bacillus licheniformis*, p. 303–314. *In* D. Schlessinger (ed.), Microbiology—1976. American Society for Microbiology, Washington, D.C.
13. **Thorne, C. B., and H. B. Stull.** 1966. Factors affecting transformation of *Bacillus licheniformis*. J. Bacteriol. **91**:1012–1020.
14. **Tyeryar, F. J., Jr., W. D. Lawton, and A. M. MacQuillan.** 1968. Sequential replication of the chromosome of *Bacillus licheniformis*. J. Bacteriol. **95**:2062–2069.
15. **Tyeryar, F. J., Jr., M. J. Taylor, W. D. Lawton, and I. D. Goldberg.** 1969. Cotransduction and co-transformation of genetic markers in *Bacillus subtilis* and *Bacillus licheniformis*. J. Bacteriol. **100**:1027–1036.
16. **Young, F. E., and G. A. Wilson.** 1972. Genetics of *Bacillus subtilis* and other gram-positive sporulating bacilli, p. 77–106. *In* H. O. Halvorson, R. Hanson, and L. L. Campbell (ed.), Spores V. American Society for Microbiology, Washington, D.C.
17. **Young, F. E., and G. A. Wilson.** 1975. Chromosomal map of *Bacillus subtilis*, p. 596–614. *In* P. Gerhardt, R. N. Costilow, and H. L. Sadoff (ed.), Spores VI. American Society for Microbiology, Washington, D.C.

Bacteriophage MP13 Transduction of *Bacillus megaterium* QM B1551

JAMES C. GARBE AND PATRICIA S. VARY

Department of Biological Sciences, Northern Illinois University, DeKalb, Illinois 60115

The *Bacillus megaterium* transducing bacteriophage MP13 was further characterized as to dimensions, buoyant density in CsCl gradients, and sensitivities to heat, salts, and organic solvents. It was also used in preliminary genetic studies of the leucine genes of this species. Nearly all the leucine genes were closely linked and showed a low cotransduction frequency with a *phe* mutation.

The genetics of *Bacillus* has been studied mainly in *B. subtilis* because of the availability of both transformation and transduction methods. Transducing phages have also been isolated for other *Bacillus* species, including *B. amyloliquefaciens, B. pumilus, B. cereus, B. anthracis, B. licheniformis,* and *B. thuringiensis* (2, 18–21), but the genetic analysis of these species has been limited. This is unfortunate since it would seem prudent to investigate the genetic and biochemical characteristics of other species of this diverse genus more fully for several reasons. First, *Bacillus* species differ markedly from one another in such areas as guanine plus cytosine content, sporulation, germination requirements, and spore morphology (4). Second, extensive genetic analysis of sporulation has been done in *B. subtilis*, but few correlations of gene function with morphogenesis have been possible, partly because of the difficulties of producing adequate rates of sporulation and germination. Third, *Bacillus* is a potentially safer host for recombinant DNA research than *Escherichia coli*.

The *Bacillus* species *B. megaterium* is not closely related to either the *B. subtilis* group or the *B. cereus* group (5) and is of particular interest because, in contrast to *B. subtilis*, it sporulates with 100% efficiency and germinates synchronously when heat activated. It should therefore be an ideal species for investigating cellular differentiation, and indeed, several laboratories have studied the properties of spore biochemistry (13–17, 22). Many mutants of interest have also been isolated. In addition to auxotrophic and antibiotic-resistant mutants, there are interesting biochemical mutants (7, 10, 12) as well as mutants defective in germination

(23, 24, 29). Unfortunately, neither transformation nor transduction techniques have been available to analyze these mutants and coordinate genetic information with the biochemical data.

Several attempts to establish genetic exchange in *B. megaterium* have been made in the past. Our laboratory tested several *Bacillus* transducing phages including PBS1, SP10, SP15, CP51, and CP53, as well as *B. megaterium* phages ϕT and MP-7, for transduction of *B. megaterium* without success (unpublished data). Fodor and Alfoldi reported that they could detect recombinants by using protoplast fusion techniques, but discovered that the physiological state of the protoplasts influenced the genetic outcome of the crosses significantly (8). Recently, workers in their laboratory and that of Carlton (3, 27) successfully transformed *B. megaterium* protoplasts with plasmid DNA from *B. subtilis* and *Staphylococcus*. These techniques do not allow exchange of chromosomal DNA, however.

A new phage, MP13, which mediates generalized transduction in *B. megaterium* QM B1551, has recently been isolated in our laboratory (25, 26). Since the phage is large, it should be very useful for mapping studies similar to those using PBS1 in *B. subtilis*. With the availability of a transducing phage, the mapping of the *B. megaterium* chromosome and the analysis of mutants both biochemically and genetically should now be possible. We have therefore attempted to characterize MP13 further, to improve its frequency of transduction, and to initiate genetic studies of the leucine genes.

RESULTS AND DISCUSSION

Phage characteristics. MP13 is a large phage with a contractile tail, several curly tail fibers, and a head that may be octahedral (26). Its dimensions are summarized in Table 1. Table 1 also lists other characteristics of the phage and its growth cycle. The latent period was fairly consistent, but wide variations in burst size were observed which might depend somewhat on the state of the cells used as inoculum. Phage particles centrifuged in CsCl exhibited two bands (Table 1), with the major band containing approximately 80% of the phages. Similar results have been observed in four separate preparations. In two independent experiments, the two bands had similar plaque-forming activity, but the lighter minor band contained in addition five transducing particles per 10^5 PFU. The DNA from either band had a density of 1.737 g/cm^3 and a melting temperature of only 65.2°C (M. Mandel, personal communication), suggesting that DNA density variations did not account for the two phage peaks and that MP13 DNA might contain an unusual base. One possible explanation for the two peaks might be that there is a fairly frequent event in which less than a headful of phage DNA is packaged. In fact, the consistent appearance of two plaque sizes (which do not breed true) could be indicative of such an event.

Optimal transducing conditions. We originally reported a transducing frequency of 1 × 10^{-7} to 8 × 10^{-7} (25). To increase this frequency, we examined conditions necessary for both phage propagation and transductional procedures. All strains and mutants used in transductions are listed in Table 2. Lysates were prepared by infecting, at a multiplicity of infection (MOI) of 1, wild-type QM B1551 cells grown in various media (M, SNB, and MC [16, 26, 30]), at several temperatures (24, 30, and 37°C), at various pH's (6.0 to 8.0), and with various extents of aeration. The resulting lysates were then tested under constant conditions for transduction of JV78 to

TABLE 1. *Summary of phage characteristics of MP13*

Characteristic	Value
Dimensions[a]	
Hexagonal head	97 nm in diam
Contractile tail	202 by 17 nm
Contracted tail	88 by 26 nm
Baseplate	88 nm
Latent period[b]	105 ± 10 min
Burst size	240–883
Sensitivity to salts[c]	
MgSO$_4$, 60 mM	12
CaSO$_4$, 60 mM	8
KCl, 60 mM	0
Sensitivity to solvents[c]	
Chloroform	90
Toluene	88
Ether	86
Sensitivity to heat for 30 min[c]	
50°C	0
55°C	0
60°C	60
Buoyant density of phage particles in CsCl[d]	
Major band	1.490 g/cm^3
Minor band	1.482 g/cm^3

[a] Phages were prepared for electron microscopy as described previously (26). Measurements were calibrated by the method of Luftig (11) using catalase crystals and were the average of 10 to 20 phage particles.

[b] Determined by the method of Adams (1).

[c] Determined by the method of Carvahlo and Vary (6). Values shown are the percent decrease in viability.

[d] Cells in SNB were infected at an MOI of 1 and incubated at 30°C and 220 rpm until lysis at 2.5 to 3 h. The lysate was spun at 2,000 × g, treated overnight at 4°C with 1 μg each of DNase and RNase (Sigma Chemical Co.) per ml, and filtered. Phages were concentrated by two rounds of high-speed (20,000 × g) and low-speed (2,000 × g) centrifugation. Two milliliters of phage suspension at approximately 10^{12} PFU/ml was added to 16 ml of CsCl (Sequanal Grade, Pierce; refractive index of 1.3820), and the mixture was centrifuged for 47 h at 34,000 rpm and 20°C in a Beckman type 65 rotor. Fractions were collected, and the density of phage particles was calculated by the method of Frampton and Mandel (9).

TABLE 2. B. megaterium *strains*[a]

Strain	Genotype	Enzyme deficiency if known[b]	Muta- genesis
PV36	*leu-20*		ICR-191
PV37	*leu-21*		ICR-191
PV39	*leu-22*		ICR-191
PV40	*leu-23*		NTG
PV41	*leu-24*		NTG
PV45	*leu-25*		NTG
JV75	*leuB1*	α-IPM isomerase	NTG
JV76	*leu-2*		NTG
JV77	*leu-3 str-3*	Pleiotropic	EMS
JV78	*leuC4 str-3*	β-IPM dehydrogenase	EMS
JV98	*glu-1 str-3*		EMS
PV30	*pheA20*		NTG

[a] Strains with the JV prefix originated from the laboratory of J. C. Vary. Those with the PV prefix originated from this laboratory. All strains have the genetic background of *B. megaterium* QM B1551. ICR-191 = 3-Chloro-7-methoxy 9-(3-chloroethyl)-aminopropylaminoacridine dihydrochloride; NTG = *N*-methyl-*N'*-nitro-*N*-nitrosoguanidine; EMS = ethyl methane sulfonate; IPM = isopropylmalate.

[b] Determined for strains JV75, JV77, and JV78 by J. C. Vary.

TABLE 3. *Two-factor crosses of leucine mutants*[a]

Donor	Recipients										
	JV98	JV75	JV76	JV77	JV78	PV36	PV37	PV39	PV40	PV41	PV45
wt	1,547	110	256	242	43	62	305	27	0	58	326
JV75	451	1	7	7	1	0	29	3	0	20	35
		(.97)	(.91)	(.90)	(.92)	(1.0)	(.67)	(.62)		(0)	(.63)
JV76	714	11	1	9	4	8	6	2	0	5	19
		(.78)	(.99)	(.92)	(.80)	(.72)	(.96)	(.84)		(.81)	(.87)
JV77	692	7	9	0	3	4	77	9	0	37	14
		(.86)	(.92)	(1.0)	(.84)	(.86)	(.44)	(.26)		(0)	(.90)
JV78	970	10	17	26	1	5	31	4	0	22	50
		(.86)	(.89)	(.83)	(.96)	(.87)	(.84)	(.76)		(.40)	(.76)
PV36	1,907	3	15	34	4	0	9	1	0	64	104
		(.98)	(.95)	(.81)	(.92)	(1.0)	(.98)	(.97)		(.10)	(.74)
PV37	1,567	13	7	15	4	8	1	0	0	2	63
		(.88)	(.97)	(.94)	(.91)	(.87)	(1.0)	(1.0)		(.97)	(.81)
PV39	1,332	8	3	4	1	1	0	0	0	0	62
		(.92)	(.99)	(.98)	(.97)	(.98)	(1.0)	(1.0)		(1.0)	(.78)
PV40	1,581	7	4	10	2	7	0	1	0	0	57
		(.94)	(.98)	(.96)	(.95)	(.89)	(1.0)	(.96)		(1.0)	(.83)
PV41	811	19	6	30	8	8	1	1	0	1	30
		(.67)	(.95)	(.76)	(.64)	(.75)	(.99)	(.93)		(.97)	(.82)
PV45	1,932	11	6	6	3	4	1	1	0	47	1
		(.92)	(.98)	(.98)	(.94)	(.95)	(1.0)	(.97)		(.35)	(1.0)

[a] Transducing lysates were propagated using the optimal conditions established (see text). The nutrient required for each mutant was added at 40 μg/ml during growth. Titers varied from 8×10^9 to 4×10^{10} PFU/ml. Recipient cells were also grown in the presence of the required nutrient, pelleted at $2,000 \times g$ for 2 min, and resuspended in an equal volume of MCT broth. Phages spread on MCT plates were irradiated for 35 s with 48 ergs/mm^2 per s of UV light at 254 nm. Final MOI = 5 to 6 PFU/colony-forming unit. Crosses were made to determine both the donor and recipient ability of the leucine mutants and the donor ability of the wild-type lysate. The expected number of prototrophs in each cross was calculated by: $E = M_d/W_d \times W_m$, where M_d is the number of prototrophs observed using each leucine mutant as donor to the glutamate recipient, W_d is the wild-type lysate crossed to the same recipient, and W_m is the number of prototrophs observed when the wild-type lysate was used as donor to each leucine mutant. The cotransduction frequency (CF) was then determined by: $CF = E - O/E$, where O = number of observed prototrophs in the leucine $\times$ leucine crosses. The top number in each cross is the average number of Leu$^+$ colonies present per plate based on four plates. The number in parentheses is the frequency of cotransduction calculated as described above.

leucine prototrophy. Lysates produced at 30 to 37°C in SNB either with reduced aeration or at pH 8 gave the highest proportion of transducing particles (data not shown). Transductional frequencies of 3×10^{-6} transductants per PFU were consistently obtained by both methods. In contrast to procedures successful for PBS1 lysates, the presence of 4 μg of chloramphenicol per ml did not enhance the proportion of transducing particles.

Transduction conditions were then varied, keeping lysate production constant. Recipient growth stage (absorbancy at 660 nm of 0.5 to 1.2 and spores), amount of UV exposure (15 to 45 s at 52 ergs/mm^2 per s), media (glucose salts with various amino acid supplements or casein or salts), pH (6.0 to 8.0), MOI (0.1 to 100), temperature (4 to 45°C), and presence or absence of antiserum were all tested for their effect on transducing frequencies. The optimal conditions for transduction included infection of a recipient in late logarithmic growth, an acidic selection medium, and incubation at 30°C. Conditions which prevented superinfection of the transductants, such as UV inactivation of phage particles, a low MOI, and the use of antiserum, all enhanced the transduction frequency. It should be pointed out that the MOI based on colony-forming units is a maximum value since most of the cells are present as chains of two to four cells. For all subsequent transductions, lysates were produced by infection of donor cells at midlogarithmic growth in SNB at an MOI of 1, and the culture was shaken at 100 rpm until clearing was observed. Optimal transduction conditions used were as follows: phages were spread on MCT (26) plates at pH 6.5 and exposed to UV light for 30 s; then enough washed cells, absorbancy at 660 nm of 1.1 (10^8 colony-forming units per ml), were spread over the phage to give an MOI of 1 to 5, and plates were incubated at 30°C for 24 to 48 h. This plate method yielded 250 to

300 colonies per plate, approximately the same number of prototrophs obtained when a broth mixture of UV-irradiated phages and cells was incubated for 15 min and then spread on a plate containing anti-MP13 serum. Frequencies of 8×10^{-6} transductants per PFU could be obtained by this method, a 10- to 80-fold increase over those first reported.

Mapping of leucine mutations. To initiate mapping of the *B. megaterium* chromosome, we crossed 10 leucine mutants (see Table 2) in two-factor crosses. Comparison with the wild-type control established the percent cotransduction of pairs of leucine mutations, as shown in Table 3. Nearly all the leucine mutations tested were cotransduced with each other at high frequency. This suggests that the leucine genes are probably clustered in *B. megaterium* as they are in *B. subtilis*. Some reciprocal crosses were not consistent, such as PV37 crossed with JV77. It was also evident that PV40 was a poor recipient, although it was sensitive to MP13 and showed high cotransduction frequencies when used as donor. Also, PV41 results were anomalous. This leucine mutation was cotransduced at fairly high frequencies as donor, but showed great variability in linkage when used as recipient. The cotransduction values found for leucine mutations are similar to the frequencies (88 to 95%) reported for *B. subtilis* by Ward and Zahler (28) using PBS1.

Cotransductional mapping. The cotransductional frequency of *leu* with *phe* in *B. subtilis* has been reported to be 60% (28). In preliminary experiments, we have tested these same markers for cotransduction in *B. megaterium* to determine whether *phe* could be used as an outside marker for three-factor crosses. PV30 (*pheA20*) was used as donor to JV78 (*leuC4*). Of a total of 562 Leu$^+$ colonies from five different experiments, 13 were also Phe$^-$, a cotransduction frequency of 2.3%. It is to be expected that the two markers would be cotransduced at a lower frequency by MP13 if near the ends of the transducing fragment, since this phage is smaller than PBS1. The *phe* marker is not close enough to use as a locus for three-factor crosses, unfortunately, but we are now testing other markers in the same area.

Preliminary results have shown that, for leucine markers at least, the chromosome of *B. megaterium* is similar to that of *B. subtilis*. For initial studies this is fortunate since the *B. subtilis* chromosome may be used as a guide for suggesting possible crosses. MP13 has also been shown to be adequate for mapping and therefore should be a valuable tool in the genetic analysis of *B. megaterium*.

ACKNOWLEDGMENTS

We thank Margaret Franzen for excellent technical assistance and Elon W. Frampton for assistance with the density gradients.

This study was supported in part by grant 4407176 awarded by the Biomedical Research Support Grant Program, Division of Research Resources, National Institutes of Health, and in part by National Science Foundation grant PCM-7922162 (P.S.V.).

LITERATURE CITED

1. **Adams, M. H.** 1959. Bacteriophages. Interscience Publishers, New York.
2. **Bramucci, M. C., K. M. Keggins, and P. S. Lovett.** 1977. Bacteriophage conversion of spore-negative mutants to spore-positive in *Bacillus pumilus*. J. Virol. **22:** 194–202.
3. **Brown, B. J., and B. C. Carlton.** 1980. Plasmid-mediated transformation in *Bacillus megaterium*. J. Bacteriol. **142:**508–512.
4. **Candeli, A., A. DeBartolomeo, V. Mastrandrea, and F. Trotta.** 1979. Contribution to the characterization of *Bacillus megaterium*. Int. J. Syst. Bacteriol. **29:**25–31.
5. **Candeli, A., V. Mastrandrea, G. Cenci, and A. De-Bartolomeo.** 1978. Sensitivity to lytic agents and DNA base composition of several aerobic spore-bearing bacilli. Zentralbl. Bakteriol. Parasitenkd. Infektionskr. Hyg. Abt. 2 **133:**250–260.
6. **Carvahlo, P. M., and J. C. Vary.** 1977. Isolation and characterization of a *Bacillus megaterium* QM B1551 bacteriophage. J. Gen. Virol. **36:**547–550.
7. **Decker, S. F., and D. R. Lang.** 1977. *Bacillus megaterium* mutant deficient in membrane-bound adenosine triphosphatase activity. J. Bacteriol. **131:**98–104.
8. **Fodor, K., and L. Alfoldi.** 1979. Polyethylene-glycol induced fusion of bacterial protoplasts. Mol. Gen. Genet. **168:**55–59.
9. **Frampton, E. W., and M. Mandel.** 1970. Properties of the deoxyribonucleic acid contained in the defective particle coliphage 15. J. Virol. **5:**8–13.
10. **Lang, D. R., and S. J. Decker.** 1977. Mutants of *Bacillus megaterium* resistant to uncouplers of oxidative phosphorylation. J. Biol. Chem. **252:**5936–5938.
11. **Luftig, R.** 1967. An accurate measurement of the catalase crystal period and its use as an internal marker for electron microscopy. J. Ultrastruct. Res. **20:**91–102.
12. **Postemsky, C. J., S. S. Dignam, and P. Setlow.** 1978. Isolation and characterization of *Bacillus megaterium* mutants containing decreased levels of spore protease. J. Bacteriol. **135:**841–850.
13. **Rossignol, D. P., and J. C. Vary.** 1979. L-Proline-triggered germination of *Bacillus megaterium* spores. J. Bacteriol. **138:**431–441.
14. **Scott, I. R., G. S. A. G. Stewart, M. A. Koncewicz, D. J. Ellar, and A. Crafts-Lighty.** 1978. Sequence of biochemical events during germination of *Bacillus megaterium* spores, p. 95–103. *In* G. Chambliss and J. C. Vary (ed.), Spores VII. American Society for Microbiology, Washington, D.C.
15. **Setlow, P.** 1978. Purification and characterization of additional low-molecular weight basic proteins degraded during germination of *Bacillus megaterium* spores. J. Bacteriol. **136:**331–340.
16. **Shay, L. K., and J. C. Vary.** 1978. Biochemical studies of glucose initiated germination in *Bacillus megaterium*. Biochim. Biophys. Acta **538:**284–292.
17. **Singh, R. P., and P. Setlow.** 1979. Purification and properties of phosphoglycerate phosphomutase from spores and cells of *Bacillus megaterium*. J. Bacteriol. **137:**1024–1027.

18. **Thorne, C. B.** 1968. Transducing bacteriophage for *Bacillus cereus.* J. Virol. **2:**657–662.

19. **Thorne, C. B.** 1968. Transduction in *Bacillus thuringiensis* and *Bacillus anthracis.* Bacteriol. Rev. **32:**358–361.

20. **Thorne, C. B.** 1978. Transduction in *Bacillus thuringiensis.* Appl. Environ. Microbiol. **35:**1109–1115.

21. **Thorne, C. B., and J. B. Kowalski.** 1976. Temperate bacteriophages for *Bacillus licheniformis,* p. 303–314. *In* D. Schlessinger (ed.), Microbiology—1976. American Society for Microbiology, Washington, D.C.

22. **Vary, J. C.** 1972. Spore germination of *Bacillus megaterium* QM B1551 mutants. J. Bacteriol. **112:**640–642.

23. **Vary, J. C.** 1975. Properties of *Bacillus megaterium* temperature-sensitive germination mutants. J. Bacteriol. **121:**197–203.

24. **Vary, J. C., and A. Kornberg.** 1970. Biochemical studies of bacterial sporulation and germination. XXI. Temperature-sensitive mutants for initiation of germination. J. Bacteriol. **101:**327–330.

25. **Vary, P. S.** 1979. Transduction in *Bacillus megaterium.* Biochem. Biophys. Res. Commun. **88:**1119–1124.

26. **Vary, P. S., and W. F. Halsey.** 1980. Host range and partial characterization of several new bacteriophages for *Bacillus megaterium* QM B1551. J. Gen. Virol. **51:** 137–146.

27. **Vorobjeva, I. P., I. A. Khmel, and I. Alfoldi.** 1980. Transformation of *Bacillus megaterium* protoplasts by plasmid DNA. FEMS Microbiol. Lett. **7:**195–198.

28. **Ward, J. B., and S. A. Zahler.** 1973. Genetic studies of leucine biosynthesis in *Bacillus subtilis.* J. Bacteriol. **116:**719–726.

29. **Wax, R., E. Freese, and M. Cashel.** 1967. Separation of two functional roles of L-alanine in the initiation of *Bacillus subtilis* spore germinations. J. Bacteriol. **94:** 522–529.

30. **Yehle, C. O., and R. H. Doi.** 1967. Differential expression of bacteriophage genomes in vegetative and sporulating cells of *Bacillus subtilis.* J. Virol. **1:**935–947.

Bacillus thuringiensis: Studies on Chromosomal and Extra-chromosomal DNA

M.-M. LECADET, D. LERECLUS, M.-O. BLONDEL, AND J. RIBIER

Institut de Recherche en Biologie Moléculaire, Centre National de la Recherche Scientifique—Université Paris VII, 75221 Paris Cedex 05, France

Phage CP-54Ber was shown to act as a generalized transducing phage for *Bacillus thuringiensis* Berliner strains, and this allowed us to demonstrate several linkage groups. Analogies with the *B. subtilis* genome in three different regions of the chromosome were found. Extrachromosomal DNA elements ranging in molecular weight from 1.5×10^6 to 60×10^6 were found in a variety of strains, some of which contained exclusively high-molecular-weight species. DNA homologies were shown for plasmids originating from different sources.

For the past 3 years the entomopathogen *Bacillus thuringiensis* and its parasporal crystal have been intensively studied. A general interest in extrachromosomal material, the rapid evolution of in vitro recombinant DNA technology, and recent promising results with genetics in this organism (12, 13, 19) have combined to trigger new research in this area.

It is well known that the parasporal crystal results from the assembly of identical subunits (3, 4, 8) which are coded for by a stable mRNA (9) and that its appearance is a consequence of the first morphological events leading to sporogenesis (15). However, a number of important aspects of crystal synthesis remain unknown. Among the basic questions to be resolved are (i) the number and location of genetic determinants involved in crystal biogenesis and (ii) the role of plasmids, if any, in crystal synthesis.

To help answer these questions, we have tried to gain basic information on the *B. thuringiensis* genome. Our approach has been to look for an efficient system of genetic exchange in *B. thuringiensis* Berliner strains. Simultaneously, we tried to characterize extrachromosomal elements in these strains and others.

RESULTS AND DISCUSSION

Development of a transduction system using phage CP-54Ber. The phage CP-54Ber originated from phage CP-54 (kindly provided by C. Thorne) and was selected on the basis of its ability to adsorb to and propagate on the Berliner strains, conditions which were not satisfactorily fulfilled by the parent phage CP-54.

In a previous paper (12), we reported the isolation and characterization of this new phage, which was shown to act as a generalized transducing phage in the Berliner 1715 strain. Transduction frequencies were generally between 10^{-5} and 10^{-6}. These results enabled us to undertake linkage studies. Results of cotransduction experiments are summarized in Table 1. Several linkage groups were found. One group showed linkage among *his-1*, *trp-1*, and *lys-1* and probably corresponds to the *hisB*, *trp*, *lys* cluster in *B. subtilis*. The mutations *arg-1* and *met-1* formed a second group. The *arg-1* mutant was able to grow on arginine, ornithine, or citrulline and thus may correspond to *argO* mutants in *B. subtilis*, whereas *met-1* and *met-2*, which have a strict requirement for methionine, may be analogous to *metC* or *metD*. These results are in good agreement with those reported by Thorne (19) and by Perlak et al. (14), who used other strains of *B. thuringiensis* and phages CP-51 or TP-13.

The third group we found concerned linkage of *ade-1* to *cys-2* and of *cys-2* to *rfm-1* and *rfm-2*; the latter markers specify resistance to rifampin. Mutant Mr15 (*rfm-1*) is an asporogenous mutant blocked at stage II/III of sporulation; M-RevG is a Spo$^+$ pseudorevertant of Mr15. Mutant Mr12 *rfm-2* is an early asporogenous mutant which is blocked before stage II. Both *rfm-1* and *rfm-2* were cotransducible with *cys-2*. Weak linkage was detected between *rfm-1* and *ade-1* or *ade-3*. The *ade-1* and *ade-3* mutations, which cause a strict requirement for adenine, may correspond to *purA* in *B. subtilis*, whereas

TABLE 1. *Cotransduction of linked markers in* B. thuringiensis *Berliner 1715*[a]

Donor		Recipient		Cotransduction (%)	Linkage group
Strain	Genotype	Strain	Genotype		
M-22-5	*his-1*	M-10-3	*trp-1*	18	
M-10-3	*trp-1*	M-22-5	*his-1*	22	
M-10-3	*trp-1*	M-11-3	*lys-1*	14–18	I
M-11-3	*lys-1*	M-10-3	*trp-1*	10	
M-22-5	*his-1*	M-11-3	*lys-1*	86	
M-1a	*met-1*	M-15	*arg-1*	36–41	
B1715	Wild type	AM-8	*arg-1 met-2*	32–36	II
M-21	*ade-1*	M-12	*cys-2*	14	
Mr15	*rfm-1*	M-12	*cys-2*	24	
M-RevG	*rfm-1*	M-12	*cys-2*	26	
Mr12	*rfm-2*	M-12	*cys-2*	17	III
M-RevG	*rfm-1*	M-32₃	*cys-3*	20	
M-RevG	*rfm-1*	M-21	*ade-1*	1	
M-RevG	*rfm-1*	M-20-1	*ade-3*	2	

[a] Growth of recipient cells, phage propagation on the donor strains, and the transduction procedure were previously reported (12). In experiments using cells bearing a single marker, transductants were first selected on the medium supplemented with the amino acid required by the donor. In a second step, double recombinants were detected by a replica procedure after individual transductants had been picked. In experiments using double auxotrophs, double recombinants were scored among transductants selected for each recipient marker. Percent cotransduction is 100 × number of double recombinants/number of transductants tested.

the *cys-2* and *cys-3* mutations, which cause a strict requirement for cysteine, may be analogous to *cysA*. Crosses involving individual markers taken from each of the three linkage groups indicate that groups I, II, and III are not linked to one another.

From the results it is clear that phage CP-54Ber, which is a large phage, is a convenient tool for chromosomal mapping in Berliner strains and possibly in other varieties of *B. thuringiensis*. Furthermore, our linkage studies point to possible analogies with the *B. subtilis* genome in three different regions of the chromosome. By using various asporogenous or crystalless species bearing a number of genetic markers, we can expect to localize crystal determinants.

Extrachromosomal elements in *B. thuringiensis* strains. Most, if not all, *B. thuringiensis* strains carry DNA elements which behave as covalently closed circular material in a CsCl gradient in the presence of ethidium bromide. Figure 1 presents a tentative scheme giving the electrophoretic pattern of extrachromosomal DNA extracted from several representative strains of known serotypes (1). It suggests three remarks:

1. Most of the strains display a complex pattern including molecules of various sizes whose molecular weights range from 1.5×10^6 to 60×10^6.

2. A common characteristic of the strains is the presence of high-molecular-weight species ($>25 \times 10^6$); these molecules are generally found in lesser amounts than the smaller ones.

3. Strains belonging to serotypes 4, 5, and 6 contain almost exclusively these high-molecular-weight species. Until now, we have not found bands in the Finitimus strain (serotype 2), but their absence has to be confirmed by other methods.

Further characterization of the predominant DNA species in some of the strains—Berliner, Sotto, Dendrolimus, Aizawai—allowed us to confirm the profiles of these DNA elements and to distinguish between covalently closed circular and open circular forms. In the Berliner 1715 strain we found at least six distinct species whose molecular weights, calculated from the lengths of the molecules, are in good agreement with gel analysis. Moreover, electron microscope observations have confirmed that these molecules (including the largest ones) are supercoiled DNA.

In view of the abundance and diversity of the extrachromosomal material, we decided to examine homologies among species. Figure 2 shows a schematic representation of results obtained when electrophoretically separated plasmids were hybridized to ^{32}P-labeled DNA molecules extracted from different species. After hybridization with radioactive plasmids of the Berliner 1715 strain (Fig. 2A), we found the usual pattern in the homologous strain and in the other Ber-

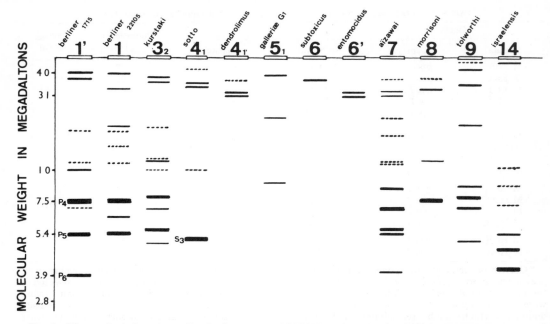

FIG. 1. *Electrophoretic pattern of extrachromosomal DNA elements prepared from strains of B. thuringiensis. Plasmids were extracted from exponentially growing cells (optical density ≤ 3.0) by two different procedures: one consisted of obtaining clear lysates according to a slightly modified version of the Clewell and Helinski procedure (5); the other consisted of lysozyme lysis followed by alkaline denaturation as described by Birnboim and Doly (2). In the first case plasmids were purified either by the usual technique of CsCl-ethidium bromide gradient centrifugation or by hydroxyapatite chromatography by the method of Colman et al. (6). The technique of Birnboim and Doly was found more convenient for extracting high-molecular-weight species. Electrophoresis was run in horizontal agarose gels (0.7%) at 130 V for 5 min and then at 70 V for 14 to 15 h. Numbers above each slot indicate the serotypes according to Barjac's classification (1). Molecular weights on the scale are calculated from measurements of contour length (11). Full lines represent superhelical forms (covalently closed circular); dotted lines represent open circular forms. P4, P5, P6, and S3 are names of particular plasmids.*

liner strain (var. *thuringiensis*); some plasmids are not detected or are represented by a weak signal in serotypes 3, 4 (plasmid S3), $4_1'$, and 9. After hybridization with radioactive giant plasmids of the Dendrolimus strain (31 and 34 megadaltons, respectively), we observed (Fig. 2B) the usual pattern in the homologous $4_1'$ and a positive reaction with most of the high-molecular-weight plasmids, except for 5_1 (and with one species of 4); no reaction at all was seen with plasmids smaller than 20 megadaltons in any of the strains examined.

To summarize, the results suggest: (i) complete homology between equivalent elements within serotype I, (ii) partial homologies from one serotype to another, and (iii) complete absence of homology between the giant plasmids, at least those of the $4_1'$ serotype, and the smaller species, thus indicating that the two size groups might have different origins. Concerning this last point, it is important to note the absence of

reaction between the giant plasmids of the $4_1'$ serotype and plasmid S3 of the 4_1 serotype.

Do these plasmids have a functional role in the cell? To answer the question of whether these plasmids have a functional role in the cell, we have taken two approaches. One approach consisted of correlating the presence of a given plasmid with the expression of a genetic marker such as resistance to antimicrobial agents. A second consisted of demonstrating that a plasmid takes part in the genetic determination of crystal formation.

With the first approach we have shown that a number of B. *thuringiensis* strains carry resistance genes to several antibiotics, among them ampicillin, neomycin, and spectinomycin; resistance to chloramphenicol and tetracycline is not so widely distributed. In the two cases that we examined, ampicillin resistance in the B.1715 strain and tetracycline resistance in a serotype 5 strain (Galleriae Gi), we failed to

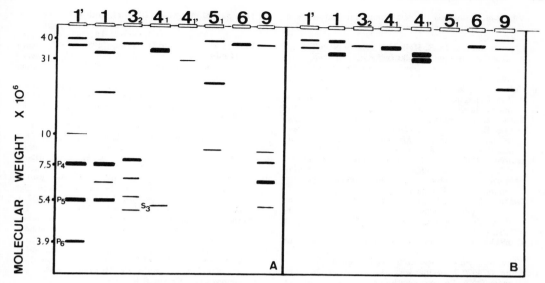

FIG. 2. *Homologies between plasmids from different strains of* B. thuringiensis. *The scheme represents results of Southern hybridization (17). Extrachromosomal DNAs transferred to nitrocellulose filters from agarose gels were hybridized to radioactive plasmids* ^{32}P *labeled by nick translation (16). In (A) the labeled probe corresponds to total plasmids of the Berliner 1715 strain. In (B) the labeled probe corresponds to high-molecular-weight species of the Dendrolimus strain (serotype $4_{1'}$). Numbers above each slot represent serotypes (1); the usual names of the strains are as follows: 1', Berliner 1715; 1, Berliner 22105 (var. thuringiensis); 3_2, Kurstaki; 4_1, Sotto; $4_{1'}$, Dendrolimus; 5_1, Galleriae Gi; 6, Subtoxicus; 9, Tolworthi. Molecular weights on the scale are calculated from measurements of contour length (11). Thickness of the lines representing plasmids is a function of the intensity of the signal as seen on autoradiograms.*

detect any clear change in the plasmid patterns after curing.

As to the question of whether or not plasmids play a role in crystal biosynthesis, there may not be a single answer. Because of the diversity of species grouped within the known serotypes, it is obvious that various situations may exist with regard to localization and functioning of the genes which govern crystal morphogenesis.

It is a fact that differences in the ability to lose their crystal are observed among serotypes or among strains within a given serotype. In the Berliner 1715 isolate, crystal production is extremely stable; this observation, together with previous results including hybridization with RNA species, pointed to a chromosomal localization of the gene encoding the protein. In another isolate of serotype I, the Berliner 22105 (var. *thuringiensis*) strain, the crystal is more easily lost, either spontaneously or after mild curing treatments. In one case we have observed the disappearance of a giant plasmid in a crystalless clone. One interesting example is that of an acrystalliferous strain of serotype 4 which is a Sotto isolate that has spontaneously lost its crystal. In this case we observed significant changes in the pattern: plasmid S3 and two giant plasmids had disappeared.

These results, taken together with data reported recently by several groups (7, 10, 18), reflect the complexity of the extrachromosomal material in *B. thuringiensis* and the possible diversity of situations regarding crystal determinism.

ACKNOWLEDGMENTS

We thank R. Dedonder, in whose laboratory this work was conducted, and A. Klier for helpful discussions. We are grateful to H. de Barjac for providing us with *B. thuringiensis* strains and to R. d'Ari for reading the manuscript.

LITERATURE CITED

1. **Barjac, H. de.** 1978. Un nouveau candidat à la lutte biologique contre les moustiques: *Bacillus thuringiensis* var. israelensis. Entomophaga **23**:309–319.

2. **Birnboim, H. C., and J. Doly.** 1979. A rapid alkaline extraction procedure for screening recombinant plasmid DNA. Nucleic Acids Res. **7**:1513–1523.

3. **Bulla, L. A., Jr., K. J. Kramer, and L. T. Davidson.** 1977. Characterization of the entomocidal parasporal crystal of *Bacillus thuringiensis*. J. Bacteriol. **130**:373–383.

4. **Chestukhina, G. G., I. A. Zalunin, L. I. Kostiwa, T. S. Kotova, S. P. Kattuka, and V. M. Stepanov.** 1980. Crystal forming proteins of *Bacillus thuringiensis*. The

limited proteolysis by endogenous proteinases as a cause of their apparent multiplicity. Biochem. J. **187:** 457–465.

5. **Clewell, B., and D. R. Helinski.** 1969. Supercoiled circular DNA protein complex in *Escherichia coli*: purification and induced conversion to an open circular DNA form. Proc. Natl. Acad. Sci. U.S.A. **62:**1159–1166.

6. **Colman, A., M. Byers, B. Primrose, and A. Lyons.** 1978. Rapid purification of plasmid DNAs by hydroxyapatite chromatography. Eur. J. Biochem. **91:**303–310.

7. **Faust, R. M., J. Spizizen, V. Gage, and R. S. Travers.** 1979. Extrachromosomal DNA in *Bacillus thuringiensis* var. *kurstaki*, var. *Finitimus,* var. Sotto and in *Bacillus popilliae*. J. Invertebr. Pathol. **33:**233–238.

8. **Glatron, M.-F., M.-M. Lecadet, and R. Dedonder.** 1972. Structure of the parasporal inclusion of *Bacillus thuringiensis* Berliner; characterization of a repetitive subunit. Eur. J. Biochem. **30:**330–338.

9. **Glatron, M.-F., and G. Rapoport.** 1972. Biosynthesis of the parasporal inclusion of *Bacillus thuringiensis*: half-life of its corresponding messenger RNA. Biochimie **54:** 1291–1301.

10. **Gonzales, J. M. J., and B. C. Carlton.** 1980. Patterns of plasmid DNA in crystalliferous and acrystalliferous strains of *Bacillus thuringiensis*. Plasmid **3:**92–98.

11. **Lang, D.** 1970. Molecular weights of coliphages and coliphage DNA. Contour length and molecular weight of DNA from bacteriophages T4, T5 and T7 and from bovine papillonna virus. J. Mol. Biol. **54:**557–565.

12. **Lecadet, M.-M., M.-O. Blondel, and J. Ribier.** 1980. Generalized transduction in *Bacillus thuringiensis* var. Berliner 1715. J. Gen. Microbiol. **121:**203–212.

13. **Martin, P. A. W., and D. H. Dean.** 1980. Genetic manipulation in the insect pathogen *Bacillus thuringiensis*, p. 155–161. *In* C. Stuttard and K. R. Rozee (ed.), Plasmids and transposons.

14. **Perlak, F. J., C. L. Mendelsohn, and C. B. Thorne.** 1979. Converting bacteriophage for sporulation and crystal formation in *Bacillus thuringiensis*. J. Bacteriol. **140:**699–706.

15. **Ribier, J., and M.-M. Lecadet.** 1973. Etude ultrastructurale et cinétique de la sporulation de *Bacillus thuringiensis* var. Berliner 1715. Remarques sur la formation de l'inclusion parasporale. Ann. Microbiol. (Paris) **124A:**311–344.

16. **Rigby, P. W. J., M. Dieckmann, C. Rhodes, and P. Berg.** 1977. Labeling deoxyribonucleic acid to high specific activity *in-vitro* by nick translation with DNA-polymerase I. J. Mol. Biol. **113:**237–251.

17. **Southern, E. M.** 1975. Detection of specific sequences among DNA fragments separated by gel electrophoresis. J. Mol. Biol. **98:**503–517.

18. **Stahly, D. P., D. W. Dingman, L. A. Bulla, Jr., and A. I. Aronson.** 1978. Possible origin and function of the parasporal crystals in *Bacillus thuringiensis*. Biochem. Biophys. Res. Commun. **84:**581–588.

19. **Thorne, C. B.** 1978. Transduction in *Bacillus thuringiensis*. Appl. Environ. Microbiol. **38:**1109–1115.

Use of Protoplast Fusion for Complementation of Sporulation Mutations in *Bacillus subtilis*

BRIAN N. DANCER AND JOEL MANDELSTAM

Microbiology Unit, Department of Biochemistry, University of Oxford, Oxford OX1 3QU, United Kingdom

Sporulation mutants (Spo⁻ and Osp) blocked at stages III, IV, and V were transferred to resuspension medium and converted to protoplasts with lysozyme-sucrose at t_3 (3 h after transfer). They were mixed with wild-type protoplasts, also prepared at t_3, and fused by use of polyethylene glycol. Incubation was continued to t_{10} to allow spore completion. The suspensions were heated at 80°C for 40 min and plated with counterselection against wild-type spores. Complementation occurred with mutations lying in 9 of 18 loci tested. The increase in colony-forming units was >1,000-fold, and the colonies had the same phenotype as the Spo⁻ or Osp parent. Eight of the complementable mutants were stage IV and one was stage V. Stage III mutants were not complemented. Complementation analysis of three alleles in the *spoIVC* locus shows that there are at least two cistrons in the locus.

Mutations in many of the genes that control the sequence of events in the sporulation of *Bacillus subtilis* have been mapped and assigned to about 30 sporulation loci presumed to function as operons (8, 10). The total number of loci controlling sporulation has been statistically estimated at between 42 and 59 (6). However, nothing definite is known, so far, about numbers of genes within loci. What information there is comes from two lines of investigation. Fine-structure mapping data indicate that recombination indices between sporulation markers in the same locus may be as high as about 0.5 (10). Since a recombination index of 0.2 would cover a gene for a protein with a molecular weight of about 30,000 (2), it seems likely that these loci probably contain more than one gene but are not very large. Moreover, mutations in the same locus sometimes lead to slightly different phenotypes, again suggesting that different functional units are being mutated. Cloning studies provide a second possible line of investigation. When the technology has developed further and the cloned sporulation loci are sequenced and their products are identified, much more information can be expected.

Until now, studies of sporulation loci have been hampered by the lack of a reliable complementation system with which to study sporulation mutations. A method for obtaining merodiploids as tandem duplications in *B. subtilis* has been described (1). However, the method is technically difficult and limited to mutations in only a segment of the chromosome. Also, the results, at least in the case of the *spo0A* locus, have been difficult to interpret unambiguously (14).

We have used a different approach for obtaining complementation of *spo* loci. It has been known for some time that protoplasts of *B. subtilis* will complete sporulation if they contain fully engulfed prespores (4). Hence, we considered that if protoplasts of sporulation mutants were fused with sporulating protoplasts of the wild-type strain it might be expected that in some cases the spores of the mutant could be "rescued" by the sporogenic strain. In practice, fusion of *B. subtilis* protoplasts can be achieved by treatment with polyethylene glycol (PEG) (12). When the cells are sporulating, the mother cells fuse but the forespores do not (5). Rescue of the Spo⁻ strain can be expected only if the mutation is expressed in the mother cell and is recessive to the wild type. It has previously been shown that some *spo* mutations seem to be expressed in the mother cell whereas others are apparently expressed solely in the spore (7, 9).

The experiments reported here show that *B. subtilis* protoplasts can be fused during sporulation and that some mutants are complemented by the wild-type strain. Topological considerations (see Discussion) limit the method to *spo* mutants blocked at stage III, IV, or V of the seven stages of sporulation (11). Complementation relationships between pairs of spo strains mutated in the same and in different loci have been studied. Preliminary experiments indicate that in one spo locus, spoIVC, there is more than one cistron.

RESULTS

Completion of sporulation in protoplasts of wild-type sporulating cells after fusion.
When portions of a culture of the wild-type strain were subjected to a protoplast fusion procedure at intervals during sporulation, there was an optimal fusion time that allowed formation of complete spores (Table 1). The yield of spores increased to a peak in samples subjected to the protoplast induction and PEG treatment at t_3 and then declined. By t_3, less than 5% of the cells had developed forespores discernible in a phase-contrast microscope. These results are in agreement with those of Fitz-James (4) in showing that only protoplasts containing spores at stage III or later can complete sporulation. We have no explanation for the drop in spore yields after t_3, but empirically the experiment indicates that PEG treatment of protoplasts at t_3 does not greatly affect the ultimate yield of spores. After

TABLE 1. *Spore yields from protoplasts of sporulating wild-type cells treated with PEG*[a]

Time of protoplast formation	Heat-resistant spores at t_{10} (colony-forming units/ml)
t_0	0 ($<10^1$)
t_1	10^3
t_2	7×10^5
t_3	6×10^7
t_4	0

[a] Cells of strain 168 (*trpC2 spo*+) were grown in hydrolyzed casein medium to a density of about 0.25 mg (dry weight equivalent) per ml and then transferred to a sporulation medium containing glutamate, inorganic salts, and the auxotrophic requirement (13). At the time of resuspension (t_0) and at hourly intervals up to 4 h (t_1 to t_4), 60-ml portions of the culture were harvested by centrifugation at room temperature, resuspended in 6 ml of sporulation medium containing 20% (wt/vol) sucrose and 2 mg of lysozyme per ml, and incubated at 42°C. When protoplast formation was complete (ca. 10 min), the preparations were centrifuged at room temperature for 5 min at 5,000 rpm. The pellets were taken up in 1 ml of sucrose medium containing in addition 40% (wt/vol) PEG. After 1 to 2 min, the PEG was diluted with 5 ml of sucrose medium, and aeration at 37°C was continued. At t_{10} the preparations were heated at 80°C for 40 min, diluted, and plated on minimal agar (3). The control culture (not subjected to lysozyme and PEG treatment) had 4×10^7 heat-resistant colony-forming units per ml at t_{10}.

t_4, an increasing number of sporangia already contained heat-resistant spores by the time the protoplast fusion was carried out (data not shown).

Fusion of Spo⁻ mutants with Spo⁺ strains. The spo mutations causing blocks at stages III, IV, and V of sporulation lie in 18 distinct loci (10, 15). When representative strains of each of these loci were fused with a sporogenic strain, heat-resistant spores giving rise to asporogenous colonies could be recovered from some of the fusions (Table 2). The Spo⁻ strains that could be rescued by this procedure carry mutations in the spoIIIC, spoIVA, spoIVB, spoIVC, spoIVD, spoIVE, spoIVF, spoIVG, and spoVF loci. In all of these, fusion with a sporogenic strain led to an increase in recovered spores of the Spo⁻ strain of 500-fold and usually more than 1,000-fold. The Spo⁻ strains that were not rescued carry mutations in the following loci: spoIIIA, spoIIIB, spoIIID, spoVA, spoVB, spoVC, spoVD, and spoVE. One strain (36, spoIIIE) may have been weakly complemented.

The yield of spores in these experiments was little affected by whether strain 168 or strain MY2016 was the Spo⁺ donor.

These experiments suggested that rescue of the asporogenous strains was due to complementation by the wild type. To establish this conclusion firmly, we had to show (i) that the colony-forming units obtained as a result of the fusion process were spores, (ii) that colonies recovered at the end of the experiment were genetically identical with the asporogenous parents, and (iii) that fusion was necessary, i.e., the rescue was not simply the result of cross-feeding.

(i) Resistance properties of spores obtained by rescue in fused protoplasts. We considered the possibility that in the fusion experiments vegetative cells or prespores of the Spo⁻ strains were less susceptible to heat killing than they would have been in less dense suspensions or in the absence of PEG. This is almost certainly ruled out by the fact that when any Spo⁻ strain was fused with itself there were no survivors except in the case of the oligosporogenous mutants 5 and 92.1 (Table 2). However, to be absolutely certain, we constructed heat-killing curves for strain 133.2 rescued by fusion with the wild type. There was rapid killing in the first 10 min at both 65 and 80°C, and the viable count fell by about half (Fig. 1). Continuation of heating up to 60 min produced no further drop in colony-forming units. The survivors of heat treatment after 60 min were then homogenized with 0.05% toluene for 60 s. This treatment, which would have killed any surviving vegetative cells, produced no reduction in

TABLE 2. *Fusion of Spo⁻ mutants with Spo⁺ strains*[a]

Spo⁻ strain	spo allele isolation no.[b]	Genotype	Spo⁺ strain	Selection on minimal medium supplemented[c] with:	Spo colonies (colony-forming units/ml)		Complementation and spore count (b/a) ratio
					a. *spo* strain fused alone	b. After fusion with Spo⁺	
59.2	NG14.7	*metC3 spoIIIA59*	168[d]	Methionine	<5	<10	− (/)
1S38	94U	*trpC2 spoIIIC94*[e]	MY2016[d]	Tryptophan	<5	2×10^4	+ (>4×10^3)
67	NG17.23	*trpC2 spoIVA67*	MY2016	Tryptophan	<5	4×10^4	+ (>8×10^3)
165.1	P7	*spoIVB165*	168	—	<5	1.5×10^4	+ (>3×10^3)
133.1	Z7	*trpC2 spoIVC133*	MY2016	Tryptophan	<5	4×10^4	+ (>8×10^3)
92.1	92	*metC3 spoIVD92*	168	Methionine	15	4.5×10^4	+ (>3×10^3)
1S40	11T	*trpC2 spoIVE11*	MY2016	Tryptophan	<5	6.5×10^4	+ (>1.3×10^4)
88	88	*trpC2 metC3 spoIVF88*	MY2016	Tryptophan, methionine	<5	5×10^3	+ (>1×10^3)
5	A8	*trpC2 spoIVG5*	MY2016	Tryptophan	25	3.4×10^4	+ (1.4×10^3)
156.1	W10	*phe-1 spoVD156*	168	Phenylalanine	<5	<10	− (/)
91	91	*trpC2 spoVB91*	MY2016	Tryptophan	<5	<10	− (/)
224	DG47	*trpC2 spoVF224*	MY2016	Tryptophan	<5	1.2×10^5	+ (>2.4×10^4)

[a] Cultures (90 ml) were grown and allowed to sporulate as described in the footnote to Table 1, except that cells were converted to protoplasts only at t_3. Protoplasts of a Spo⁻ strain were mixed with Spo⁺ protoplasts prepared in parallel (3 ml of each). The mixture and the two parent protoplast preparations were then subjected to the PEG treatment. At t_{10}, the suspensions were heated at 80°C for 40 min and plated with counterselection against the Spo⁺ strain. The data for strains carrying mutations in *spoIIIB*, *spoIIID*, *spoVA*, *spoVB*, and *spoVC* were very similar to those for strains 59.2, 91, and 156.1; i.e., they were not rescued by Spo⁺ strains.

[b] Isolation numbers are those given by Piggot and Coote (10).

[c] L-Tryptophan was added at 20 µg/ml; other amino acids were added at 200 µg/ml.

[d] Strain 168 is *trp spo⁺*; strain MY2016 is *lys his spo⁺*.

[e] *spoIIIC94* should probably be classified as a stage IV mutation (10).

viable counts. It therefore appeared that the plateau level reached after heating at 80°C for 10 min represented the survival of heat-resistant spores and that the standard heating time of 40 min (see Table 1) was ample.

It is, however, worth noting that spores formed in fusion experiments were slowly killed by prolonged treatment at 85°C (Fig. 1), possibly as a result of slower development of heat resistance under these experimental conditions.

(ii) Genetic identification of rescued colonies. Asporogenous colonies rescued by fusion of strain 133.2 (*spo-133 phe*) with the wild-type strain were tested for their genotype. The cells were transformed to phenylalanine independence by a method that ensures congression (8), i.e., the transfer of a second, unlinked marker in addition to the marker selected for. First, the cells were transformed with DNA from strain 133.1 (*spo-133 trpC2*), carrying the identical *spo* mutation. Phe⁺ recombinant colonies were obtained, but all of the 18,600 examined were Spo⁻; i.e., there was no recombination between the *spo* mutations. Second, rescued colonies of strain 133.2 were crossed with strain 23.1 (*spoIVC23 trpC2*), which carries a linked *spo* mutation, and with strain 36.1 (*spoIIIE36 trpC2*), which carries an unlinked *spo* mutation. In each cross more

than 2,000 *phe⁺* transformants were examined. The recombination index between *spo-133* and *spo-23* (0.199) was the same as when the post-fusion isolates were used in place of strain 133.2 (0.205).

(iii) Necessity for fusion. Protoplasts of strain 133.2 and of the wild-type strain, both prepared at t_3, were separately treated with PEG and mixed only after the PEG had been diluted out: no rescue occurred. However, if the protoplasts were mixed before addition of PEG, there was good recovery of spores (Table 2).

Complementation between *spo* loci. Since the results described so far were apparently due to complementation, we considered it likely that strains complemented by the wild-type strain would also be complemented by other Spo⁻ mutants. Conversely, Spo⁻ mutants not complemented by the wild-type strain should also be not complemented by other Spo⁻ strains. To test this, we carried out fusions between pairs of Spo⁻ mutants (Table 3). We found that *spoIVC* and *spoIVG* strains (both complementable) reciprocally complemented one another, whereas *spoV* mutations (not complementable) complemented *spoIVC* but were not themselves complemented. Fusions of strains carrying two non-

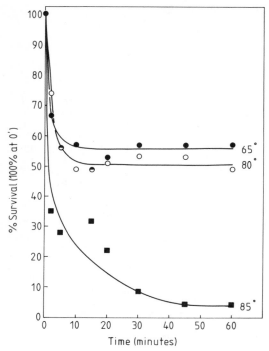

FIG. 1. *Heat-killing curves for spores formed by an asporogenous mutant when fused with the wild-type strain. Cells of strain 133.2 (phe-1 spoIVC) and the wild type were fused as described in the footnote to Table 1. Portions of the mixed culture at t_{10} were heated at 65, 80, or 85°C. Samples were taken at intervals and plated on minimal medium with counterselection against the wild-type strain. All colonies obtained were asporogenous and the counts (colony-forming units per milliliter) at 0 min were 3.5×10^4.*

complementable *spoV* mutations resulted in no colonies of either parental type.

Complementation analysis of mutations in the same *spo* locus. A further prediction of the results shown in Table 2 is that complementable mutations in the same *spo* locus should complement one another unless they are in the same cistron. On this basis the fusion procedure can be used for complementation analysis.

The *spoIVC* locus contains three mapped alleles that have been ordered as follows: 133 ——— 149 ——— 23 (3). When pairs of strains carrying these alleles were fused, there was reciprocal complementation between *spo-133* and *spo-23* and between *spo-149* and *spo-23* (Table 3). However, no colonies of either parental type were recovered from fusions containing *spo-133* and *spo-149*. This suggests strongly that *spo-133*

and *spo-149* are located in one cistron and *spo-23* is located in another.

DISCUSSION

The results described here indicate that when protoplasts of sporulating cells are fused with those of a Spo⁻ mutant it is possible, in some cases, to rescue the Spo⁻ strain. The method is limited for topological reasons to mutants blocked at stages III, IV, and V of sporulation, and it is most successful if applied at t_3, i.e., after many of the prespores have been engulfed. The survivors of the fusion are resistant to extended heating at 80°C followed by toluene treatment, so it is exceedingly unlikely that they are persistent vegetative cells or prespores. Even so, it could still be maintained that they were derived from recombination rather than complementation. Thus, if on fusion the wild-type chromosome recombines with the mutant chromosome, there would be a genetically wild-type mother cell containing a prespore with a Spo⁻ genotype. The wild-type chromosome could then correct the Spo⁻ defect and allow complementation of sporulation with concomitant loss of the mother cell and its wild-type chromosome. Recombination of this type can occur if asporogenous mutants are transformed with wild-type DNA at t_0 (7, 9). This interpretation is unlikely in view of the experiments with strains containing mutations in the same *spo* locus. Thus, *spoIVC133* and *spoIVC23* are roughly equidistant from *spoIVC149* but on opposite sides (3). If recombination had been responsible for the observed degree of sporulation, the numbers of spores recovered in fusion crosses between *spo-133* and *spo-149* and between *spo-23* and *spo-149* should have been similar. In fact the first cross yielded no spores, and the second cross yielded both parental types (Table 3). It is thus very unlikely that the observed spores derive from recombination rather than complementation.

From Table 2 it can be seen that nine of the loci examined showed complementation. Most of these are concerned with stage IV of sporulation. The notable exception is *spoVF224*, with which a high degree of recovery was obtained. This mutation is phenotypically cured by exogenous dipicolinic acid. Presumably, the block in the synthesis of this compound is overcome in fused protoplasts where the wild-type strain would provide the necessary enzyme.

We have no explanation for the lack of complementation of other *spoV* strains or of the *spoIII* strains. It is possible, but unlikely, that these mutations are all either dominant or expressed in the spore compartment. It follows

TABLE 3. *Complementation between* spo *strains*[a]

A		B		Spore formation (colony-forming units/ml), Spo⁻ colonies	
Strain	Genotype	Strain	Genotype	A selected for	B selected for
133.2	*phe spoIVC* (C)[b]	156.2	*trpC2 spoVD* (NC)[b]	5.6×10^3	0 (<5)
133.2	*phe spoIVC* (C)	91	*trpC2 spoVB* (NC)	5.3×10^3	0 (<5)
149.1	*phe spoIVC* (C)	156.2	*trpC2 spoVD* (NC)	8.0×10^3	0 (<5)
149.1	*phe spoIVC* (C)	91	*trpC2 spoVB* (NC)	2.5×10^3	0 (<5)
133.2	*phe spoIVC* (C)	5	*trpC2 spoIVG* (C)	1.1×10^4	3.0×10^3
156.1	*phe spoVD* (NC)	5	*trpC2 spoIVG* (C)	0 (<5)	4.0×10^4
156.1	*phe spoVD* (NC)	91	*trpC2 spoVB* (NC)	0 (<5)	0 (<5)
133.2	*phe spoIVC133* (C)	23.1	*trpC2 rif spoIVC23* (C)	1.3×10^3	5.6×10^3
149.1	*phe spoIVC149* (C)	23.1	*trpC2 rif spoIVC23* (C)	1.5×10^3	5.2×10^3
149.1	*phe spoIVC149* (C)	133.1	*trpC2 spoIVC133* (C)	0 (<10)	0 (<10)

[a] Pairs of strains carrying different *spo* mutations were fused together with PEG as described in the footnote to Table 1. After the heat treatment at t_{10}, the survivors were plated with separate selection for each parent strain. No survivors were found if any of the strains were fused with themselves.

[b] C or NC indicates that the strain was complementable or not complementable by fusion with wild type (Table 2).

that some alteration in experimental procedure might allow some of them to be complemented. However, even without this, it should now be possible to carry out a detailed complementation analysis of the nine sporulation loci that are complementable. Such an analysis should establish the numbers of cistrons in each locus. We have already shown that the *spoIVC* region contains at least two cistrons, but the proper analysis of this region requires the testing of many more alleles.

ACKNOWLEDGMENTS

This work was supported by grants from the Agricultural Research Council and the Science Research Council.

LITERATURE CITED

1. **Audit, C., and C. Anagnostopoulos.** 1973. Genetic studies relating to the production of transformed clones diploid in the tryptophan region of the *Bacillus subtilis* genome. J. Bacteriol. **114**:18–27.
2. **Carlton, B. C.** 1966. Fine-structure mapping by transformation in the tryptophan region of *Bacillus subtilis.* J. Bacteriol. **91**:1795–1803.
3. **Coote, J. G.** 1972. Sporulation in *Bacillus subtilis.* Genetic analysis of oligosporogenous mutants. J. Gen. Microbiol. **71**:17–27.
4. **Fitz-James, P. C.** 1964. Sporulation in protoplasts and its dependence on prior forespore development. J. Bacteriol. **87**:667–675.
5. **Frehel, C., A.-M. Leheritier, C. Sanchez-Rivas, and P. Schaeffer.** 1979. Electron microscopic study of *Bacillus subtilis* protoplast fusion. J. Bacteriol. **137**:1354–1361.
6. **Hranueli, D., P. J. Piggot, and J. Mandelstam.** 1974. Statistical estimate of the total number of operons specific for *Bacillus subtilis* sporulation. J. Bacteriol. **119**:684–690.
7. **Lencastre, H. de, and P. J. Piggot.** 1979. Identification of different sites of expression for *spo* loci by transformation of *Bacillus subtilis.* J. Gen. Microbiol. **114**:377–389.
8. **Piggot, P. J.** 1973. Mapping of asporogenous mutations of *Bacillus subtilis*: a minimum estimate of the number of sporulation operons. J. Bacteriol. **114**:1241–1253.
9. **Piggot, P. J.** 1978. Organization of *spo* locus expression during sporulation of *Bacillus subtilis*: evidence for different loci being expressed in the mother cell and in the forespore, p. 122–126. *In* G. Chambliss and J. C. Vary (ed.), Spores VII. American Society for Microbiology, Washington, D.C.
10. **Piggot, P. J., and J. G. Coote.** 1976. Genetic aspects of bacterial endospore formation. Bacteriol. Rev. **40**:908–962.
11. **Ryter, A.** 1965. Étude morphologique de la sporulation de *Bacillus subtilis*. Ann. Inst. Pasteur Paris **108**:40–60.
12. **Schaeffer, P., B. Cami, and R. D. Hotchkiss.** 1976. Fusion of bacterial protoplasts. Proc. Natl. Acad. Sci. U.S.A. **73**:2151–2155.
13. **Sterlini, J. M., and J. Mandelstam.** 1969. Commitment to sporulation in *Bacillus subtilis* and its relationship to development of actinomycin resistance. Biochem. J. **113**:29–37.
14. **Trowsdale, J., S. M. H. Chen, and J. A. Hoch.** 1978. Evidence that *spo0A* mutations are recessive in *spo0A⁻/spo0A⁺* merodiploid strains of *Bacillus subtilis.* J. Bacteriol. **135**:99–113.
15. **Young, M., and J. Mandelstam.** 1980. Early events during bacterial endospore formation. Adv. Microb. Physiol. **20**:103–162.

Defective Sporulation of a Spore Germination Mutant of *Bacillus subtilis* 168

R. J. WARBURG

Genetics Department, Birmingham University, Birmingham B15 2TT, United Kingdom

The sporulation properties of a spore germination mutant containing the *gerJ50* mutation were studied. The mutant was delayed in the acquisition of resistances to both chemicals and heat during sporulation, but no morphological defect was detected in thin sections. The levels of total extracellular protease, intracellular serine protease, succinate dehydrogenase, and alkaline phosphatase activities were normal, as was that of dipicolinic acid.

The very low level of metabolism of a spore (13) makes it likely that the processes occurring during its germination are programmed during sporulation or the stages preceding sporulation (6). This means that the determination of germination characteristics of spores can be regarded as an integral part of sporulation and that any defects occurring during sporulation may well affect germination of spores. This view accords with the properties of strains containing the *gerE36* mutation, which form coat-defective spores and are blocked in germination (15a, 16) and also by the occurrence of dipicolinic acid-requiring mutants of several *Bacillus* spp. which do not germinate in normal germinants (25). Spores of strains containing the *gerJ50* mutation differ from those of the wild type in that they only partially lose absorbance and release hexosamine-containing fragments during germination, whereas their release of dipicolinic acid and loss of resistance properties are normal (R. J. Warburg and A. Moir, J. Gen. Microbiol., in press). That these *gerJ50* spores are also sensitive to heating at 90°C, with this resistance being attained only late in the process, may well point to a defect in sporulation (14). This hypothesis was tested with two isogenic strains, one of which contained the *gerJ50* mutation.

RESULTS AND DISCUSSION

The two isogenic strains, 9014 (*trpC2*) and 9015 (*trpC2 gerJ50*), were derived from a phage SPP1 transduction cross using an *aroB2 trpC2* recipient and a *trpC2 gerJ50* donor, selecting for Aro+ (Warburg and Moir, in press).

During sporulation of strain 9015, the following properties (data not presented) were similar to those of 9014: (i) increase in the number of phase-bright spores with time; (ii) timing of the various stages that could be observed in thin sections of spores (19, 21), such as septum formation, forespore engulfment, and the presence of cortical and coat material, and the proportion of spores at each stage during sporulation (coat structure in thin sections and freeze fractures was also normal [Warburg and Moir, in press]); and (iii) levels of intracellular serine protease activity (5), total extracellular protease activity, alkaline phosphatase activity, and dipicolinic acid (11) during sporulation. Thus, 9015 differs from previously isolated mutants of *B. subtilis* which are defective in germination and either will not respond to germinants (16) or have a defective spore coat and form lysozyme-sensitive spores (15a, 16) and from similar mutants of *B. cereus* (3, 4, 22, 23) and *Clostridium perfringens* (18). Some of these mutants also had a defective intracellular protease (3), whereas the activity of this enzyme is normal in 9015. The level of dipicolinic acid in strain 9015 was also normal, unlike that in the mutants of several *Bacillus* spp. (25). Finally, the level of succinate dehydrogenase activity during sporulation was also measured because *gerJ50* is located close to a mutation, *cit-74*, which affects the level of this enzyme (17); this too was normal (data not shown).

Strain 9015 differed from 9014 in that the onset of resistances to chemicals and heating during sporulation was delayed (Fig. 1). This delay was approximately 60 min for resistances to chloroform or heat and 90 min for toluene, xylene, and octanol. The resistances to the latter

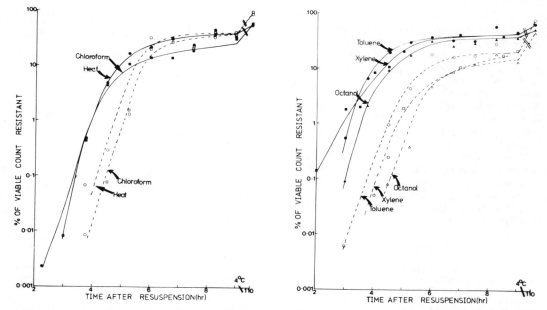

FIG. 1. *Acquisition of resistances to chemicals and heat (70°C for 30 min). Fresh single colonies from nutrient agar were inoculated into the growth medium of Sterlini and Mandelstam (24), shaken for 15 h at 26°C, transferred to 42°C for 2 h, and then suspended in minimal glutamate medium. After 9 h, the cultures were transferred to 4°C. Samples (0.1 ml) were taken and diluted into 10 ml of minimal salts (1); 2-ml portions of these dilutions were treated with 0.25 ml of the appropriate chemical, and blended on a Vortex mixer. After separation of the two layers, dilutions were plated in soft nutrient agar (1.1%, wt/vol, with NaCl, 0.5%) onto Difco nutrient agar (2.8%). Colonies were counted after incubation at 37°C for 15 h. An untreated control was also included to monitor the viable count. Open symbols and broken lines for the gerJ50 mutant; closed symbols and solid lines for the wild type.*

chemicals did not reach wild-type levels until the cultures had remained at 4°C for 5 days, at which time cell lysis was complete. Such a delay in the acquisition of resistance properties during sporulation has also been observed in strains containing *sco* mutations (2, 15). Some *sco* strains differ from 9015 in that they are also delayed in the progress of sporulation, remaining at stage III or V for longer than the wild type (20). At some stage in sporulation, all *sco* mutants had elevated levels of extracellular protease activity (7), and some also had elevated levels of alkaline phosphatase activity (12). The levels of these activities were normal in 9015 (data not presented).

The results obtained demonstrate that the *gerJ50* mutation results in sporulation abnormalities. It has also been found to affect the release of hexosamine-containing fragments during germination and to cause the production of heat-sensitive spores (Warburg and Moir, in press). All of these properties are likely to derive from a single pleiotropic mutation, since their simultaneous acquisition by recombinants from both transformation and transduction (by

phages PBS1 and SPP1) crosses using a series of genetic markers has been clearly demonstrated (Warburg and Moir, in press).

It has been shown that resistances of wild-type spores to organic chemicals and heat during sporulation are not attained until stage III or later (14) and that effects upon forespore ultrastructure of both chemicals and heat are lessened after this time (19). The appearance of cortex material was correlated with resistance to xylene, as was cortical maturation or coat development, or both, with resistance to toluene, octanol, and chloroform. Heat resistance did not appear until after the coat was fully formed. Both chemical and heat resistances are dependent upon the presence of cortex (9, 10), the amount required for xylene or octanol resistance and for heat resistance (80°C for 10 min) being 25% and 90% of the maximum, respectively. Thus, one explanation for the delay in the acquisition of sporulation-related resistances by strain 9015 could be that either the cortex or its development, or both, is defective. If its peptidoglycan structure was also altered, then the other properties of the mutant can be accounted for. The heat sensitivity of the spores could result from the cortex not being able to dehydrate the core to the same extent as in the wild type (8), and the incomplete release of peptidoglycan fragments during germination could be explained by the inability of the normal lytic enzymes to break down the defective cortex. That only part of the peptidoglycan is released may reflect the existence of at least two pools of peptidoglycan that are involved in normal spore germination.

The validity of these deductions could be tested by following the deposition of cortical fragments during sporulation, e.g., by measuring the muramic lactam and by testing the sensitivity of the mutant cortical peptidoglycan to wild-type lytic enzymes.

ACKNOWLEDGMENTS

This work was supported by a Research Studentship from the Medical Research Council.

I am extremely grateful to Anne Moir, who has been a constant source of information, suggestions, and valuable comments. I am also indebted to both her and D. A. Smith for their careful reading of and comments on the manuscript. My thanks go to G. Gould, G. Dring, D. Walton, and S. Taylor for valuable discussions, as well as to S. Haley, J. Yeomans, and C. Davies for expert technical assistance and to M. Johnson for skillful typing.

LITERATURE CITED

1. **Anagnostopoulos, C., and J. Spizizen.** 1961. Requirements for transformation in *Bacillus subtilis*. J. Bacteriol. **81**:741–746.
2. **Balassa, G., B. Dod, V. Jeannoda, P. Milhaud, J. Zucca, J. C. F. Sousa, and M. T. Silva.** 1978. Pleiotropic control mutations affecting the sporulation of *Bacillus subtilis*. Ann. Microbiol. (Paris) **129B**:537–549.
3. **Cheng, Y. E., and A. I. Aronson.** 1977. Alterations of spore coat processing and protein turnover in a *Bacillus cereus* mutant with a defective postexponential intracellular protease. Proc. Natl. Acad. Sci. U.S.A. **74**:1254–1258.
4. **Cheng, Y. E., P. FitzJames, and A. I. Aronson.** 1978. Characterization of a *Bacillus cereus* protease mutant defective in an early stage of spore germination. J. Bacteriol. **133**:336–344.
5. **Dancer, B. N., and J. Mandelstam.** 1975. Production and possible function of serine protease during sporulation of *Bacillus subtilis*. J. Bacteriol. **121**:406–410.
6. **Dion, P., and J. Mandelstam.** 1980. Germination properties as marker events characterizing later stages of *Bacillus subtilis* spore formation. J. Bacteriol. **141**:786–792.
7. **Dod, B., G. Balassa, E. Raulet, and V. Jeannoda.** 1978. Spore control (Sco) mutations in *Bacillus subtilis*. II. Sporulation and the production of extracellular proteases and α-amylase by Sco mutants. Mol. Gen. Genet. **163**:45–56.
8. **Gould, G. W., and G. J. Dring.** 1975. Role of an ex-

panded cortex in resistance of bacterial endospores, p. 541–546. *In* P. Gerhardt, R. N. Costilow, and H. L. Sadoff (ed.), Spores VI. American Society for Microbiology, Washington, D.C.
9. **Imae, Y., and J. L. Strominger.** 1976. Relationship between cortex content and properties of *Bacillus sphaericus* spores. J. Bacteriol. **126**:907–913.
10. **Imae, Y., and J. L. Strominger.** 1976. Cortex content of asporogenous mutants of *Bacillus subtilis*. J. Bacteriol. **126**:914–918.
11. **Janssen, F. W., A. J. Lund, and L. E. Anderson.** 1958. Colorimetric assay for dipicolinic acid in bacterial spores. Science **127**:26–27.
12. **Jeannoda, V., and G. Balassa.** 1978. Spore control (Sco) mutations in *Bacillus subtilis*. IV. Synthesis of alkaline phosphatase during sporulation of Sco mutants. Mol. Gen. Genet. **163**:63–73.
13. **Lewis, J. C.** 1969. Dormancy, p. 301–358. *In* G. W. Gould and A. Hurst (ed.), The bacterial spore. Academic Press, Inc., New York.
14. **Milhaud, P., and G. Balassa.** 1973. Biochemical genetics of bacterial sporulation. IV. Sequential development of resistances to chemicals and physical agents during sporulation of *Bacillus subtilis*. Mol. Gen. Genet. **125**:241–250.
15. **Milhaud, P., G. Balassa, and J. Zucca.** 1978. Spore control (Sco) mutations in *Bacillus subtilis*. I. Selection and genetic mapping of Sco mutants. Mol. Gen. Genet. **163**:35–44.
15a.**Moir, A.** 1981. Germination properties of a spore coat-defective mutant of *Bacillus subtilis*. J. Bacteriol. **146**:1106–1116.
16. **Moir, A., E. Lafferty, and D. A. Smith.** 1979. Genetic analysis of spore germination mutants of *Bacillus subtilis* 168: the correlation of phenotype with map location. J. Gen. Microbiol. **111**:165–180.
17. **Ohne, M., B. Rutberg, and J. A. Hoch.** 1973. Genetic and biochemical characterization of mutants of *Bacillus subtilis* defective in succinate dehydrogenase. J. Bacteriol. **115**:738–745.
18. **Sacks, L. E., and R. S. Thomas** 1979. High yields of coatless spores of *Clostridium perfringens* strain 8-6 in a defined medium. Can. J. Microbiol. **25**:642–645.
19. **Sousa, J. C. F., M. T. Silva, and G. Balassa.** 1978. Ultrastructure effects of the chemical agents and moist heat on *Bacillus subtilis*. II. Effect on sporulating cells. Ann. Microbiol. (Paris) **129B**:377–390.
20. **Sousa, J. C. F., M. T. Silva, and G. Balassa.** 1978. Spore control (Sco) mutations in *Bacillus subtilis*. V. Electron microscope study of delayed morphogenesis. Mol. Gen. Genet. **163**:285–291.
21. **Spurr, A. R.** 1969. A low-viscosity epoxyresin embedding medium for electron microscopy. J. Ultrastruct. Res. **26**:31–43.
22. **Stelma, G. N., Jr., A. I. Aronson, and P. FitzJames.** 1978. Properties of *Bacillus subtilis* temperature-sensitive mutants altered in spore coat formation. J. Bacteriol. **134**:1157–1170.
23. **Stelma, G. N., Jr., A. I. Aronson, and P. C. FitzJames.** 1980. A *Bacillus cereus* mutant defective in spore coat deposition. J. Gen. Microbiol. **116**:173–185.
24. **Sterlini, J. M., and J. Mandelstam.** 1969. Commitment to sporulation in *Bacillus subtilis* and its relationship to actinomycin resistance. Biochem. J. **113**:29–37.
25. **Zytkovicz, T. H., and H. O. Halvorson.** 1971. Some characteristics of dipicolinic acid-less mutant spores of *Bacillus cereus*, *Bacillus megaterium*, and *Bacillus subtilis*, p. 49–52. *In* H. O. Halvorson, R. Hanson, and L. L. Campbell (ed.), Spores V. American Society for Microbiology, Washington, D.C.

Specialized Transduction with *Bacillus subtilis* Bacteriophage SPβ

STANLEY A. ZAHLER AND RUTH Z. KORMAN

Division of Biological Sciences, Cornell University, Ithaca, New York 14853

Phage SPβ of *Bacillus subtilis* gives specialized transduction of bacterial genes close to its prophage. We have produced a collection of bacterial strains carrying the SPβ prophage in various chromosomal locations. Induced phage lysates of these strains can be screened to identify those containing transducing particles for any particular selectable marker. We can then construct heterogenotes with two copies of the selected marker and of nearby genes. Cistron relationships and *cis-trans* interactions of mutations in the duplicated region can be determined.

Spizizen's original transformable strain of *Bacillus subtilis*, strain 168, and most of its derivatives are lysogenic for SPβ, a large temperate phage (6). The SPβ prophage lies in its normal attachment site on the *B. subtilis* chromosome near the terminus of replication between the *ilvA* and *kauA* genes (7) and makes up about 4% of the chromosome. The genome size is about 126 kilobases (83 megadaltons) (P. S. Fink, R. Z. Korman, J. M. Odebralski, and S. A. Zahler, Mol. Gen. Genet., in press). The phage is related to, and freely recombines with, the phages φ3T and ρ11 (1a, 6).

When an SPβ lysogen is induced, among the particles released are a few that contain nearby bacterial genes—the *ilvA* and *kauA* genes, for example, and a few others. Such particles also contain some phage DNA, but usually they do not contain the whole phage genome. When these particles infect a new bacterium, they can transduce it, e.g., from IlvA⁻ to IlvA⁺. Such a particle is called a *defective specialized transducing phage* (TP): *defective* because it lacks some phage genes and therefore cannot make plaques, *transducing* because it can carry genes from one bacterial strain to another, and *specialized* because it can carry only genes close to the prophage from which it originated. The SPβ system is quite similar to specialized transduction in the *Escherichia coli* phage λ (1).

A bacterium that is transduced by a TP frequently becomes diploid for the bacterial DNA carried by the phage. Since two copies of the bacterial DNA are present in the transductant, it is possible to study cistron relationships within the piece of bacterial DNA and to determine

cis- and *trans*-dominant alleles within the region. We estimate that the bacterial DNA in a single TP from SPβ may be as long as 40 kilobases (26 megadaltons).

If we transduce a nonlysogen with a TP, the entire TP genome may integrate into the bacterial DNA in the region of homology.

If we transduce a lysogen with a TP, the TP genome may integrate into the bacterial DNA in the region of homology (the B or *bacterial* configuration), or it may integrate into the resident prophage DNA (the P or *prophage* configuration). In either of these two cases, the result is a *double lysogen* carrying a TP and a whole SPβ prophage. When a culture of such a double lysogen is induced, it releases many TP particles and many SPβ plaque-forming particles, although the ratio is never greater than about 1:100.

EXPERIMENTAL METHODS AND RESULTS

Phage and bacteria. The methods used for growing phages and bacteria are given in reference 3. Methods for curing bacterial strains of the prophage by cotransducing the unoccupied SPβ attachment site with the *trpC* gene (or other linked markers) by phage PBS1 are given in references 3 and 7.

Most of the SPβ phages described here carry the *c2* mutation (3), which permits lysogens to be induced by a brief heat shock followed by incubation at 37°C. The heat shock consists of heating a logarithmically growing culture at 50°C for 5.5 min if a single copy of *c2* is present or for 10 min if two copies of *c2* are present in a double lysogen.

SP*βc2int5* is a mutant of SP*βc2* that is deficient in the enzyme needed for integration into the normal SP*β* attachment site (R. Rosenthal et al., submitted for publication). Fewer than 1 cell in 10^4 infected with SP*βc2int5* becomes a lysogen. The rare lysogens can be isolated on agar plates because they excrete a bacteriocin-like substance (betacin) that kills nonlysogens (1a). Most lysogens of SP*βc2int5* carry the prophage at sites other than the normal attachment site.

B. subtilis strain CU156 has the genotype *trpC2 ilvA3 citD2*. The spontaneous *citD2* mutation is a deletion of most or all of the SP*β* prophage, plus the adjacent bacterial genes *kauA* and *citK*. CU156 and its derivatives that carry *citD2* lack a functional SP*β* attachment site. When infected by SP*βc2*, only a small fraction of the cells become lysogens; their prophages are scattered around the *B. subtilis* chromosome. The lysogens can be isolated on agar plates because of their production of betacin.

Determination of B or P configuration in double lysogens. When a lysogen is transduced by a TP, the TP prophage may integrate into the bacterial chromosome in the region of the selected marker (the B configuration) or into the resident prophage (the P configuration). In studying dominance relationships between alleles of the duplicated bacterial genes, it is important to use double lysogens in which recombination between the heterogenotic regions is minimized; that is, we must use double lysogens in the P configuration.

To determine the configuration, we produce PBS1 transducing phage in the double lysogen. This phage is used to learn whether the selectable marker on the TP is cotransducible with *metB* or *citK*, two genes close to the resident prophage in the normal attachment site. Cotransduction indicates that the double lysogen is in the P configuration.

In the case of TPs carrying *leu* genes, it was also possible to look for cotransduction of *leu* with *phe*, a marker close to *leu* on the *B. subtilis* chromosome. Cotransduction of *leu* and *phe* indicates that the double lysogen is in the B configuration. In general, about equal numbers of double lysogens are in the B and P configurations. In analyzing the controls of the *leu* region, all double lysogens were in the P configuration.

Collection of lysogens. Strain CU156 or one of its *citD2* derivatives was grown in antibiotic medium no. 3 (Difco) to a concentration of about 10^8 cells per ml. One-tenth milliliter of the culture was added to a tube of overlay agar and

poured on a tryptose blood agar base (TBAB) plate containing 0.2% glucose. After the overlay dried, drops of SP*βc2* or SP*βc2int5* containing 10^7 to 10^8 phage per ml were placed on the overlays and allowed to dry, and the plates were incubated for 72 to 96 h at 22 to 24°C. Bacteria from the partly cleared regions of phage lysis were purified on TBAB + 0.2% glucose + anti-SP*β* antiserum and were tested for betacin production. Lysogens (identified by betacin production) were heat induced (3).

A few lysogens in our collection were isolated by infection of sensitive bacteria by SP*βc2int5* in broth, followed by plating on TBAB in the presence of a clear-plaque mutant of SP*βc2int5* (to kill sensitive cells). Rare lysogens were isolated from the TBAB plate after overnight incubation at 37°C.

We now have a collection of more than 200 *B. subtilis* strains carrying the SP*βc2* or SP*βc2int5* prophage at chromosomal locations other than the normal attachment site. We have heat induced these strains to produce phage lysates. When we want to find a prophage close to a selectable gene of interest to us, we screen the lysates, looking for transductants. In this way we can find strains that give TPs for many, if not all, *B. subtilis* genes.

We will briefly describe two chromosome regions we have examined with SP*β*, to indicate some of its possibilities.

***dal-sup* region.** We have isolated a strain of *B. subtilis* that carries SP*βc2int5* between the *dal* and *purB* genes, close to *dal*. The external suppressor genes *sup-3* and *sup-44* lie close to *dal* (2) on the side away from the prophage. We have constructed strains carrying each of these suppressors and the prophage. From them we isolated TPs that carry both the *dal+* gene (which encodes alanine racemase) and the *sup* gene. The TPs were used to construct double lysogens.

Lysates of the double lysogens contain many TPs ($>10^5$/ml) for *sup-3 or sup-44*. We can use these lysates to determine whether any auxotrophic strain is suppressible. A culture of the auxotroph is spread on the surface of a minimal agar plate lacking the needed supplement. A diluted drop of the TP lysate is placed on the spread plate. If the auxotrophic mutation is suppressible, a solid mass of transductant growth appears in the region of the drop after incubation of the plate. The method works whether or not the auxotroph is lysogenic for SP*β* (R. H. Lipsky and S. A. Zahler, unpublished data).

***leu* region.** A detailed study of the *ilvBC-leu* (= "*leu*") region of the *B. subtilis* chromosome is under way. We have found a strain in which

SPβc2 has been inserted into the *citF* gene (which encodes succinate dehydrogenase), close to the *leu* region (C. J. Mackey, R. Z. Korman, and S. A. Zahler, manuscript in preparation). From this prophage we have isolated TPs that carry all of the genes of interest to us. The order of genes is *citF-ilvB-ilvC-leuA-leuC-leuB-leuD* (4).

We have isolated TPs carrying different amounts of bacterial DNA from this region. One carries *ilvB* and ends in *ilvC*. Other individual TPs end in *leuA*, *leuC*, *leuB*, or *leuD*, and one contains all of the genes from *ilvB* through *leuD*. Spot tests for complementation, like the one described for suppressibility, permit us to assign new *ilv* or *leu* mutations to their appropriate cistrons.

By starting with a TP that carries the entire *leu* region, we have by a relatively simple set of manipulations produced double lysogens that contain two copies of, for example, the *leuA5* negative allele. Lysates of such a double lysogen contained TPs that complemented *ilvB*, *ilvC*, *leuB*, *leuC*, and *leuD* auxotrophs, but not *leuA* auxotrophs, in spot tests. In fact, the existence of the *leuD* cistron was discovered when we found leucine auxotrophs that were complemented by lysates from *leuA*, *leuB*, and *leuC* double lysogens (Mackey, Korman, and Zahler, unpublished data).

One can select *B. subtilis* mutants that overproduce leucine and thus are resistant to the leucine analog 4-azaleucine. Such mutants are constitutive for the *ilvBC-leu* enzymes. One such mutation, *azlA102*, lies very close to the left end of *ilvB* (5). We have tested this mutation in heterogenotes that carry one copy of *azlA102* and one copy of *azlA*$^+$. The *azlA102* mutation is *cis* dominant and thus seems likely to be an operator-constitutive mutation, a promoter-up mutation, or an attenuator mutation. Another leucine-overproducing mutation, *azlD1*, which is also closely linked to *ilvBC-leu*, is recessive in this test; it acts as though it is a repressor-constitutive mutation.

CONCLUSION

The prospects seem good that phage SPβ will allow us to carry out in *B. subtilis* the sorts of cistron analyses and control mechanism analyses that have been possible in *E. coli* for two decades. We hope to find new kinds of controls; *E. coli* and *B. subtilis* have probably not had a common ancestor for 10^9 years (10^{11} generations?), which should be sufficient time for the evolution of interesting variations.

The methods that we are developing for the analysis of the *ilvBC-leu* region will also work for genetic regions that control developmental properties such as sporulation and germination, provided that a selectable marker can be found within about 40 kilobases of the genes of interest.

ACKNOWLEDGMENTS

Three excellent technicians (Richard Rosenthal, Patryce A. Toye, and Janice M. Odebralski), four graduate students (Pamela S. Fink, Irene Gage, Catherine J. Mackey, and Robert H. Lipsky), and three undergraduate students (Robert Herenstein, Richard A. Rosenberg, and Cathy Vocke) were responsible for most of the work described here. We are very grateful to them. H. E. Hemphill of Syracuse University has freely exchanged ideas, bacteria, and phages with us; we thank him.

This work was supported by grants from the National Science Foundation.

LITERATURE CITED

1. **Campbell, A.** 1971. Genetic structure, p. 13–44. *In* A. D. Hershey (ed.), The bacteriophage lambda. Cold Spring Harbor Laboratory, Cold Spring Harbor, N.Y.
1a. **Hemphill, H. E., I. Gage, S. A. Zahler, and R. Z. Korman.** 1980. Prophage-mediated production of a bacteriocinlike substance by SPβ lysogens of *Bacillus subtilis*. Can. J. Microbiol. **26:**1328–1333.
2. **Henner, D., and W. Steinberg.** 1979. Genetic location of the *Bacillus subtilis sup-3* suppressor mutation. J. Bacteriol. **139:**668–670.
3. **Rosenthal, R., P. A. Toye, R. Z. Korman, and S. A. Zahler.** 1979. The prophage of SPβcdcitK$_1$, a defective specialized transducing phage of *Bacillus subtilis*. Genetics **92:**721–739.
4. **Ward, J. B., Jr., and S. A. Zahler.** 1973. Genetic studies of leucine biosynthesis in *Bacillus subtilis*. J. Bacteriol. **116:**719–726.
5. **Ward, J. B., Jr., and S. A. Zahler.** 1973. Regulation of leucine biosynthesis in *Bacillus subtilis*. J. Bacteriol. **116:**727–735.
6. **Warner, F. D., G. A. Kitos, M. P. Romano, and H. E. Hemphill.** 1977. Characterization of SPβ: a temperate bacteriophage from *Bacillus subtilis* M168. Can. J. Microbiol. **23:**45–51.
7. **Zahler, S. A., R. Z. Korman, R. Rosenthal, and H. E. Hemphill.** 1977. *Bacillus subtilis* bacteriophage SPβ: localization of the prophage attachment site, and specialized transduction. J. Bacteriol. **129:**556–558.

Cloning of a Sporulation Gene in *Bacillus subtilis*

E. DUBNAU, N. RAMAKRISHNA, K. CABANE, AND I. SMITH

Department of Microbiology, The Public Health Research Institute of the City of New York, Inc., New York, New York 10016

A 0.8-megadalton *Bgl*II restriction fragment of *Bacillus licheniformis* cloned into the *Bgl*II site of the plasmid pBD64 can complement *spo0H* mutations of *B. subtilis*. Using the helper system described by T. J. Gryczan, S. Contente, and D. Dubnau (Mol. Gen. Genet. **177**:459–467, 1980), we isolated the clone by selecting for the Spo$^+$ phenotype and antibiotic resistance. The insert is functional in both orientations and thus presumably has its own promoter.

There is an extensive body of descriptive information on sporulation in *Bacillus subtilis*. About 50 mutations, blocked at various stages of sporulation, have been described and mapped (11). Recently, the cloning of a number of genes which are necessary for normal sporulation (13; also this volume) has expanded tremendously the possibility for studying the regulation of such genes. The requirements for proper transcription of these genes, such as modified RNA polymerases (7), can now be determined, and also the isolation and study of their gene products is now possible.

In this paper we describe the cloning of a segment of the *B. licheniformis* chromosome onto a plasmid vector, pBD64, forming a plasmid which specifically complements *spo0H* mutations of *B. subtilis*.

RESULTS

Cloning of a DNA fragment which complements *spo0H* mutations. The strategy employed was to clone *spo*$^+$ genes from *B. licheniformis* into a *B. subtilis spo* recipient strain. Since there is no detectable transformation for *spo*$^+$ by *B. licheniformis* DNA with *B. subtilis spo* recipients (E. Dubnau and I. Smith, unpublished data), we were able to use the host plasmid rescue system of Gryczan et al. (4). DNA taken up by competent cells of *B. subtilis* is generally nicked or otherwise damaged during uptake (2). When the host cell carries a plasmid that is homologous to the vector plasmid, damaged DNA of a vector carrying cloned DNA can be "rescued" by recombination between the recipient plasmid and the homologous portion of the donor DNA. This greatly increases the probability of obtaining transformants with cloned DNA (4).

The plasmid vector used was pBD64, which confers resistance to kanamycin and chloramphenicol. It has a unique *Bgl*II restriction site. A cloning event in this site inactivates kanamycin resistance (6). (Plasmids pUB110, pBD9, and pBD64 were obtained from D. Dubnau [5]. Restriction, ligation, and transformation techniques were as described by Gryczan et al. [3].)

DNA from *B. licheniformis* (strain FD02) and DNA from pBD64 were ligated together after treatment with restriction endonuclease *Bgl*II. The ligated mixture was used to transform IS1007, a *B. subtilis* strain with the *spo0H75* (9) marker carrying pUB110. The latter is homologous with a portion of the vector pBD64. Recombinant colonies were selected on minimal medium containing chloramphenicol, and Spo$^+$ recombinants were selected by treatment with chloroform after 2 days of incubation (8). Pigmented colonies appeared after 2 days of additional incubation, following the chloroform treatment, at a frequency of 10^{-3} per Cm^r transformant. The yield of Cm^r transformants from the ligated mixture was 5.9×10^4/μg of vector DNA.

Several clones were obtained after single-colony purification which were Spo$^+$, Cm^r, and Kan^s. Plasmid DNA prepared from these recombinants could transform IS171, a *spo0H75 recE4* strain, simultaneously to Cm^r and Spo$^+$, and these transformants were also Kan^s. Plasmid DNA from these clones had a molecular weight of about 4×10^6, whereas the molecular weight of the vector is 3.2×10^6 (Fig. 1). The pBD64 vector containing the insert DNA is called pIS1.

Sporulation phenotype of strains carry-

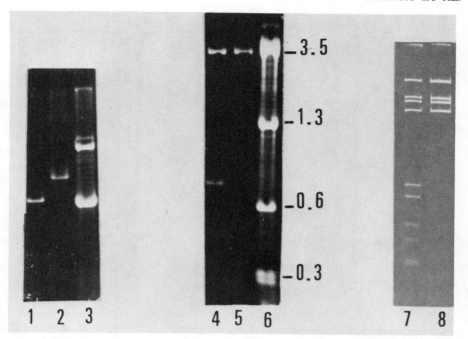

FIG. 1. *Size of pIS1 and the cloned insert. The leftmost panel illustrates the electrophoretic analysis on a 0.8% agarose gel of the covalently closed circular forms of: pBD64 (3.2 megadaltons), lane 1; pIS1 (4.0 megadaltons), lane 2; and pUB110 (3.0 megadaltons), lane 3. In the center panel, BglII-restricted pIS1 DNA (lane 4) and pBD64 DNA (lane 5) were analyzed by means of electrophoresis on a 0.8% agarose gel. HindIII-restricted PM2 DNA (lane 6) was also electrophoresed to provide molecular-weight standards. The values on the right of this panel show the size (in megadaltons) of the four largest PM2 HindIII fragments. The right panel shows the electrophoretic separation of an MboI digest of pIS1 (lane 7) and pBD64 (lane 8) on a 5% acrylamide gel. In all of the gels, the DNA was visualized by staining with ethidium bromide, and photographs were taken as previously described (3).*

ing pIS1. Colonies of *spo0H75* strains carrying pIS1 became brown after 2 days of incubation at 37°C, whereas colonies of the otherwise isogenic *spo0H75* strain carrying pBD64 remained white and eventually lysed. Sporulating cultures of strains carrying pIS1 were resistant to heat (75°C) and to chloroform treatment. Spores formed by such strains looked the same under the phase-contrast microscope as those from IS230, a *spo*$^+$ strain which is otherwise isogenic with the *spo0H* strains used. Spores were present at about the same frequency (50 to 80%) in sporulating cultures of IS230 and *spo0H75* strains carrying pIS1.

Complementation specificity of pIS1. Transductional crosses were performed with the use of AR9 transducing phage (3) grown on strains carrying either pIS1 or pBD64. Strains bearing several *spo0H* markers, including *spo0H12-3* (9), *spo0H46* (10), and *spo0H17* (10), were used as recipients. Cm^r transductants were selected and treated with chloroform for the

Spo$^+$ selection. All *spo0H* recipient strains could be transduced to Spo$^+$ with phage grown on the strain carrying pIS1, but not with phage grown on the strain carrying pBD64, with 100% linkage between Cm^r and Spo$^+$. The pIS1 phage could not complement *spoIIE20* or *spoIIIA53* mutations (10). Other experiments, in which pBD64 or pIS1 DNA was used to transform 19 different *spo* strains that carried mutations from the classes *spo0*, *spoII*, *spoIII*, *spoIV*, and *spoV*, also showed that only *spo0H* mutations could be complemented by pIS1.

Restriction map of pIS1. The molecular weight of the cloned insert was found to be 0.82 × 10^6, as determined by the electrophoretic mobility on agarose gels of the fragment obtained after restriction of pIS1 with *BglII* (Fig. 1).

A restriction map of the insert was constructed by single or multiple digestion of pIS1 DNA with various enzymes. The size of various fragments obtained with double digestion by *BglII* and a series of other restriction endonu-

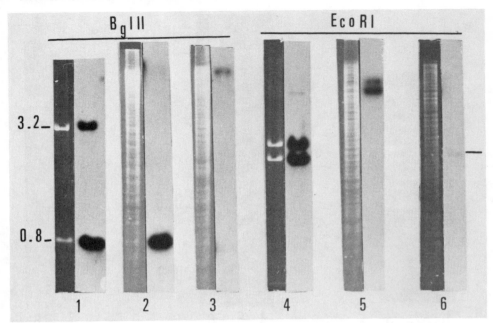

F_IG_. 2. *Chromosomal origin of cloned insert. DNA from pIS1 (lanes 1 and 4), B. licheniformis FD02 (lanes 2 and 5), and B. subtilis 168 (lanes 3 and 6) was restricted with BglII or EcoRI and electrophoresed on 0.8% agarose gels. The DNA was then transferred to diazobenzyloxymethyl papers (1) and hybridized with purified ^{32}P nick-translated cloned insert DNA (12). Each lane shows the ethidium bromide-stained gel on the left and the autoradiogram on the right. Lanes 1 to 3 show the results of BglII digestion, and lanes 4 to 6 are from EcoRI restriction. The larger band shown in the BglII digestions is the linearized pBD64 (3.2 × 10⁶), and the smaller band is the excised insert (0.8 × 10⁶) (lane 1). Contamination of the ^{32}P nick-translated insert by pBD64 DNA accounts for the radioactivity seen at the level of the pBD64 linearized DNA in this hybridization. The line in lane 6 points to an extremely faint radioactive band present in the EcoRI restriction pattern of B. subtilis DNA.*

cleaves (*Mbo*I, *Hae*III, *Hpa*II, *Eco*RI, and *Bcl*I) was determined. From these data, the distance of the restriction site or sites from one of the *Bgl*II cloning sites could be established (Fig. 1, Table 1).

Chromosomal origin of the cloned insert. To study the chromosomal origin of the cloned *B. licheniformis* insert in pIS1, we used DNA: DNA hybridization. pIS1, *B. subtilis*, and *B. licheniformis* DNA were digested with *Bgl*II and *Eco*RI restriction enzymes. The restriction fragments were separated by electrophoresis, transferred to diazobenzyloxymethyl derivatized paper (1), and hybridized to ^{32}P nick-translated (12) insert DNA (Fig. 2). The 0.8-megadalton *Bgl*II fragment in pIS1, which has the *spo$^+$* complementing activity, has a counterpart of the same molecular weight in *Bgl*II-digested *B. licheniformis* DNA, whereas a much fainter band, corresponding to a molecular weight of ~2 × 10⁷, is observed in *Bgl*II-restricted *B. subtilis*. We have observed that *spo0H$^+$* transforming activ-

ity from *B. subtilis Bgl*II-restricted DNA also migrated electrophoretically in this region of 0.8% agarose gels (unpublished data). These results are not conclusive, as the upper parts of agarose gels do not resolve DNA fragments very well, but they suggest a low level of homology between the *spo0H* region of *B. subtilis* and the cloned fragment of *B. licheniformis*. There are two *Eco*RI fragments of *B. licheniformis* DNA which hybridize with the insert DNA (Fig. 2). This was expected from the fact that there is one *Eco*RI cleavage site on the insert DNA (Table 1). These *Eco*RI fragments are quite large, approximately 10⁷ megadaltons.

The cloned insert carries a promoter for the Spo$^+$ function. If the insert carries its own promoter, its function would be independent of the insertional orientation into the vector. If the function were dependent on a vector promoter, however, the insert would be less likely to function in both orientations, since that would re-

TABLE 1. *Restriction map of the cloned insert[a]*

BglII↓		BcIIA		BclI↓		BcIIB	BglII↓

	MboA		MboB		MboE	MboD	

Mbo fragment	Mol wt	Restriction sites
A	0.22×10^6	HindIII, HaeIII
B	0.19×10^6	TacI, HpaII
C	0.13×10^6	?
D	0.11×10^6	HpaII, HhaI
E	$\sim 0.07 \times 10^6$	EcoRI, HhaI
F	$\sim 0.07 \times 10^6$	?
	0.79×10^6	

[a] Additional restriction sites not mapped on *Mbo* fragments are *Tac*I (one site), *Hin*dIII (one site), *Hin*fI (one site), *Cla*I (two sites), *Rsa*I (several cuts), and *Dde*I (several cuts). Enzymes which do not have sites are *Xba*I, *Bam*HI, *Pvu*II, *Pst*I, *Xho*I, *Sal*I, and *Hpa*I. *Mbo*C and *Mbo*F fragments have not yet been placed on the restriction map.

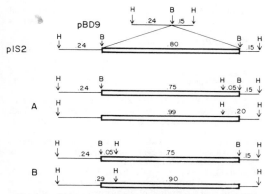

FIG. 3. *Orientation of the cloned insert in pIS2. The insertion of the cloned fragment into the* BglII *site of pBD9 is illustrated in the upper part of the figure. A and B refer to the two possible orientations the insert may have. In each alternative a linear map showing the location of* BglII *and* HaeIII *restriction sites is placed above the* HaeIII *digestion map as a reference. The numbers above the line refer to the calculated size of fragments resulting from single or double restriction digests.*

quire a vector promoter properly located on either side of the insert.

To test the dependence of the Spo[+] function of the insert upon its orientation, we recloned it from pIS1 into the *Bgl*II site of PBD9, a plasmid determining kanamycin and erythromycin resistance (6). Ery[r] transformants were selected after transformation of IS160, a *spo0H75* strain, with DNA from a ligated mixture of *Bgl*II-restricted pBD9 and pIS1. Seventeen of 300 Ery[r] transformants were Kan[s] and therefore were

carrying pBD9 with some alteration in the *kan* gene. Fifteen of the 17 were Spo[+], and they all carried plasmid DNA of the same size, which was larger than pBD9. Of the two Spo[-] Kan[s] clones obtained, one was Cm[r] and probably carried an insert derived from pBD64. The other carried a plasmid much smaller than those carried by the Spo[+] clones, apparently a deletion. Therefore, all clones which obtained the entire insert became Spo[+]. pBD9 containing the insert in the *Bgl*II site is called pIS2.

The two possible orientations are distinguishable, as shown in Fig. 3. We analyzed the restriction enzyme patterns of plasmid DNA from nine pIS2 clones. Seven of the nine had one type of *Hae*III digestion pattern, and the other two had another pattern. The two types fit the predicted *Hae*III restriction patterns expected with the cloned fragment inserted in each possible orientation (Fig. 3). These results indicate that the insert is functional in both orientations and presumably carries its own promoter.

Stability of pIS1. When a *spo0H75* strain carrying pIS1 is grown in the absence of chloramphenicol, thereby removing selective pressure, the plasmid is lost at a rapid rate. Brown *spo[+]* colonies have white sectors when grown in the absence of antibiotic. White Spo[-] colonies were found to be sensitive to chloramphenicol, so the antibiotic resistance segregates together with the Spo[+] phenotype. In contrast, an isogenic strain (*spo0H75*) carrying pBD64, when grown in the absence of antibiotic, retains its antibiotic resistance. In one experiment, an isogenic set of strains, one infected with pBD64 and the other with pIS1, were grown in culture overnight in the absence of chloramphenicol. The following day they were plated on rich, nonselective media. Single colonies were picked and checked for resistance to chloramphenicol. All (100 of 100 colonies) of the pBD64-bearing strain were Cm[r], and 53% (53 of 100 colonies) of the pIS1-bearing strain were Cm[r]. In all cases the Cm[r] and Spo[+] phenotype segregated together.

The copy numbers of pIS1 and pBD64, as well as pIS2a, pIS2b, and pBD9, were determined by the method of Weisblum et al. (14). The ratio of [*methyl*-³H]thymidine incorporated into chromosomal and plasmid DNA was measured after separation of the DNAs on agarose gels. This value was corrected by the ratio of the molecular weights of chromosome and plasmid DNA. The copy number of plasmids carrying the insert was always two- to threefold lower than that of the parent vector, regardless of insertional orientation.

The decreased copy number of the plasmid with the cloned insert and its instability compared with the vector plasmid may reflect growth inhibition of those cells in the culture with a high copy number of the cloned insert.

DISCUSSION

The cloning of the *B. licheniformis* gene fragment which complements *spo0H* mutations in *B. subtilis* provides us with a handle to study the expression and control of a gene with activity required for sporulation. As yet, we do not know the chromosomal location of this *B. licheniformis* gene or whether it performs the same functions in that organism as in *B. subtilis*.

The *spo0H* function or functions are required for the expression of at least one sporulation-specific mRNA (0.4-kilobase RNA of Segall and Losick [13]), so it seems likely to be a gene which is functional early in sporulation. It is now possible to look for proteins specified by the complementing gene and also for specific mRNA transcription in sporulating cells.

ACKNOWLEDGMENTS

We greatly appreciate the many helpful discussions we had with D. Dubnau, T. Gryczan, G. Grandi, and Y. Kozlov. We especially thank A. Howard for expert secretarial assistance.

This work was supported by Public Health Service research grant GM-19693 from the National Institutes of Health to I.S.

LITERATURE CITED

1. **Alwine, J. C., D. J. Kemp, B. A. Parker, J. Reiser, J. Renart, G. R. Stark, and G. M. Wahl.** 1979. Detection of specific RNAs or specific fragments of DNA by fractionation in gels and transfer to diazobenzyloxymethyl paper. Methods Enzymol. **68:**220–242.

2. **Contente, S., and D. Dubnau.** 1979. Characterization of plasmid transformation in *Bacillus subtilis*: kinetic properties and the effect of DNA conformation. Mol. Gen. Genet. **167:**251–258.

3. **Gryczan, T. J., S. Contente, and D. Dubnau.** 1978. Characterization of *Staphylococcus aureus* plasmids introduced into *Bacillus subtilis*. J. Bacteriol. **134:**318–329.

4. **Gryczan, T., S. Contente, and D. Dubnau.** 1980. Molecular cloning of heterologous chromosomal DNA by recombination between a plasmid vector and a homologous resident plasmid in *Bacillus subtilis*. Mol. Gen. Genet. **177:**459–467.

5. **Gryczan, T. J., and D. Dubnau.** 1978. Construction and properties of chimeric plasmids in *Bacillus subtilis*. Proc. Natl. Acad. Sci. U.S.A. **75:**1428–1432.

6. **Gryczan, T., A. G. Shivakumar, and D. Dubnau.** 1980. Characterization of chimeric plasmid cloning vehicles in *Bacillus subtilis*. J. Bacteriol. **141:**246–253.

7. **Haldenwang, W. G., and R. Losick.** 1979. A modified RNA polymerase transcribes a cloned gene under sporulation control in *Bacillus subtilis*. Nature (London) **282:**256–260.

8. **Hoch, J. A.** 1971. Selection of cells transformed to prototrophy for sporulation. J. Bacteriol. **105:**1200–1201.

9. **Hoch, J. A., and J. L. Mathews.** 1973. Chromosomal location of pleiotropic negative sporulation mutations in *Bacillus subtilis*. Genetics **73:**215–228.

10. **Piggot, P.** 1973. Mapping of asporogenous mutations of *Bacillus subtilis*: a minimum estimate of the number of sporulation operons. J. Bacteriol. **114:**1241–1253.

11. **Piggot, P. J., and J. G. Coote.** 1976. Genetic aspects of bacterial endospore formation. Bacteriol. Rev. **40:**908–962.

12. **Rigby, P. W. J., M. Dieckmann, C. Rhodes, and P. Berg.** 1977. Labeling deoxyribonucleic acid to high specific activity *in vitro* by nick translation with DNA polymerase I. J. Mol. Biol. **113:**237–251.

13. **Segall, J., and R. Losick.** 1977. Cloned *Bacillus subtilis* DNA containing a gene that is activated early during sporulation. Cell **11:**751–761.

14. **Weisblum, B., M. Y. Graham, T. Gryczan, and D. Dubnau.** 1979. Plasmid copy number control: isolation and characterization of high-copy-number mutants of plasmid pE194. J. Bacteriol. **137:**635–643.

Cloning of *spo0* Genes with Bacteriophage and Plasmid Vectors in *Bacillus subtilis*

FUJIO KAWAMURA, HIDENORI SHIMOTSU, HIUGA SAITO, HIROHIKO HIROCHIKA,[1] AND YASUO KOBAYASHI[2]

Institute of Applied Microbiology, University of Tokyo, Bunkyo-ku, Tokyo 113, Japan, and Department of Biology, Faculty of Science, Hiroshima University, Hiroshima 730, Japan

Sporulation genes *spo0F*[+] and *spo0B*[+] were cloned in *Bacillus subtilis* by "prophage transformation" using the temperate phages $\rho11$ and $\phi105$. When an *Eco*RI fragment harboring *spo0F*[+] was cloned with pUB110 as vector, it neither restored the sporulation of a *spo0F* host nor allowed the sporulation of a Spo[+] host. Another sporulation gene, *spo0B*[+], did not exhibit such inhibitory effect.

Numerous genetic studies on sporulation in *Bacillus subtilis*, especially on *spo0* genes which determine the earliest stage of sporulation, have been carried out (2, 11). Mutations in these genes are highly pleiotropic and seem to cause defects in factors required for the initiation of the sporulation process (2, 4–6, 11, 15). Studies on these gene functions are, therefore, of particular interest and may be advanced by the introduction of gene cloning techniques in *B. subtilis*.

We recently developed a novel method called prophage transformation for constructing specialized transducing phages of $\rho11$ (8) and $\phi105$ (3). DNA molecular weights of the former (11) and the latter (3) are 78×10^6 and 26×10^6, respectively. Each phage has its own advantages as a vector. We have used these phages to clone the early sporulation genes *spo0B*[+] and *spo0F*[+].

RESULTS AND DISCUSSION

Construction and characterization of $\rho11spo0F$[+]. We used the phage $\rho11$ since it was found to be the most efficient vector for shotgun experiments with various *B. subtilis* genes. Procedures for cloning of the *spo0F*[+] gene are shown in Fig. 1. DNAs of $\rho11$ and sporeforming *B. subtilis* were digested with *Eco*RI and then ligated with T4 DNA ligase. The mixture was used to transform competent cells of a *spo0F* strain lysogenic for $\rho11$, and Spo[+] transformants were selected. The transformation may have occurred by recombination with host chromo-

somal or prophage sequences. All the Spo[+] transformants were mixed, incubated, and treated with mitomycin C to induce the phage $\rho11$, and transducers were selected by transduction of *spo0F* cells to Spo[+]. One of the Spo[+] transductants, UOT 0182, was further examined.

By CsCl density gradient analysis, phage particles prepared from UOT 0182 showed two distinct bands at buoyant densities of 1.532 and 1.498 g/cm[3] (7). The particles that banded at the higher density had lost their tails and exhibited neither plaque-forming nor transducing ability, whereas DNA extracted from these particles showed Spo[+] and Thy[+] transforming activity. Many other artificially constructed transducing phages have been shown to lose tails frequently (3, 7). On the other hand, the phages that banded at the lower density exhibited all phage functions, although these phages were heterogeneous and composed of a small number of $\rho11spo0F$[+] and a large number of plaque-forming phages which did not have the *spo0F*[+] marker. The origin of the plaque-forming phages is unknown but might be ascribed to a chromosomal rearrangement which is frequently observed in the artificially constructed transducing phage $\rho11$ (9). In the prophage transformation method, repeated infection of these primary, unusual transducers was found generally to be effective in obtaining more efficient transducers which were plaque forming.

The *Eco*RI fragment carrying the *spo0F*[+] gene in $\rho11spo0F$[+] was determined by transformation assay. After gel electrophoresis, DNA fragments were extracted from gel slices and tested for *spo0F*[+] transforming activity. The

[1] Present address: Mitsubishi-Kasei Institute of Life Science, Machida-shi, Tokyo 194, Japan.

[2] Present address: Department of Applied Biochemistry, Hiroshima University, Fukuyama 720, Japan.

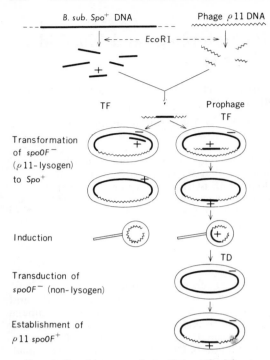

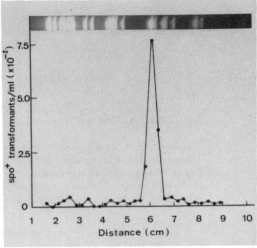

FIG. 2. *Detection of EcoRI fragment exhibiting* spo0F+ *transformation activity. ρ11spo0F+ DNA (6 μg) was digested with EcoRI and fractionated on a 1.5% cylindrical agarose gel. After electrophoresis, the gel was longitudinally cut into two equal portions. One of them was stained with ethidium bromide and used to identify band positions; the other was cut into 2-mm slices, and DNA was extracted by a freeze-squeeze method and used for transformation assay.*

FIG. 1. *Prophage transformation method for construction of specialized transducing phage ρ11spo0F+. Equal amounts (10 μg) of ρ11 DNA and sporeforming B. subtilis DNA prepared from the defective phage PBSX, both digested with restriction endonuclease EcoRI, were mixed and ligated. A 10-μg amount of the ligated DNA in 0.1 ml of ligation mixture was added to 2 ml of competent cells of B. subtilis SPO77 (trpC2 thyA thyB spo0F77) lysogenic for ρ11, and the cells were incubated at 37°C for 1 h. After addition of 8 ml of Schaeffer's sporulation medium (13), cells were further incubated at 37°C for 3 days with shaking until spore formation was completed. In total, 120 Spo+ transformants were obtained, which were then mixed together, grown in L-broth, and treated with 0.5 μg of mitomycin C per ml to induce phage ρ11. One milliliter of this lysate (10^8 PFU/ml) was added to 2 ml of B. subtilis SPO77 cells (10^8 cells/ml) in L-broth and incubated for 30 min at 37°C. After addition of 7 ml of Schaeffer's sporulation medium (13), cells were incubated for 3 days at 37°C with shaking. Heat-resistant spores were obtained at a frequency of 3 × 10^3/ml.*

activity was found in a 1.3-megadalton *Eco*RI fragment (Fig. 2). As can be seen in Fig. 2, however, the isolation of the 1.3-megadalton fragment was difficult because of the neighboring fragments.

Construction and characterization of φ105spo0F+ and φ105spo0B+. Subsequently, we tried to transfer the *spo0F+* gene to another temperate phage, φ105, because the DNA of

φ105 is relatively small and is convenient for isolation of the individual DNA fragments. The procedure was almost the same as that shown in Fig. 1, except that we used *ρ11spo0F+* DNA and φ105 DNA instead of *B. subtilis* DNA and ρ11 DNA, respectively. After DNA cleavage, ligation, and transformation of a *spo0F* recipient (JH649, *spo0F221 trpC2 phe-1*) lysogenic for φ105, Spo+ transformants were obtained at a relatively high efficiency (ca. 10^4/ml). The lysate obtained by inducing the mixed culture of the transformants gave a high titer of transducing activity, and φ105spo0F+ was isolated from a resulting transductant.

Specialized transducing phage φ105spo0B+ was also obtained by the successive prophage transformation procedures with ρ11 and φ105 as described above (H. Hirochika et al., in preparation). The recipient used for selection of *spo0B+* marker was JH648 (*spo0B136 trpC2 phe-1*).

The *Eco*RI cleavage maps of φ105spo0B+ and φ105spo0F+ in comparison with the map of wild-type φ105 are shown in Fig. 3. Agarose-gel electrophoretic patterns of the three phage DNAs are shown in Fig. 4. The fragments carrying *spo0B+* and *spo0F+* genes could be isolated, as is clearly shown in the right lanes of the gel. Since the *spo0B+* and *spo0F+* genes are func-

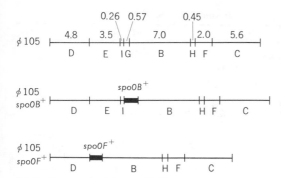

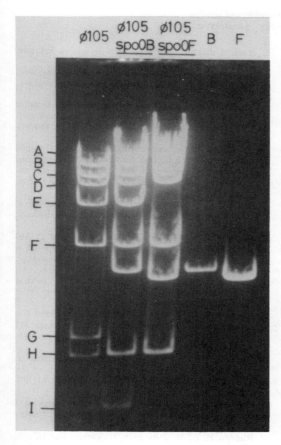

FIG. 3. EcoRI cleavage maps of φ105, φ105spo0B⁺, and φ105spo0F⁺ genomes. Details will be reported elsewhere (H. Hirochika et al., in preparation).

FIG. 4. Agarose gel electrophoresis of EcoRI digest of φ105, φ105spo0F⁺, and φ105spo0B⁺. DNA (0.1 to 1.5 μg) was electrophoresed at a constant voltage of 80 V for 2.5 h at room temperature. Lanes F and B indicate 1.3-megadalton (spo0F) and 1.4-megadalton (spo0B) EcoRI fragments, respectively.

tional in both ρ11 and φ105 prophages, the 1.4- and 1.3-megadalton EcoRI fragments seem to contain complete coding sequences for the respective genes.

spo0 gene cloning with pUB110. In contrast to the phage vector, which exists as a single prophage genome in each of the host cells, plasmid vectors such as pUB110 are known to reside as multiple copies (1). We therefore examined the copy number effect of spo0 genes on the sporulation process by using these vector systems. The 1.3- and 1.4-megadalton fragments harboring spo0F⁺ and spo0B⁺ genes were isolated from slices of low-melting-point agarose gel after electrophoresis of transducing φ105 DNAs. The isolated DNA fragments and pUB110 DNA (Kmʳ) were cleaved with EcoRI, ligated, and used to transform spo0F and Spo⁺ recipients to kanamycin resistance. Both recipients carried the recE4 mutation to stabilize the recombinant plasmid (Table 1).

The upper half of Table 1 shows the transformation with spo0F recipients. As shown in the second line, half of the Kmʳ transformants had recombinant plasmids carrying the spo0F⁺ gene; nevertheless, the phenotype of these transformants was defective in sporulation. A similar result was reported by Rhaese et al. (12). The lower half of the table shows the transformation of Spo⁺ recipients. In spite of the Spo⁺ genotype, none of the tested transformants carrying spo0F⁺ plasmids sporulated well under conventional conditions. On the other hand, a recombinant plasmid carrying spo0B⁺ gene restored the sporulation of a spo0B host (data not shown) and did not inhibit the sporulation of a Spo⁺ host.

The inhibitory effect of spo0F⁺ carried by plasmid was further examined with four Kmʳ transformant clones. All clones exhibited little sporulation under conventional conditions (Table 2). Among the rare sporulating cells in each population, however, the plasmid was always found to be more or less deleted as compared with the parental recombinant plasmid. One of the deleted plasmids, pUBΔSF-12, as well as the original pUB110 did not inhibit sporulation. The sizes of deletions in comparison with the physical map of the spo0F⁺ 1.3-megadalton fragment were studied, and all the deleted plasmids were found to have less than the entire spo0F⁺ gene (data not shown).

In conclusion, we have cloned spo0F⁺ and spo0B⁺ genes successfully with temperate phage vectors and found that the ρ11 vector works well for the first shotgun experiment and the φ105 vector is useful for isolation and fine analysis of the cloned genes. In addition, we noticed the

TABLE 1. *Cloning of* spo0 *genes with pUB110 in* spo0F *and* Spo⁺ *hosts*[a]

Recipient	DNA (µg)			Transformant		
	pUB110	spo0F⁺	spo0B⁺	Km^r/ml	Recombinant plasmid/Km^r	Spo^{+b}/Km^r
spo0F221 recE4	0.6	0.2	—	10	—	0
	5	0.2	—	1.8×10^2	6^c/12	0
	0.6	—	0.2	10	—	0
	5	—	0.2	2.5×10^2	—	0
Spo⁺ *recE4*	0.6	0.2	—	80	—	+ and −
	5	0.2	—	7.4×10^2	18^c/40	+ and −
	0.6	—	0.2	1.8×10^2	—	All +
	5	—	0.2	7.1×10^2	10^d/39	All +

[a] Indicated amounts of *Eco*RI fragments were joined by T4 ligase and added to the competent cells of UOT 0294 (*spo0F221 recE4 nonB1*) or UOT 0277 (*recE4 hisA1 metB5 nonB1*). After 1.5 h of incubation at 37°C, cells were plated on tryptose blood agar base (Difco) containing 5 µg of kanamycin per ml and 0.1% glucose. Plasmids in kanamycin-resistant (Km^r) transformants were examined by agarose gel electrophoresis.

[b] Judged from the appearance of colored colony.

[c] Contained *spo0F⁺* DNA; the phenotype was Spo⁻ regardless of the host genotype.

[d] Contained *spo0B⁺* DNA; there was no effect on the Spo⁺ phenotype.

TABLE 2. *Inhibition of sporulation by* spo0F⁺ *plasmid*[a]

Strain	Viable cells/ml	Spores/ml
Spo⁺ *recE4*(pUB110)	1.8×10^8	1.2×10^8
Spo⁺ *recE4*(pUBSF-8)	6.2×10^7	3.6×10^4
Spo⁺ *recE4*(pUBSF-10)	3.6×10^7	5.6×10^3
Spo⁺ *recE4*(pUBSF-13)	6.4×10^7	1.3×10^3
Spo⁺ *recE4*(pUBSF-19)	6.8×10^7	3.6×10^3
Spo⁺ *recE4*(pUBΔSF-12)	1.8×10^8	1.6×10^8

[a] Spo⁺ transformants obtained from experiments in Table 1 were incubated in Schaeffer's sporulation medium (13) for 26 h at 37°C. Heat-resistant spores were counted as described in the legend of Fig. 1. pUB110 and pUBΔSF-12 did not carry the *spo0F⁺* gene.

usefulness of the temperate phage vectors for cloning of a gene such as *spo0F⁺*, which inhibits sporulation when cloned with the multicopy plasmid pUB110. A similar phenomenon was observed by R. Losick and colleagues (personal communication). They found that plasmid pUB110 carrying the promoter of the 0.4-kb gene (14) inhibited the sporulation of a Spo⁺ host.

ACKNOWLEDGMENTS

This work was supported in part by a Grant-in-Aid for Scientific Research and by a Grant-in-Aid for Special Project Research from the Ministry of Education, Science, and Culture of Japan.

LITERATURE CITED

1. **Gryczan, T. J., and D. Dubnau.** 1978. Construction and properties of chimeric plasmids in *Bacillus subtilis*. Proc. Natl. Acad. Sci. U.S.A. **75:**1428–1432.

2. **Hoch, J. A.** 1976. Genetics of bacterial sporulation. Adv. Genet. **18:**69–98.

3. **Iijima, T., F. Kawamura, H. Saito, and Y. Ikeda.** 1980. A specialized transducing phage constructed from *Bacillus subtilis* phage φ105. Gene **9:**115–126.

4. **Ito, J.** 1973. Pleiotropic nature of bacteriophage tolerant mutants obtained in early blocked asporogenous mutants of *Bacillus subtilis* 168. Mol. Gen. Genet. **124:**97–106.

5. **Ito, J., G. Mildner, and J. Spizizen.** 1971. Early blocked asporogenous mutants of *Bacillus subtilis* 168. I. Isolation and characterization of mutants resistant to antibiotic(s) produced by sporulating *Bacillus subtilis* 168. Mol. Gen. Genet. **112:**104–109.

6. **Ito, J., and J. Spizizen.** 1971. Abortive infection of sporulating *Bacillus subtilis* 168 by φ2 bacteriophage. J. Virol. **7:**515–523.

7. **Kawamura, F., H. Saito, H. Hirochika, and Y. Kobayashi.** 1980. Cloning of sporulation gene, *spo0F*, in *Bacillus subtilis* with ρ11 phage vector. J. Gen. Appl. Microbiol. **26:**363–373.

8. **Kawamura, F., H. Saito, and Y. Ikeda.** 1979. A method for construction of specialized transducing phage ρ11 of *Bacillus*. Gene **5:**87–91.

9. **Mizukami, T., F. Kawamura, and H. Saito.** 1980. Genetic instability of artificially constructed phage ρ11 carrying histidine A gene of *Bacillus subtilis*. J. Gen. Appl. Microbiol. **26:**307–310.

10. **Mizukami, T., F. Kawamura, H. Takahashi, and H. Saito.** 1980. A physical map of the genome of the *Bacillus subtilis* temperate phage ρ11. Gene **11:**157–162.

11. **Piggot, P. J., and J. G. Coote.** 1976. Genetic aspects of bacterial endospore formation. Bacteriol. Rev. **40:**908–962.

12. **Rhaese, H. J., R. Groscurth, R. Vetter, and H. Gilbert.** 1979. Regulation of sporulation by highly phosphorylated nucleotides in *Bacillus subtilis*, p. 145–159. *In* G. Koch and D. Richter (ed.), Regulation of macromolecular synthesis by low molecular weight mediators. Academic Press, Inc., New York.

13. **Schaeffer, P., H. Ionesco, A. Ryter, and G. Ballasa.** 1963. La sporulation de *Bacillus subtilis*: etude gene-

tique et physiologique. Colloq. Int. C.N.R.S. **124:**553–563.

14. **Segall, J., and R. Losick.** 1977. Cloned *Bacillus subtilis* DNA containing a gene that is activated early during sporulation. Cell **11:**751–761.

15. **Trowsdale, J., S. M. H. Chen, and J. A. Hoch.** 1979. Genetic analysis of a class of polymyxin resistant partial revertants of stage 0 sporulation mutants of *Bacillus subtilis*: map of the chromosome region near the origin of replication. Mol. Gen. Genet. **173:**61–70.

Cloning and Characterization of Sporulation Genes in *Bacillus subtilis*

YASUO KOBAYASHI,[1] HIROHIKO HIROCHIKA,[2] FUJIO KAWAMURA, AND HIUGA SAITO

Department of Biology, Faculty of Science, Hiroshima University, Hiroshima 730, Japan, and Institute of Applied Microbiology, University of Tokyo, Bunkyo-ku, Tokyo 113, Japan

Specialized transducing phages carrying sporulation genes, *spo0B* and *spo0F*, were constructed by use of *Bacillus subtilis* temperate phage ρ11. The *spo0B* and *spo0F* genes reside on 1.4- and 1.3-megadalton *Eco*RI fragments of the phage DNA, respectively. Sodium dodecyl sulfate-polyacrylamide gel analysis of *spo0B* proteins synthesized in UV-irradiated cells infected with ρ11p*spo0B* showed that the 1.4-megadalton fragment codes for at least one protein with a molecular weight of 39,000, which is synthesized in both vegetative and sporulating cells.

Sporulation of *Bacillus subtilis* is a series of well-defined morphological stages. A large number of sporulation-defective mutants ave been isolated and their mutations have been mapped (12). These mutants were classified into five types (*spo0*, *spoII*, *spoIII*, *spoIV*, and *spoV*) according to the stage at which sporulation of the mutant is blocked. Among these mutants, the *spo0* type is particularly interesting because it shows the earliest block. It might be expected that the mechanism controlling initiation of sporulation would be clarified by the analysis of *spo0* mutants. Nine *spo0* genes, *spo0A*, *spo0B*, etc., are known, and these genes have been mapped to different loci on the *B. subtilis* chromosomes. Almost no details of the function and the structure of these genes and their gene products are known. A gene cloning technique developed by Kawamura et al. (6) makes it possible to clone sporulation genes in *B. subtilis*. Using this technique, we have succeeded in cloning the *spo0B* and *spo0F* genes.

RESULTS AND DISCUSSION

Construction of the specialized transducing phages ρ11p*spo0B* and ρ11*spo0F*. Using *B. subtilis* temperate phage ρ11, we constructed specialized transducing phages ρ11p*spo0B* and ρ11*spo0F* by the method of Kawamura et al. (6). *B. subtilis* DNA (*spo*+) was prepared from defective phage PBSX, which was induced from a

[1] Present address: Department of Applied Biochemistry, Hiroshima University, Fukuyama 720, Japan.

[2] Present address: Mitsubishi-Kasei Institute of Life Sciences, Michida-shi, Tokyo 194, Japan.

sporeforming strain, UOT-0110-1 (*trpC2 met-14 su3*+; University of Tokyo). Equal amounts (10 µg/ml) of *B. subtilis* DNA and ρ11 DNA were cleaved by endonuclease *Eco*RI, mixed, and joined by bacteriophage T4 ligase. Transformation was carried out by the method of Wilson and Bott (18). Ten micrograms of the ligated DNA in 0.1 ml of the ligation mixture was added to 2 ml of the competent cells of strain JH648 (*trpC2 phe-1 spo0B136*), JH649 (*trpC2 phe-1 spo0F221*), or Spo077 (*trpC2 thyA thyB spo0F77*), all lysogenic for ρ11. Spo+ transformants were selected on Schaeffer's sporulation medium (14) by the method of Hoch (1). These Spo+ transformants would consist of true Spo+ transformants and lysogens of Spo+ tranducing phage. The latter were selected by mixing all Spo+ colonies and inducing the transducing phages with mitomycin C (0.5 µg/ml). The lysate was used to transduce *spo0* strains to Spo+. Spo+ transductants of strains JH648 (*spo0B*), JH649 (*spo0F*), and Spo077 (*spo0F*) were purified and used for the following experiments.

Characterization of the transducing phages. Phages induced from Spo+ transductants of JH648 had the same plaque-forming activity as ρ11 (1.33 × 10^10 PFU/ml). However, phages induced from Spo+ transductants of JH649 or Spo077 were not plaque formers. When phages were analyzed by CsCl density gradient centrifugation, ρ11p*spo0B* formed a single band, but ρ11*spo0F* formed two distinct bands at buoyant densities of 1.532 and 1.498 g/cm^3. The particles that banded at the higher density might have lost their tails, since they exhibited neither

TABLE 1. *Spo⁺ transducing activity of the transducing phage[a]*

Strain[b]	Relevant genotype	Infected with	Spores/ml	Viable cells/ml
JH648	spo0B136	—	10	5.7×10^8
		$\rho 11pspo0B$	6.9×10^7	ND[c]
		$\rho 11wt$	10	ND
JH649	spo0F221	—	10	1.4×10^9
		$\rho 11spo0F$	1.9×10^7	ND
		$\rho 11wt$	10	ND
Spo077	spo0F77	—	8.0×10	1.3×10^8
		$\rho 11spo0F$	2.9×10^{5d}	1.0×10^8
		$\rho 11wt$	2.0×10	1.0×10^8

[a] The overnight culture of the recipient strain was inoculated into Schaeffer's sporulation medium (14). After phage addition, incubation was continued for 48 h at 37°C. Heat-resistant spores were counted by plating the cells after heating the cell suspension for 10 min at 80°C.

[b] Strains JH648 and JH649 were obtained from J. Hoch; strain Spo077 was a spontaneous mutant of 168TT (*trpC2 thyA thyB*).

[c] ND, Not determined.

[d] In this experiment, crude phage lysate was used instead of purified transducing phage. This might be a reason why sporulation efficiency is low in this case.

TABLE 2. *Genetic complementation test[a]*

Tester strain[b]	$\rho 11pspo0B$	$\rho 11pspo0B$-m15	$\rho 11pspo0B$-m18
JH648	+	−	−
3Y	+	−	−
ts32	+	−	−
6Z	+	−	−
17NA	+	−	−
9NA	+	−	−

[a] Phage solution was spotted on a lawn of the tester strain which was spread on Schaeffer's sporulation medium. Incubation was at 37°C for 2 days. Symbols: +, spores produced on the spot; −, no spores.

[b] Strain 3Y (*trpC2 spo03Y*) was from the Bacillus Genetic Stock Center. Strain ts32 [*trpC2 thyA thyB spo0B32*(Ts)] was a spontaneous mutant of 168TT. The strain cannot sporulate at 42°C but does sporulate at 37°C. Strains 6Z, 17NA, and 9NA were obtained from A. Uchida.

plaque-forming nor Spo⁺ transducing activity. On the other hand, the phages of low density were capable of plaque formation and transduction, although the plaque-forming phages among them had lost the ability to transduce *spo0F* to Spo⁺.

Table 1 shows the Spo⁺ transducing activity of the transducing phages. These results indicate that the *spo0B* and *spo0F* mutations are recessive to the wild-type allele.

Genetic complementation test. Strain 3Y (= strain 1S18 [*trpC2 spo03Y*]) of the Bacillus Genetic Stock Center is a *spo0* mutant isolated by Ionesco et al. (3). Its mutation was suggested to be an allele of *spo0F* (12). However, we found that the *spo03Y* mutation is located between *spo0B136* and *phe-1* (H. Hirochika et al., in preparation). Moreover, strain 3Y lysogenic for

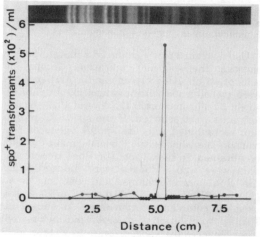

FIG. 1. *Detection of EcoRI fragment containing spo0B⁺ transforming activity. ρ11pspo0B DNA (6 μg) was digested with EcoRI and fractionated on a 1.5% cylindrical agarose gel (0.7 by 11 cm). The electrophoresis was carried out at 40 mA for 6 h at room temperature with the use of a Tris-phosphate buffer system (8). After electrophoresis, the gel was cut longitudinally into two equal parts. One of them was stained with ethidium bromide and used to identify band positions. The other was sliced (2 mm thick, each), and each slice was placed in 0.2 ml of Spizizen mineral salts (15) supplemented with 30 mM MgCl₂. Then DNA was extracted by a freeze-squeeze method (17). The DNA solution (0.2 ml) and the competent cell culture (0.3 ml) of strain JH648 were mixed and incubated at 37°C for 30 min; then Spo⁺ transformants were selected by the method of Hoch (1).*

$\rho 11pspo0B$ formed spores, but strain 3Y lysogenic for $\rho 11spo0F$ did not form spores. These results suggest that *spo03Y* is an allele of *spo0B*.

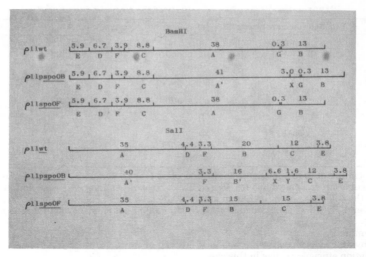

FIG. 2. *Cleavage map of ρ11, ρ11pspo0B, and ρ11spo0F DNA. The cleavage map of ρ11wt DNA was taken from Mizukami et al. (9). The molecular weight of the fragments was determined from the electrophoretic mobility relative to φ29 DNA, φ29 EcoRI fragments (4), ρ11 DNA, and ρ11 SalI and BamHI fragments (9). Fragment size is expressed in megadaltons.*

Using strain 3Y, we produced a mutant phage carrying the *spo0B3Y* mutation. Competent cells of strain JH648 lysogenic for ρ11p*spo0B*$^+$ were transformed with a saturating amount of strain 3Y chromosomal DNA, and Phe$^+$ transformants were selected. If the *spo0B3Y* mutation recombined into the *spo0B*$^+$ gene of the transducing phage, Spo$^-$ transformants would be obtained. In fact, 1% of the Phe$^+$ transformants were Spo$^-$. Phages were induced from the Spo$^-$ transformants, and two of them (ρ11p*spo0B*-m15 and ρ11p*spo0B*-m18) were used for complementation tests. As shown in Table 2, sporulation of all *spo0B* mutants tested was recovered by lysogenization by ρ11p*spo0B*, but not by ρ11p*spo0B*-m15 or ρ11p*spo0B*-m18. These results suggest that all *spo0B* mutants examined are allelic. This result is in good agreement with those of Hoch et al. (2). A similar result was obtained for the *spo0F* locus (data not shown).

Identification of *Eco*RI fragment carrying *spo0B* and *spo0F* genes. *Eco*RI fragments carrying *spo0B* and *spo0F* genes were identified by transformation experiments. Transducing phage DNA was cut into about 30 fragments by *Eco*RI and fractionated by agarose gel electrophoresis. Some differences in the cleavage pattern were observed for ρ11DNA, ρ11p*spo0B* DNA, and ρ11*spo0F* DNA (data not shown; 5; H. Hirochika et al., in preparation). The gel was sliced, and DNA was extracted from each slice and used for the assay of the transforming activity. As shown in Fig. 1, *spo0B*$^+$ activity is found

in a 1.4-megadalton fragment of ρ11p*spo0B* DNA. Similarly, *spo0F*$^+$ transforming activity was found in a 1.3-megadalton fragment of ρ11*spo0F* DNA (5; F. Kawamura et al., this volume).

Cleavage map of ρ11p*spo0B* and ρ11*spo0F*. To learn the location of *spo0B* and *spo0F* genes on each transducing phage genome, we prepared the cleavage map of the phage genomes, using endonucleases *Bam*HI and *Sal*I (Fig. 2). Details of the mapping procedure will be published elsewhere (5; Hirochika et al., in preparation). As can be seen from Fig. 2, the molecular weight of ρ11p*spo0B* DNA is 4 × 10^6 to 6 × 10^6 higher than that of ρ11 DNA. This means that, in addition to the 1.4-megadalton *Eco*RI fragment, several other *Eco*RI fragments may be inserted into the ρ11p*spo0B* phage genome. In contrast, the molecular weight of ρ11*spo0F* is the same as that of ρ11, but the sizes of the *Sal*I-B and *Sal*I-C fragments of ρ11 DNA have been changed in ρ11*spo0F*. This result suggests that the fragment carrying the *spo0F* gene may be inserted in this region.

To determine more precisely the location of *spo0B* and *spo0F* genes in each transducing phage genome, we carried out transformation experiments using *Bam*HI and *Sal*I fragments of each phage as donor DNA. We found that *Bam*HI-A' and *Sal*I-X fragments of ρ11p*spo0B* DNA have a transforming activity of *spo0B*$^+$. Similarly, *Bam*HI-A and *Sal*I-B fragments of ρ11*spo0F* have a *spo0F*$^+$ transforming activity

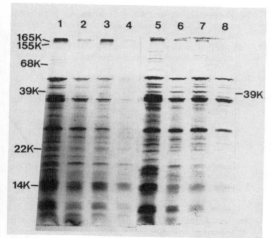

FIG. 3. *Phage-directed protein synthesis in UV-irradiated cells at various stages of growth and sporulation. Strain UVS-1 (trpC2 uvr-114) was used as a host strain. This strain sporulates normally and is sensitive to UV light. The UV-sensitive strain was used because UV-sensitive mutation reduces the background level of protein synthesis in UV-irradiated cells. Cells were grown in Sterlini and Mandelstam's resuspension medium (16) supplemented with 0.2% glucose, spun down, and suspended in the spent medium. The cell suspension (4 ml) in a 9-cm petri dish was irradiated for 5 min with stirring at a distance of 25 cm from two 10-W germicidal lamps, and then cells (0.2 ml) were infected with purified phage at a multiplicity of infection of 100. After incubation at 37°C for 5 min, 2.5 µCi of ^{14}C-amino acid mixture (New England Nuclear) was added, and incubation was continued for 20 min; then a large excess of nonradioactive amino acid mixture (0.1 mM, each) was added. Cells were washed once and suspended in 30 µl of the lysing buffer containing 30 mM Tris-hydrochloride, pH 6.8, lysozyme (200 µg/ml), DNase (10 µg/ml), and phenylmethylsulfonyl fluoride (300 µg/ml). The suspension was incubated at 37°C for 15 min, mixed with 20 µl of 4× sample buffer, and heated at 100°C for 2 min. The samples containing about 10^5 cpm of radioactivity were subjected to sodium dodecyl sulfate-polyacrylamide gel electrophoresis as described by Laemmli and Favre (7). Gels were stained with Coomassie brilliant blue, dried, and exposed to X-ray film for autoradiography. Marker proteins used for the molecular-weight estimation were β' (165,000), β (155,000), and α (39,000) subunits of E. coli RNA polymerase; bovine serum albumin (68,000); trypsin inhibitor (21,500); and lysozyme (14,400). (1 and 5) Vegetative cells; (2 and 6) t_1 cells; (3 and 7) t_2 cells; (4 and 8) t_3 cells; (1–4) ρ11 infection; (5–8) ρ11pspo0B infection.*

(data not shown; 5; Hirochika et al., in preparation).

Identification of *spo0B* gene product. To identify the *spo0B* gene product, we analyzed

proteins coded for by ρ11pspo0B. Cells at various stages of sporulation were irradiated with UV light and infected with ρ11 or ρ11pspo0B. Proteins were labeled with ^{14}C-amino acid mixture and fractionated by sodium dodecyl sulfate-polyacrylamide gel electrophoresis (Fig. 3). Comparison of the results obtained with ρ11 and ρ11pspo0B infections showed that a 39-kilodalton protein is coded for by ρ11pspo0B, but not by ρ11. Other differences were not detected, although minor differences might be present. In all stages examined (log phase to t_3, 3 h after the initiation of sporulation), the 39-kilodalton protein was synthesized.

To examine whether this protein is coded for by the 1.4-megadalton *Eco*RI fragment, we analyzed proteins coded for by phage deletion mutants. These mutants were obtained by heat treatment of the phage in the presence of sodium pyrophosphate (11). All of them had a deletion in the 1.4-megadalton *Eco*RI fragment. They had no *spo0B*+ transducing activity and did not synthesize the 39-kilodalton protein (data not shown; Hirochika et al., in preparation). This result suggests that the 1.4-megadalton fragment codes for the 39-kilodalton protein.

Cloning of sporulation genes in a temperate phage. Recently, two sporulation genes, *spo0F* (13) and *spoVC* (10), have been cloned by using a plasmid as a vector. In the present paper, we have presented a method for the cloning of sporulation genes by using a temperate phage as vector. Temperate phages have several advantages over plasmid vectors in the study of sporulation. First, the analysis of the time of gene expression and the identification of gene products are performed easily in UV-irradiated cells, since the cloned gene can be inserted into cells at any time of sporulation. Second, gene dosage effects can be avoided, since, in general, only one copy of the phage genome is in the lysogen. Third, phage deletion mutants, which are useful for the analysis of gene organization, are easily obtained (11). Since our method is applicable to any sporulation gene, it will provide a powerful tool for future analysis of sporulation mechanisms.

ACKNOWLEDGMENTS

We thank J. Hoch, A. Uchida, E. Freese, and the Bacillus Genetic Stock Center (D. Dean) for providing us bacterial strains.

LITERATURE CITED

1. **Hoch, J. A.** 1971. Selection of cells transformed to prototrophy for sporulation markers. J. Bacteriol. **105:** 1200–1201.

2. **Hoch, J. A., M. A. Schiflett, J. Trowsdale, and S. M. H. Chen.** 1978. Stage 0 genes and their products, p. 127–130. *In* G. Chambliss and J. C. Vary (ed.), Spores VII. American Society for Microbiology, Washington, D.C.

3. **Ionesco, H., J. Michel, B. Cami, and P. Schaeffer.** 1970. Genetics of sporulation in *Bacillus subtilis* Marburg. J. Appl. Bacteriol. **33**:13–24.

4. **Ito, J., F. Kawamura, and S. Yanofsky.** 1976. Analysis of φ29 and φ15 genomes by bacterial restriction endonucleases, *Eco*RI and *Hpa*I. Virology **70**:37–51.

5. **Kawamura, F., H. Saito, H. Hirochika, and Y. Kobayashi.** 1980. Cloning of sporulation gene, *spo0F*, in *Bacillus subtilis* with ρ11 phage vector. J. Gen. Appl. Microbiol. **26**:363–373.

6. **Kawamura, F., H. Saito, and Y. Ikeda.** 1979. A method for construction of specialized transducing phage ρ11 of *Bacillus subtilis*. Gene **5**:87–91.

7. **Laemmli, U. K., and M. Favre.** 1973. Mutation of the head of bacteriophage T4. I. DNA packing events. J. Mol. Biol. **80**:575–599.

8. **Loening, U. E.** 1968. Molecular weights of ribosomal RNA in relation to evolution. J. Mol. Biol. **38**:355–365.

9. **Mizukami, T., F. Kawamura, H. Takahashi, and H. Saito.** 1980. A physical map of the genome of the *Bacillus subtilis* temperate phage ρ11. Gene **11**:157–162.

10. **Moran, C. P., Jr., R. Losick, and A. L. Sonenshein.** 1980. Identification of a sporulation locus in cloned *Bacillus subtilis* deoxyribonucleic acid. J. Bacteriol. **142**:331–334.

11. **Parkinson, J. S., and R. J. Huskey.** 1971. Deletion mutants of bacteriophage lambda. J. Mol. Biol. **56**:369–384.

12. **Piggot, P. J., and J. G. Coote.** 1976. Genetic aspects of bacterial endospore formation. Bacteriol. Rev. **40**:908–962.

13. **Rhaese, H. J., R. Groscurth, R. Vetter, and H. Gilbert.** 1979. Regulation of sporulation by highly phosphorylated nucleotides in *Bacillus subtilis*, p. 145–159. *In* G. Koch and D. Richter (ed.), Regulation of macromolecular synthesis by low molecular weight mediators. Academic Press, Inc., New York.

14. **Schaeffer, P., H. Ionesco, A. Ryter, and G. Balassa.** 1963. La sporulation de *Bacillus subtilis*: étude génétique et physiologique. Coll. Int. C.N.R.S. **124**:553–563.

15. **Spizizen, J.** 1958. Transformation of biochemically deficient strains of *Bacillus subtilis* by deoxyribonucleate. Proc. Natl. Acad. Sci. U.S.A. **44**:1072–1078.

16. **Sterlini, J. M., and J. Mandelstam.** 1969. Commitment to sporulation in *Bacillus subtilis* and its relationship to development of actinomycin resistance. Biochem. J. **113**:29–37.

17. **Thuring, R. W. J., J. P. M. Sanders, and P. Borst.** 1975. A freeze-squeeze method for recovering long DNA from agarose gels. Anal. Biochem. **66**:213–220.

18. **Wilson, G. A., and K. F. Bott.** 1968. Nutritional factors influencing the development in the *Bacillus subtilis* transformation system. J. Bacteriol. **95**:1439–1449.

Characterization of *Bacillus subtilis* rRNA Genes

KENNETH F. BOTT, FRANCES E. WILSON, AND GEORGE C. STEWART

University of North Carolina School of Medicine, Department of Bacteriology, Chapel Hill, North Carolina 27514

Bacillus subtilis rRNA labeled with ^{32}P was used as a hybridization probe to detect the DNA fragments carrying rRNA genes. A physical map of the rRNA genes was constructed by using this probe against Southern blots of restriction digests of whole *B. subtilis* DNA. By probing *Escherichia coli* clone banks carrying fragments of *B. subtilis* DNA, we recovered cloned fragments derived from individual rRNA gene sets. These clones were used to correlate cloned DNA with the hybridizing restriction fragments of total *B. subtilis* DNA. Since the spacer sequences adjacent to the rRNA gene sets consist of unique DNA, each set of genes can be mapped by integrating a cloned spacer fragment with its plasmid vector into its homologous site on the chromosome. PBS1-mediated transduction was used to establish the locus of one such inserted antibiotic resistance determinant. At least one rRNA unit maps between *purA* and *cysA*, outside the ribosomal cluster adjacent to the SPO2 attachment site.

Our laboratory has embarked on a project to isolate and characterize the rRNA genes from *Bacillus subtilis* with the aim of discerning the structural or functional differences between the individual operons which might relate to control of endospore formation. This investigation would also allow a study of the regulation of this multigene system in a manner analogous to that used by Nomura and his colleagues for *Escherichia coli* (12). We would then be able to characterize directly the differences between a grampositive and a gram-negative species in this regard.

A strong foundation for these studies was provided by Chow and Davidson (3) with their heteroduplex analysis of *B. subtilis* ribosomal DNA. Their study suggested that there are 7 to 10 copies of the rRNA genes on the chromosome and that each is homologous and separated from the others by heterologous spacer DNA. The Chow and Davidson study enabled the measurement of several other spacer regions and demonstrated a clustering of the gene sets on the chromosome. This is in contrast to the situation found with *E. coli*, in which there is not a clustering of the rRNA operons (12). In addition, Colli et al. (4, 5, 18) have demonstrated a 16S, 23S, 5S transcriptional order for the *B. subtilis* rRNA.

To characterize the genes, we probed the chromosomal restriction fragments with purified species of 16S or 23S rRNA in the presence of a 10-fold excess of the other unlabeled RNA species.

rRNA was isolated from cells grown for four to six generations in the presence of ^{32}P as described (9, 13). In addition, we have made use of several collections of cloned *B. subtilis* DNA fragments in *E. coli* vectors (8, 14; E. Ferrari, D. J. Henner, and J. A. Hoch, and W. Steinberg, personal communications).

Table 1 lists the patterns of hybridization obtained when *Bam*HI, *Sma*I, *Eco*RI, and *Bgl*I digests of *B. subtilis* chromosomal DNA were probed with radioactive rRNA. It is important to note that a unique multiple-band pattern was obtained with each enzyme. The fewest bands in any digest seen to date is seven, as shown in the *Bgl*I digest.

All the evidence we have obtained indicates that each of the gene sets is homologous with regard to the restriction sites within the coding sequence for 16S and 23S RNA. A composite map of these sites is shown in Fig. 1. Only the *Hind*III site distal to the 5S rRNA coding sequence is variable. On the basis of the 5S RNA sequence reported by Sogin et al. (16), the *Hind*III site is one base outside of the coding sequence for mature 5S RNA. The nucleotide sequences of two distinct precursors of 5S RNA have been reported for *B. subtilis* (12, 16). Our computer analysis indicates that only one of these would possess a *Hind*III site. The *B. subtilis* operons also differ from those of *E. coli* (1) by not having tRNA sequences in the junction between the 16S and 23S coding regions.

Recombinant plasmids and Charon 4A phage

119

TABLE 1. *Molecular size and number of chromosomal restriction fragments which hybridize to rRNA[a]*

EcoRI		BamHI		SmaI		BglII[b]	
16S RNA	23S RNA	16S RNA	23S RNA	16S RNA	23S RNA	16S RNA	23S RNA
10		27		30		42	42
	9.2		25	18[c]	18[c]	19	19
	8.0	23	23	14		18	18
	6.3		14		7.3	15	15
	4.5		11		5.4	14.5	14.5
	4.3	8.0		3.3		13.5	13.5
4.0	4.0	6.4		3.0[c]	3.0	10	10
3.5	3.5		6.3[c]	2.8	2.8		
3.0	3.0	6.2[c]			2.2		
	2.4	5.4	5.4	2.1	2.1		
2.0		5.0	5.0	2.0			
1.8		4.4		0.4[c]	0.4[c]		
1.5			3.5				
1.2[c]	1.2[c]						
	0.8[c]						

[a] Sizes are given in kilobases, resolved on 0.7 to 0.8% agarose.
[b] Requires 0.35% agarose for resolution.
[c] Intensity on autoradiographs suggests these represent multiple copies.

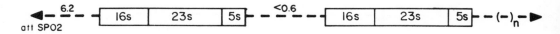

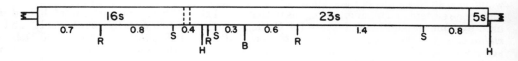

No Bgl I sites

FIG. 1. *Generalized map of* B. subtilis *rRNA genes. The top half of this figure represents the ribosomal gene sets located adjacent to the SPO2 attachment site as identified by Chow and Davidson (3). The bottom half represents the physical map of an rRNA gene set. Sizes are given in kilobases. The vertical dashed lines represent the junction (size yet undetermined) between the 16S and 23S rRNA genes. B, BamHI; H, HindIII; R, EcoRI; S, SmaI.*

clones containing rRNA sequences were identified by using either [³²P]rRNA or nick-translated DNA (15) from clones previously obtained by use of the labeled RNA (9, 10; C. P. Moran, Ph.D. thesis, University of North Carolina, Chapel Hill, 1979) (Fig. 2). The spacers of plasmids p14B1 and p14B8 do not contain *Bgl*I sites, and since rRNA coding regions also lack *Bgl*I sites (see Fig. 1), two gene sets must be represented on each of two chromosomal *Bgl*I fragments. Since there are seven *Bgl*I fragments which hybridize to rRNA (Table 1), there must be at least nine rRNA gene sets in *B. subtilis*. This supersedes the estimate of eight suggested

by Moran and Bott (9). An estimate of nine is also consistent with the finding of 11 *Eco*RI chromosomal fragments hybridizing to 23S rRNA (Table 1). Two of these represent internal fragments (0.8 and 1.2 kilobases) common to all of the gene sets, and the remainder represent the terminal sequences from nine 23S rRNA genes.

In every case we have attempted to verify: that the cloned fragments represent true segments of the chromosome, whether the restriction site map of each cloned ribosomal fragment corresponds with the generalized site map, and that the spacer DNAs are heterologous as as-

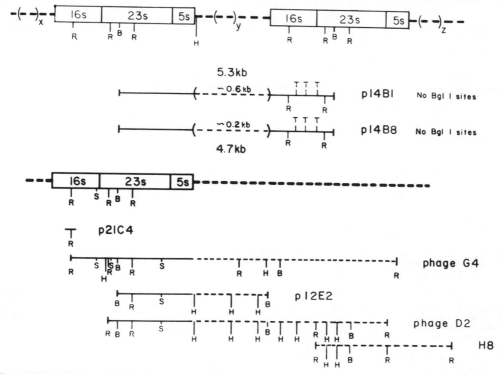

FIG. 2. *Plasmid and Charon 4A clones containing ribosomal DNA sequences from* B. subtilis. *Dashed lines represent unique spacer sequences, and solid lines represent sequences corresponding to an rRNA gene set. Unique DNA sequences contained in the HindIII fragments located to the right of the rRNA coding block in the plasmid p12E2 were used as a radioactive probe to identify overlapping DNA sequences from phage in the Charon 4A bank. The distal EcoRI fragments from the resulting phage (D2) was used as a probe to isolate the adjacent DNA sequence (phage H8). Phage D2 and H8 represent 21.2 kilobases of DNA adjacent to the portion of the rRNA gene set contained on p12E2. No homology to rRNA was detected in these spacer sequences. Phage G4 represents part of a different rRNA operon as indicated by the physical map of its spacer sequence. P21C4 was obtained from K. Hutchison (8). Restriction endonuclease abbreviations are as in Fig. 1 with the addition of T = TaqI.*

sayed by hybridization. Our objective is to make use of the unique nature of these linked spacer sequences to map the chromosomal site of each gene set. For these mapping experiments we used the chimeric plasmid pCS540 constructed by Chang and Cohen (2). This plasmid cannot replicate in *B. subtilis* but has a chloramphenicol marker that will function if introduced into the cell by transformation.

A 14.2-kilobase *Eco*RI restriction fragment from the plasmid p12E2 (9) was ligated into the *Eco*RI site of pCS540. This recombinant plasmid contains a portion of the 23S coding sequence, the 5S gene, and adjacent spacer DNA. This plasmid (designated pE2A) was purified from *E. coli* and then introduced into a recombination-proficient *B. subtilis* 168 strain by the protoplast transformation procedure of Chang and Cohen (2). Chloramphenicol-resistant *B. subtilis* transformants were selected.

Physical mapping has shown that this plasmid integrates into the *Bgl*I fragment which contains the p12E2 rRNA gene set (unpublished data). Genetic mapping of the integrated chloramphenicol resistance marker from pE2A was accomplished by means of PBS1-mediated transduction. Analogous plasmids have been used to map a gene of *B. subtilis* under sporulation control (6, 11).

The rRNA operon represented by the p12E2 plasmid maps near the guanine marker located between *purA* and *cysA* on the *B. subtilis* chromosome (Fig. 3). It now appears that this region, which has been reported to contain certain ribosome-associated functions (7, 17), contains at least one rRNA gene set (this study). Mapping of other unique sequences associated with different ribosomal gene sets from our clone collection is currently under way in our laboratory.

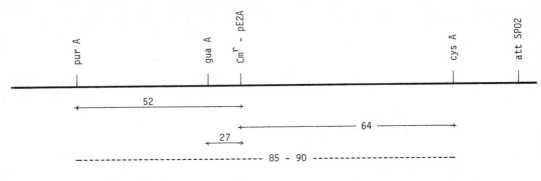

Recipient strains:

    K - 1   pur A,   cys A,   trp C
    B4   gua A,   trp C

FIG. 3. *Genetic map constructed by PSB1 transduction of chloramphenicol resistance. Map distances represent 100 − % cotransduction of the two markers. B. subtilis 168 derivative K-1 was obtained from R. Dedonder; B4, from J. Hoch.*

ACKNOWLEDGMENTS

We thank E. R. Allen, L. M. Anderson, M. F. Lampe, and M. A. Hollis for valuable discussion, technical assistance, and enthusiastic criticism during all phases of this study. P. Zuber generously exchanged plasmid clones and unpublished data from his dissertation.

This project was supported by grant PCM78-09814 from the National Science Foundation and Public Health Service grant GM26399 from the National Institute of General Medical Sciences. G.C.S. is the recipient of Public Health Service postdoctoral fellowship GM07900 from the National Institutes of Health.

LITERATURE CITED

1. **Chambliss, G. H., et al. (ed.).** 1980. Ribosomes: structure, function and genetics. University Park Press, Baltimore.
2. **Chang, S., and S. N. Cohen.** 1979. High frequency transformation of *Bacillus subtilis* protoplasts by plasmid DNA. Mol. Gen. Genet. **168:**111–115.
3. **Chow, L. T., and N. Davidson.** 1973. Electron microscope mapping of the distribution of ribosomal genes of the *Bacillus subtilis* chromosome. J. Mol. Biol. **75:**265–279.
4. **Colli, W., and M. Oishi.** 1969. Ribosomal RNA genes in bacteria: evidence for the nature of the physical linkage between 16S and 23S RNA genes in *Bacillus subtilis.* Proc. Natl. Acad. Sci. U.S.A. **64:**642–649.
5. **Colli, W., I. Smith, and M. Oishi.** 1971. Physical linkage between 5S, 16S, and 23S ribosomal RNA genes in *Bacillus subtilis.* J. Mol. Biol. **56:**117–127.
6. **Haldenwang, W. G., C. D. B. Banner, J. F. Ollington, R. Losick, J. A. Hoch, M. B. O'Connor, and A. L. Sonenshein.** 1980. Mapping a cloned gene under sporulation control by insertion of a drug resistance marker into the *Bacillus subtilis* chromosome. J. Bacteriol. **142:**90–98.
7. **Henner, D. J., and J. A. Hoch.** 1980. The *Bacillus subtilis* chromosome. Microbiol. Rev. **44:**57–82.
8. **Hutchison, K. W., and H. O. Halvorson.** 1980. Cloning of randomly sheared DNA fragments from a φ105 ly-

sogen of *Bacillus subtilis.* Identification of prophage containing clones. Gene **8:**267–278.
9. **Moran, C. P., and K. F. Bott.** 1979. Restriction enzyme analysis of the ribosomal RNA genes of *Bacillus subtilis.* J. Bacteriol. **140:**99–105.
10. **Moran, C. P., and K. F. Bott.** 1979. Organization of transfer RNA genes and ribosomal RNA genes in *Bacillus subtilis.* J. Bacteriol. **140:**742–744.
11. **Moran, C. P., R. Losick, and A. L. Sonenshein.** 1980. Identification of a sporulation locus in cloned *Bacillus subtilis* deoxyribonucleic acid. J. Bacteriol. **142:**331–334.
12. **Nomura, M., and L. E. Post.** 1980. Organization of ribosomal genes and regulation of their expression in *E. coli,* p. 671–691. *In* G. H. Chambliss et al. (ed.), Ribosomes: structure, function, and genetics. University Park Press, Baltimore.
13. **Potter, S. S., K. F. Bott, and J. E. Newbold.** 1977. Two-dimensional restriction analysis of the *Bacillus subtilis* genome: gene purification and ribosomal ribonucleic acid gene organization. J. Bacteriol. **129:**492–500.
14. **Rapoport, G., A. Klier, A. Billault, F. Fargette, and R. Dedonder.** 1979. Construction of a colony bank of *E. coli* containing hybrid plasmids representative of the *Bacillus subtilis* 168 genome. Mol. Gen. Genet. **176:**239–245.
15. **Rigby, P. W. J., M. Dieckmann, C. Rhodes, and P. Berg.** 1977. Labeling deoxyribonucleic acid to high specific activity in vitro by nick translation with DNA polymerase I. J. Mol. Biol. **113:**237–251.
16. **Sogin, M. L., N. R. Pace, M. Rosenberg, and S. M. Weissman.** 1976. Nucleotide sequence of a 5s ribosomal RNA precursor from *Bacillus subtilis.* J. Biol. Chem. **251:**3480–3488.
17. **Trowsdale, J. S., M. H. Chen, and J. A. Hoch.** 1979. Genetic analysis of a class of polymyxin resistant partial revertants of stage 0 sporulation mutants of *Bacillus subtilis*; map of the chromosome region near the origin of replication. Mol. Gen. Genet. **173:**61–70.
18. **Zingales, B., and W. Colli.** 1977. Ribosomal RNA genes in *Bacillus subtilis.* Evidence for a contranscription mechanism. Biochim. Biophys. Acta **474:**562–577.

Instability of Cloned *Bacillus subtilis* DNA Associated with a Transposon-Like Structure

KEITH W. HUTCHISON, ELAINE SACHTER, AND HARLYN O. HALVORSON

Rosenstiel Basic Medical Sciences Research Center, Brandeis University, Waltham, Massachusetts 02254

pBS16H7 is a recombinant plasmid carrying seven kilobase pairs of *Bacillus subtilis* DNA. The plasmid is unstable, and spontaneous deletions of the insert result in a series of smaller stable plasmids. The deletions of the insert are associated with a stem-loop structure.

Bacterial sporulation is a developmental process involving a sequential set of morphological and biochemical changes. To understand the molecular events which control this process, we need DNA templates encompassing a defined gene or set of genes involved in spore formation for the in vitro analysis of gene expression. As a first step in this process, a gene bank was constructed using randomly sheared DNA fragments from *Bacillus subtilis* cloned in *Escherichia coli* on the plasmid pMB9 (8). Statistically, the gene bank contained 78% of the *B. subtilis* chromosome (8).

It was our initial intention to probe the gene bank for plasmids containing the *spo0B* region of the *B. subtilis* chromosome. This region is not represented in the bank. In addition, only 40% of the φ105 prophage genome is in the bank. A number of other investigators have screened the gene bank for other regions of the chromosome and have also found them lacking (A. Aronson, H. Paulus, and L. Brown, personal communications). The only reported success has been the finding of clones containing the "*0.4 kb*" gene identified by Segall and Losick (5, 13), and the rRNA coding sequences (K. F. Bott, F. E. Wilson, and G. C. Stewart, this volume).

Several other observations have led us to believe that regions of the *B. subtilis* chromosome may be difficult to clone in *E. coli*. During construction of the gene bank, it was noted that of 2,700 plasmids analyzed by "toothpick assay" (1) less than 1,000 contained plasmids larger than pMB9 (8). In addition, the average insert size of the cloned DNA was 7 kilobases, whereas the DNA used to construct the bank was 20 kilobases in size. In this regard it should be noted that one of the rRNA clones is smaller than pMB9, yet contains part of the *B. subtilis* 16S

rRNA coding sequence (Bott et al., this volume). Finally, preliminary experiments hybridizing *B. subtilis* DNA to DNA from the gene bank indicated that less than 50% of the genome is represented (K. Hutchison, unpublished data).

The process generating adenine-thymine (AT) tails is known to result in deletions at the site of tail formation during construction of the recombinant plasmid (7). Such a process does not seem likely to cause extensive loss of the cloned DNA after the plasmid has been established. It was therefore of interest to isolate a plasmid in which

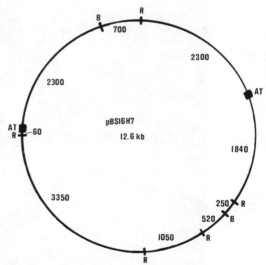

FIG. 1. *Restriction map of pBS16H7. Numbers on the inside of the circle are the number of base pairs for each fragment. Symbols: ■, AT tails (approximately 110 base pairs); R, EcoRI restriction endonuclease sites; B, BamHI restriction endonuclease sites.*

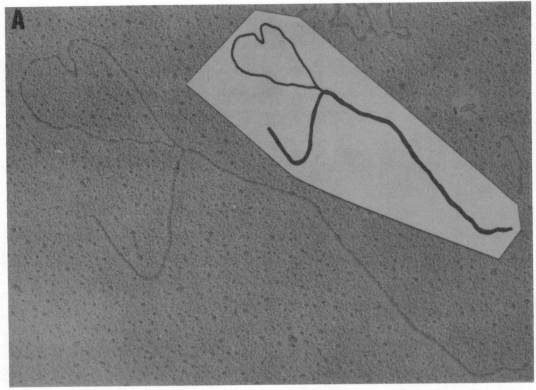

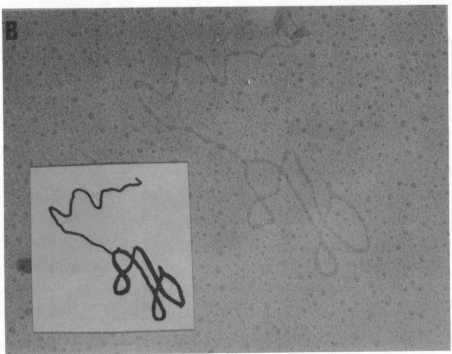

the loss of inserted DNA could be investigated. Plasmid pBS16H7 was originally identified by colony hybridization (4) with [^{32}P]RNA from sporulating cells, though the relationship of the plasmid with sporulation has not been confirmed. The intact plasmid is 12.6 kilobase pairs in size. During growth, a series of smaller stable plasmids are formed by loss of varying amounts of inserted DNA. We present here an analysis of the plasmid pBS16H7 and its instability.

RESULTS AND DISCUSSION

The restriction map of pBS16H7 was determined by partial and double digests with EcoRI and BamHI restriction endonucleases (Fig. 1). During analysis of pBS16H7, it was noted that more than one plasmid could be found in preparations of plasmid DNA. All additional plasmids were smaller than pBS16H7. Individual colonies could be obtained carrying only the smaller plasmids, and all clones carrying pBS16H7 also carried a smaller plasmid. A heteroduplex between pBS16H7 and a smaller plasmid showed the smaller plasmid to be completely homologous to pBS16H7 (Hutchison, unpublished data). It appeared, therefore, that pBS16H7 was an unstable plasmid, generating smaller, stable plasmids of various sizes. The instability of pBS16H7 could not be related to its size, cultural conditions, or the use of chloramphenicol to amplify the plasmid. In addition, the host strain, E. coli HB101, was shown to have retained its RecA phenotype (Hutchison, unpublished data).

To determine whether the instability of pBS16H7 was related to a particular area of the plasmid, we formed heteroduplexes between pBS16H7 and several smaller plasmids. pBS16H7 was linearized by partial digestion with BamHI restriction endonuclease. All the small plasmids isolated so far have lost the BamHI site within the insert. Heteroduplexes between pBS16H7 and the smaller plasmids therefore give two types of structures, depending

on whether pBS16H7 was cleaved in the pMB9 portion of the molecule (type I) or in the insert portion of the molecule (type II). These two possible structures were seen in a heteroduplex between pBS16H7 and one of the smaller plasmids, PBS16H7-6 (Fig. 2). Figure 2A represents a typical deletion loop structure. The apparent stem was not seen in other type I molecules and is believed to be a consequence of the extended nature of this particular molecule. Figure 2B shows only one single-stranded arm coming from the double-strand circle. This is because the deletion in generating pBS16H7-6 terminates at the BamHI site within the insert.

A surprising structure was seen when pBS16H7 was heteroduplexed to another small plasmid, pBS16H7-5 (Fig. 3A and B). In this case, not only was inserted DNA lost, but also the remaining portion of B. subtilis DNA was inverted in the plasmid (Fig. 3C). Measurements from the heteroduplexes place the retained portion of B. subtilis DNA as coming from the center of the insert (Fig. 3C).

The results of the analysis of several plasmids are shown in Fig. 4. The smallest plasmid isolated so far is pBS16H7-8, which is at or near the size of pMB9. It has not been determined whether some or all of the AT tails are still present. It is obvious from Fig. 4 that the endpoints of excision are random and that there is a central portion of the molecule which is always lost. The exception to this is pBS16H7-5, in which the central portion of the molecule is retained in an inverted position.

A possible reason for the instability of the plasmid was seen when pBS16H7 was allowed to reanneal under conditions favoring intrastrand hybridization. "Snapback" structures formed by pBS16H7 are shown in Fig. 5. The pMB9 portion of the molecule is separated from the insert by the double-stranded neck formed by the AT tails. An obvious stem-loop structure was seen near the BamHI site of the insert (Fig. 5B).

It should be noted that the molecule shown in

FIG. 2. *Heteroduplex of pBS16H7 DNA. pBS16H7 was linearized by partial digestion with BamHI restriction endonuclease. The full-sized linear molecule was purified by electrophoresis on a 0.7% agarose gel. pBS16H7-6 was linearized by complete digestion with BamHI restriction endonuclease. Amounts of 1 µg of each DNA were mixed (total volume, 75 µl); and 125 µl of 0.1 N NaOH–0.02 M EDTA was added. The DNA was incubated for 10 min at room temperature, and 25 µl of 2 M Tris, pH 7.4, and 250 µl of formamide were added. The DNA was incubated at 19°C for 2 h. The DNA was spread for electron microscopy by the method of Kleinschmidt (10). (A) Type I heteroduplex, pBS16H7 cleaved at the BamHI site in pMB9. (B) Type II heteroduplex, pBS16H7 cleaved at the BamHI site in the insert. Measurements of the heteroduplexes were made by tracing electron micrographs with an electronic graphics calculator (Numonics Instruments). At least 10 heteroduplexes were measured (percent σ < 2).*

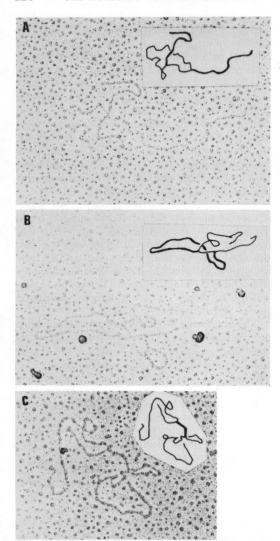

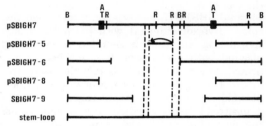

FIG. 4. *Location of deletions and stem-loop structure of pBS16H7 DNA. Symbols: as in Fig. 1, except for* – – – –, *border of stem structure, and* –·–·–, *border of loop structure.*

pMB9 site. The secondary spur has not been seen in molecules cut at the insert site. In these molecules the loop does not increase in size. To account for this, we believe that the spur at the end of the molecule is folded back on itself.

The correlation of the stem-loop structure with the loss of cloned DNA is intriguing. The stem-loop structure is similar to that of transposons in *E. coli*. To date, no indigenous transposons have been reported in *B. subtilis*. Insertion sequences, transposable elements, and bacteriophage Mu are known to generate deletions in the chromosome upon curing (3, 9, 12). Any given deletion is generally to one side or the other of the transposon (9). Deletions generated in pBS16H7 extend to both sides of the stem-loop structure.

Movable regions of the genome, involving site-specific recombinational events and the regulation of gene expression, are not uncommon in biological systems (2, 6, 11, 13, 14, 16). Another possibility is that the stem-loop structure represents an invertible region of the chromosome. Such a region, with a similar stem-loop structure, has been shown to exist in *Salmonella*, for the control of flagellum synthesis (15, 16).

B. subtilis DNA appears to be quite unstable when cloned on plasmids in *E. coli*. Plasmid pBS16H7 is the first unstable plasmid to be analyzed. The instability is associated with a particular stem-loop structure. It is possible that pBS16H7 may represent a unique region of the chromosome. However, given the instability of cloned *B. subtilis* DNA in *E. coli*, it may represent a unique opportunity to study a general phenomenon.

FIG. 3. *Heteroduplex of pBS16H7 DNA and pBS16H7-5 DNA. See Fig. 2 for details. (A) Type I heteroduplex. (B) Type II heteroduplex. (C) Type I heteroduplex with hybridization in the inverted region.*

Fig. 5B has not formed the AT stem. Snapback structures can also be found (approximately 25%) in which the AT stem has formed, but not the stem-loop structure. It has not been determined whether the AT stem or the stem-loop structure is more stable. The stem-loop structure has a stem of approximately 120 base pairs, a loop of 1.5 kilobases, and a secondary spur of 240 base pairs. This secondary spur was seen in 75% of the molecules which had been cut at the

ACKNOWLEDGMENT

This work was supported by Public Health Service research grant GM 18904 from the National Institute of General Medical Sciences.

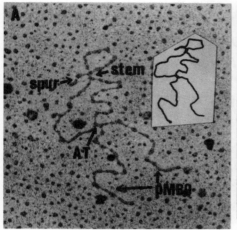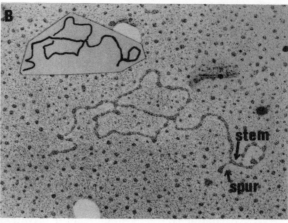

FIG. 5. *Snapback of pBS16H7. (A) Type I molecule. (B). Type II molecule. Snapback analysis was as described for Fig. 2 except that no small plasmid DNA was added, the formamide concentration was 30% (vol/vol), and the incubation time for reannealing was reduced to 30 min.*

LITERATURE CITED

1. **Barnes, W. N.** 1977. Plasmid detection and siting in single colony isolates. Science **195**:393–394.
2. **Bernard, O., N. Hozumi, and S. Tonegawa.** 1978. Sequences of mouse immunoglobin light chain genes before and after somatic changes. Cell **15**:1133–1144.
3. **Bukhari, A. I.** 1975. Reversal of mutator phage Mu integration. J. Mol. Biol. **96**:87–99.
4. **Gergen, J. P., R. H. Stern, and P. C. Wensink.** 1979. Filter replicas and permanent collections of recombinant DNA plasmids. Nucleic Acids Res. **7**:2115–2136.
5. **Haldenwang, W. G., C. D. B. Banner, J. F. Ollington, R. Losick, J. A. Hoch, M. B. O'Connor, and A. L. Sonenshein.** 1980. Mapping a cloned gene under sporulation control by insertion of a drug resistance marker into the *Bacillus subtilis* chromsome. J. Bacteriol. **142**:90–98.
6. **Hicks, J. B., J. N. Strathern, and I. Herskowitz.** 1977. The cassette model of mating-type interconversion, p. 457–462. *In* A. I. Bukhari, J. A. Shapiro, and J. L. Adhya (ed.), DNA insertion elements, plasmids and episomes. Cold Spring Harbor Laboratory, Cold Spring Harbor, N.Y.
7. **Humphries, P., R. Old, L. W. Coggins, T. McShane, C. Watson, and J. Paul.** 1978. Recombinant plasmids containing *Xenopus laeris* globin structural genes derived from complementary DNA. Nucleic Acids Res. **5**:905–924.
8. **Hutchison, K. W., and H. O. Halvorson.** 1980. Cloning of randomly sheared fragments from a φ 105 lysogen of *Bacillus subtilis*. Identification of prophage containing clones. Gene **8**:267–278.
9. **Kleckner, N., R. K. Chan, B. K. Tye., and D. Botstein.** 1975. Mutagenesis by insertion of a drug-resistance element carrying an inverted repetition. J. Mol. Biol. **97**:561–575.
10. **Kleinschmidt, A. K.** 1968. Monolayer techniques in electron microscopy of nucleic acid molecules. Methods Enzymol. **12B**:361–377.
11. **McClintock, B.** 1957. Chromsome organization and genic expression. Cold Spring Harbor Symp. Quant. Biol. **21**:197–216.
12. **Reif, H.-J., and H. Saedler.** 1975. IS1 is involved in deletion formation in the *gal* region of *E. coli* K12. Mol. Gen. Genet. **137**:17–28.
13. **Segall, J., and R. Losick.** 1977. Cloned *Bacillus subtilis* DNA containing a gene that is activated early during sporulation. Cell **11**:751–761.
14. **Seidman, J. G., E. E. Max, and P. Leder.** 1979. A kappa-immunoglobulin gene is formed by site-specific recombination without further somatic mutation. Nature (London) **280**:370–375.
15. **Zieg, J., M. Hilmen, and M. Simon.** 1978. Regulation of gene expression by site-specific inversion. Cell **15**:237–244
16. **Zieg, J., M. Silvermen, M. Hilmen, and M. Simon.** 1977. Recombinational switch for gene expression. Science **196**:170–172.

B. Initiation of Sporulation and Regulation of Metabolism

Initiation of *Bacillus subtilis* Sporulation Caused by the Stringent Response

JUAN M. LOPEZ, KOZO OCHI, AND ERNST FREESE

Laboratory of Molecular Biology, National Institutes of Neurological and Communicative Disorders and Stroke, Bethesda, Maryland 20205

Sporulation of *Bacillus subtilis* can be initiated by the stringent response to partial or transient amino acid deprivation. Relaxed (*relA*) mutants sporulate only when the synthesis of guanine nucleotides is decreased by other means, e.g., by addition of decoyinine. Decoyinine can also counteract the inhibition of sporulation by fusidic acid and kasugamycin.

Experiments from our laboratory had demonstrated that a partial decrease in the synthesis of guanine nucleotides initiates sporulation (3, 4). For example, sporulation can be induced, in the presence of excess ammonium ions, glucose, and phosphate, by the addition of decoyinine, a specific inhibitor of GMP synthetase. In addition, all conditions of nutritional starvation initiating sporulation, including nitrogen or phosphate limitation or the step-down from Casamino Acids/glutamate medium to glutamate medium, were accompanied by a decrease in GTP (and GDP) (8). As the latter step-down condition involves the transient deprivation of amino acids, it produces a stringent response, which includes a decrease in the rate of RNA synthesis, an increase in ppGpp and pppGpp, and a decrease in the synthesis of guanine nucleotides (5; J. M. Lopez, A. Dromerick, and E. Freese, submitted for publication). We show here that the sporulation resulting from this step-down is mainly caused by the stringent response because it does not occur in a relaxed strain (gene symbol, *relA*), which lacks the symptoms of the stringent response. To verify that the stringent response alone can initiate sporulation, we have also produced it in an auxotroph by controlling the supply of an amino acid.

Several groups have found that various antibiotics which bind to ribosomes inhibit sporulation at concentrations at which they affect growth only little (e.g., fusidic acid [2], kasuga-mycin [16; E. R. Allen and K. F. Bott, Abstr. Annu. Meet. Am. Soc. Microbiol. 1980, I102, p. 101]). In addition, many mutants resistant to such antibiotics do not sporulate at either normal or elevated temperatures (for reviews, see 10, 12). These effects have been interpreted to imply that certain properties of the translation machinery (ribosomes) are specifically needed for sporulation and not for growth. We show here that the inhibition of sporulation by fusidic acid or kasugamycin can be counteracted by decoyinine, an inhibitor of GMP synthetase.

RESULTS

Sporulation and nucleotide changes after transfer from Casamino Acids/glutamate to glutamate medium. The sporulation of stringent (*rel+*) and *relA* strains observed 10 and 20 h after transfer from Casamino Acids/glutamate medium to glutamate medium (13) is shown in Table 1. Whereas the *rel+* strain (61831) had already sporulated well 10 h after the transfer, the *relA* strain (61852) produced a much lower spore titer even 20 h after the transfer. The changes in the intracellular concentrations of the nucleoside triphosphates in the stringent strain have been published before (8); only GTP decreased, whereas the other nucleoside triphosphates increased for at least 1 h. Figure 1 shows the rapid decrease of GTP which corresponded to the increase of ppGpp (and pppGpp; not shown); later, these changes were

TABLE 1. *Sporulation after cell exposure to different step-down conditions*

Strain	Genotype except *rel*	*relA* property	Condition	Spores/ml after	
				10 h	20 h
61831	*lys trpC2*	+	CAA-Glu → Glu[a]	8×10^6	6×10^7
61852	*lys trpC2*	−	CAA-Glu → Glu	8×10^3	7×10^5
		−	CAA-Glu → Glu + Dec[b]	2×10^6	5×10^7
61886	*ilv-del1 kauA1*	+	0 mM Omv[c]	2×10^2	ND[d]
			0.4 mM Omv	4×10^7	ND
			1.5 mM Omv	3×10^2	ND
61885	*ilv-del1 kauA1*	−	0 mM Omv	1×10^3	ND
			0.4 mM Omv	1×10^3	ND
			0.4 mM Omv + Dec[e]	5×10^6	ND
			1.5 mM Omv	9×10^2	ND
			1.5 mM Omv + Dec[e]	1.2×10^7	ND

[a] Cells were grown in Casamino Acids/glutamate medium (CAA-Glu) (13) containing 2 mM K_2HPO_4, 0.186 mM $CaCl_2$, and 20 mM morpholinopropane sulfonate, adjusted to pH 7.0 with KOH. When the optical density at 600 nm was 0.8, the cells were rapidly collected on a membrane filter and suspended in the same volume of a glutamate replacement medium (Glu) (15) containing the same additions as above.

[b] The replacement medium also contained 1.8 mM decoyinine (Dec).

[c] Cells were grown and transferred as in Fig. 2a. The titer of heat-resistant spores (20 min at 75°C) was determined 10 h after the transfer.

[d] ND, Not determined.

[e] The transfer medium also contained 1.0 mM decoyinine (Dec).

reduced, presumably because the rates of RNA and protein synthesis had then adapted to the new nutrient condition. In the relaxed strain, ppGpp (and pppGpp; not shown) increased only little, while GTP decreased only slowly and to a level similar to that eventually observed for the stringent strain. This limited GTP decrease was not sufficient to initiate sporulation (Table 1). However, if the concentration of GTP in this strain was further decreased by the addition of decoyinine (Fig. 1), the relaxed strain also sporulated well (Table 1). (GTP later increased again, probably because RNA [and protein] synthesis adapted to the reduced supply of GTP resulting from the partial inhibition of the synthesis of guanine nucleotides.)

Sporulation and nucleotide changes after limitation of one amino acid. Since amino acids are actively transported, with K_m values in the micromolar range, we could control their intracellular concentration by the extracellular concentration only if we avoided the active transport. We have done this in several ways, only one of which is described here. We used an *ilv kau* double mutant (61886) deficient in isoleucine and valine synthesis and in the active transport of the oxo-acid precursors of these amino acids (6). We grew this strain in synthetic medium (containing excess glucose) in the presence of isoleucine and valine (Ile medium) and transferred the cells early in exponential growth to a medium containing valine and different concentrations of oxomethylvalerate (Omv). Without Omv or with excess (>1.5 mM) Omv, the mutant did not sporulate. However, at intermediate Omv concentrations good sporulation was observed which was optimal at 0.4 mM Omv (Table 1). In contrast, the isogenic relaxed strain (*ilv kau rel* = 61885) did not sporulate at any Omv concentration, but it did sporulate upon addition of decoyinine (with both 0.4 and 1.5 mM Omv). We used the condition which allowed optimal sporulation of the stringent strain (61886; 0.4 mM Omv = Omv medium) to measure the nucleotide changes after cell transfer. In the stringent strain, both ppGpp and pppGpp increased while GTP decreased rapidly; eventually, the concentrations of these nucleotides leveled off at a less extreme value (Fig. 2). In the relaxed strain, ppGpp and pppGpp increased only little and GTP decreased slowly to values similar to those eventually reached in the stringent strain. Although the limited GTP decrease did not suffice to initiate sporulation of the relaxed strain, this strain did sporulate (Table 1) when the concentration of GTP was further reduced (Fig. 2) by addition of decoyinine at the time of cell transfer to Omv medium. (GTP decreased below the critical concentration of about 150 pmol/AM [see Fig. 1 legend] only slowly Fig. 2], and sporulation was correspondingly delayed by about 2 h [not shown].)

Inhibition of sporulation by fusidic acid and kasugamycin. Kasugamycin was a much

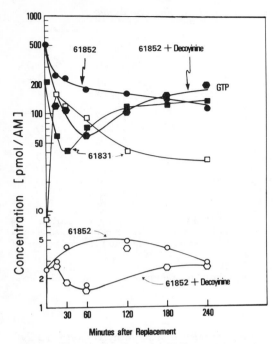

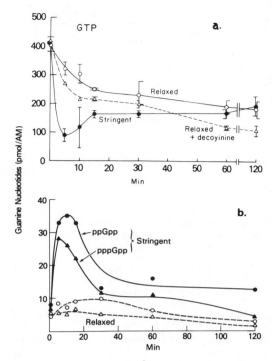

FIG. 1. *Changes in the intracellular concentration of guanine nucleotides after cell transfer of a stringent (61881) or a relaxed (61852) strain from a Casamino Acids/glutamate medium to a glutamate medium. Cells were grown (at 37°C from an optical density at 600 nm of 0.1) in Casamino Acids/glutamate medium (15) containing 2 mM (final concentration) $K_2H^{32}PO_4$ (100 µCi/ml), 20 mM morpholinopropane sulfonate, adjusted to pH 7.0 by KOH, 50 µg of L-lysine per ml, and 70 µg of L-tryptophan per ml. When the optical density at 600 nm was 0.8, the cells were washed on a membrane filter and transferred to a glutamate medium (15) containing the same concentrations of L-lysine, L-tryptophan, and $H^{32}PO_4$. Decoyinine (1.8 mM) was added at the time of transfer to a parallel culture of the relaxed strain. Just before and at different times after the transfer, 100-µl amounts of culture were transferred to 3 µl of 20 M formic acid on ice. After 60 min, the extracts were freed of all debris and frozen. They were later chromatographed and electrophoresed together with standards, and the radioactivity of relevant spots was measured (10). Solid symbols, GTP; open symbols, ppGpp. The intracellular concentration of pppGpp was always lower than that of ppGpp. The results are expressed in picomoles per AM, where AM measures the amount of cells. One AM unit is that amount of cells that would give an optical density at 600 nm of 1 if it were suspended in 1 ml. No pppAppp was detected, which means that its concentration must have been below 4 pmol/AM.*

more potent inhibitor of growth in the Casamino Acids/glutamate medium than in the Omv medium, whereas fusidic acid inhibited growth with similar potency in both media (see legends to Fig. 3 and 4). When cells (strains 61886) were

FIG. 2. *Changes in the intracellular concentrations of guanine nucleotides after cell transfer of a stringent (61886) or a relaxed (61885) ilv kau auxotroph to limiting Omv concentrations. (a) Cells were grown in synthetic medium [5 mM K_2HPO_4, 100 mM morpholinopropane sulfonate, adjusted to pH 7.0 by KOH, 10 mM $(NH_4)_2SO_4$, 2 mM $MgCl_2$, 0.7 mM $CaCl_2$, 50 µM $MnCl_2$, 5 µM $FeCl_3$, 1 µM $ZnCl_2$, 2 µM thiamine, 55 mM D-glucose, and 20 mM L-glutamate, adjusted to pH 7.0 by KOH] containing 1 mM L-isoleucine and 2 mM DL-valine. When the optical density at 600 nm was 0.5, the cells were rapidly washed with synthetic medium on membrane filters and transferred to several flasks with 100 ml of synthetic medium containing 0.4 mM Omv and 0.6 mM DL-valine. For the relaxed strain (61885) one transfer culture (○) contained only the stated additions whereas a parallel culture (△) also contained decoyinine (2.0 mM). Just before and at the stated times after the transfer, the cells were rapidly collected on membrane filters (10-cm diameter, Schleicher & Schuell Co.) and extracted by 1.5 ml of ice-cold formic acid (0.5 M). The extracts were freed from debris and dried by centrifugation under a vacuum. Nucleotide amounts were determined by high-pressure chromatography (Ochi, Kandala, and Freese, submitted for publication). Each point represents an average of one to four experiments; the horizontal bars represent the values for two experiments or the standard deviation for more than two experiments. (b) Cells were grown as in (a), but both the initial and the transfer medium contained 2 mM $K_2H^{32}PO_4$ (100 µCi/ml). At different times, 1-ml portions were filtered and the cells were extracted by 0.1 ml of 0.5 M formic acid. The nucleotides in the extracts were chromatographed and*

transferred from Ile to Omv medium containing different amounts of an antibiotic, the spore titer decreased with increasing antibiotic concentration; fusidic acid was more potent than kasugamycin (Fig. 3). In the absence of an antibiotic, when sporulation was initiated by the addition of decoyinine (to a culture of strain 61886 in Ile medium), the spore titer was a little higher than that observed after Omv deprivation (Fig. 3). This was typical for the more efficient sporulation initiation caused by decoyinine addition than that caused by partial deprivation of one amino acid. Addition of one of the antibiotics together with decoyinine, either to a culture in Ile medium or after cell transfer to the Omv medium, inhibited sporulation much less than when the antibiotic was added alone to the culture in Omv medium (Fig. 3).

The sporulation initiated by the transfer of cells (60015) from Casamino Acids/glutamate medium to glutamate medium was not much inhibited (less than 65%) by concentrations of fusidic acid (0.13 µg/ml) which inhibited growth in the Casamino Acids/glutamate medium (by 60%). But kasugamycin strongly inhibited sporulation, and this effect was again counteracted by decoyinine (Fig. 4).

For both antibiotics, the nucleotide changes resulting from cell transfer to the replacement media were measured. When kasugamycin (75 µg/ml) was present after cells had been transferred from Casamino Acids/glutamate to glutamate medium, both the height and the duration of the increase of ppGpp and pppGpp were reduced (maximal values reached were 6 and 9 pmol/AM, respectively); correspondingly, GTP decreased only for a short time (15 min) to 33% and then increased back to 78% of its original value (which was 462 pmol/AM). In the presence of both kasugamycin and decoyinine (1.8 mM), the GTP concentration decreased to 20% of its original value and then increased only slowly (in 90 min) to 50% of this value. The effects in the Omv medium were much more drastic: the increases of ppGpp and pppGpp were very small (maximal values of 6.5 and 7.5 pmol/AM, respectively, reached after 15 min with fusidic acid [0.14 µg/ml], and 6.5 and 5.0 pmol/AM reached after 5 min with kasugamycin [400 µg/ml]). Instead of decreasing, the concentration of GTP increased (from 440 to a maximum of 1,970 pmol/AM in 10 min with fusidic acid and from 484 to a maximum of 1,200 pmol/AM in 5 min with kasugamycin). However, addition of decoyinine (optimal levels of 0.3 mM

to counteract fusidic acid and 0.5 mM to counteract kasugamycin) again caused the normal GTP decrease (by 70 to 80%).

DISCUSSION

We have shown that sporulation can be initiated by the stringent response to partial amino

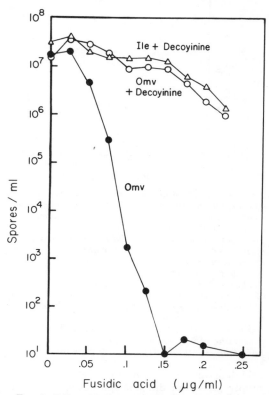

FIG. 3. Effects of fusidic acid on sporulation initiated by partial Omv deprivation or by decoyinine addition. Cells of the stringent ilv kau strain (61886) were grown in synthetic medium (see Fig. 2) containing 0.5 mM each L-isoleucine and L-valine. When the optical density at 600 nm was 0.5, the cells were washed and transferred to the same medium (triangles) or to limiting Omv conditions as in Fig. 2 (circles); the transfer flasks contained different amounts of fusidic acid (●) or the same plus decoyinine (0.3 mM; open symbols). The titer of heat-resistant spores was determined 13 h after the transfer. Much higher concentrations of kasugamycin were needed to inhibit sporulation; the inhibition was prevented by decoyinine. For example, 300 µg of kasugamycin per ml reduced the spore titer to 80/ml during Omv deprivation but only to 2.5×10^6/ml with Omv + decoyinine (0.5 mM) and to 3.0×10^6/ml with isoleucine + decoyinine (0.5 mM). The interpolated antibiotic concentrations producing 50% growth inhibition in synthetic medium containing isoleucine and valine were 0.2 µg of fusidic acid per ml and 4 mg of kasugamycin per ml.

quantitated (8). Circles, ppGpp; triangles, pppGpp; solid symbols, stringent strain (61886); open symbols, relaxed strain (61885).

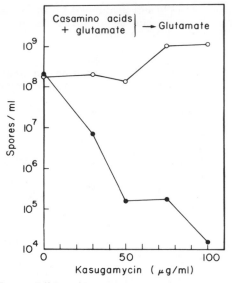

FIG. 4. *Effects of kasugamycin and decoyinine on sporulation initiated by transfer from Casamino Acids/glutamate medium to glutamate medium. Cells of the stringent strain 60015 (metC7 trpC2) were grown in Casamino Acids/glutamate medium to an optical density at 600 nm of 0.5, collected on a membrane filter, and suspended in glutamate medium as in Fig. 1. The suspended cells were distributed to flasks containing different amounts of kasugamycin (●) or the same plus 1.8 mM decoyinine (○). The titer of heat-resistant spores was determined 10 h after the transfer. The growth in Casamino Acids/glutamate medium was inhibited 67% by 0.13 µg of fusidic acid per ml and 58% by 75 µg of kasugamycin per ml. Only kasugamycin inhibited sporulation by more than 90%.*

acid deprivation. Without deprivation, the cells simply grew and did not sporulate; when the required amino acid (isoleucine here) was completely missing (no Omv added), sporulation could not proceed, presumably because the necessary macromolecules could not be synthesized. The sporulation resulted from the stringent response, which included an extensive decrease of GTP, because it did not occur in a *relA* mutant. But the stringent response is not necessary for sporulation because *relA* strains can sporulate in nutrient sporulation medium (9, 11) or in synthetic medium plus decoyinine (which decreases the synthesis of guanine nucleotides but does not produce a stringent response). Thus, all examined conditions of nutritional deprivation which initiate sporulation have been correlated with a decrease of GTP (and GDP), whereas other nucleotides have increased under some and decreased under other conditions (3, 4, 8).

The increase of (p)ppGpp and the decrease of GTP were most pronounced very soon after the

step-down; eventually, the cells adapted to the new nutrient condition. When only one amino acid was deficient, the changes of guanine nucleotides were less pronounced and the adaptation occurred faster (20 min; Fig. 2) than when many amino acids were removed (120 min; Fig. 1). We have generally found that the partial deprivation of a single amino acid does not produce as efficient sporulation as the longer-lasting effect resulting from the deficiency of several amino acids.

The antibiotics fusidic acid and kasugamycin, which affect ribosomal function, inhibited the sporulation initiated by the stringent response to partial Omv deprivation (Fig. 3). This inhibition could be counteracted by the addition of decoyinine. Kasugamycin also inhibited sporulation after cell transfer from Casamino Acids/glutamate to glutamate medium (Fig. 4), and decoyinine again restored sporulation completely. These results show that at least these two antibiotics do not inhibit sporulation by preventing the translation of sporulation-specific genes but only by interfering with the metabolic changes needed for the initiation of sporulation under a particular nutrient condition, i.e., one producing a stringent response. They do not inhibit sporulation initiated by another condition, i.e., deprivation of guanine nucleotides by decoyinine. As will be analyzed elsewhere, the effect of decoyinine on the inhibition by other antibiotics varies from complete to no restoration of sporulation.

At concentrations at which they inhibited sporulation almost completely, the antibiotics used here reduced the stringent response, as was shown by a reduced increase of (p)ppGpp and either a reduced decrease of GTP (after transfer to glutamate) or even a strong increase of GTP (in Omv medium). Therefore, the antibiotics may inhibit sporulation by reducing the stringent response. This effect could be produced in two ways: (i) the binding of the antibiotic to the ribosomes may directly interfere with the function of the stringent factor which binds to the 50S ribosome and, in the presence of uncharged tRNA, produces ppGpp by using ATP to phosphorylate GTP (2), or (ii) the inhibition of protein synthesis by the antibiotic may keep more tRNA in the charged state and thereby indirectly prevent the stringent response, which depends on the presence of uncharged tRNA. Fusidic acid (15) binds to elongation factor G of the 50S subunit and inhibits the GTP hydrolysis involved in the translocation step of protein synthesis. Sporulation eventually becomes resistant to fusidic acid (2). This is understandable, at least under the conditions used here, because

2 to 3 h after sporulation has been initiated, its continuation no longer depends on the presence of the stringent response (5; Ochi, Kandala, and Freese, submitted for publication). Kasugamycin, like other aminoglycosides, binds to the 30S ribosomal subunit. In contrast to other aminoglycosides, it exclusively inhibits initiation of protein synthesis (7, 15).

LITERATURE CITED

1. **Cashel, M.** 1975. Regulation of bacterial ppGpp and pppGpp. Annu. Rev. Microbiol. **29**:301–318.
2. **Fortnagel, P., and R. Bergmann.** 1973. Alteration of the ribosomal fraction of *Bacillus subtilis* during sporulation. Biochim. Biophys. Acta **299**:136–141.
3. **Freese, E., J. E. Heinze, and E. M. Galliers.** 1979. Partial purine deprivation causes sporulation of *Bacillus subtilis* in the presence of excess ammonia, glucose and phosphate. J. Gen. Microbiol. **115**:193–205.
4. **Freese, E., J. M. Lopez, and K. Ochi.** 1981. Role of guanine nucleotides and of the stringent response to amino acid deprivation in the initiation of bacterial sporulation, p. 11–16. *In* D. Schlessinger (ed.), Microbiology—1981. American Society for Microbiology, Washington, D.C.
5. **Gallant, J. A.** 1979. Stringent control in *E. coli*. Annu. Rev. Genet. **13**:393–415.
6. **Goldstein, B. J., and S. A. Zahler.** 1976. Uptake of branched-chain α-keto acids in *Bacillus subtilis*. J. Bacteriol. **127**:667–670.
7. **Helser, T. L., J. E. Davies, and J. E. Dahlberg.** 1972. Mechanism of kasugamycin resistance in *Escherichia coli*. Nature (London) New Biol. **235**:6–9.
8. **Lopez, J. M., C. L. Marks, and E. Freese.** 1979. The decrease of guanine nucleotides inititiates sporulation of *Bacillus subtilis*. Biochim. Biophys. Acta **587**:238–252.
9. **Nishino, T., J. Gallant, P. Shalit, L. Palmer, and T. Wehr.** 1979. Regulatory nucleotides involved in the *rel* function of *Bacillus subtilis*. J. Bacteriol. **140**:671–679.
10. **Piggot, P., and J. Coote.** 1976. Genetic aspects of bacterial endospore formation. Bacteriol. Rev. **40**:908–962.
11. **Rhaese, H. J., R. Grade, and H. Dichtelmueller.** 1976. Studies on the control of development: correlation of initiation of differentiation with synthesis of highly phosphorylated nucleotides in *Bacillus subtilis*. Eur. J. Biochem. **64**:205–213.
12. **Sonenshein, A. L., and K. M. Campbell.** 1978. Control of gene expression during sporulation, p. 179–192. *In* G. Chambliss and J. C. Vary (ed.), Spores VII. American Society for Microbiology, Washington, D.C.
13. **Sterlini, J., and J. Mandelstam.** 1969. Commitment to sporulation in *Bacillus subtilis* and its relationship to development of actinomycin resistance. Biochem. J. **113**:29–37.
14. **Tai, P.-C., B. J. Wallace, and B. D. Davis.** 1973. Actions of aurintricarboxylate, kasugamycin and pactamycin on *Escherichia coli* polysomes. Biochemistry **12**:616–620.
15. **Tanaka, N.** 1975. Fusidic acid, p. 436–447. *In* J. W. Corcoran and F. E. Hahn (ed.), Antibiotics III. Springer Verlag, New York.
16. **Tominaga, A., and Y. Kobayashi.** 1978. Kasugamycin-resistant mutants of *Bacillus subtilis*. J. Bacteriol. **135**:1149–1150.

Role of the *spo0F* Gene in Differentiation in *Bacillus subtilis*

H. J. RHAESE, R. GROSCURTH, E. AMANN, H. KÜHNE, AND R. VETTER

Institut für Mikrobiologie/Molekulare Genetik, Fachbereich Biologie, Universität Frankfurt, Frankfurt/M., Federal Republic of Germany

Synthesis of the highly phosphorylated nucleotide adenosine-bis-triphosphate (p_3Ap_3) was measured by using various growth and replacement media. Its regulation was determined in an in vitro system which allowed us to measure sugar transport and p_3Ap_3 synthesis simultaneously. Cloning of the *spo0F* gene, which seems to code for p_3Ap_3 synthetase in a λ vector, has been achieved.

In a series of papers (5–8, 12), we have shown that sporulation in *Bacillus subtilis* apparently requires synthesis of the unusual highly phosphorylated nucleotide adenosine-5′,3′(2′)bis-triphosphate (abt or p_3Ap_3). This nucleotide is produced when growing cells are deprived of carbon sources which trigger sporulation. A mutant (JH649) with a mutation in the *spo0F* gene is unable to synthesize p_3Ap_3 (12). Spontaneous revertants of this mutant regain the ability both to sporulate and to synthesize p_3Ap_3 (12). In vitro synthesis of the nucleotide is achieved (10) by using membrane vesicles from the sporogenous wild-type strain, which always synthesize p_3Ap_3 regardless of the developmental stage; membranes from the asporogenous (*spo0F*) mutant JH649 do not. We were able to isolate a DNA fragment with a molecular weight of 1.3 × 10^6 which is able to transform mutant JH649 to sporogeny and to p_3Ap_3-synthesizing capacity (11). Cloning of this gene in the *B. subtilis* plasmid pBS161-1 (2) did not cause mutant JH649 to sporulate after transfection, indicating that expression of this gene is somehow affected (11).

In a recent paper Gallant and co-workers (4) confirmed our discovery of two highly phosphorylated nucleotides, p_2Ap_2 and p_3Ap_2, synthesized in *B. subtilis* after nutrient shift down (9). However, they claimed that p_3Ap_3, which seems to correlate with sporulation, is not present in this organism. Forced to repeat these experiments, we found that we were also unable to detect p_3Ap_3 when using their conditions, for the following reasons. (i) Nutrient sporulation medium (1) contains 0.1 M phosphate buffer. With this medium it is not possible to reach a sufficiently high specific activity to detect p_3Ap_3. (ii) The chromatography system used does not allow migration of p_3Ap_3 in formic acid/LiCl. Thus, p_3Ap_3 remains at the origin and hides under other substances when chromatographed in the second dimension. (iii) Nutrient sporulation medium does not contain glucose (1). Therefore, the glucose analog α-methylglucoside cannot mimic a downshift and never induces sporulation in this medium. In contrast, it inhibits p_3Ap_3 synthesis in vivo and in vitro (this communication). Thus, Gallant and co-workers (4) unintentionally confirmed our experiments, because their conclusions are based on wrong assumptions. Downshifts may be necessary but are not sufficient to induce sporulation in *B. subtilis*. Sporulation is much more complex.

RESULTS AND DISCUSSION

To test further our model of initiation of sporulation, we measured synthesis of p_3Ap_3 in *B. subtilis* cells grown in different rich (mostly glucose-containing) media and shifted to poor media lacking glucose, thereby allowing sporulation to occur. This latter condition is important, since p_3Ap_3 synthesis is correlated with sporulation, not with any trivial effect involved in starvation or connected with the *rel* function (6); some investigators seem to overlook this mandatory condition for the synthesis of p_3Ap_3 (4). As can be seen in Fig. 1, p_3Ap_3 was detectable (spot near the origin, O) within 15 min after shift from a rich to a poor medium. In Fig. 2, the incorporation of ^{32}P into p_3Ap_3 was quantitatively measured for cells grown in different media (SYM, nutrient broth, Mandelstam growth medium) and shifted to sporulation media (SSM or Mandelstam replacement medium). (See legend to Fig. 1 for composition of SYM and SSM.) There were some differences in the amount of

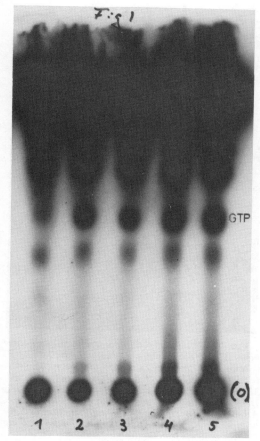

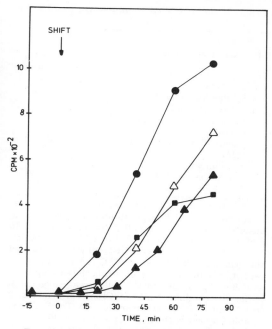

FIG. 2. *Quantitative determination of p_3Ap_3 synthesized by* B. subtilis *grown in SYM and shifted to SSM* (●), *grown in nutrient broth and shifted to SSM* (■), *grown in Mandelstam (13) growth medium and shifted to replacement sporulation medium* (▲), *or grown in Mandelstam's medium containing 1 mM glucose and shifted to replacement sporulation medium* (△). *Conditions were the same as those described in the legend to Fig. 1. Spots corresponding to p_3Ap_3 were cut out from the PEI plates and counted in a liquid scintillation counter.*

FIG. 1. *Autoradiogram of formic acid extracts of* B. subtilis *grown in SYM in the presence of $H_3{}^{32}PO_4$ and shifted to SSM also containing ${}^{32}P_i$. Extracts were prepared 1 min after shift (column 1) and every 15 min (columns 2, 3, 4, and 5) thereafter. The spot near the origin (O) was shown to be p_3Ap_3. SYM is composed of: (part 1) 0.2% yeast extract, 6 mM KCl, 40 mM NH_4Cl, 20 mM NaCl; (part 2) 12 mM $MgSO_4$, 5 mM sodium citrate; (part 3) 0.5 mM $CaCl_2$, 0.1 mM $MnCl_2$; (part 4) 10 mM glucose; (part 5) 3 mM potassium phosphate, pH 7; (part 6) 30 mM Tris-hydrochloride, pH 7. Only final concentrations are given. All parts were autoclaved separately. In ${}^{32}P$ experiments part 5 was omitted. The replacement medium SSM consists of SYM but without parts 1 and 4 and with an amino acid mixture (5). Extraction with formic acid, thin-layer chromatography, and autoradiography have been described previously (5).*

p_3Ap_3 produced; shift from SYM and SSM seemed to be most efficient, but the difference did not seem to be significant. We therefore assume that p_3Ap_3 synthesis is medium independent so long as cells are starved for carbon sources and sporulation is allowed to occur or at least to be initiated. It is important that in

labeling experiments unlabeled P_i be omitted and that $H_3{}^{32}PO_4$ be added to both growth *and* replacement media. The specific activity in SYM (minus part 5) is approximately 3.3×10^5 cpm/nmol of phosphate. Replacement media, on the other hand, contain only carrier-free $H_3{}^{32}PO_4$. Using this method together with our standard methods of chromatography (5), we were always able to measure p_3Ap_3. However, if too much salt or protein (resulting from lysis, etc.) is present in the extracts, p_3Ap_3 may be covered by the large spot at the origin.

From the data shown in Fig. 2 and the quantitative determination of GTP present in the same extracts, we were able to roughly calculate the number of p_3Ap_3 molecules per *B. subtilis* cell. Assuming that labeling of p_3Ap_3 is comparable to that of GTP, we found that the GTP concentration is higher than that of p_3Ap_3 by a factor of 10^2 to 2×10^2. According to Gallant et al. (4), GTP is present at 1 nmol/10^9 cells, which corresponds to 10^{-18} mol per cell or 10^5 molecules

TABLE 1. *Sugar transport and p_3Ap_3 synthesis in vitro*[a]

Components of the in vitro system	% Activity	
	Transport	p_3Ap_3 synthesis
Basic system	0	100
+ soluble fraction (S)	0	40
+ glucose (10 mM)	0	90
+ methyl-α-D-glucopyranoside (10 mM)	0	50
+ glucose-6-phosphate (10 mM)	0	60
+ S + glucose	80	10
+ S + glucose − PEP	0	30
+ S + glucose − ATP	100	0

[a] Membranes of *B. subtilis* were isolated by growing strain 60015 in SYM in a 16-liter fermentor to an absorbancy at 600 nm of ≈1.0. After centrifugation, 10 g of wet cells was resuspended in 1,500 ml of buffer. DNase (5 mg) and lysozyme (100 mg) were added, and the cells were incubated at 37°C for 45 min. After centrifugation at $5,000 \times g$ for 10 min and subsequently at $40,000 \times g$ for 35 min, the pellet was suspended in buffer (9) containing 1 mM $MgCl_2$ and sonicated at 2 to 3 A four times for 30 s in an ice bath. To obtain the soluble fraction S, the supernatant of the $40,000 \times g$ centrifugation was centrifuged at $100,000 \times g$, and the supernatant was used as S. The basic system for in vitro synthesis of p_3Ap_3 and measurement of glucose transport contained, in a volume of 100 μl: 10 mM $MgCl_2$, 2 mM dithiothreitol, 20 mM potassium fluoride, 4 mM ATP, 5 mM phosphoenolpyruvate (PEP), 0.1 M glycerate-3-phosphate, and 20 mg of membranes per ml.

of GTP per cell. Therefore, we find between 5×10^2 and 1×10^3 molecules of p_3Ap_3 per cell after about 60 min of induction of sporulation by nutrient starvation.

Since excess of glucose or other sugars in growth and replacement media prevent p_3Ap_3 synthesis and sporulation, we developed an in vitro system which allowed us to investigate regulation of p_3Ap_3 synthesis by components of the phosphoenolpyruvate:sugar phosphoryltransferase system.

Table 1 shows that the basic system (see footnote to table) can synthesize p_3Ap_3. When a soluble fraction (HPν ~ P and factor III) was added, p_3Ap_3 synthesis was reduced to 40%. A somewhat lower reduction was observed when glucose, α-methylglucoside, or glucose-6-phosphate was added to the basic system. Synthesis of p_3Ap_3 can occur as long as sugar is omitted and therefore no transport takes place. However, when glucose and soluble fraction were added to the basic system, transport took place at a high rate, whereas p_3Ap_3 synthesis was almost completely prevented. Whereas transport depends

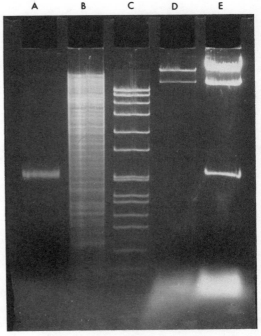

FIG. 3. *Cloning procedure for the p_3Ap_3 synthetase gene. All DNAs shown were cut with EcoRI. (A) B. subtilis DNA fraction 13 containing the spo0F gene. (B) B. subtilis DNA used to isolate fraction 13. (C) Marker DNA from phage SPP1. (D) λ NM607b_{538} DNA. (E) Cloned spo0F gene cut from the λ vector. All methods for isolating DNA and fraction 13 have been described (11), as has the method for cloning DNA fragments in λ (3).*

on phosphoenolpyruvate, p_3Ap_3 synthesis was absolutely dependent on ATP.

From the results shown in Table 1, we conclude that the PTS is involved in regulation of p_3Ap_3 synthesis but only when phosphorylation of sugars can take place, because omission of phosphoenolpyruvate from the basic system containing soluble fraction and glucose inhibited transport whereas p_3Ap_3 synthesis did occur (Table 1, line 7).

Cloning of the *spo0F* gene. To study further the effect of p_3Ap_3 synthesis on sporulation, we tried to clone the *spo0F* gene, presumed to code for the p_3Ap_3 synthetase, in a *B. subtilis* plasmid. Expression of this gene on an extrachromosomal element should allow the asporogenous mutant JH649 to synthesize p_3Ap_3 and to sporulate. Since for unknown reasons the DNA fragment carrying this gene, which was cloned in the plasmid pBS161-1, was not causing strain JH649 to sporulate (11), we tried to clone the *spo0F* gene in a λ vector in order to have it expressed in *Escherichia coli*. As can be seen in Fig. 3, isolated fragments from *Eco*RI-digested *B. subtilis* DNA were used for cloning. We were able to isolate a clone (clear plaque) containing a fragment of 1.3×10^6 daltons (Fig. 3E). Reisolation of the fragment from this clone by *Eco*RI digestion allowed proof by transformation that the cloned fragment is indeed carrying the *spo0F* gene, as we have shown previously (11). Expression of this gene in *E. coli* is currently in progress. A fragment of similar size carrying the *spo0F* gene was recently cloned by Saito and co-workers in a *B. subtilis* phage (H. Saito, personal communication; F. Kawamura et al., this volume).

LITERATURE CITED

1. **Fortnagel, P., and E. Freese.** 1968. Analysis of sporulation mutants. II. Mutants blocked in the citric acid cycle. J. Bacteriol. **95:**1431–1438.
2. **Kreft, J., K. Bernhard, and W. Goebel.** 1978. Recombinant plasmids capable of replication in B. subtilis and E. coli. Mol. Gen. Genet. **162:**59–67.
3. **Murray, N. E., W. J. Brammar, and K. Murray.** 1977. Lambdoid phages that simplify the recovery of *in vitro* recombinants. Mol. Gen. Genet. **150:**53–61.
4. **Nishino, T., J. Gallant, P. Shalit, L. Palmer, and T. Wehr.** 1979. Regulatory nucleotides involved in the *rel* function of *Bacillus subtilis*. J. Bacteriol. **140:**671–679.
5. **Rhaese, H. J., H. Dichtelmüller, and R. Grade.** 1975. Studies on the control of development. Accumulation of guanosine tetraphosphate and pentaphosphate in response to inhibition of protein synthesis in Bacillus subtilis. Eur. J. Biochem. **56:**385–392.
6. **Rhaese, H. J., H. Dichtelmüller, R. Grade, and R.**
Groscurth. 1975. High phosphorylated nucleotides involved in regulation of sporulation in *Bacillus subtilis*, p. 335–340. *In* P. Gerhardt, R. N. Costilow, and H. L. Sadoff (ed.), Spores VI. American Society for Microbiology, Washington, D.C.
7. **Rhaese, H. J., R. Grade, and H. Dichtelmüller.** 1976. Studies on the control of development. Correlation of initiation of differentiation with synthesis of highly phosphorylated nucleotides in Bacillus subtilis. Eur. J. Biochem. **64:**205–213.
8. **Rhaese, H. J., and R. Groscurth.** 1976. Control of development. Role of regulatory nucleotides synthesized by membranes of Bacillus subtilis in initiation of sporulation. Proc. Natl. Acad. Sci. U.S.A. **73:**331–335.
9. **Rhaese, H. J., and R. Groscurth.** 1978. Studies on the control of development. Synthesis of two highly phosphorylated nucleotides depends on changes in the composition of ribosomes at the beginning of sporulation in Bacillus subtilis. Eur. J. Biochem. **85:**517–528.
10. **Rhaese, H. J., and R. Groscurth.** 1979. Apparent dependence of sporulation on synthesis of highly phosphorylated nucleotides in Bacillus subtilis. Proc. Natl. Acad. Sci. U.S.A. **76:**842–846.
11. **Rhaese, H. J., R. Groscurth, R. Vetter, and H. Gilbert.** 1979. Regulation of sporulation by highly phosphorylated nucleotides in Bacillus subtilis, p. 145–159. *In* G. Koch and D. Richter (ed.), Regulation of molecular synthesis by low molecular weight mediators. Academic Press, Inc., New York.
12. **Rhaese, H. J., J. A. Hoch, and R. Groscurth.** 1977. Studies on the control of development. Isolation of Bacillus subtilis mutants blocked early in sporulation and defective in synthesis of highly phosphorylated nucleotides. Proc. Natl. Acad. Sci. U.S.A. **74:**1125–1130.
13. **Sterlini, J. M., and J. Mandelstam.** 1969. Commitment to sporulation in Bacillus subtilis and its relationship to development of actinomycin resistance. Biochem. J. **113:**29–37.

Regulation of Nitrogen Metabolism and Sporulation in *Bacillus licheniformis*

H. J. SCHREIER, T. M. SMITH, T. J. DONOHUE,[1] AND R. W. BERNLOHR

Department of Microbiology, Cell Biology, Biochemistry, and Biophysics, The Pennsylvania State University, University Park, Pennsylvania 16802

The intracellular concentrations of NH_4^+, glutamine, glutamate, and α-ketoglutarate and the levels of the enzymes glutamate synthase, glutamine synthetase, and glutamate dehydrogenase were measured in batch and continuous cultures of *Bacillus licheniformis* growing in the presence of a variety of sole nitrogen sources. Our results indicate that the levels of these enzymes are regulated by the nature of the nitrogen source, but the levels of the above metabolites, individually, are not related to the levels of the enzymes. Furthermore, neither pools of the above metabolites nor levels of the three enzymes are directly related to the control of the initiation of sporulation.

The classic study by Schaeffer et al. (17) clearly showed that bacilli can be induced to sporulate during growth if the culture is deprived of nitrogen in the presence of excess carbon source. Because of that observation, work in this laboratory has been directed toward understanding the regulation of nitrogen metabolism and its role in the induction of sporulation (4). Particularly, we have been concerned with establishing the identity of the so-called nitrogen signal in *Bacillus licheniformis*. An approach to studying this is to examine the control of the nitrogen assimilatory pathway in *B. licheniformis* grown under various conditions of nitrogen limitation. At the same time, the effect of these conditions on sporulation frequency can be noted and compared with effects on nitrogen metabolism. In this study we determined the levels of the assimilatory enzymes glutamine synthetase, glutamate synthase, and glutamate dehydrogenase and the intracellular pool levels of their substrates, ammonia, glutamine, α-ketoglutarate, and glutamate. By doing this, we hoped to determine (i) whether the level of any of these metabolites changes in a manner consistent with a role in regulating the enzymes of nitrogen assimilation and (ii) whether any of these potential regulatory metabolites or enzymes play a role in controlling the process of sporulation.

[1] Present address: Department of Microbiology, University of Illinois, Urbana, IL 61801.

RESULTS

Table 1 summarizes the results obtained when *B. licheniformis* was grown in the presence of 11 different nitrogen sources with glucose as the carbon source. As can be seen, different nitrogen sources produced cells which grew at significantly different rates. Similarly, as the doubling time increased with respect to the nitrogen source being used, the number of heat-resistant spores formed during growth increased. Cultures utilizing glutamine, ammonia, urea, or sodium glutamate as sole nitrogen sources did not contain a measurable fraction of spores when determined at one generation before the end of exponential growth. Those utilizing arginine, ornithine or nitrate as sole sources of nitrogen yielded a 5- to 10-fold increase in heat-resistant spores. Cultures of cells utilizing alanine, γ-aminobutyrate, and potassium glutamate contained a significant number of heat-resistant spores (approximately 1 to 30% of the total colony-forming units).

The levels of the assimilatory enzymes in cells grown in the presence of various nitrogen sources are shown in Table 1. The activity of glutamate dehydrogenase varied 10- to 100-fold. The amount of enzyme was elevated in cells grown in the presence of glutamine or ammonia, as the nitrogen source and was very low under the other conditions. The activity of glutamine synthetase varied 10-fold, being low when cells were grown in the presence of glutamine or ammonia and high under the other conditions.

TABLE 1. *Effect of nitrogen source on growth, sporulation, enzyme levels, and ammonia, glutamine, glutamate, and α-ketoglutarate pools in B.* licheniformis *grown in batch culture*[a]

N source	Doubling time[b] (min)	Specific activity[c] of:			Intracellular pool[d] of:				Sporulating fraction[e]
		GDH	GS	GltS	Amm	Gln	Glu	α-Kg	
Glutamine	50	350	22	72	2	3.7	69	0.8	$<1 \times 10^{-6}$
(NH$_4$)$_2$SO$_4$	60	480	81	267	3	2.1	59	0.8	$<1 \times 10^{-6}$
Urea	60	ND[f]	280	330	ND	ND	62	0.8	$<1 \times 10^{-6}$
Glutamate (Na$^+$)	60	ND	300	32	ND	ND	59	1.0	$<1 \times 10^{-6}$
Arginine	65	20	263	150	3	5.7	88	0.3	5×10^{-5}
Proline	70	ND	207	9	ND	ND	60	1.1	ND
Ornithine	70	20	300	15	ND	ND	65	0.4	5×10^{-5}
Nitrate	90	50	291	164	0.8	4.5	71	<0.1	1×10^{-5}
Alanine	120	30	296	107	0.8	4.7	59	<0.1	1×10^{-2}
γ-Aminobutyrate	150	5	ND	13	ND	ND	ND	ND	3×10^{-1}
Glutamate (K$^+$)	180	10	310	16	0.8	5.2	ND	ND	3×10^{-1}

[a] *B. licheniformis* A5 was germinated and grown in the minimal salts A medium previously described (18). All samples were taken at one generation before the end of exponential growth. In all cases 15 mM glucose was used as the carbon source along with the indicated nitrogen source at 20 mM.

[b] Values are given in minutes per doubling time.

[c] Expressed in nanomoles per minute per milligram of protein. Preparation of cell extracts for glutamine synthetase (GS) assays was done as previously described (4). Cell extracts for the assay of glutamate synthase (GltS) and glutamate dehydrogenase (GDH) were prepared in the following manner. Harvested cells were washed four times in 50 mM Tris-hydrochloride (pH 7.9) buffer containing 5 mM 2-mercaptoethanol, 1 mM EDTA, and 200 mM KCl. The cells were then suspended (at 1 to 2% of the original cell volume) in 50 mM Tris-hydrochloride (pH 7.9)–5 mM 2-mercaptoethanol–1 mM EDTA and were broken by passing twice through an Aminco French pressure cell at 20,000 lb/in². Cell debris was removed by centrifugation at 18,000 × g for 20 min at 4°C. The supernatant solution was used as the source of all enzyme activities and was used within 12 h after French press treatment. Protein was determined by the 230/260 method of Kalb and Bernlohr (8). GS was assayed by a modification of the radiochemical method of Prusiner and Milner (16). Standard assay conditions have been described (4). GDH and GltS were both assayed by recording the rate of oxidation of NADPH at 340 nm at 37°C with a Cary 118 recording spectrophotometer. GDH standard assay conditions were described by Phibbs and Bernlohr (15). GltS was assayed by a modification of the method described by Meers et al. (12). The standard reaction mix contained 50 mM Tris-hydrochloride (pH 8.0), 5 mM 2-mercaptoethanol, 35 mM α-ketoglutarate (K$^+$), 20 mM glutamine, and 0.3 mM NADPH. Reactions were initiated by the addition of glutamine, and the activity was determined after subtracting endogenous NADPH oxidase and GDH activities.

[d] Cell extracts for pool studies were prepared as described (18). Ammonia (Amm) and glutamine (Gln) were determined by the radiochemical method of Kalb et al. (9). Glutamate (Glu) and α-ketoglutarate (α-Kg) were determined with the fluorometric assay described by Lowry and Passonneau (11) using a Turner model III fluorometer. Assays were done at 37°C. Standard solutions of glutamate and α-ketoglutarate were standardized with a Cary 118 recording spectrophotometer. Pool sizes are expressed as millimolar intracellular concentrations. Results are the averages of at least duplicate analyses of extracts prepared from two separate sets of cultures. The variation between the result of an individual assay and the final average was less then ±30%.

[e] Sporulating fraction is defined as the fraction of total viable cells which are spores and was determined as described by Donohue and Bernlohr (4).

[f] ND, Not determined.

Glutamate synthase levels varied approximately 30-fold and were at their highest activity when ammonia and urea were used as the nitrogen source, at an intermediate level when cells were grown in the presence of glutamine, arginine, nitrate, or alanine, and at their lowest levels under the other conditions.

The intracellular pool sizes of the metabolites of interest are shown in Table 1. Ammonia varied from 3- to 4-fold, the highest levels being obtained when cells were grown in the presence of nitrogen sources which yielded very low sporulating fractions and the lowest being seen for those nitrogen sources which yielded high spor-

ulating fractions. The glutamine pool varied approximately 1.5- to 3-fold, the lowest value being obtained when ammonia was the nitrogen source. Glutamate remained in the 60 to 90 mM range under all conditions. The intracellular pool of α-ketoglutarate varied 10-fold; the low values were obtained for cells utilizing sodium glutamate, ornithine, nitrate, or urea.

A comparison of the pool size of any one metabolite, or of combinations of these pools, with any one of the enzyme levels shows that the pools of these metabolites do not change in a manner consistent with their use in the regulation of any of the enzymes. Similarly, neither pools nor enzyme levels seem related to the control of sporulation. Because enzyme levels seen at one generation before stationary phase (the time of sampling in this study) may be due to physiological conditions present one to two generations earlier, we decided to examine the levels of the enzymes and metabolites in cells grown in continuous culture. This approach would assure steady-state conditions under which we can measure pools, enzyme levels, and sporulation.

Table 2 summarizes the results obtained when *B. licheniformis* was grown in continuous culture. The observed dilution rates were dependent on the individual nitrogen source used for growth. The sporulating fraction varied at least 10^7-fold. The activities of glutamate dehydrogenase, glutamine synthetase, and glutamate synthase remained the same as those seen in batch culture (Table 1).

The ammonia pool varied approximately 6-fold. As in batch culture, the low concentrations were obtained from cells grown in the presence of a nitrogen source which produced a slow dilution rate (e.g., γ-aminobutyrate). The level of glutamine varied approximately 13-fold, and as in batch culture (Table 1), the cells grown in the presence of ammonia had the lowest glutamine pool. The glutamate pool remained high under all conditions except for the cultures grown in the presence of urea and glutamine. These growth conditions yielded pool sizes which were in the 40 mM range and were approximately 1.5-fold lower than the respective pool sizes seen in batch culture (Table 1). The α-ketoglutarate pools varied over a range of approximately 14-fold, and the values obtained were in some cases different from those seen in batch culture (e.g., glutamine, arginine, and nitrate). However, α-ketoglutarate pools, as well as the other pools, change during the life cycle in batch culture (1, 4, 5), and the values obtained

TABLE 2. *Effect of nitrogen source on growth, sporulation, enzyme levels, and ammonia, glutamine, glutamate, and α-ketoglutarate pools in* B. licheniformis *grown in continuous culture*[a]

N source	Dilution rate (ml/ min)	Specific activity[b] of:			Intracellular pool[c] of:				Sporulating fraction[d]
		GDH	GS	GltS	Amm	Gln	Glu	α-Kg	
Urea	7.0 ± 0.2	174	240	260	5.7 ± 1.8	10.5 ± 3.5	44 ± 11	0.8 ± 0.2	$<1 \times 10^{-8}$
$(NH_4)_2SO_4$	6.3 ± 0.5	350	85	400	10.8 ± 2.3	1.6 ± 0.5	70 ± 22	1.2 ± 0.4	$<1 \times 10^{-8}$
Glutamine	6.0 ± 0.2	300	11	160	5.5 ± 2.0	17.0 ± 4.3	42 ± 8	1.6 ± 0.3	$<1 \times 10^{-8}$
Arginine	5.0 ± 0.1	30	250	160	6.1 ± 1.6	21.1 ± 4.9	63 ± 7	1.2 ± 0.2	$<1 \times 10^{-8}$
Glutamate (Na^+)	4.6 ± 0.6	6	240	11	7.3 ± 2.8	11.2 ± 4.2	150 ± 30	2.9 ± 0.6	$<1 \times 10^{-8}$
Nitrate	3.7 ± 0.4	11	220	180	1.7 ± 0.8	11.1 ± 2.7	68 ± 10	1.2 ± 0.2	2×10^{-6}
Ornithine	3.5 ± 0.1	10	240	14	2.7 ± 1.1	11.9 ± 3.1	60 ± 8	0.5 ± 0.1	5×10^{-5}
γ-Aminobutyrate	1.6 ± 0.1	13	247	26	3.2 ± 1.0	11.4 ± 2.4	111 ± 11	0.2 ± 0.1	4×10^{-3}
Glutamate (K^+)	1.9 ± 0.1	10	200	10	4.2 ± 1.2	6.5 ± 4.6	130 ± 10	0.5 ± 0.1	1×10^{-1}

[a] Continuous culture experiments were done by use of a modified 750-ml F-1000 culture vessel (New Brunswick Scientific Co., New Brunswick, N.J.). Agitation was maintained at 600 rpm by using the vessel's stirring apparatus. Temperature was controlled at 37°C, and the medium input was regulated by a Buchler peristaltic pump. An 8-liter aspirator flask was used as a medium reservoir and was connected to a vessel intake tube which allowed medium to be pumped to the bottom of the vessel. Aeration was obtained by first passing air through a glass-fiber column, through two water traps (the second heated to 35°C), and finally through the stirring mechanism. Air flow was maintained at 16,000 ml/min at 1 atm, 21°C. A two-way stopcock was attached to the bottom of the vessel to allow fast sampling. A typical experiment was run as follows. A small inoculum (<2% by volume) of *B. licheniformis* from a mid-exponential-phase culture was transferred to the culture vessel, which contained 500 ml of the minimal salts A medium (18) adjusted to pH 6.8 and supplemented with 15 mM glucose and the indicated nitrogen source at 20 mM. The culture was grown to a cell density of 10^8 cells per ml, which corresponded to the mid-exponential phase in batch culture. Cell density was measured turbidimetrically with a Klett-Summerson colorimeter equipped with a no. 54 (green) filter. The culture was maintained in this steady state for the equivalent of 10 doubling times and was subsequently harvested.

[b] As described in Table 1.

[c] Expressed as millimolar intracellular concentration. The preparation of cell extracts for pool studies was done as described (18) except that 100 ml of cells was rapidly emptied into a graduated cylinder and filtered on a 142-mm membrane filter apparatus (Millipore Corp.). The filtered cells were washed twice with 50 ml of prewarmed (37°C) medium lacking the carbon and nitrogen sources. Total time elapsed from the start of sample collection to immersion of the filter paper in 30 ml of 0.3 N $HClO_4$ was less than 20 s. $KHCO_3$ (2 M) was used for neutralization. Results are the averages of at least triplicate analyses of extracts prepared from two separate sets of cultures. Ammonia (Amm), glutamine (Gln), glutamate (Glu), and α-ketoglutarate (α-Kg) were assayed as described in Table 1.

[d] As described in Table 1.

here when continuous culture was used may indicate that the continuous culture data are indicative of steady-state conditions.

DISCUSSION

Our experiments were designed to determine whether the nitrogen signal which regulates the control of sporulation is an enzyme(s) or metabolite(s) of the nitrogen assimilation pathway. Our results indicate that the levels of the enzymes studied are regulated by the nature of the nitrogen source utilized for growth; however, no single metabolite pool or combination of pools seems to be related to the regulation of any of the enzymes. Several laboratories (2, 3, 7, 14, 15) have proposed that specific pools of metabolites play a role in the regulation of the levels of certain enzymes, but our results do not support these hypotheses. Elmerich (6) stated that the glutamate synthase of *B. megaterium* is repressed by glutamate. Phibbs and Bernlohr (15) proposed that the glutamate dehydrogenase of *B. licheniformis* is repressed by glutamate. The data on pool sizes reported here show that glutamate levels are not related to the levels of glutamate synthase or glutamate dehydrogenase when cells are grown in the presence of various nitrogen sources. A recent study by Deshpande et al. (2) suggested that the level of glutamate synthase of *B. subtilis* is regulated by the glutamine pool size. We have observed no direct relationship between the levels of glutamate synthase and glutamine from cultures growing in the presence of different nitrogen sources.

Our data indicate that neither the pools nor the levels of the enzymes, alone or in simple combinations, are related to the control of sporulation in *B. licheniformis*. Since these are apparently not involved in the regulation of sporulation, other factors should be sought. Possibilities such as a purine nucleotide (7, 13) or a charged tRNA (10) should be considered.

ACKNOWLEDGMENT

This study was supported in part by grant PCM77-27548 from the National Science Foundation.

LITERATURE CITED

1. Clark, V. L., D. E. Peterson, and R. W. Bernlohr. 1972. Changes in free amino acid production and intracellular amino acid pools of *Bacillus licheniformis* as a function of culture age and growth media. J. Bacteriol. 112:715–725.
2. Deshpande, K. L., J. R. Katze, and J. F. Kane. 1980. Regulation of glutamate synthase from *Bacillus subtilis* by glutamine. Biochem. Biophys. Res. Commun. 95:55–60.
3. Deuel, T. F., and S. Prusiner. 1974. Regulation of glutamine synthetase from *Bacillus subtilis* by divalent cations, feedback inhibitors and L-glutamine. J. Biol. Chem. 249:257–264.
4. Donohue, T. J., and R. W. Bernlohr. 1978. Carbon and nitrogen catabolite repression, metabolite pools, and the regulation of sporulation in *Bacillus licheniformis*, p. 293–298. In G. Chambliss and J. C. Vary (ed.), Spores VII. American Society for Microbiology, Washington, D.C.
5. Donohue, T. J., and R. W. Bernlohr. 1978. Effect of cultural conditions on the concentrations of metabolic intermediates during growth and sporulation of *Bacillus licheniformis*. J. Bacteriol. 135:363–372.
6. Elmerich, C. 1972. Le cycle du glutamate, point de depart du metabolisme de l'azote, chez *Bacillus megaterium*. Eur. J. Biochem. 27:216–224.
7. Elmerich, C., and J.-P. Aubert. 1975. Involvement of glutamine synthetase and the purine nucleotide pathway in repression of bacterial sporulation, p. 385–390. In P. Gerhardt, R. N. Costilow, and H. L. Sadoff (ed.), Spores VI. American Society for Microbiology, Washington, D.C.
8. Kalb, V. F., and R. W. Bernlohr. 1977. A new spectrophotometric assay for protein in cell extracts. Anal. Biochem. 82:362–371.
9. Kalb, V. F., T. J. Donohue, M. G. Corrigan, and R. W. Bernlohr. 1978. A new and specific assay for ammonia and glutamine sensitive to 100 pmol. Anal. Biochem. 90:47–57.
10. Lapointe, J. 1975. Role of glutamyl-transfer ribonucleic acid in regulation of glutamate metabolism and assimilation of ammonia in *Escherichia coli* cells: analogies with metabolism of glutamate and glutamine during initiation of sporulation in *Bacillus megaterium* cells, p. 381–384. In P. Gerhardt, R. N. Costilow, and H. L. Sadoff (ed.), Spores VI. American Society for Microbiology, Washington, D.C.
11. Lowry, O. H., and J. V. Passonneau. 1972. A flexible system of enzymatic analysis. Academic Press, Inc., New York.
12. Meers, J. L., D. W. Tempest, and C. M. Brown. 1970. Glutamine (amide):2-oxoglutarate amine transferase oxido-reductase (NADP); an enzyme involved in the synthesis of glutamate by some bacteria. J. Gen. Microbiol. 64:187–194.
13. Mitani, T., J. E. Heinze, and E. Freese. 1977. Induction of sporulation in *B. subtilis* by decoyinine or hadacidin. Biochem. Biophys. Res. Commun. 77:1118–1125.
14. Pan, F. L., and J. G. Coote. 1979. Glutamine synthetase and glutamate synthase activities during growth and sporulation of *Bacillus subtilis*. J. Gen. Microbiol. 112:373–377.
15. Phibbs, P. V., and R. W. Bernlohr. 1971. Purification, properties, and regulation of glutamic dehydrogenase of *Bacillus licheniformis*. J. Bacteriol. 106:375–385.
16. Prusiner, S., and L. Milner. 1970. A rapid radioactive assay for glutamine synthetase, glutaminase, asparagine synthetase and asparaginase. Anal. Biochem. 37:429–438.
17. Schaeffer, P., J. Millet, and J.-P. Aubert. 1965. Catabolic repression of bacterial sporulation. Proc. Natl. Acad. Sci. U.S.A. 54:704–711.
18. Siegel, W. H., T. Donohue, and R. W. Bernlohr. 1977. Determination of pools of tricarboxylic acid cycle and related acids in bacteria. Appl. Environ. Microbiol. 34:512–517.

Nitrofurans Can Partially Overcome the Glucose Repression of Sporulation

PETER FORTNAGEL

Abteilung für Mikrobiologie, Institut für Allgemeine Botanik, Universität Hamburg, Hamburg, Germany

Nitrofurantoin was found to induce the stringent response in *Bacillus subtilis*. The glucose repression of sporulation could be partially overcome by nitrofurans. The induction of spore formation required a defined reduction of growth. In nutrient sporulation medium sporulation was not influenced by low concentrations of nitrofurantoin.

Nitrofurans are a group of synthetic antibiotics which are reported to act at the initiation of translation (4). In *Escherichia coli* these antibiotics interfere with the synthesis of inducible enzymes such as β-galactosidase or galactokinase much more than with the synthesis of most cell proteins or of phage protein (10), possibly suggesting regulation of protein synthesis to a certain extent at the level of translation.

The analysis of the action of the translational inhibitors and the analysis of resistant mutants in *Bacillus subtilis* have led to the postulate that effective sporulation requires a precise translational control of protein synthesis (2, 6, 7). It was therefore tempting to study the action of nitrofurans on the growth and development of this organism.

RESULTS AND DISCUSSION

Inhibition of bacterial growth by NF. Growth of *B. subtilis* stopped immediately after 80 μM nitrofurantoin (NF) was added to a culture in S_6 minimal medium in the presence of 100 mM glucose and 0.1% Casamino Acids (Fig. 1A). Eventually, the culture recovered from the inhibition (after 1 h), and growth resumed at a reduced rate. Concentrations of 100 μM blocked growth permanently.

In contrast, when 100 μM NF was added to cells in nutrient sporulation medium, growth did not stop immediately (Fig. 1B), but continued at a reduced rate. The final cell titer reached at T_5 was 2.0×10^8 colony-forming units/ml compared with 3.9×10^8 colony-forming units/ml in the untreated control.

Care was taken to treat both cultures with NF at the same cell density (optical density at 578 nm [OD_{578}], 0.75), since it is known that the effectiveness of NF may decrease as the size of the inoculum decreases (1).

Nitrofuran influence on sporulation. Sporulation in nutrient sporulation medium in the presence of NF is shown in Fig. 2. Although the growth rate was reduced after NF addition and the final cell titer reached was low, sporulation was normal with respect to the frequency

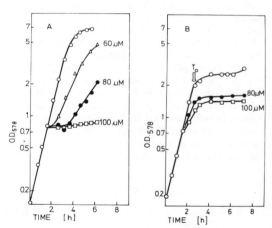

FIG. 1. *Growth inhibition by NF [N-(5-nitro-2-furfurylidene)-1-aminohydantoin].* B. subtilis *strain 60 015 (trpC metC) was grown (A) at 37°C in S_6 minimal medium (9) containing 100 mM glucose, 0.1% Casamino Acids, 20 mM NH_4^+, and 5 mM PO_4^{3-} or (B) at 40°C in nutrient sporulation medium (3). The cultures were inoculated from tryptose blood agar base plates at an initial OD_{578} of 0.05 or less. When the OD_{578} was 0.75, the cultures were divided, and 80 μM (●) or 100 μM (□) NF was added. (○) Untreated control culture.*

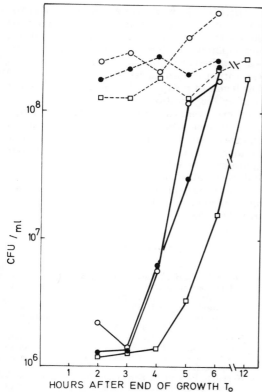

FIG. 2. *Influence of NF on sporulation in nutrient sporulation medium. Strain 60 015 was grown in nutrient sporulation medium at 40°C as shown in Fig. 1B. At an OD$_{578}$ of 0.75, NF was added (○, none; ●, 80 μM; or □, 100 μM). The total cell titer (dashed lines) and the spore titer (after incubation at 75°C for 20 min) (solid lines) were measured at intervals starting at T$_2$, 2 h after the end of growth of the untreated culture (T$_0$; see Fig. 1B).*

of the spores. Although growth was reduced by 80 μM NF, the sporulation pattern did not change. The major increase of spores started 3 h after the end of growth (at 40°C), as in the untreated control culture. In the presence of 100 μM NF, spores occurred with a delay of 1 h. At T$_{12}$, this culture contained normal amounts of spores.

In S$_6$ minimal medium sporulation normally remains repressed as a result of the presence of 100 mM glucose (9). However, it can be induced. The critical step for a successful induction of sporulation is a decrease of the guanosine nucleotide pools, especially GTP (8, 9; J. M. Lopez, A. Dromerick, and E. Freese, personal communication). Such a reduction of GTP and concomitant induction of sporulation can be accomplished with inhibitors of guanosine nucleotide

biosynthesis such as decoyinine (9). In addition, this can be achieved with a limited stringent response (8; Lopez et al., personal communication), since this response includes an increase in the intracellular concentrations of ppGpp and pppGpp and a corresponding decrease of GTP (see J. M. Lopez, K. Ochi, and E. Freese, this volume).

As can be seen from Fig. 3, the nitrofurans NF and urfadyn induced sporulation in S$_6$ minimal medium. The frequency of spores was 10% or more 18 h after nitrofuran addition (80 μM) as compared with less than 0.3% in the untreated control.

NF induces the stringent response. The pool concentrations of GTP, ppGpp, and pppGpp were determined after addition of 80 μM NF to a culture growing in S$_6$ minimal medium with 100 mM glucose and 0.1% Casamino

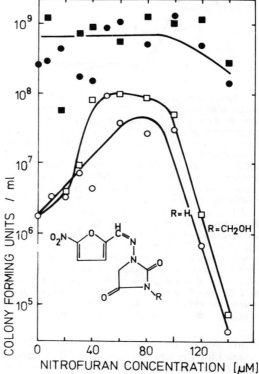

FIG. 3. *Sporulation induction by nitrofurans in S$_6$ minimal medium. The cultural conditions were the same as described for Fig. 1A. Spore titers (after exposure of the culture to 75°C for 20 min) (○, □) and total viable cell titers (●, ■) were measured 18 h after addition of the inhibitors NF (○, ●) or urfadyn (□, ■). Both inhibitors were a gift from RÖHM-pharma, Darmstadt, Germany. Stock solutions were 1 mg/ml in acetone.*

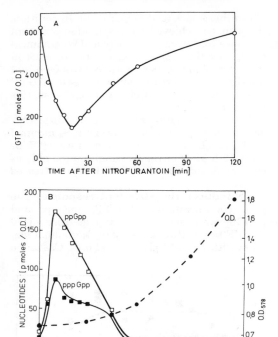

FIG. 4. *NF-caused changes of (A) GTP and (B) ppGpp and pppGpp concentrations. Nucleotide pools were measured as described in reference 8. During exponential growth, 40 μCi of $^{32}PO_4{}^{3-}$ per ml was added to a culture growing in S_6 minimal medium as described for Fig. 1A. When the OD_{578} was 0.75, 80 μM NF was added. At the intervals indicated, 2-ml portions were withdrawn and filtered through 0.45-μm-pore size membrane filters. The filters were immediately extracted with 300 μl of 0.4 M formic acid at 0°C. The sampling procedure took less than 15 s. The extracts were chromatographed on polyethyleneimine cellulose thin-layer plates with 1.5 M potassium phosphate (pH 3.4) to separate GTP, ppGpp, and pppGpp from all other nucleotides. After autoradiography, the radioactive spots were cut out for quantitative measurement of the nucleotide pool concentration. The results are expressed as picomoles per OD_{578} unit of the culture. Symbols: ○, GTP; □, ppGpp; ■, pppGpp; ---●---, growth of the culture (OD_{578}).*

Acids. As can be seen from Fig. 4A, GTP decreased for the next 30 min after NF addition. The GTP pool recovered thereafter, and 2 h after NF addition normal amounts of GTP were measured.

The GTP decrease was a consequence of the stringent response initiated by NF. Figure 4B shows that ppGpp and pppGpp were formed immediately after NF addition. A maximal value was reached after 15 min. Thereafter, the concentration of both nucleotides declined. Normal low levels of ppGpp and pppGpp were found 1 h after NF addition. At the same time, cells recovered from the arrest of growth (Fig. 4B).

Conditions which initiate the stringent response to partial or transient deprivation of one or more amino acids are sufficient to initiate sporulation (see Lopez et al., this volume). The induction of sporulation by nitrofurans may, therefore, be explained by a similar mechanism, since nitrofurans induced the stringent response in *B. subtilis*.

LITERATURE CITED

1. Carey, W. F., H. E. Russell, and J. R. O'Connor. 1961. The effect of nitrofurans on the survival of bacteria in vivo. Antimicrob. Agents Annu. 1960, p. 152–158.
2. Fortnagel, P., and R. Bergmann. 1973. Alteration of the ribosomal fraction of *Bacillus subtilis* during sporulation. Biochim. Biophys. Acta **299**:136–141.
3. Freese, E., and P. Fortnagel. 1967. Analysis of sporulation mutants. I. Response of uracil incorporation to carbon sources and other mutant properties. J. Bacteriol. **94**:1957–1969.
4. Herrlich, P., and M. Schweiger. 1976. Nitrofurans, a group of synthetic antibiotics, with a new mode of action: discrimination of specific messenger RNA classes. Proc. Natl. Acad. Sci. U.S.A. **73**:3386–3390.
5. Kobayashi, H., K. Kobayashi, and Y. Kobayashi. 1977. Isolation and characterization of fusidic acid-resistant, sporulation-defective mutants of *Bacillus subtilis*. J. Bacteriol. **132**:262–269.
6. Kobayashi, Y., and T. Domoto. 1975. Role of ribosomes in bacterial sporulation, p. 115–131. *In* T. Ishikowa, Y. Maruyama, and H. Matsumiya (ed.), NRI symposia on molecular biology: growth and differentiation in microorganisms. University of Tokyo Press, Tokyo.
7. Kobayashi, Y., H. Kobayashi, and H. Hirochika. 1978. Role of elongation factor G in sporulation of *Bacillus subtilis*, p. 242–249. *In* G. Chambliss and J. C. Vary (ed.), Spores VII. American Society for Microbiology, Washington, D.C.
8. Lopez, J. M., C. L. Marks, and E. Freese. 1979. The decrease of guanine nucleotides initiates sporulation of *Bacillus subtilis*. Biochim. Biophys. Acta **587**:238–252.
9. Mitani, T., J. E. Heinze, and E. Freese. 1977. Induction of sporulation in *Bacillus subtilis* by decoyinine and hadacidin. Biochem. Biophys. Res. Commun. **77**:1118–1125.
10. Wagner, E., M. Schweiger, H. Ponta, and P. Herrlich. 1977. Messenger-selective inhibitor for the initiation of translation in *Escherichia coli*: nitrofurantoin. FEBS Lett. **83**:337–340.

Biosynthesis of RNA Polymerase in *Bacillus subtilis* During Induction of the Stringent Response

RANDELL T. LIBBY, SARAH J. KUHL, AND LYLE R. BROWN

Department of Microbiology, Oregon State University, Corvallis, Oregon 97330

The biosynthesis of RNA polymerase (nucleosidetriphosphate:RNA nucleotidyltransferase, EC 2.7.7.6) was examined in isogenic *relA⁺/relA* strains of *Bacillus subtilis* during induction of the stringent response. Induction of the stringent response by amino acid limitation of the *relA⁺* strain resulted in a 1.5-fold stimulation in σ-subunit synthesis; ββ' subunits were immediately shut off, and assembly (or synthesis, or both) of the α-subunit was stimulated 2.5-fold. No stimulation in RNA polymerase synthesis was observed: (i) in the *relA* strain after amino acid limitation or (ii) after addition of O-methylthreonine to *relA⁺/relA* strains.

The synthesis of bacterial rRNA and protein components of the ribosome is subject to the stringent control mechanism (5, 6, 23). In *relA⁺* strains of *Bacillus subtilis* (17, 22) or *Escherichia coli*, tRNA charging imbalances due to amino acid limitation or inactivation of an aminoacyl tRNA synthetase result in the rapid accumulation of (p)ppGpp as well as a concomitant reduction (i) in the production of stable RNA molecules (8) and (ii) in ribosomal protein (r-protein) mRNA. During amino acid limitation of *relA⁺* (wild type) strains of *E. coli*, ppGpp is produced (4) through an idling reaction of uncharged tRNA bound in the ribosomal A site (2, 7, 19). The reduction in stable RNA synthesis has been observed after ppGpp accumulates during conditions of amino acid starvation, and upon addition of the limiting amino acid, synthesis of RNA resumes only after ppGpp disappears (3). Recently, physiological levels of ppGpp have been found to inhibit the in vitro synthesis of r-proteins, elongation factors G and Tu, and the alpha subunit of RNA polymerase (10, 11, 14, 18), but synthesis of the β and β' subunits of RNA polymerase continues unabated (12, 15). In vivo, elicitation of the stringent response in *relA⁺* (but not *relA*) strains of *E. coli* results in the rapid accumulation of ppGpp and subsequent specific inhibition of genes coding for rRNA and r-proteins. Transcription of the r-proteins and *rpoA* gene coding for the alpha subunit of RNA polymerase, located near 72 min, as well as the four 50S r-proteins located adjacent to the *rpoBC* operon at 88.5 min, is specifically inhibited. Transcription of the *rpoBC* operon coding for the β and β' subunits

of RNA polymerase is unaffected under these conditions (15). In contrast, *lac* and *trp* operons are stimulated to produce increased levels of mRNA ater the addition of ppGpp (21).

In *B. subtilis*, ppGpp and pppGpp have been found to accumulate during amino acid limitation of *relA⁺* strains (22, 24). Recent experiments in which the reported isoleucyl tRNA synthetase inhibitor O-methylthreonine was used have demonstrated that a variety of polyphosphorylated nucleotides resembling pppApp, ppApp, pppGpp, ppGpp, pGpp, and ppGp accumulate in wild-type *B. subtilis* cells (17). In addition, a transient accumulation of ppGpp and pppGpp has been reported in wild-type cells of *B. subtilis* upon temperature upshift and after addition of rifampin and lipiarmycin (V. L. Price, Ph.D. thesis, Oregon State University, Corvallis, 1979). A transient increase in (p)ppGpp has been observed in *B. subtilis* during initial stages of sporulation. This is accompanied by a sharp decrease in RNA polymerase synthesis (R. Libby and L. R. Brown, unpublished data). In this investigation we examined the biosynthesis of RNA polymerase subunits after induction of the stringent response by amino acid starvation and O-methylthreonine addition in an isogenic pair of *relA⁺/relA* strains of *B. subtilis*. We observed a discoordinate rate of RNA polymerase subunit biosynthesis in the *relA⁺* strain after induction of the stringent response.

RESULTS AND DISCUSSION

Biosynthesis of RNA polymerase during the stringent response. To determine whether RNA polymerase biosynthesis in *B. subtilis* is subject to stringent control, we simultaneously monitored: (i) the production of (p)ppGpp and (ii) the synthesis or assembly, or both, of the $\beta\beta'$, σ, and α subunits into the core and holoenzyme (i.e., $\alpha_2\beta\beta'$, $\alpha_2\beta\beta'\sigma$) after induction of the stringent response. Isogenic lysine-requiring *relA*[+] (BR16) and *relA* (BR17) strains of *B. subtilis* were grown in a modification of Gallant's Tris-glucose medium to an optical density at 650 nm (OD_{650}) of 0.15, at which time the culture was divided, and 200 μCi of $^{32}P_i$ per ml was added to one portion. The nucleotide pools were allowed to equilibrate for one generation as described above. At an OD_{650} of 0.30, the unlabeled culture was concentrated fivefold and resuspended in the same medium lacking lysine. Samples were pulse-labeled with 10 μCi of L-[4,5-^{3}H(N)]leucine per ml and chased for 2 min with 100 μg of unlabeled leucine per ml. Monitoring of (p)ppGpp accumulation and RNA polymerase subunit biosynthesis was synchronized after the transfer of the unlabeled and ^{32}P-equilibrated cultures to lysine-deficient medium as described above.

Induction of the stringent response resulted in the rapid accumulation of (p)ppGpp in the *relA*[+], but not the *relA*, strain within 3 to 5 min after shift to the lysine-deficient medium (Table 1). Synthesis of total protein was immediately affected in the *relA*[+] strain (Fig. 1), whereas a more gradual decline in the rate of total protein synthesis was observed in the *relA* strain (Fig. 2). However, by 20 min after the shift to lysine-deficient medium, both *relA*[+] and *relA* strains had reduced their rate of total protein synthesis to approximately 15% of its preshift level. In the *relA*[+] strain an immediate depression in the rate of synthesis of the RNA polymerase $\beta\beta'$ subunits was observed. Synthesis of the σ subunit was specifically stimulated 1.5-fold during the peak of (p)ppGpp accumulation, and accumulation of α into an immunoprecipitable form was stimulated 2.5-fold (Fig. 1). Lysine deprivation of the *relA* strain resulted in a different pattern of RNA polymerase synthesis. Synthesis of the $\beta\beta'$ and σ subunits continued unabated until the level of total protein synthesis fell to approximately 50% of its preshift rate, at which time the biosynthesis of these subunits dropped precipitously. An immediate reduction in the rate of assembly (or synthesis) of α into holoenzyme was observed (Fig. 2).

Therefore, control of RNA polymerase subunit biosynthesis in *B. subtilis* is subject to stringent control and results in the discoordinate change in the rate of subunit synthesis observed in the *relA*[+] strain. This change is coincident with the transient accumulation of the polyphosphorylated nucleotides ppGpp and pppGpp. Hence, induction of the stringent response, as judged by the accumulation of (p)ppGpp in *B. subtilis*, has an effect on RNA polymerase biosynthesis opposite to that previously observed in *E. coli* (14, 15). We have made independent observations on early sporulating cells and have demonstrated a correlation between the accumulation of polyphosphorylated guanosine, but not adenosine nucleotides (17), and immediate inhibition of $\beta\beta'$-subunit synthesis (R. Libby, Ph.D. thesis, Oregon State University, Corvallis, 1980).

O-Methylthreonine-induced synthesis of polyphosphorylated nucleotides. The proposed isoleucyl tRNA synthetase inhibitor, *O*-methylthreonine (17), has also been used to stimulate production of the highly phosphorylated nucleotides by creating an amino acid-deficient state. The addition of *O*-methylthreonine (2 mg/ml) to the *relA*[+] strain was found to stimulate the accumulation of ppGpp and pppGpp by 4 min after *O*-methylthreonine addition, but no such stimulation was observed in the *relA* strain (Table 1). Biosynthesis of RNA polymerase subunits in both the *relA*[+] and *relA* strains was very similar: after addition of *O*-

TABLE 1. *Nucleotide levels (picomoles per optical density unit)*

| | After lysine starvation[a] | | | | | After addition of O-methylthreonine[b] | | | |
| | BR16 (*relA*[+]) | | BR17 (*relA*) | | | BR16 (*relA*[+]) | | BR17 (*relA*) | |
Time (min)	ppGpp	pppGpp	ppGpp	pppGpp	Time (min)	ppGpp	pppGpp	ppGpp	pppGpp
0	15	<5	7	<5	0	5	<5	11	<5
3	77	36	5	<5	2	21	28	18	<5
5	93	57	5	<5	4	76	56	16	<5
7	20	15	5	<5	7	23	26	8	<5
10	10	6	5	<5	10	7	9	10	<5
14	13	10	5	<5	14	10	6	<5	<5
20	17	8	5	<5	20	9	6	<5	<5

[a] See legend for Fig. 1.
[b] See legend for Fig. 3.

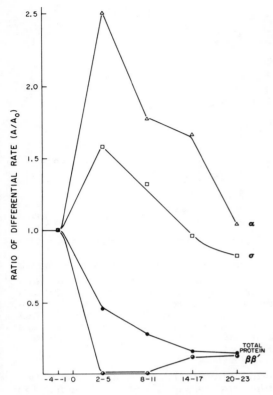

FIG. 1. *Effect of stringent response induction on RNA polymerase biosynthesis in* B. subtilis *BR16.* B. subtilis *BR16 (relA⁺) was grown in a modification of Gallant's Tris-glucose medium (17) which contained: 100 mM Trizma-HCl, pH 7.4, 0.1 mM KH_2PO_4, 1.32 mM $FeCl_3$, 10 mM $(NH_4)_2SO_4$, 1.43 mM $Na_3C_6H_5O_7$·$2H_2O$, 0.2% glucose, 100 µg each of proline, glycine, alanine, glutamic acid, aspartic acid, and arginine per ml, 40 µg each of cysteine, methionine, tyrosine, tryptophan, and phenylalanine per ml, and 20 µg of lysine per ml. When the OD_{650} of the culture reached 0.15, 200 µg of $^{32}P_i$ per ml was added to one-half of the culture, and the nucleotide pools were allowed to equilibrate for one generation. At an OD_{650} of 0.30, the cultures were centrifuged (3,000 × g) and simultaneously resuspended in lysine-deficient Tris-glucose medium. The accumulation of (p)ppGpp was monitored in the ^{32}P-equilibrated cultures as previously described (17). Biosynthesis of RNA polymerase subunits was simultaneously monitored by pulse-labeling for 3 min with 10 µCi of L-[4,5-^{3}H(M)]leucine (specific activity, 40 to 60 Ci/mmol), followed by 2 min of chase with 100 µg of unlabeled leucine. Reactions were terminated on ice after the addition of NaN₃ to 1 mM. Total protein levels were determined by addition of a 250-µl sample to an equal volume of ice-cold 10% trichloroacetic acid and precipitation for 2 to 3 h on ice before counting. Samples were sonicated in a Bronwill Biosonik IV Sonifier cell disrupter for 1 min and subsequently centrifuged (100,000 × g) at*

methylthreonine to exponentially growing cultures, synthesis of the $\beta\beta'$ subunits was immediately depressed in both *relA⁺* and *relA* strains to approximately 5% and 20%, respectively, and was found to continue thereafter at a nearly constant rate for the duration of the experiment (Fig. 3 and 4). Although a slight deviation in the pattern of σ subunit synthesis was noted between the *relA⁺* and *relA* strains, synthesis of the σ subunit continued at a relatively high level in both strains (Fig. 3 and 4). The synthesis and assembly of α subunit into an immunoprecipitable form continued at a nearly constant rate after an initial depression noted upon O-meth-

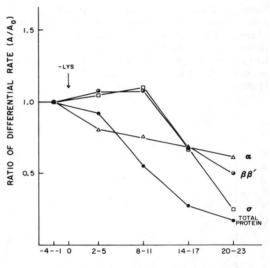

FIG. 2. *Effect of lysine starvation on RNA polymerase biosynthesis in* B. subtilis *BR17.* B. subtilis *BR17 (relA) was grown in the modified Tris-glucose medium and was monitored for synthesis of RNA polymerase subunits and accumulation of (p)ppGpp as described in the legend to Fig. 1. Symbols as in Fig. 1.*

2°C for 1 h. *Resulting high-speed supernatants were concentrated (125 µl) in a negative-pressure protein dialysis concentrator (Biomolecular Dynamics, Beaverton, Ore.), and RNA polymerase subunits were precipitated with antisera to RNA polymerase β, β', and σ subunits. Precipitated subunits of RNA polymerase were fractionated along with the internal standard, ^{35}S-labeled RNA polymerase, on sodium dodecyl sulfate-polyacrylamide gels as previously described (Libby, Ph.D. thesis, Oregon State University, 1980). RNA polymerase subunit biosynthesis was calculated as a ratio of the postshift (A) to the preshift (A₀) rate of synthesis and expressed as the ratio of the differential rate (i.e., A/A₀). Symbols: ○, $\beta\beta'$ subunits; □, σ subunit; △, α subunit.*

ylthreonine addition. The α subunit appears to be more sensitive to the effects of O-methylthreonine in both *relA*⁺ and *relA* strains.

Thus, induction of the stringent response by amino acid deprivation and O-methylthreonine addition do not affect RNA polymerase biosynthesis in the same fashion as would have been predicted from previous data (17). The initial rate of depression observed in synthesis of the σ subunit, after addition of O-methylthreonine, at present cannot be explained. With this exception, RNA polymerase biosynthesis in both *relA*⁺ and *relA* strains responds in nearly the same fashion after O-methylthreonine addition. The lack of any stimulation of the rate of synthesis and assembly of the σ and α subunits, in spite of the accumulation of the polyphosphorylated nucleotides after O-methylthreonine addition, may be the result of a generalized inhibitory effect on protein synthesis which may antagonize the effects of the normal stringent response.

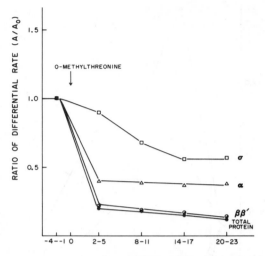

FIG. 4. *Effect of O-methylthreonine on RNA polymerase biosynthesis in* B. subtilis *BR17.* B. subtilis *BR17* (relA) *was grown in the modified Tris-glucose medium and monitored for synthesis of RNA polymerase subunits and accumulation of (p)ppGpp as described in the legend to Fig. 3. Symbols as in Fig. 1.*

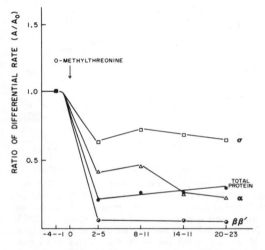

FIG. 3. *Effect of O-methylthreonine on RNA polymerase biosynthesis in* B. subtilis *BR16.* B. subtilis *BR16* (relA⁺) *was prelabeled with 200 μCi of* ³²Pᵢ *per ml in modified Tris-glucose medium for one generation as described in the legend to Fig. 1. At an OD₆₅₀ of 0.30, both labeled and unlabeled cultures were centrifuged (3,000 × g) and resuspended in fresh Tris-glucose medium containing lysine (100 μg/ml) and O-methylthreonine (2 mg/ml). Biosynthesis of RNA polymerase subunits was monitored by pulse-labeling with 5 μCi of* L-[4,5-³H]leucine *per ml (specific activity, 125 Ci/mmol) while simultaneously monitoring the accumulation of (p)ppGpp as described in the Fig. 1 legend. Quantitation of RNA polymerase was performed as described in the Fig. 1 legend. Symbols as in Fig. 1.*

In contrast to the effects observed in *B. subtilis*, *relA*⁺ cells of *E. coli* have been found to respond in a different manner after induction of the stringent response. These differences include: (i) the specific inhibition of transcription of the *rpoA* gene, encoding the α subunit of RNA polymerase, and (ii) the unabated transcription from *rpoBC* coding for the β and β′ subunits (15). The coincident inhibition of r-proteins S11, S4, and L17, located adjacent to the *rpoA* gene, and of r-proteins L10 and L7/L12, located adjacent to *rpoBC*, is consistent with the view that *rpoA* (α) is cotranscribed along with *rpsK* (S11) and *rpsD* (S4), and that transcription of *rplJL* (L10, L7/L12) and *rpoBC* (ββ′) originates from separate promoters which respond to different signals (1).

Based on the accelerated rates of synthesis (and assembly) of the σ and α subunits observed in our studies, we speculate that either the promoters for *rpoA* and the putative *rpoD* gene are under separate promoter control or transcription from these promoters may be specifically stimulated after induction of the stringent response. Inhibition in the rate of synthesis of the ββ′ subunits, after induction of the stringent response, could result if the *rpoBC* genes lie within an r-protein gene transcription unit which itself is inhibited by the effects of the stringent response. Alternatively, a separate promoter from *rpoBC* which is sensitive to stringent response induction could also be postulated.

A transient accumulation of (p)ppGpp has been observed during early sporulation, and this response is accompanied by a decline in *rpoBC* gene expression. This reaction, similar in all respects to that observed during amino acid limitation, makes it tempting to postulate an effector role in some early step in sporulation for the highly phosphorylated nucleotides, but no evidence is available to support this speculation. However, one could predict that a general decline in *rpoBC* synthesis coupled with an increased synthesis of sporulation-specific RNA polymerase subunits could readily facilitate the adaptive response needed by the sporulating cell.

ACKNOWLEDGMENTS

We are grateful to S. Halling, K. Burtis, and R. Doi for providing us with the RNA polymerase antisera. We also thank Virginia Price for advice on nucleotide analysis and Jonathan Gallant for the isogenic pair of *relA*⁺/*relA lys* strains of *B. subtilis*.

This work was supported by Public Health Service grant GM 19617-07 from the National Institute of General Medical Sciences.

LITERATURE CITED

1. **Barry, G., C. L. Squires, and C. Squires.** 1979. Control features within the rplJL-rpoBC transcription unit of *Escherichia coli.* Proc. Natl. Acad. Sci. U.S.A. **76:**4922–4926.

2. **Block, R., and W. A. Haseltine.** 1974. In vitro synthesis of ppGpp and pppGpp, p. 747–761. *In* M. Nomura, A. Tissieres, and P. Lengyel (ed.), Ribosomes. Cold Spring Harbor Laboratory, Cold Spring Harbor, N.Y.

3. **Cashel, M.** 1969. The control of ribonucleic acid synthesis in *Escherichia coli.* J. Biol. Chem. **244:**3133–3141.

4. **Cashel, M., and J. Gallant.** 1974. Cellular regulation of guanosine tetraphosphate and guanosine pentaphosphate, p. 733–745. *In* M. Nomura, A. Tissieres, and P. Lengyel (ed.), Ribosomes. Cold Spring Harbor Laboratory, Cold Spring Harbor, N.Y.

5. **Dennis, P., and M. Nomura.** 1974. Stringent control of ribosomal protein gene expression in *Escherichia coli.* Proc. Natl. Acad. Sci. U.S.A. **71:**3819–3823.

6. **Dennis, P., and M. Nomura.** 1975. Stringent control of the transcriptional activities of ribosomal protein genes in *E. coli.* Nature (London) **225:**460–465.

7. **Haseltine, W. A., and R. Block.** 1973. Synthesis of guanosine tetra and pentaphosphate requires the presence of a codon-specific uncharged transfer ribonucleic acid in the acceptor site of ribosomes. Proc. Natl. Acad. Sci. U.S.A. **70:**1564–1568.

8. **Ikemura, T., and J. E. Dahlberg.** 1973. Small ribonu-

cleic acids of *Escherichia coli.* J. Biol. Chem. **248:**5033–5041.

9. **Iwakura, Y., K. Ito, and A. Ishihama.** 1974. Biosynthesis of RNA polymerase in *Escherichia coli.* I. Control of RNA polymerase content at various growth rates. Mol. Gen. Genet. **133:**1–23.

10. **Jaskunas, S. R., R. R. Burgess, and M. Nomura.** 1975. Identification of a gene for the α-subunit of RNA polymerase at the str-spc region of the *Escherichia coli* chromosome. Proc. Natl. Acad. Sci. U.S.A. **72:**5036–5040.

11. **Jaskunas, S. R., L. Lindahl, M. Nomura, and R. R. Burgess.** 1975. Identification of two copies of the gene for the elongation factor EF-Tu in *E. coli.* Nature (London) **257:**458–462.

12. **Johnsen, M., and N. P. Fiil.** 1976. Synthesis of ribosomal protein L7/L12 *in vitro*, p. 221–225. *In* N. O. Kjeldgaard and O. Maaløe (ed.), Alfred Benzon Symposium IX: Control of ribosome synthesis. Academic Press, Inc., New York.

13. **Kirschbaum, J. B., and J. Scaife.** 1974. Evidence for a λ transducing phage carrying the genes for the β and β' subunits of *Escherichia coli* RNA polymerase. Mol. Gen. Genet. **132:**193–201.

14. **Lindahl, L., L. Post, and M. Nomura.** 1976. DNA-dependent *in vitro* synthesis of ribosomal proteins, protein elongation factors, and RNA polymerase subunit α: inhibition by ppGpp. Cell **9:**439–448.

15. **Maher, D. L., and P. P. Dennis.** 1977. *In vivo* transcription of *E. coli* genes coding for rRNA, ribosomal proteins and subunits of RNA polymerase: influence of the stringent control system. Mol. Gen. Genet. **155:**208–211.

16. **Miller, J. H.** 1972. Experiments in molecular genetics. Cold Spring Harbor Laboratory, Cold Spring Harbor, N.Y.

17. **Nishino, T., J. Gallant, P. Shalit, L. Palmer, and T. Wehr.** 1979. Regulatory nucleotides involved in the *rel* function of *Bacillus subtilis.* J. Bacteriol. **140:**671–679.

18. **Nomura, M., and S. R. Jaskunas.** 1976. Organization of genes for ribosomal RNA, ribosomal proteins, protein elongation factors and RNA polymerase subunits in *Escherichia coli*, p. 191–204. *In* N. O. Kjelgaard and O. Maaløe (ed.), Alfred Benzon Symposium IX: Control of ribosome synthesis. Academic Press, Inc., New York.

19. **Pedersen, F. S., E. Lund, and N. O. Kjelgaard.** 1973. Codon specific, tRNA dependent *in vitro* synthesis of ppGpp and pppGpp. Nature (London) New Biol. **243:**13–15.

20. **Primakoff, P., and P. Berg.** 1970. Stringent control of transcription of phage φ80 psu₃. Cold Spring Harbor Symp. Quant. Biol. **35:**391–396.

21. **Reiness, B., H. L. Yang, G. Zubay, and M. Cashel.** 1975. Effects of guanosine tetraphosphate on cell-free synthesis of *Escherichia coli* ribosomal RNA and other gene products. Proc. Natl. Acad. Sci. U.S.A. **72:**2881–2885.

22. **Smith, I., P. Paress, K. Cabane, and E. Dubnau.** 1980. Genetics and physiology of the *rel* system of *Bacillus subtilis.* Mol. Gen. Genet. **178:**271–279.

23. **Stent, G. S., and S. Brenner.** 1961. A genetic locus for the regulation of ribonucleic acid synthesis. Proc. Natl. Acad. Sci. U.S.A. **47:**2005–2014.

24. **Swanton, M., and G. Edlin.** 1972. Isolation and characterization of an RNA relaxed mutant of *Bacillus subtilis.* Biochem. Biophys. Res. Commun. **46:**583–588.

Effect of Citric Acid Cycle Intermediates on *Bacillus subtilis* Sporulation: a Reexamination

TERRANCE LEIGHTON

Department of Microbiology and Immunology, University of California, Berkeley, California 94720

The effects of citric acid cycle intermediates on *Bacillus subtilis* mass doubling times, rates of protease accumulation, and rates of sporulation were reexamined.

The role of proteolytic enzymes in *Bacillus subtilis* sporulation is unclear. Serine protease and metal protease are synthesized and excreted by sporulating *B. subtilis* cells concomitant with the initiation of sporulation (6–8). Circumstantial evidence suggests that the metal protease may not be required for sporulation (7) and that the serine protease may be more directly involved with developmental functions (1, 3, 6). None of these studies is particularly compelling as a result of the lack of a large collection of structural gene lesions in the relevant polypeptides. Ohne and Rutberg (9) reported that citric acid cycle intermediates apparently reduce the level of extracellular serine and metal proteases without affecting the sporulation processes. Their results seem to support the conclusion that at least the extracellular form of the serine protease is not required for sporulation. However, several experimental variables were not rigorously controlled by these investigators. The aeration of their cultures was suboptimal, resulting in reduced rates of sporulation (compare their Fig. 1F and the data in Fig. 1). Their cultures were not in a state of balanced exponential growth prior to the initiation of sporulation, especially when citric acid cycle intermediates were present. The kinetics of growth, protease accumulation, and sporulation were not followed in several instances. The possible effects of divalent cation chelation, by citric acid cycle intermediates, on protease activity and sporulation were not examined. To gain a better appreciation of the relevance of Ohne and Rutberg's results, I have repeated a number of their experiments under more exacting conditions.

METHODS AND RESULTS

W168 (prototrophic *B. subtilis*) was grown in one of the following media: 1XS contained 8 g of nutrient broth (Difco) per liter, 1.2 mM MgSO$_4$·7H$_2$O, 1 mM Ca(NO$_3$)$_2$, 0.1 mM MnCl$_2$, and 0.001 mM FeSO$_4$; 1XS-pH 6.5 was 1XS plus 0.05 M potassium phosphate buffer, pH 6.5; and 2XS-pH 6.5 was 1XS-pH 6.5 plus an additional 8 g of nutrient broth per liter. Citric acid cycle intermediates were prepared as described (9) and were present at a final concentration of 80 mM. Cells were grown in the appropriate medium for at least 10 exponential mass doublings prior to the initiation of an experiment (4, 5). Optimal conditions for cell culture aeration have been described (4). Serine protease activity, metal protease activity, and forespore refractility were quantitated as described (5, 6).

The citric acid cycle intermediates citric acid and oxalacetic acid were chosen for intensive study since Ohne and Rutberg (9) had reported that these compounds caused large decreases in protease activity while allowing significant levels of sporulation. To assess possible direct effects of citrate on serine and metal protease activity, I incubated (8 h, 37°C) pure serine protease (150 μg of protein per ml, 0.05 M phosphate buffer, pH 6.5, containing 1 mM CaCl$_2$) or t$_2$ culture supernatant (t$_2$ = 2 h after the deviation from exponential growth phase) in the presence or absence of 80 mM citrate. Pure serine protease activity and the activity of culture supernatant serine and metal protease was decreased approximately 25% relative to control samples (data not shown). These effects were much smaller than the 70 to 80% inhibition of proteolytic activity observed by Ohne and Rutberg (9). Hence, a direct effect of citrate on protease stability or activity did not suffice to explain their results.

To characterize further the in vivo consequences of citrate and oxalacetate addition, I compared the effects of 80 mM levels of these compounds on vegetative growth rates, protease accumulation (an early sporulation event), and refractile body accumulation (a late sporulation event). In 1XS-pH 6.5, the medium used by Ohne and Rutberg, citrate affected vegetative growth rates (45-min mass doubling time versus 30-min mass doubling time in untreated cultures) and the rates of protease and refractile body accumulation (Fig. 1 and 2). The citrate effects were partially ameliorated by the addition of excess calcium nitrate (10 mM). The mass doubling time of a culture growing in 1XS-pH 6.5 medium containing 80 mM citrate and 10 mM calcium was approximately 37 min (Fig. 1). Citrate addition to 1XS-pH 6.5 medium resulted in kinetically complex effects on protease synthesis; at early sporulation stages, serine and metal proteases were underproduced, whereas at later time periods serine protease was overproduced (Fig. 2). The underproduction of protease activity caused by citrate (t_0 to t_2) could be overcome by the addition of 10 mM calcium; however, the later overproduction of serine protease was not affected (Fig. 2). These results were complicated further by the fact that citrate affected the balance of protease activities; that is, metal protease accumulation occurred at 50% of the normal rate. The differential inhibition of metal protease accumulation was prevented by the addition of 10 mM calcium. These data could be reconciled with those of Ohne and Rutberg (9) if one assumes that they selected a time point for protease measurement when enzyme accu-

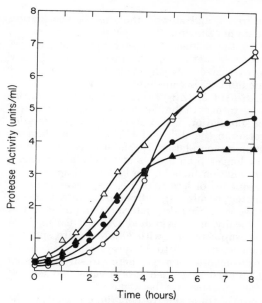

FIG. 2. *Effect of citric acid cycle intermediates on the accumulation of protease activity. Cultures were grown in 1XS-pH 6.5 medium. Symbols: ●, control culture; ○, culture plus 80 mM citrate; △, culture plus 80 mM citrate and 10 mM Ca(NO₃)₂; ▲, culture plus 80 mM oxalacetate. Each data point represents an average of four independent experiments (standard deviation, ±10%). One unit of proteolytic activity hydrolyzes 1 mg of azocasein per h at 37°C. At any point in time, approximately 80% of the proteolytic activity is due to serine protease and 20% of the proteolytic activity is due to metal protease, except in the case of the 80 mM citrate cultures, where the balance of activity is 90% serine protease and 10% metal protease. t_0 = 2-h time point.*

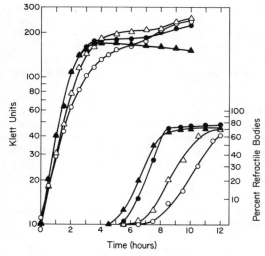

FIG. 1. *Effect of citric acid cycle intermediates on growth (left portion of graph) and sporulation (right portion of graph). Cultures were grown in 1XS-pH 6.5 medium. Symbols: ●, control culture; ○, culture plus 80 mM citrate; △, culture plus 80 mM citrate and 10 mM Ca(NO₃)₂; ▲, culture plus 80 mM oxalacetate. Klett readings were recorded with the use of a no. 54 filter; t_0 = 2-h time point.*

mulation was *transiently* depressed.

Oxalacetate had no significant effect on vegetative growth rates, the initiation of protease accumulation, or the efficiency of sporulation (Fig. 1 and 2). Serine and metal protease levels at t_6 were approximately 25% lower than those in untreated cultures. Postexponential cell mass levels were also lowered approximately 25% when oxalacetate was present. Hence, there was no apparent selective effect of oxalacetate on protease accumulation or sporulation. The effects of oxalacetate were not influenced by the addition of 10 mM calcium ions or 1 mM manganese ions. These results are not in agreement with those of Ohne and Rutberg (9), who observed an inhibition of protease accumulation and a stimulation of sporulation by oxalacetate.

The presence of 80 mM malate, in 1XS-pH 6.5 medium, caused an exaggerated postexponential mass increase (i.e., a twofold greater than normal cell mass level at t_1 to t_2) and delayed the appearance and slowed the rate of sporulation (data not shown).

In the course of these experiments, it was observed that 1XS-pH 6.5 medium did not allow as extensive or synchronous sporulation as 1XS medium. 2XSG medium (4, 5) was superior to 1XS-pH 6.5, 2XS-pH 6.5, 1XS, and 2XS media with respect to the synchrony and absolute level of sporulation. I also observed a rapid deterioration in the ability of 1XS-pH 6.5 medium to support sporulation. If this medium was stored for longer than 5 days at 20°C (in the dark), sporulation did not occur (maximum sporulation frequency of less than 1%, i.e., fewer than 10^6 spores per ml). Vegetative growth rates were not affected during this deterioration period. Paradoxically, if deteriorated 1XS-pH 6.5 medium was supplemented with 80 mM citrate, sporulation levels of 20 to 30% were observed. In all experiments reported here, freshly prepared 1XS-pH 6.5 was used. It is possible that some of the discrepancies between these results and those of Ohne and Rutberg (9), especially in the case of oxalacetate addition, could be related to medium deterioration anomalies. The use of a suboptimal medium would provide an explanation for their finding that oxalacetate addition *stimulated* sporulation above the level of "control" cultures (9). Consistent with this interpretation is the fact that maximal control culture sporulation levels obtained in the present study (1×10^8 to 2×10^8 spores per ml) are substantially higher than those obtained by Ohne and Rutberg (3×10^7 spores per ml; 9) in the same medium.

DISCUSSION

The results reported here and those of Ohne and Rutberg (9) indicate that the effects of citric acid cycle intermediates on *B. subtilis* growth and sporulation parameters are complex. We both found that malate addition disturbs postexponential cell physiology, effecting a delay in the time of appearance and rate of accumulation of spores (see also Freese et al. [2]). Likewise, citrate addition causes a decrease in the rate of cell growth and delays the appearance of mature spores. This treatment also causes an alteration in the balance of serine and metal proteases accumulated by sporulating cultures. My results clarify the kinetic effects of citrate addition on serine and metal protease accumulation. Consistent with the Ohne and Rutberg single measurement of protease activity (9), I found that citrate addition does cause a *transient* underproduction of proteolytic activity during early sporulation time periods. However, after this period of decreased protease accumulation,

there is a rapid increase in the rate of protease elaboration, resulting in greater amounts of protease activity being accumulated by citrate-treated cultures than are found in untreated culture supernatants (see Fig. 2). Hence, citrate addition has a limited and transient effect on protease accumulation similar to its effects on the accumulation of refractile prespores. The citrate-induced disturbance of sporulation-associated events is due in part to chelation of divalent cations. Increasing the level of calcium ions in citrate-treated cultures increases the rate of protease and refractile body accumulation (Fig. 1 and 2).

Using the experimental conditions described in Methods and Results, I have been unable to repeat the observation that oxalacetate addition causes an increase in sporulation efficiency and concomitantly restricts the accumulation of serine and metal proteases (9). It is likely that the more optimal conditions utilized in the present study are responsible for the observed differences. This inference is supported by the finding reported here that oxalacetate addition has little effect on sporulation frequency (Fig. 1), whereas Ohne and Rutberg found that this compound caused an increase in the *absolute efficiency* of sporulation compared with an unsupplemented "control" culture (9). Citrate addition was observed by Ohne and Rutberg (9) to produce similar effects, suggesting that their experimental conditions did not allow a substantial fraction of the total cell population to complete the sporulation process (see Methods and Results). From the results of the Ohne and Rutberg study, it was not clear what fraction of the total cell population was able to sporulate under various experimental conditions and how the levels of proteolytic enzymes correlated with sporulation efficiency. The results reported here define the effects of citric acid cycle intermediates on sporulation-associated events under conditions in which a majority of the total cell population is undergoing synchronous development.

ACKNOWLEDGMENTS

This research was supported by grants from the National Science Foundation and the National Institutes of Health.

LITERATURE CITED

1. Dancer, B. N., and J. Mandelstam. 1975. Production and possible function of serine protease during sporulation of *Bacillus subtilis*. J. Bacteriol. **121**:406–410.
2. Freese, E., Y. K. Oh, E. B. Freese, M. D. Diesterhaft, and C. Prasad. 1972. Suppression of sporulation of *Bacillus subtilis*, p. 212–221. *In* H. O. Halvorson, R. Hanson, and L. L. Campbell (ed.), Spores V. American Society for Microbiology, Washington, D.C.
3. Geele, G., E. Garrett, and J. H. Hageman. 1975. Effect

of benzeneboronic acids on sporulation and on production of serine proteases in *Bacillus subtilis* cells, p. 391–396. *In* P. Gerhardt, R. N. Costilow, and H. L. Sadoff (ed.), Spores VI. American Society for Microbiology, Washington, D.C.

4. **Leighton, T.** 1974. Further studies on the stability of sporulation messenger ribonucleic acid in *Bacillus subtilis*. J. Biol. Chem. **246:**7808–7812.

5. **Leighton, T. J., and R. H. Doi.** 1971. The stability of messenger ribonucleic acid during sporulation in *Bacillus subtilis*. J. Biol. Chem. **246:**3189–3195.

6. **Leighton, T. J., R. H. Doi, R. A. J. Warren, and R. A.**

Kelln. 1973. The relationship of serine protease activity to RNA polymerase modification and sporulation of *Bacillus subtilis*. J. Mol. Biol. **76:**103–122.

7. **Michel, J., and J. Millet.** 1970. Physiological studies on early-blocked sporulation mutants of *Bacillus subtilis*. J. Appl. Bacteriol. **33:**220–227.

8. **Millet, J.** 1970. Characterization of proteinases excreted by *Bacillus subtilis* Marburg strain during sporulation. J. Appl. Bacteriol. **33:**207–219.

9. **Ohne, M., and B. Rutberg.** 1976. Repression of sporulation in *Bacillus subtilis* by L-malate. J. Bacteriol. **125:**453–460.

Relationship Between Glutamine Pools and Enzymes of Nitrogen Metabolism in *Bacillus subtilis*

KATHRYN L. DESHPANDE,[1] RANDY S. FISCHER, JON R. KATZE, AND JAMES F. KANE

Department of Microbiology and Immunology, University of Tennessee Center for the Health Sciences, Memphis, Tennessee 38163

We examined the relationship in *Bacillus subtilis* between glutamine pools and the activities of glutamate synthase and glutamine synthetase. We observed an inverse relationship between intracellular glutamine concentration and glutamate synthase activity and a direct relationship between the glutamine pool and the activity of glutamine synthetase. We propose that glutamine is an important regulatory element in the control of glutamate synthase activity; it does not appear to be a corepressor of glutamine synthetase.

In *Bacillus subtilis*, the assimilation of inorganic nitrogen into an organic form occurs primarily through the coupled reactions of glutamate synthase (8) (GOGAT, EC 2.6.1.53) and glutamine synthetase (GS, EC 6.3.1.2).

GS: NH_3 + glutamate

$$+ ATP \xrightarrow{Mn^{2+} \text{ or } Mg^{2+}} \text{glutamine}$$

$$+ ADP + P_i$$

GOGAT: glutamine + α-ketoglutarate

$$+ NADPH \rightarrow 2 \text{ glutamate} + NADP^+$$

Glutamate dehydrogenase (GDH, EC 1.4.1.4) also occurs in *B. subtilis* (7), but its physiological function is unknown at present.

Recently, we reported that the activity of GOGAT in *B. subtilis* is inversely proportional to the internal concentration of glutamine (3). We have examined the sequence of intracellular events involved in this regulatory process and have found that a loss of GOGAT activity is associated with an increase in both the activity of GS and the glutamine pool.

RESULTS AND DISCUSSION

Previous results (3, 4) indicated (i) that there is an inverse relationship between the activity of GOGAT and the intracellular concentration of glutamine and (ii) that there is no consistent correlation between GOGAT activity and the intracellular levels of aspartate, glutamate, or alanine. The inverse relationship between

[1] Present address: Department of Biochemistry, University of Missouri at Columbia, Columbia, MO 65212.

GOGAT and the glutamine pool was investigated further by shifting growing cultures of a *trpC* mutant, strain NP19, from an NH_4^+ plus glutamate medium (high GOGAT activity) to glutamate medium (low GOGAT activity) and vice versa. The specific activities of GOGAT and GS, as well as the intracellular levels of aspartate, glutamate, glutamine, and alanine, were monitored at various times after the shifts. The intracellular levels of the four amino acids are expressed as ratios: that is, the concentration of aspartate, glutamate, glutamine, or alanine to the concentration of serine. Serine was chosen as a reference since it is not directly involved in nitrogen metabolism and its concentration is generally around 1 mM. Since serine was used as an internal control, we believed that variations in sample time, cell volume, and amino acid analyses could be minimized. As an additional control, the ratio of threonine plus glycine to serine is shown also.

The results of the shift from NH_4^+ plus glutamate to glutamate are shown in Table 1. Glutamate was the most abundant intracellular amino acid, constituting 88 to 93% of the total of aspartate, glutamate, glutamine, and alanine. There was, however, little variation in its level relative to serine. Although the pools of aspartate and alanine varied about 2-fold, there does not appear to be a correlation between these changes and GOGAT activity. The glutamine concentration, on the other hand, increased about 5-fold in 30 min. This increased level of glutamine results from a 2.5-fold derepression of GS activity. The level of GS protein was monitored by rocket immunoelectrophoresis and was found to increase 2.9-fold. These results demonstrate that after the removal of NH_4^+ from the medium the *glnA* locus is derepressed. The increased level of GS results in a rapid build-up of intracellular glutamine, and the level of both GS and glutamine remained 2- to 3-fold greater than at zero time for the duration of the experiment.

The level of GOGAT decreased after the increase in the glutamine pool. This result suggests a close correlation between the activity of GOGAT and the intracellular pool of gluta-

TABLE 1. *Effect of shifting from NH_4^+ plus glutamate to glutamate medium on the activity of GOGAT, GS, and the intracellular amino acid pools[a]*

Time[b] (min)	A_{600}[c]	Intracellular level[d] of:					Relative activity[e] of:		
		Aspartate	Glutamate	Glutamine	Alanine	Threonine-glycine	GOGAT	GS	GS[f] antigen
0	0.35	3.2	78	1.1	1.4	0.5	1.00	1.0	1.0
30	0.47	2.8	77	5.5	1.3	1.0	0.91	2.5	2.9
60	0.53	3.7	76	2.6	1.8	0.8	0.64	3.0	3.7
90	0.61	3.5	90	2.7	0.9	0.7	0.47	3.0	3.1
120	0.69	2.2	104	3.7	2.2	0.8	0.34	2.6	3.9

[a] The *trpC* mutant NP19 was grown in minimal glucose medium (4) containing 0.2% $(NH_4)_2SO_4$ and 0.2% glutamate.

[b] At zero time, the culture was shifted to the minimal glucose medium containing only 0.2% glutamate.

[c] The absorbance change of one of these experiments is shown. Prior to the shift, the doubling times of the NH_4^+ plus glutamate cultures were 55 to 60 min.

[d] The internal amino acid pools were measured by high-pressure liquid chromatography (3) and are expressed as a ratio of the concentration of the indicated amino acid to the concentration of serine. If it is assumed that the cell volume remains constant and 1 mg of protein is equivalent to 8 μl of internal volume, then the concentrations of serine are 1.1, 1.0, 1.1, 1.1, and 1.1 mM at the indicated times. The concentrations of serine ranged from a low of 0.6 mM to a high of 1.8 mM. All ratios represent an average of three separate experiments. Threonine and glycine elute together under these conditions.

[e] The relative activity of 1.0 is equal to 5.8 μmol of NADPH oxidized per h per mg of protein for GOGAT (2) and 65 nmol of γ-glutamyl hydroxamate formed per min per mg of protein for GS with both Mg^{2+} (20 mM) and Mn^{2+} (0.2 mM) as the activating cations. These values were an average of eight and five samples, respectively.

[f] GS antibody was prepared as described (1). Rocket immunoelectrophoresis was carried out in a 1.2% agarose gel containing a 1:40 dilution of antibody. A 2-μl sample of crude extract was used as the source of antigen. These values were an average of two samples.

TABLE 2. *Effect of shifting from glutamate to glutamate plus NH_4^+ on the activities of GOGAT and GS and the intracellular amino acid pools[a]*

Time[b] (min)	A_{600}[c]	Intracellular level[d] of:					Relative activity[e] of:		
		Aspartate	Glutamate	Glutamine	Alanine	Threonine-glycine	GOGAT	GS	GS[f] antigen
0	0.34	5.3	84	5.5	2.0	0.6	1.0	1.00	1.00
30	0.36	4.6	55	6.7	0.8	1.0	2.2	0.88	—
60	0.41	3.8	62	6.1	0.4	0.7	2.4	0.69	0.57
90	0.49	2.3	58	3.0	0.8	0.4	2.3	0.37	0.39
120	0.57	2.0	52	1.5	0.5	0.3	2.7	0.25	0.18
180	0.98	1.9	54	0.5	0.7	0.2	4.6	0.29	0.18

[a] The *trpC* mutant NP19 was grown in minimal glucose medium containing only 0.2% glutamate as nitrogen source.

[b] At zero time, the culture was shifted from glutamate (0.2%) to a fresh medium containing both $(NH_4)_2SO_4$ (0.2%) and glutamate (0.2%).

[c] The absorbance change of one of three experiments is shown. Prior to the shift, the doubling times of the cultures were 140 to 150 min.

[d] See Table 1. The concentrations of serine at the indicated times were: 0.8, 1.0, 0.9, 1.0, 1.3, and 1.6 mM, respectively. The values ranged from 0.6 to 2.0 mM.

[e] The relative activity of 1.0 for GOGAT is equal to 1.2 μmol of NADPH oxidized per h per mg of protein. An activity of 1.0 for GS is equal to 320 nmol of γ-glutamyl hydroxamate formed per min per mg of protein with both Mg^{2+} (20 mM) and Mn^{2+} (0.2 mM) as the activating cations.

[f] See Table 1.

mine. This is consistent with our previous results with batch cultures, which showed an inverse relationship between glutamine and activity of GOGAT. Furthermore, when chloramphenicol was included in the resuspension medium, there was no increase in GS activity and no decrease in the activity of GOGAT. We propose that the accumulation of intracellular glutamine plays an important role in bringing about the loss of GOGAT activity and is dependent upon an increased activity of GS.

In the reverse experiment (Table 2), several pool changes occurred after the shift to NH_4^+ plus glutamate medium. At zero time, the glutamine pool was 5.5-fold greater than the serine pool, GOGAT was repressed, and GS was derepressed. After 30 min, there was a reproducible 2.5-fold increase in GOGAT activity. During this time, however, the concentration of glutamine increased relative to serine, and there was a 2.5-fold decrease in the concentration of alanine, a 50% decrease in the glutamate concentration, and a small drop in the aspartate concentration. It is not clear what change was responsible for the increase in GOGAT activity, however. For the next 90 min there was little change in the activity of GOGAT, but the pools of aspartate and glutamine and the activity of GS decreased with time. An increase in the activity of GOGAT was observed in the 180-min sample, and this increase was associated with a 3-fold drop in the concentration of glutamine. No other amino acid changed significantly during the final hour of the experiment.

We draw two conclusions from these experiments. First, the activity of GOGAT is inversely related to the glutamine pool. We do not know whether this glutamine effect is at the transcriptional or posttranslational level. Second, there is a direct correlation between the intracellular glutamine pool and the activity of GS. This is not consistent with glutamine functioning as a

corepressor (5, 6, 9) of GS activity in *B. subtilis*. We propose that some other metabolite is the primary effector molecule controlling the expression of the *glnA* locus.

ACKNOWLEDGMENTS

We thank Pat Moore for excellent technical assistance and Rose Larkins for excellent secretarial assistance.

This research was supported by Public Health Service grants AM19443 and AM2696301 from the National Institute of Arthritis and Metabolic Diseases.

LITERATURE CITED

1. **Dean, D. R., J. A. Hoch, and A. I. Aronson.** 1977. Alteration of the *Bacillus subtilis* glutamine synthetase results in overproduction of the enzyme. J. Bacteriol. **131**:981–987.
2. **Deshpande, K. L., and J. F. Kane.** 1980. Glutamate synthase from *Bacillus subtilis*: in vitro reconstitution of an active amidotransferase. Biochem. Biophys. Res. Commun. **93**:308–314.
3. **Deshpande, K. L., J. R. Katze, and J. F. Kane.** 1980. Regulation of glutamate synthase from *Bacillus subtilis* by glutamine. Biochem. Biophys. Res. Commun. **95**:55–60.
4. **Deshpande, K. L., J. R. Katze, and J. F. Kane.** 1981. Effect of glutamine on enzymes of nitrogen metabolism in *Bacillus subtilis*. J. Bacteriol. **145**:768–774.
5. **Deuel, T. F., and S. Prusiner.** 1974. Regulation of glutamine synthetase from *Bacillus subtilis* by divalent cations, feedback inhibitors, and L-glutamine. J. Biol. Chem. **249**:257–264.
6. **Deuel, T. F., and E. R. Stadtman.** 1970. Some kinetic properties of *Bacillus subtilis* glutamine synthetase. J. Biol. Chem. **245**:5206–5213.
7. **Kane, J. F., and K. L. Deshpande.** 1979. Properties of glutamate dehydrogenase from *Bacillus subtilis*. Biochem. Biophys. Res. Commun. **88**:761–767.
8. **Meers, J. L., D. W. Tempest, and C. M. Brown.** 1970. Glutamine (amide):2-oxoglutarate amino transferase oxido-reductase (NADP), an enzyme involved in the synthesis of glutamate by some bacteria. J. Gen. Microbiol. **64**:187–194.
9. **Pan, F. L., and J. G. Coote.** 1979. Glutamine synthetase and glutamate synthase activities during growth and sporulation in *Bacillus subtilis*. J. Gen. Microbiol. **112**:373–377.

Association Between DNA Replication and Interference by Bromodeoxyuridine with Sporulation in *Bacillus subtilis*

J. G. COOTE AND C. BINNIE

Department of Microbiology, University of Glasgow, Glasgow, Scotland

5-Bromo-2′-deoxyuridine prevented sporulation in *Bacillus subtilis thyA thyB but-32* at concentrations that had little effect on growth. The inhibitory effect on sporulation was confined to the period in which DNA replication was in progress, and recovery of sporulation capacity on removal of 5-bromo-2′-deoxyuridine required further DNA replication. Termination of replication during sporulation was delayed by 5-bromo-2′-deoxyuridine.

In many eucaryotic cells 5-bromo-2′-deoxyuridine (BUdR), an analog of thymidine (TdR), has diverse and selective effects on processes associated with differentiation without significantly altering cell growth or gross RNA and protein synthesis (1, 11, 17). The precise mode of action of BUdR has not been satisfactorily resolved. There is often a positive correlation between the effects on differentiation and the degree of incorporation of BUdR into DNA in place of TdR. The inhibitory action of BUdR can often be reversed by subsequent DNA replication in its absence (13), although a reversal of BUdR inhibition of differentiation has been noted in some instances without a concomitant alteration in the BUdR content of the DNA (5, 19). An explanation of its inhibitory action via alterations in transcription has been suggested (10, 17), but it is clear that BUdR can alter the surface properties and membrane fluidity of cells (8, 16), which may in turn interfere with differentiation. Whether cell surface alterations are a direct effect of BUdR or are a consequence of its incorporation into DNA is not clear.

Preliminary work (4) indicated that incorporation of BUdR into the DNA of a thymidine-requiring, BUdR-tolerant strain of *Bacillus subtilis*, at a level that did not significantly affect growth, was associated with a marked reduction in spore formation. This report suggests that the inhibitory effect of BUdR on sporulation is associated with the requirement for DNA replication during the onset of the process.

RESULTS AND DISCUSSION

BUdR inhibited the growth of *B. subtilis thyA thyB trpC* when supplied in place of TdR. A BUdR-tolerant derivative was isolated (4), and the tolerance mutation (*but-32*) was subsequently transferred by DNA-mediated transformation back into the *thyA thyB* background. This strain, *B. subtilis thyA thyB but-32*, grew normally with a BUdR to TdR ratio of between 7.5 and 15.0 µg/ml to 1.0 µg/ml, but only between 1 and 5% of the cells, compared with control cells using TdR alone, were able to form spores. Presence of the analog allowed normal, or slightly increased, production of two early marker events of sporulation, exoprotease and alkaline phosphatase, but the bulk of the cells became blocked before spore septum formation and did not produce dipicolinic acid, a later sporulation event.

DNA replication during sporulation and the effect of BUdR. It was necessary to determine whether the inhibitory effect of BUdR was associated with the period during the onset of sporulation when DNA replication occurs (12). Use of 6-(*p*-hydroxyphenylazo)uracil (HPUra), a specific inhibitor of DNA synthesis, had indicated that successful sporulation depends on the completion of chromosome replication (6). In these experiments cells of *B. subtilis trpC* began to escape the inhibitory effect of HPUra on sporulation about 35 min after the initiation of the process, and this correlated with the termination of DNA replication. Unlike the wild type, cells of strain *but-32* began to escape the inhibitory effect of HPUra at a slightly later time, about 75 min after initiation (Fig. 1a), and whereas escape was essentially complete 2 h after initiation in the wild type (6), the spore incidence with strain *but-32* continued to rise steadily until about 5 h after initiation (Fig. 1b). This difference may relate to the suggestion that

157

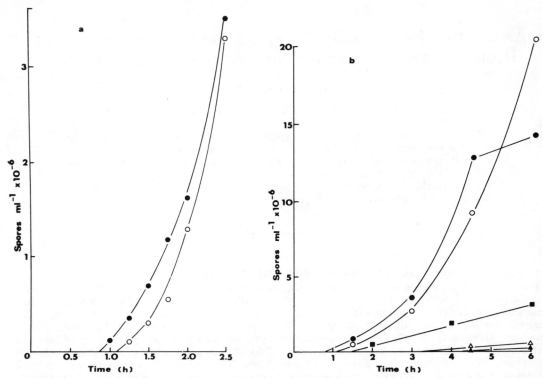

FIG. 1. *Escape of* B. subtilis but-32 *from HPUra and BUdR inhibition of sporulation. Cells growing exponentially in a rich casein hydrolysate medium were resuspended in a poor minimal salts medium to initiate sporulation (20). (a) Cells were grown and resuspended in media containing TdR (8.5 µg/ml). At intervals, two 10-ml portions were removed; the cells of one were treated with HPUra (50 µg/ml) (○) and the cells of the other were collected on a membrane filter, washed, and suspended in sporulation medium (10 ml) containing TdR (1.0 µg/ml) + BUdR (15 µg/ml) (●). (b) Three separate cultures were grown and resuspended with (i) TdR (8.5 µg/ml), (ii) TdR (8.5 µg/ml) + HPUra (1.0 µg/ml), and (iii) TdR (1.0 µg/ml) + BUdR (15 µg/ml). At intervals, 10-ml portions of cultures i and ii were resuspended as above with TdR (1.0 µg/ml) + BUdR (15 µg/ml). Culture i treated with HPUra (○) or BUdR (●). Culture ii treated with HPUra (△) or BUdR (▲). Culture iii treated with HPUra (■). Heat-resistant spores were determined at t_{20} by heating a cell sample at 85°C for 15 min followed by dilution and plating on nutrient agar.*

the rate of replication is slower in TdR-requiring strains (7, 9) or the possibility that the replication essential for sporogenesis, although proceeding at the same rate as in the wild type, is initiated over a wider time span under starvation conditions (18). The pattern of escape of cells from BUdR inhibition of sporulation was similar to that obtained with HPUra (Fig. 1), although several experiments indicated that BUdR escape preceded that of HPUra by some 15 min.

DNA synthesis in the wild type can be slowed down by partially inhibitory concentrations of HPUra, and escape of sporulation from the effect of a high level of HPUra is consequently delayed (22). Addition of a subinhibitory amount of HPUra to sporulating cells of *but-32* delayed the escape of sporulation from both HPUra and

BUdR inhibitions to a similar extent (Fig. 1b). Similarly, cells grown and resuspended with BUdR showed delayed escape of sporulation from HPUra inhibition (Fig. 1b). It should be noted that the inclusion of uridine or deoxycytidine at 50 µg/ml in the sporulation medium had no marked effect on BUdR inhibition of sporulation.

Recovery from BUdR inhibition of sporulation. Recovery from inhibition of sporulation by BUdR was obtained if the analog was removed from the sporulation medium and replaced by TdR (4). This recovery in sporulation capacity on removal of BUdR was not obtained if DNA synthesis was prevented by starvation for TdR or treatment with HPUra (Table 1). Addition of HPUra to cells at intervals after

Table 1. *Recovery from BUdR inhibition of sporulation*[a]

Determination	Colony-forming units per ml at indicated time (h) after shifting cell sample to TdR medium					
	t_3	t_4	t_5	t_6	t_7	t_8
Viable cells at time of transfer	950×10^5	720×10^5	670×10^5	970×10^5	$1,080 \times 10^5$	250×10^5
Spores at t_{20}	17×10^5 (55×10^5)	52×10^5	29×10^5	19×10^5	12×10^5	10×10^5
Spores at t_{20}; HPUra added at time of transfer	(0.4×10^5)					
Spores at t_{20}; HPUra added 20 min after transfer	(0.8×10^5)					
Spores at t_{20}; HPUra added 45 min after transfer	1.0×10^5 (0.7×10^5)		2.0×10^5			0.7×10^5
Spores at t_{20}; HPUra added 60 min after transfer	(1.7×10^5)					
Spores at t_{20}; HPUra added 90 min after transfer	4.0×10^5 (6.0×10^5)		5.0×10^5			2.0×10^5

[a] Cells were grown and resuspended in media containing TdR (1.0 µg/ml) plus BUdR (15 µg/ml). At intervals, 10-ml samples were removed, and the cells were collected on a membrane filter, washed, and suspended in sporulation medium (10 ml) containing TdR (8.5 µg/ml). At t_3, t_5, and t_8 two additional samples were transferred in the same way, and HPUra (50 µg/ml) was added to one 45 min and to the other 90 min after the transfer. Viable-cell counts were determined after dilution and plating of a portion of the culture onto nutrient agar. Figures in parentheses represent the results of a separate experiment in which HPUra was added at more frequent intervals after transfer of cells at t_3. In this experiment a value of 2.2×10^5 spores per ml at t_{20} was obtained for cells incubated in TdR plus BUdR medium without transfer.

transfer to medium with TdR showed that the cells only began to regain their capacity to sporulate after about 60 min. This delay apparently was not caused by a lag in DNA synthesis, as this proceeded normally after transfer to TdR medium but ceased immediately when HPUra was added (Fig. 2a). Cells exhibited the capacity to sporulate after removal of BUdR at any time up to t_8 in BUdR medium (t_1, t_2, etc., are hours after initiation of sporulation), but the capacity declined somewhat from t_4 to t_8 (Table 1). By t_8, the number of viable cells was reduced slightly; in addition, the rate of incorporation of [¹⁴C]-TdR into acid-insoluble material was less at t_7 than at t_3 on transfer of cells from BUdR to TdR medium (Fig. 2a). The rate of [¹⁴C]TdR incorporation at t_3 was, in turn, less than that shown by cells simply resuspended in TdR medium from t_0 (Fig. 2a).

Thus, the recovery of cells from BUdR inhibition tended to decline the longer the cells were exposed to the analog, and this correlated to some extent with a decrease in the rate of DNA replication on transfer to TdR medium. This raised the possibility that the effect of BUdR was to prevent DNA replication in the majority of cells such that termination of replication was not possible and successful sporulation was therefore prevented. However, [¹⁴C]TdR incorporation appeared to proceed normally in cells exposed to BUdR at various times after initiation of sporulation (Fig. 2b). It should be noted

that, whereas [¹⁴C]TdR incorporation leveled off between t_4 and t_5 in cells given TdR from t_0 (Fig. 2a), replication continued in a linear manner up to t_7 in cells given BUdR from t_0 (Fig. 2b). This agreed with the later escape time from HPUra inhibition of sporulation exhibited by BUdR-treated cells (Fig. 1b). Cells exposed to BUdR after 3 h with TdR continued replication almost linearly for a further 3 to 4 h and, somewhat surprisingly, cells exposed to BUdR after 6 h with TdR (when replication had almost ceased [Fig. 2a]) also incorporated [¹⁴C]TdR at a linear rate for almost 2 h (Fig. 2b).

Marker frequency analysis of sporulating cells. It seemed, therefore, that DNA replication was not adversely affected by exposure of cells to BUdR. This was confirmed by marker frequency analysis (Fig. 3). In these experiments DNA was prepared at intervals after initiation of sporulation from cells exposed to BUdR, and the ratio of pur^+ to met^+ transformants was compared with that given by DNA prepared from cells sporulated with TdR alone. The $purA16$ marker is located near the origin of replication, and the $metB5$ marker is located near the terminus. The ratio of these markers will give an indication of the average number of replication positions per chromosome (n), using the relationship pur^+/met^+ ratio $= 2^n$ (21). DNA prepared from cells grown for several generations in casein hydrolysate medium with TdR (1.0 µg/ml) plus BUdR (15 µg/ml) gave a con-

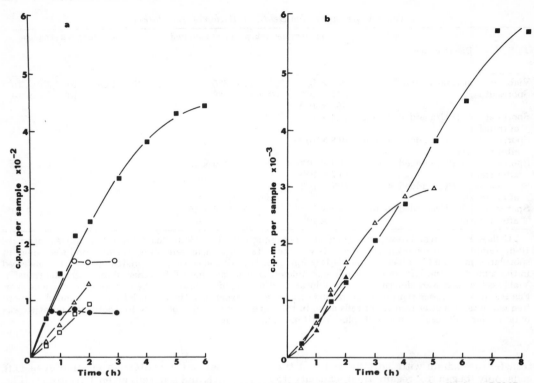

FIG. 2. *Incorporation of radioactivity into acid-insoluble material after transfer of cells from BUdR to TdR medium or vice versa. (a) Cells were grown and resuspended in media containing TdR (1.0 μg/ml) + BUdR (15 μg/ml). At t_0 (■), t_3 (△), and t_7 (□), portions (10 ml) of the culture were removed; the cells were collected on a membrane filter, washed, and suspended in sporulation medium (10 ml) containing TdR (8.5 μg/ml) + [^{14}C]TdR (0.025 μCi/ml). In a separate experiment two additional samples at t_3 were transferred in the same way; HPUra (50 μg/ml) was added to one 45 min (●) and to the other 90 min (○) after the transfer. (b) Cells were grown and resuspended in media with TdR (8.5 μg/ml), and at t_0 (■), t_3 (△), and t_6 (▲) samples were transferred, as above, to sporulation medium with TdR (1.0 μg/ml) + BUdR (15 μg/ml) + [^{14}C]TdR (0.025 μCi/ml). Samples (0.2 ml) were removed at intervals and added to 1 ml of 1 N KOH. After incubation at 37°C for 2 h to degrade RNA, 2.4 ml of ice-cold trichloroacetic acid (10%, wt/vol) containing TdR (100 μg/ml) was added. The precipitates were collected on Whatman GF/C glass fiber filters, washed with ice-cold trichloroacetic acid (10%, wt/vol), dried, and the radioactivity was determined in a scintillation counter.*

sistently higher pur^+/met^+ transformant ratio than DNA from cells grown with TdR (8.5 μg/ml) alone. In four experiments in which this ratio was compared, an average pur^+/met^+ ratio of 9.9 was obtained for BUdR-grown cells, compared with an average ratio of 5.9 for TdR-grown cells. Spores of *B. subtilis* contain completed chromosomes (15, 18), and the pur^+/met^+ ratio obtained with spore DNA can be used as a correction factor to standardize the ratios obtained during growth to a value of 1.0 for spore DNA. DNA prepared from *B. subtilis* 168 *trpC* spores (by the method of Sargent [18]) gave a pur^+/met^+ ratio of 0.98 (average of three determinations), whereas DNA from *thyA thyB but-32* spores gave a ratio of 1.38. The discrepancy between the ratios from the wild type and TdR-

requiring strains is probably due to a lack of isogenicity in the *metB* region of the latter strains, which lowers transformation efficiency at the *metB* locus (3). If the average marker ratios for TdR- and BUdR-grown cells are standardized with respect to spore DNA, then values of 4.27 and 7.17, respectively, are obtained. The chromosome replication time (C) can be calculated from the relationship $n = C/r$, where n is the average number of replication positions per chromosome and r is the generation time of the bacteria (7). With a generation time of 35 min and n values of 2.1 and 2.8 (derived from the standardized pur^+/met^+ ratios), DNA replication times of 74 min and 98 min for TdR- and BUdR-grown cells are obtained. A replication time of 53 min was calculated for the wild type

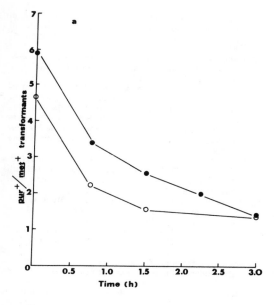

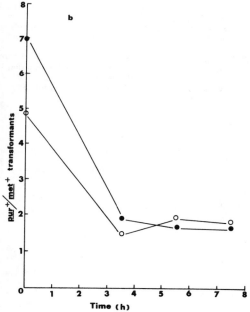

FIG. 3. *Marker frequency analysis during sporu-lation of* B. subtilis *but-32. Cells were grown and resuspended in media containing TdR (8.5 µg/ml) (○) or TdR (1.0 µg/ml) + BUdR (15 µg/ml) (●). Immediately (t₀) and at intervals thereafter, samples (10 ml) were removed, treated with 0.1 ml of 0.1 M NaN₃, and centrifuged; the pellets were suspended in 1.0 ml of 0.15 M NaCl–0.1 M EDTA (pH 8.0) containing lysozyme (500 µg/ml). The cells were incubated for 90 min at 37°C with RNase (50 µg/ml) added after 30 min and protease (200 µg/ml) added after 60 min of incubation. The lysate was then shaken at 4°C for*

(7). The effect of BUdR in slowing replication has been noted previously (14).

During sporulation, the pur^+/met^+ ratio from cells given TdR fell over an initial 90-min period (Fig. 3a) and thereafter did not alter significantly up to $t_{7.5}$ (Fig. 3b). For BUdR-treated cells the ratio only leveled off about 180 min after initiation and again seemed to remain constant up to $t_{7.5}$. In neither case did the plateau value fall to the 1.38 ratio obtained from spore DNA. This might indicate either that chromosomes were not always completely replicated during sporulation of the *but-32* strain or that replication required for sporulation was terminated in some cells at later times during the sporulation period. The latter possibility would agree with the prolonged escape of *but-32* cells from HPUra inhibition of sporulation (Fig. 1b) and would account for the continued incorporation of [^{14}C]TdR beyond t_4 (Fig. 2a). Other work has indicated that termination of replication required for sporulation reached a peak of t_2, but then declined slowly over the next 2 h (18).

In summary, the inhibitory effect of BUdR on sporulation was associated with the requirement for DNA replication. A delay in the achievement of termination of replication, while other changes necessary for sporulation continue normally, may set up an imbalance which blocks the process. However, BUdR has an inhibitory effect on sporulation up to about 15 min before termination of replication, and reversal of the inhibition by TdR apparently required a period of DNA synthesis roughly equivalent to a complete chromosome replication time. This might suggest that BUdR has a more direct effect on some chromosome-associated function such that replication specifically required for sporulation is either not initiated or not terminated in the required manner.

ACKNOWLEDGMENTS

We thank M. Achison for skilled technical assistance. The work was supported by the Science Research Council.

LITERATURE CITED

1. Biswas, D. K., K. T. Abdullah, and B. A. Brennessel.

30 min after the addition of 0.2 ml of 5 M KCl, twice the volume of cooled 95% ethanol was added, and the DNA was collected. Strain Mu8u5u16 (purA16 leu-8 metB5) was used as recipient for transformation, and pur⁺ and met⁺ transformants were selected separately after addition of DNA (0.1 to 0.2 µg/ml) to competent cells (2). Transformation was performed three times for each DNA sample, and the mean values are presented. Results shown are for two experiments (a and b).

1979. On the mechanism of 5-bromodeoxyuridine induction of prolactin synthesis in rat pituitary tumor cells. J. Cell Biol. **81**:1–9.

2. **Bott, K. F., and G. A. Wilson.** 1967. Development of competence in the *Bacillus subtilis* transformation system. J. Bacteriol. **94**:562–570.

3. **Callister, H., and R. G. Wake.** 1974. Completed chromosomes in thymine-requiring *Bacillus subtilis* spores. J. Bacteriol. **120**:579–582.

4. **Coote, J. G.** 1977. Interference by bromodeoxyuridine with differentiation in a prokaryote. Nature (London) **267**:635–637.

5. **Davidson, R. C., and E. R. Kaufman.** 1977. Deoxycytidine reverses the suppression of pigmentation caused by 5-bromodeoxyuridine without changing the amount of 5-bromodeoxyuridine in DNA. Cell **12**:923–929.

6. **Dunn, G., P. Jeffs, N. H. Mann, D. M. Torgersen, and M. Young.** 1978. The relationship between DNA replication and the induction of sporulation in *Bacillus subtilis*. J. Gen. Microbiol. **108**:189–195.

7. **Ephrati-Elizur, D., and S. Borenstein.** 1971. Velocity of chromosome replication in thymine-requiring and independent strains of *Bacillus subtilis*. J. Bacteriol. **106**:58–64.

8. **Evans, I., P. Distefano, K. R. Case, and H. B. Bosmann.** 1977. Cell surface changes caused by growth of B16 melanoma cells in bromodeoxyuridine. FEBS Lett. **78**:109–112.

9. **Gillin, F. D., and A. T. Ganesan.** 1975. Control of chromosome replication in thymine-requiring strains of *Bacillus subtilis*. J. Bacteriol. **123**:1055–1067.

10. **Gordon, J. S., G. I. Bell, H. C. Martinson, and W. J. Rutter.** 1976. Selective interaction of 5'-bromodeoxyuridine substituted DNA with different chromosomal proteins. Biochemistry **15**:4778–4786.

11. **Levitt, D., and A. Dorfman.** 1973. Control of chondrogenesis in limb-bud cell cultures by bromodeoxyuridine. Proc. Natl. Acad. Sci. U.S.A. **70**:2201–2205.

12. **Mandelstam, J., J. M. Sterlini, and D. Kay.** 1971.

Sporulation of *Bacillus subtilis*. Effect of medium on the form of chromosome replication and on initiation of sporulation in *Bacillus subtilis*. Biochem. J. **125**:635–641.

13. **Muira, Y., and F. H. Wilt.** 1971. The effects of bromodeoxyuridine on yolk sac erythropoiesis in the chick embryo. J. Cell Biol. **48**:523–532.

14. **Nagley, P., and R. G. Wake.** 1969. Effect of 5-bromouracil on the pattern of DNA replication in germinating *Bacillus subtilis* spores. J. Mol. Biol. **43**:619–630.

15. **Oishi, M., H. Yoshikawa, and N. Sueoka.** 1964. Synchronous and dichotomous replications of the *Bacillus subtilis* chromosome during spore germination. Nature (London) **204**:1069–1073.

16. **Rosenthal, S. L., A. H. Parola, E. R. Blout, and R. L. Davidson.** 1978. Membrane alterations associated with "transformation" by BUdR in BUdR-dependent cells. Exp. Cell Res. **112**:419–429.

17. **Rutter, W. J., R. C. Pictet, and P. W. Morris.** 1973. Toward molecular mechanisms of developmental processes. Annu. Rev. Biochem. **42**:601–646.

18. **Sargent, M. G.** 1980. Chromosome replication in sporulating cells of *Bacillus subtilis*. J. Bacteriol. **142**:491–498.

19. **Schubert, D., and F. Jacob.** 1970. 5-Bromodeoxyuridine-induced differentiation of a neuroblastoma. Proc. Natl. Acad. Sci. U.S.A. **67**:246–254.

20. **Sterlini, J. M., and J. Mandelstam.** 1969. Commitment to sporulation in *Bacillus subtilis* and its relationship to development of actinomycin resistance. Biochem. J. **113**:29–37.

21. **Sueoka, N., and H. Yoshikawa.** 1965. The chromosome of *Bacillus subtilis*. I. Theory of marker frequency analysis. Genetics **52**:747–757.

22. **Young, M., and P. Jeffs.** 1978. Effect of delayed chromosome termination on the timing of sporulation in *Bacillus subtilis*, p. 201–204. *In* G. Chambliss and J. C. Vary (ed.), Spores VII. American Society for Microbiology, Washington, D.C.

Microcycle Sporulation in *Bacillus subtilis*

M. J. CLOUTIER, J. H. HANLIN, AND R. A. SLEPECKY

Biological Research Laboratories, Department of Biology, Syracuse University, Syracuse, New York 13210

By separating the germination and outgrowth phases and by systematic reduction of the concentrations of ingredients of a defined outgrowth medium, the minimal concentrations allowing germination and limited outgrowth of spores of *Bacillus subtilis* NCTC 3610 (the Marburg strain) were determined. Additions at specific times of Casamino Acids, amino acids (histidine, threonine, proline, homoserine), or diaminopimelic acid gave 8, 12, or 30% sporulation, respectively. Since no increase in number was observed, sporangia could only have arisen from primary cells; thus, microcycle sporulation was achieved in *B. subtilis.*

The conversion of an outgrowing cell to a sporulating cell without any intervening cell division, termed microcycle sporulation, was first described by Vinter and Slepecky (18), who induced outgrowing cells of *Bacillus cereus* NCIB 8122 and *B. megaterium* ATCC 19213 to sporulate by diluting the outgrowth medium. Microcycle sporulation has been demonstrated in *B. megaterium* QMB1551 (1, 6), *B. cereus* strain T (7, 9, 13), *B. cereus* 504 (3), and *B. brevis* AG4 (12), but not in *B. subtilis* (8). Microcycle sporulation has been a useful tool in several species other than *B. subtilis* in delineating various aspects of the sporulation process and its relationship with the cell cycle. This technique was used in demonstrating the necessity of a round of DNA replication for sporulation (6, 9–11) and in experiments which led to the suggestion that sporulation potential may be related to cell growth (10, 16). A microcycle sporulation system in *B. subtilis* would be advantageous because of the vast genetic and gene-splicing capabilities of this species.

RESULTS

The inoculation of 5 ml of a modified defined medium described by Schaeffer et al. (15) (to which 1.12 mM alanine was added to allow germination) with heat-shocked (70°C, 30 min) spores (final concentration, about 6×10^7/ml), followed by incubation in 250-ml Klett flasks with shaking (110 rpm) at 37°C, resulted in good germination, outgrowth, cell multiplication, and sporulation. Step-by-step limitation of the major ingredients of that medium (glucose, phosphates, and nitrogen) and modifications of culture conditions allowed the germinated spores

to elongate to the singlet stage only. These conditions were a medium devoid of NH_4Cl consisting of: 7.5 mM K_2HPO_4, 3.25 mM KH_2PO_4, 0.56 mM alanine, and 1.11 mM glucose, with the other ingredients remaining at the same concentration; a lower temperature of incubation (30°C); and elimination of shaking of the culture. To limit phosphate concentrations even further, we substituted Tris (40 mM, pH 7.3) as the buffering ingredient of the medium; however, reduction of the phosphate levels did not increase the level of sporulation, whereas it had been found to have this effect on *B. cereus* microcycle sporulation (9).

Separating the germination phase from the outgrowth phase gave some slight increase in primary-cell sporulation. Germination of heat-shocked spores, measured by optical density loss, total counts, and cell type determination, was typically 90% and occurred after incubation for 2.5 h at 30°C in 5 ml of a medium consisting of 0.56 mM glucose, 40 mM Tris, pH 7.3, 134 mM KCl, and 0.56 mM alanine. After germination, the cell suspensions were centrifuged and suspended in 5 ml of the following basal medium (hereafter referred to as the outgrowth medium): 1 mM glucose, 40 mM sodium acetate, 7.5 mM K_2HPO_4, 3.25 mM KH_2PO_4, 40 mM Tris, pH 7.3, 0.2 mM $MgSO_4 \cdot 7H_2O$, 0.018 mM $FeSO_4 \cdot 7H_2O$, 0.45 mM $CaCl_2$, and 0.025 mM $MnCl_2 \cdot 4H_2O$.

The addition of Casamino Acids to the outgrowth medium at 3 h of outgrowth gave 8% sporulation of primary cells (Fig. 1). The 90% germinated spores elongated to ovals, and 28% of these developed to singlets, sporangia, and a few doublets. Since no increase in number of cells was observed, sporangia could only have arisen from primary cells; thus microcycle sporulation was achieved in *B. subtilis.*

The addition of 15 μg of a five-pool mix of amino acids per ml at 3 h as a substitute for Casamino Acids gave 12% sporulation (Fig. 2). However, higher concentrations resulted in less sporulation, more doublet formation, and an increase in total numbers of cells.

The elimination of pools I and IV from the total pool mix resulted in fewer spores than omission of the other pools (Fig. 3A). Of these two pools, pool IV appeared to be more effective when added alone (Fig. 3B). Small differences in spore numbers were found upon elimination of

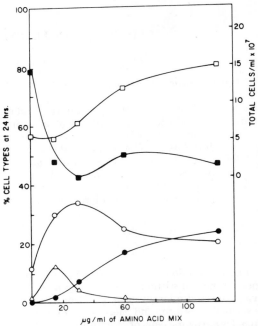

FIG. 2 *Cell types and total cell counts at 24 h in outgrowth medium to which the indicated amount of total amino acid mix (pools I through V) was added at 3 h. Symbols: □, total cell count; ■, ovals; ○, singlets; △, sporangia; ●, doublets. The amino acid pools were constructed as follows. Pool I (mg/ml): L-lysine, 0.26; L-methionine, 0.104; L-cysteine, 0.104; L-arginine 1.04. Pool II (mg/ml): L-leucine, 0.214; L-isoleucine, 0.214; L-valine, 0.214; L-glutamic acid, 0.856. Pool III (mg/ml): L-tryptophan, 0.5; L-tyrosine, 0.5; L-phenylalanine, 0.5. Pool IV (mg/ml): L-histidine, 0.376; L-threonine, 0.376; L-proline, 0.376; L-homoserine, 0.376. Pool V (mg/ml): L-alanine, 0.234; glycine, 0.234; L-serine, 0.094; L-aspartic acid, 0.936. Pools II and V were dissolved in double-distilled water and autoclaved; pools I, III, and IV were dissolved in 0.05 N HCl, adjusted to pH 7.0 with 1 N NaOH, and sterilized by filtration. The five-pool and four-pool mixes consisted of equal volumes of each individual pool. The total amino acid concentration of the stock pools and pool mixes was 1,500 μg/ml. Appropriate volumes of these stocks were added to the culture tubes to give the indicated total amino acid concentration.*

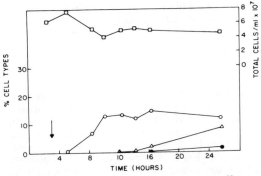

FIG. 1. *Total cell numbers and percent cell types during spore-to-spore development in B. subtilis NCTC 3610 (the Marburg strain) as a function of time in outgrowth medium. Arrow indicates the time of addition of 20 μg of Casamino Acids (Difco) per ml. Symbols: □, total cell count; ○, singlets; △, sporangia; ●, doublets. As inoculum, a water suspension of an isolated colony from a 4-day-old potato dextrose agar (PDA, Difco) plate supplemented with MnSO₄·H₂O (100 mg/liter) was spread on other PDA-Mn plates. After 4 days of incubation at 37°C, the confluent growth, largely spores, was suspended in distilled water and cleaned as described previously (4), diluted with distilled water to an optical density of 400 Klett units (no. 54 filter), distributed into 5.0-ml volumes, and stored at −20°C until used. Cell numbers and cell types were determined in a Petroff-Hausser chamber using a Zeiss microscope equipped with a phase-contrast condenser and 100× objective. In all cases at least 200 cells were counted per sample. Refractile spores (the ungerminated starting population of free spores) germinated to nonrefractile forms which swelled and elongated to cells referred to in this paper as "ovals" (Fig. 2). Some ovals remained in the culture while others elongated further to singlets, which are the primary cells. The primary cell, depending on conditions, became either a doublet (the result of symmetric division) or a sporangium, a cell containing a refractile spore within (the result of an asymmetric division prior to engulfment). Periodic dilution of cells in 50% (wt/vol) trichloroacetic acid was made to allow visualization of the symmetric septa to assure that singlets were true singlets. All the experiments reported were repeated at least once.*

individual amino acids from the pool (Fig. 3C). However, the addition of each amino acid separately revealed that proline was slightly better than homoserine in promoting sporulation of primary cells (Fig. 3D). Addition of histidine gave levels of sporulation lower than that of controls which received no addition of amino acids. Lysine up to 20 μg/ml was without effect. No doublets were detected in any experiments depicted in Fig. 3, and the total number of cells

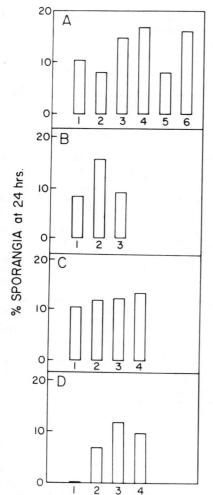

in the number of spores. Higher concentrations resulted in doublet formation and increases in total numbers of cells. DAP appeared to be the most important ingredient in the mixture; its omission resulted in no sporulation. A series of experiments indicated that either addition of DAP with amino acids or multiple additions of DAP alone at low concentrations were effective. The addition of 10 µM DAP at 1-h intervals was effective during the first 5 h (Fig. 4). Multiple additions of greater than 10 µM DAP gave increases in numbers and in doublet formation. Fewer additions gave fewer spores, and more additions gave increased numbers and doublet formation. Thirty percent of the original population developed into sporangia. To emphasize the reproducibility of the system and the ability to quantitate the developmental changes, we have presented data obtained with two separate cultures utilizing triplicate counts.

FIG. 3. *Percent sporangia after 24 h of incubation in outgrowth medium in various cultures to which the indicated amino acid pools or amino acids were added at 3 h. Panel A: 1, amino acid pools I through V; 2, all pools minus pool I; 3, minus pool II; 4, minus pool III; 5, minus pool IV; 6, minus pool V. Total amino acid final concentration was 15 µg/ml. Panel B: 1, amino acid pool I; 2, pool IV; 3, pools I and IV. Final concentration of each addition was 12 µg/ml. Panel C: 1, pool IV minus proline; 2, minus homoserine; 3, minus threonine; 4, minus histidine. Final concentration of each addition was 12 µg/ml. Panel D: 1, histidine; 2, threonine; 3, proline; 4, homoserine. Final concentration of each amino acid was 6 µg/ml.*

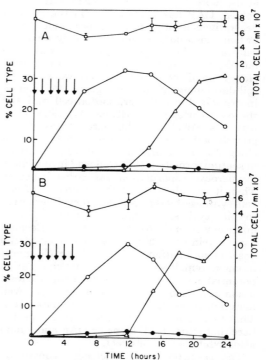

FIG. 4. *Total cell number and percent cell types during spore-to-spore development as a function of time in outgrowth medium upon the addition of DAP. Arrows indicate the times of addition of 10 µM DAP. Symbols: □, total cell count (bars give the standard deviation of triplicate counts; where none exists, counts were essentially identical); ○, singlets; △, sporangia; ●, doublets. Cell type percentages are the average of triplicate counts. Panels A and B represent separate cultures.*

at 24 h was approximately the same as at 0 h in all of those experiments.

The addition at 3 h of a mixture of suspected precursors of cell surface layer—diaminopimelic acid (DAP), N-acetylglucosamine, glucosamine, and glycerol—resulted in only a slight increase

DISCUSSION

As with other microcycle sporulation systems employing medium limitation (6, 7, 9) or shifts to minimal replacement medium (10, 18), elongation to primary cells was achieved only in extremely low concentrations of some major medium ingredients, but in contrast to the others, complete elimination of nitrogen did not lead to sporulation; only upon the addition of amino acids did sporulation of the primary cells ensue. Either a particular amino acid or group of amino acids might be required for a key pathway or the amino acids are required for protein synthesis. Sporulation levels were highest if one of those amino acids was DAP, added either alone in a certain way or in conjunction with other amino acids (Fig. 4). That DAP can be effective alone precludes an amino acid requirement for protein synthesis; alternatively, since it was effective when coupled with addition of other amino acids, more than one mechanism may be involved. DAP has limited fates in the growing vegetative cell, either being a precursor for lysine or being incorporated into peptidoglycan of the cell wall. In these experiments lysine was without effect, which suggests a cell wall role during elongation of the primary cell. However, in the cell developing into a sporangium, DAP can be incorporated into the primordial cell wall and the spore cortex peptidoglycan (17). Since DAP was most effective during early stages of outgrowth, perhaps its role here is either to promote elongation of the primary cell or to be involved in septum formation (17). Cell wall involvement in microcycle sporulation was earlier indicated when it was shown that the inhibition of cell wall synthesis by the antibiotic D-cycloserine led to microcycle sporulation in an otherwise adequate environment for cell growth (13). Because vancomycin and chloramphenicol delayed the onset of sporulation in a non-microcycle sporulation system, it was suggested that sporulation potential developed as a result of the accumulation of a fixed amount of protein or cell wall material, or both, in outgrowing spores (14). That finding may explain the enigma reported here that either DAP alone or DAP and amino acids exhibit a similar effect.

The reason for the low efficiency of conversion of ovals to primary cells is unknown. Either other required metabolites are still limiting or alterations are needed in the conditions employed.

The need for small concentrations of DAP added incrementally to obtain high levels of sporulation (Fig. 4) and the resulting decrease in sporulation and doublet formation with high levels of DAP reflect the proposed need for proper balance of metabolites in spore development (2, 5).

ACKNOWLEDGMENTS

This investigation was supported by grant DAR-7924785 from the National Science Foundation.

LITERATURE CITED

1. **Freer, J. H., and H. S. Levinson.** 1967. Fine structure of *Bacillus megaterium* during microcycle sporulation. J. Bacteriol. **94**:441–457.
2. **Freese, E.** 1977. Metabolic control of sporulation, p. 1–32. *In* A. N. Barker, J. Wolf, D. J. Ellar, G. J. Dring, and G. W. Gould (ed.), Spore research 1976. Academic Press, London.
3. **Gashinskii, V. V., and V. G. Voitsekhovskii.** 1978. Repeated microcycle of *Bacillus cereus*. Mikrobiologiya **47**:373–375.
4. **Hitchins, A. D., R. A. Greene, and R. A. Slepecky.** 1972. Effect of carbon source on size and associated properties of *Bacillus megaterium* spores. J. Bacteriol. **110**:392–401.
5. **Hitchins, A. D., and R. A. Slepecky.** 1969. Bacterial sporulation as a modified procaryotic cell division. Nature (London) **223**:804–807.
6. **Holmes, P. K., and H. S. Levinson.** 1967. Metabolic requirements for microcycle sporogenesis of *Bacillus megaterium*. J. Bacteriol. **94**:434–440.
7. **Hoyem, T., S. Rodenberg, H. A. Douthit, and H. O. Halvorson.** 1968. Changes in the pattern of proteins synthesized during outgrowth and microcycle in *Bacillus cereus* T. Arch. Biochem. Biophys. **125**:964–974.
8. **Keynan, A., A. A. Berns, G. Dunn, M. Young, and J. Mandelstam.** 1976. Resporulation of outgrowing *Bacillus subtilis* spores. J. Bacteriol. **128**:8–14.
9. **MacKechnie, I., and R. S. Hanson.** 1968. Microcycle sporogenesis of *Bacillus cereus* in a chemically defined medium. J. Bacteriol. **95**:355–359.
10. **Mychajlonka, M., A. M. Slee, and R. A. Slepecky.** 1975. Requirements for microcycle sporulation in outgrowing *Bacillus megaterium* cells, p. 434–440. *In* P. Gerhardt, H. L. Sadoff, and R. N. Costilow (ed.), Spores VI. American Society for Microbiology, Washington, D.C.
11. **Mychajlonka, M., and R. A. Slepecky.** 1974. Requirement of deoxyribonucleic acid synthesis for microcycle sporulation in *Bacillus megaterium*. J. Bacteriol. **120**:1331–1338.
12. **Pandey, N. K., and K. G. Gollakota.** 1977. Induction of microcycle sporulation in *Bacillus brevis* spp. AG4 by amethopterin. Arch. Microbiol. **114**:189–191.
13. **Rodenberg, S. D., D. J. O'Kane, R. A. Hackle, and E. Cocklin.** 1972. Factors regulating cellular development in *Bacillus cereus* T spores, p. 197–202. *In* H. O. Halvorson, R. Hanson, and L. L. Campbell (ed.), Spores V. American Society for Microbiology, Washington, D.C.
14. **Sandler, N., and A. Keynan.** 1979. Changes in sporulation potential during the growth cycle of *Bacillus subtilis*. Arch. Microbiol. **123**:9–14.
15. **Schaeffer, P., J. Millet, and J. P. Aubert.** 1965. Catabolic repression of bacterial sporulation. Proc. Natl. Acad. Sci. U.S.A. **54**:704–711.
16. **Slee, A. M., M. Mychajlonka, S. C. Holt, and R. A. Slepecky.** 1977. Characterization of patterns of symmetrical and asymmetrical division in outgrowing cells of *Bacillus megaterium*, p. 85–92. *In* D. Schlessinger

(ed.), Microbiology—1977. American Society for Microbiology, Washington, D.C.

17. **Tipper, D. J., I. Pratt, M. Guinard, S. C. Holt, and P. E. Linnett.** 1977. Control of peptidoglycan synthesis during sporulation in *Bacillus sphaericus*, p. 50–68. *In*

D. Schlessinger (ed.), Microbiology—1977. American Society for Microbiology, Washington D.C.

18. **Vinter, V., and R. A. Slepecky.** 1965. Direct transition of outgrowing bacterial spores to new sporangia without intermediate cell division. J. Bacteriol. **90:**803–807.

C. Macromolecular Synthesis During Sporulation

Two-Dimensional Polyacrylamide Gel Electrophoresis Analyses of Cytoplasmic and Membrane Proteins from Sporulating Cells and Forespore and Mother-Cell Compartments of *Bacillus cereus*

ANTHONY J. ANDREOLI, MICHAEL KAO, RICKY CHUI, JORGE CABRERA, AND SAM K. S. WONG

Department of Chemistry, California State University, Los Angeles, California 90032

Cytoplasmic and membrane proteins from late exponential and sporulating cells of *Bacillus cereus* and from forespore and mother-cell compartments of sporulating cells were recovered in the presence of proteolytic inhibitors and resolved by two-dimensional isoelectric focusing–sodium dodecyl sulfate-polyacrylamide gel electrophoresis. More than 300 cytoplasmic and 260 membrane proteins were revealed on the two-dimensional gels by Coomassie blue staining. Cytoplasmic protein profiles of late exponential and sporulating cells, through the first three stages of sporulation, showed successive changes involving increases and decreases in the relative abundance of some proteins and the appearance of new proteins. The results indicate the extent to which cytoplasmic and membrane protein profiles change during the initial four stages of sporulation. Protein degradation and differential gene expression occurred in both forespore and mother-cell compartments.

Numerous studies have made it apparent that the differentiation process which occurs during the transition of the vegetative bacterial cell through the developmental stages of sporulation in *Bacillus* species involves a series of integrated biochemical events which are believed to be controlled by sequential changes in the pattern of gene expression (5, 7, 9, 18). Thus, extensive qualitative and quantitative alterations of the enzyme patterns is typical of the changes which occur during sporogenesis. Awareness of these changes has depended primarily on the measurement of more than 70 enzymes in crude protein preparations obtained from sporulating cells and dormant spores and, to a more limited extent, their detection in isolated forespore and mother-cell compartments (2–4, 13). Consequently, we have only a limited appreciation for the impact made by the 40 to 60 clusters of genes which code for sporulation functions (8) on the profiles of cytoplasmic and membrane proteins and the timing with which this occurs during sporulation.

This study was undertaken to (i) develop an appreciation for the types of changes both cytoplasmic and membrane proteins may undergo during the four initial stages of sporulation, (ii) establish the chronology of these protein changes and correlate them with specific stages of development, (iii) ascertain the extent to which differential gene expression is manifested in early sporulating cells and in forespore and mother-cell compartments, and (iv) validate the suitability of the experimental approach adopted for future studies. Toward this end, we have exploited two highly effective and reproducible contemporary techniques: isoelectric focusing-sodium dodecyl sulfate two-dimensional gel electrophoresis for the high resolution of complex mixtures of cytoplasmic (12) and membrane (1) proteins and a method for disrupting sporulating cells which permits the release and recovery of homogeneous preparations of forespore and mother-cell components at specific stages of sporulation (2, 4).

RESULTS

Two-dimensional gel electrophoresis of cytoplasmic proteins. The two-dimensional isofocusing–sodium dodecyl sulfate-polyacrylamide gel electropherograms of the cytoplasmic proteins from exponential, T_0, and T_2 cells of *B. cereus* are shown in Fig. 1a, b, and c, respectively. At T_2, *B. cereus* sporulating cells have completed spore septum formation and have initiated prespore engulfment (stage III) (1, 2). A detailed comparison of the three electropherograms shows that specific proteins among these revealed by Coomassie blue staining underwent

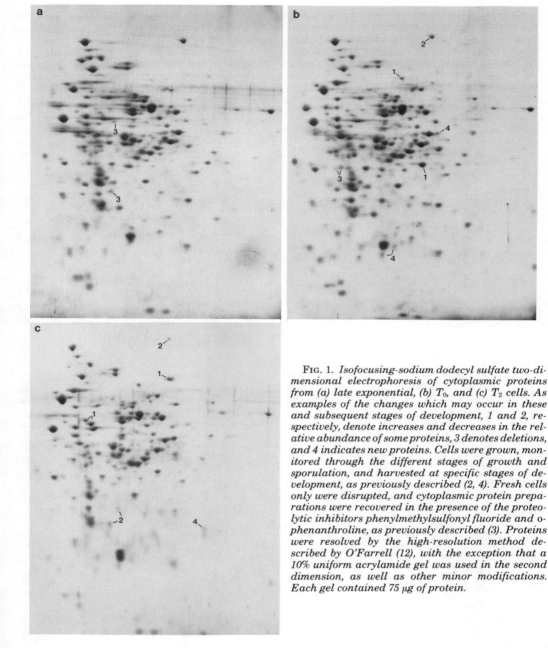

FIG. 1. *Isofocusing-sodium dodecyl sulfate two-dimensional electrophoresis of cytoplasmic proteins from (a) late exponential, (b) T_0, and (c) T_2 cells. As examples of the changes which may occur in these and subsequent stages of development, 1 and 2, respectively, denote increases and decreases in the relative abundance of some proteins, 3 denotes deletions, and 4 indicates new proteins. Cells were grown, monitored through the different stages of growth and sporulation, and harvested at specific stages of development, as previously described (2, 4). Fresh cells only were disrupted, and cytoplasmic protein preparations were recovered in the presence of the proteolytic inhibitors phenylmethylsulfonyl fluoride and o-phenanthroline, as previously described (3). Proteins were resolved by the high-resolution method described by O'Farrell (12), with the exception that a 10% uniform acrylamide gel was used in the second dimension, as well as other minor modifications. Each gel contained 75 µg of protein.*

selective net changes during the 3-h interval of development studied. These changes involved increases or decreases in the relative abundance of some proteins, the disappearance of others, and the appearance of new proteins. Examples of such protein changes are illustrated in Fig. 1.

The profiles shown in the electropherograms of cytoplasmic proteins from $T_{3.5}$ forespore and mother-cell compartments (Fig. 2a and c) had strong similarities, with primarily quantitative differences observed in approximately 25% of the proteins revealed. This similarity is not surprising since $T_{3.5}$ forespore and mother-cell cytoplasmic proteins were isolated at the close of prespore engulfment, i.e., the end of stage III (1). In contrast, the differences between the profiles of cytoplasmic proteins from $T_{4.5}$ forespore and mother-cell compartments were more pronounced (Fig. 2b and d). Qualitative and quantitative differences were observed in fully 30% of the forespore proteins revealed. In addition, it can be seen that 15% of the forespore proteins were "new" in the sense that they were not evident in the mother cell or in preengulfment sporulating cells.

A comparison of $T_{3.5}$ and $T_{4.5}$ forespore cytoplasmic protein electropherograms (Fig. 2a and b) reveals that 20% of the protein profile under-

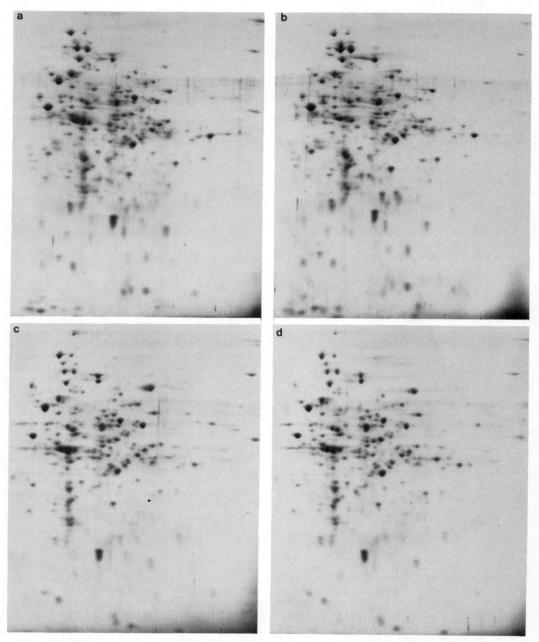

FIG. 2. *Two-dimensional, isofocusing-sodium dodecyl sulfate cytoplasmic protein profiles of (a) $T_{3.5}$ forespore, (b) $T_{4.5}$ forespore, (c) $T_{3.5}$ mother cell, and (d) $T_{4.5}$ mother cell. The methods used are the same as those described in the Fig. 1 legend. Each gel contained 75 μg of protein.*

went change during the 1-h interval of development, as a result of increased synthesis or degradation of existing proteins and de novo synthesis. In contrast, approximately 5% of the mother-cell protein profile underwent change during the interval $T_{3.5}$ to $T_{4.5}$ (Fig. 2c and d). Thus, it appears that changes in protein profiles occur at a faster rate in the forespore than the mother-cell compartment. Also, the changes which occur in both compartments involve the degradation and synthesis of existing proteins as well as the synthesis of new proteins.

Two-dimensional gel electrophoresis of membrane proteins. The two-dimensional electropherograms of membrane proteins from late exponential and sporulating cells at $T_{2.75}$ are shown in Fig. 3a and b, respectively. A comparison of the electropherograms helps to illustrate the extent to which the profile of the late exponential cell membrane changed as cells undertook the first three stages of sporulation. It should be noted that $T_{2.75}$ cells were at a midpoint in the engulfment process. The protein profiles show that changes in the relative abundance of some proteins and the deletion of others occurred. It is also evident that $T_{2.75}$ cell membranes (Fig. 3b) contained approximately 50 proteins which were not found in cells prior to the onset of sporulation.

The protein profiles of mother-cell and forespore membrane preparations isolated at the end of stage III ($T_{3.5}$) are shown in Fig. 4a and b, respectively. A close comparison of the two electropherograms shows that almost all (more than 98%) $T_{3.5}$ mother-cell membrane proteins were evident in counterpart forespore membrane preparations. However, the forespore membranes contained about 60 additional proteins which were not detectable in mother-cell membrane preparations. Thus, it appears that the majority of the 60 added forespore membrane proteins are synthesized during the approximately 2.75-h interval from T_0 to $T_{2.75}$, and only a few are synthesized in the 45-min interval from $T_{2.75}$ to $T_{3.5}$. Although most of the new proteins are present in relatively low abundance, at least 8 proteins are major membrane components in the forespore.

DISCUSSION

In this study cytoplasmic and membrane proteins on two-dimensional electropherograms were revealed with Coomassie blue, in lieu of the alternative technique involving the pulse-labeling of proteins and subsequent detection by autoradiography. One disadvantage of the latter method is that only proteins undergoing active

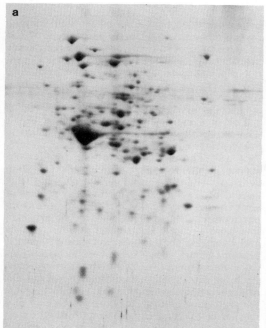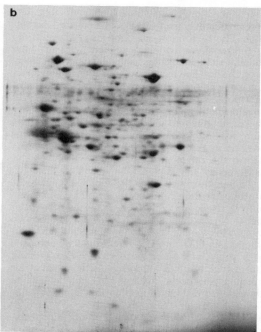

FIG. 3. *Two-dimensional protein profiles of membrane proteins from (a) late exponential and (b) $T_{2.75}$ cells. Fresh cells were disrupted and membrane preparations were recovered in the presence of proteolytic inhibitors as described in the Fig. 1 legend. Membranes were treated and proteins were extracted as described by Ames and Nikaido (1). The extracted proteins were resolved as described in the Fig. 1 legend. Each gel contained 75 μg of protein.*

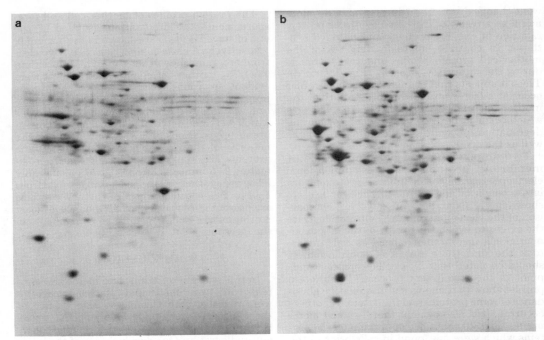

FIG. 4. *Two-dimensional protein profiles of membrane proteins from (a) $T_{3.5}$ mother cell and (b) $T_{3.5}$ forespore. Procedures used are the same as those described for Fig. 3. Each gel contained 75 μg of protein.*

and sufficiently prolonged synthesis are detectable. The technique has been used successfully to follow the chronology of synthesis of some specific proteins during sporulation (10). However, the method assumes that growing and sporulating cells and forespore and mother-cell compartments can be pulsed at different stages of development with comparable effectiveness. By choosing to reveal gel proteins with Coomassie blue, all cytoplasmic and membrane proteins present at a given stage of development and within the range of sensitivity of the stain can be detected.

To ascertain the extent to which cytoplasmic proteins remain undetected when Coomassie blue is used, we conducted a series of experiments (unpublished data) in which one of two duplicate two-dimensional gels was stained with Coomassie blue and the second gel was treated with the silver stain method of Switzer et al. (16), which is 100-fold more sensitive than other staining methods. Coomassie blue revealed about 300 proteins in gels containing a total of 75 μg of protein from preparations clarified to remove ribosomes and small membrane fragments. The silver stain method revealed an additional 33%, or a total of approximately 400 proteins. None of the additional 100 proteins revealed was a major constituent. It should be noted that the silver stain is at least as sensitive as autoradiographic methods (11). Thus, we

have concluded that Coomassie blue staining reveals a surprisingly high percentage of the proteins present in cytoplasmic and membrane preparations described in this report. Consequently, the method of choice is well suited to provide reasonably comprehensive protein profiles which reflect the array of proteins at specific stages of growth and sporulation.

The progressive changes observed in Fig. 1 were expected and are in keeping with the many enzymatic changes reported in earlier studies (5, 7, 9, 18). The changes in protein profiles observed in the forespore which reflect protein synthesis and differential gene expression (Fig. 2a and b) are consistent with earlier reports of active protein synthesis in the forespore compartment (2, 3, 6, 13). However, a comparison of forespore $T_{3.5}$ and $T_{4.5}$ protein profiles (Fig. 2a and b) shows that the degradation of specific proteins occurs, as is also evident in the mother-cell compartment (Fig. 2c and d). More recently, we have confirmed in two-dimensional gel autoradiography experiments that specific proteins undergo degradation in both forespore and mother-cell compartments (unpublished data). The finding that forespore proteins are degraded is contrary to the view that degradation of all proteins synthesized at all times during vegetative growth and sporulation is confined to the mother-cell compartment (6). However, the degradation of forespore proteins was assumed to play an important role in the turnover model proposed by Spudich and Kornberg (14, 15). The similarities evident between forespore and mother-cell profiles (Fig. 2a–d) and the profile from late exponential cells (Fig. 1a) are consistent with the finding that many proteins present during growth persist during sporulation (5, 7, 9).

The two-dimensional gels of membrane protein profiles presented in Fig. 3a and b show a series of protein increases, decreases, and deletions and the appearance of new proteins. These types of changes continue to the completion of engulfment (stage III), as is evident in Fig. 4a and b. Clearly, significant protein degradation and synthesis occurs, resulting in extensive changes in the complement of proteins found in membranes during the transition from late exponential growth through the initial three stages of sporulation.

Although forespore and mother-cell membrane protein profiles show strong similarities, forespore membranes contain 60 proteins not evident in mother-cell preparations. The similarities found are probably due in part to the fact that a portion of the mother-cell membrane undergoes synthesis, as does the outer forespore membrane, during the process of prespore engulfment (stage III). Similarities between membrane profiles from forespores, mother cells, and late exponential cells probably result from the fact that part of the inner forespore membrane and much of the mother-cell membrane already exist at the time when spore septum formation is initiated (17).

When forespore and mother-cell membrane profiles are compared with those from preceding stages (Fig. 3a and b,), it is evident that many proteins which help to distinguish forespore and mother-cell profiles from that of late exponential cells appear by $T_{2.75}$. This suggests that the majority of genes involved in the process may be expressed in a relatively concerted fashion at or around the time of spore septum formation. Current efforts are concerned with a more detailed study of the timing of this process.

ACKNOWLEDGMENTS

This investigation was supported in part by grant PCM78-06147 from the National Science Foundation, the Minority Biomedical Support and MARC Programs of the National Institutes of Health, and grant NSG-2199 from the National Aeronautics and Space Administration.

LITERATURE CITED

1. **Ames, G. F.-L., and K. Nikaido.** 1976. Two-dimensional gel electrophoresis of membrane proteins. Biochemistry **15**:616–623.

2. **Andreoli, A. J., J. Saranto, P. A. Baecker, S. Suehiro, E. Escamilla, and A. Steiner.** 1975. Biochemical properties of forespores isolated from Bacillus cereus, p. 418–424. In P. Gerhardt, R. N. Costilow, and H. L. Sadoff (ed.), Spores VI. American Society for Microbiology, Washington, D.C.

3. **Andreoli, A. J., J. Saranto, N. Caliri, E. Escamilla, and E. Piña.** 1978. Comparative study of proteins from forespore and mother-cell compartments of Bacillus cereus, p. 260–264. In G. Chambliss and J. C. Vary (ed.), Spores VII. American Society for Microbiology, Washington, D.C.

4. **Andreoli, A. J., S. Suehiro, D. Sakiyama, J. Takemoto, E. Vivanco, J. C. Lara, and M. C. Klute.** 1973. Release and recovery of forespores from Bacillus cereus. J. Bacteriol. **115**:1159–1166.

5. **Doi, R. H., and F. J. Sanchez-Anzaldo.** 1976. Complexity of protein and nucleic acid synthesis during sporulation of bacilli, p. 145–163. In D. Schlessinger (ed.), Microbiology—1976. American Society for Microbiology, Washington, D.C.

6. **Eaton, M. W., and D. J. Ellar.** 1974. Protein synthesis and breakdown in the mother-cell and forespore compartments during spore morphogenesis in Bacillus megaterium. Biochem. J. **144**:327–337.

7. **Hanson, R. S., J. A. Peterson, and A. A. Yousten.** 1970. Unique biochemical events in bacterial sporulation. Annu. Rev. Microbiol. **24**:53–90.

8. **Hoch, J. A.** 1978. Developmental genetics at the beginning of a new era, p. 119–121. In G. Chambliss and J. C. Vary (ed.), Spores VII. American Society for Microbiology, Washington, D.C.

9. **Kornberg, A., J. A. Spudich, D. L. Nelson, and M. P. Deutscher.** 1968. Origin of proteins in sporulation. Annu. Rev. Biochem. **37**:51–78.

10. **Linn, T., and R. Losick.** 1976. The program of protein synthesis during sporulation in Bacillus subtilis. Cell **8**:103–114.

11. **Merril, C. R., R. C. Switzer, and M. L. Van Keuren.** 1979. Trace polypeptide in cellular extracts and human body fluids detected by two-dimensional electrophoresis and a highly sensitive silver stain. Proc. Natl. Acad. Sci. U.S.A. **76**:4335–4339.

12. **O'Farrell, P. H.** 1975. High resolution two-dimensional electrophoresis of proteins. J. Biol. Chem. **250**:4007–4021.

13. **Singh, R. P., B. Setlow, and P. Setlow.** 1977. Levels of small molecules and enzymes in the mother cell compartment and the forespore of sporulating Bacillus megaterium. J. Bacteriol. **130**:1130–1138.

14. **Spudich, J. A., and A. Kornberg.** 1968. Biochemical studies of bacterial sporulation and germination. VI. Origin of spore core and coat proteins. J. Biol. Chem. **243**:4588–4599.

15. **Spudich, J. A., and A. Kornberg.** 1968. Biochemical studies of bacterial sporulation and germination. VII. Protein turnover during sporulation of Bacillus subtilis. J. Biol. Chem. **243**:4600–4605.

16. **Switzer, R. C., C. R. Merril, and S. Shifrin.** 1979. A highly sensitive silver stain for detecting proteins and peptides in polyacrylamide gels. Anal. Biochem. **98**:231–237.

17. **Young, E. I., and P. C. Fitz-James.** 1959. Chemical and morphological studies of bacterial spore formation. I. The formation of spores in Bacillus cereus. J. Biophys. Biochem. Cytol. **6**:467–481.

18. **Young, M.** 1978. Bacterial endospore development—an ordered sequence of gene transcription. Trends Biochem. Sci. **3**:55–59.

Solubility of Parasporal Crystals of *Bacillus thuringiensis* and Presence of Toxic Protein During Sporulation, Germination, and Outgrowth[1]

ROBERT E. ANDREWS, JR., DONALD B. BECHTEL, BARBARA S. CAMPBELL, LOREN I. DAVIDSON, AND LEE A. BULLA, JR.

U.S. Grain Marketing Research Laboratory, Agricultural Research, Science and Education Administration, U.S. Department of Agriculture, Manhattan, Kansas 66502, and Division of Biology, Kansas State University, Manhattan, Kansas 66506

The alkali-soluble parasporal crystal toxin of *Bacillus thuringiensis* was monitored throughout growth, sporulation, germination, and outgrowth by use of rocket immunoelectrophoresis, sodium dodecyl sulfate-polyacrylamide gel electrophoresis, insect bioassay, and light microscopy. Maximal synthesis of crystal antigen occurred between t_3 and t_6 of sporulation. The appearance of completed spores within the mother cell followed synthesis of the crystal toxin. Vegetative cells and early stationary-phase cells did not produce antigen, and they were not toxic to insects. The crystal toxin which is contained in spore coats persisted throughout germination, outgrowth, and first cell division. Gel filtration and insect bioassay procedures revealed that the smallest toxic component isolated from crystals treated with alkali and KSCN had an apparent molecular weight of 68,000.

Bacillus thuringiensis is a rod-shaped, aerobic, sporeforming bacterium that forms a parasporal crystalline inclusion adjacent to the endospore during postexponential cellular development (2). This parasporal crystal is highly toxic to a variety of insect pests and is the major toxic agent in insecticides formulated with *B. thuringiensis* (5). The crystal is composed of a glycoprotein protoxin (apparent molecular weight, 1.34×10^5) that is converted to a toxin (apparent molecular weight, 6.8×10^4) after ingestion by a susceptible insect (4). Apparently, synthesis of the crystalline protein toxin is a sporulation-specific event, and the protein has been shown to be present in the spore coats of *B. thuringiensis* (12).

Crystal toxin can be measured by electrophoresis in sodium dodecyl sulfate (SDS)-polyacrylamide gels (3), by rocket immunoelectrophoresis (1), and by insect bioassay (10). In the present study, we used these techniques to examine the pattern of crystal protein synthesis and to determine whether the toxic protein contained within

the spore is degraded during germination and outgrowth. Also, we report on some of the solubility properties of the crystal protein.

RESULTS AND DISCUSSION

Crystal toxin during sporulation. Figure 1 shows the pattern of growth, crystal toxin formation, and sporulation of *B. thuringiensis* subsp. *kurstaki* in 0.2% GYS medium (8). Vegetative growth occurred within the first 6 h, after which the cells entered stationary phase (t_0). Using rocket immunoelectrophoresis (1), we first detected crystal antigen at about t_3 of sporulation and observed a maximum level by t_5. At this time, completed phase-light spores began to appear, and by t_8, approximately 99% of all cells had sporulated. Insect toxicity was associated only with those cells in which antigen expression could be measured. The LC_{50} (50% lethal concentration) values for intracellular crystal toxin, beginning at about t_3, were constant throughout sporulation. No crystal antigen was discerned in vegetative and early stationary-phase cells, nor were these cells toxic to insects at levels 1,000 times the LC_{50} established for spores (9). SDS-polyacrylamide gel electrophoresis (electropherogram not shown) gave similar results. Evidently, the crystal protein toxin of *B. thurin-*

[1] Contribution No. 81-366-j, Division of Biology, Kansas Agricultural Experiment Station, Manhattan. Cooperative investigation between Agricultural Research, Science and Education Administration, U.S. Department of Agriculture, and the Kansas Agricultural Experiment Station.

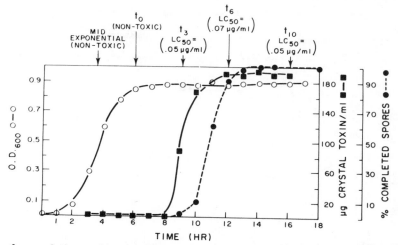

FIG. 1. *Growth, sporulation, and insecticidal toxin production of* B. thuringiensis *subsp.* kurstaki. *Vegetative cells were phased in 0.2% GYS medium (8). Growth at 28°C in 0.2% GYS medium (100 ml in 250-ml Erlenmeyer flasks shaken at 250 rpm) was measured by absorbance at 600 nm, alkali-extractable toxin antigen was assayed by rocket immunoelectrophoresis (1), toxicity was determined by insect bioassay of the tobacco hornworm (10), and sporulation was measured by phase-contrast microscopy.*

giensis is synthesized only during a 2-h period prior to the appearance of spores.

Crystal toxin during germination and outgrowth. Because spore coats of *B. thuringiensis* contain crystal protein (12), we were interested in knowing whether the protein was degraded or turned over during spore germination and outgrowth. Figure 2 depicts the crystal antigen profile in Renografin-purified spores (11) that had been heat shocked at 80°C for 30 min and allowed to germinate in 0.2% GYS medium. As shown, the optical density of such a culture decreased for the first 30 min and then gradually increased throughout outgrowth and first cell division (t_{120}). The amount of extractable crystal toxin antigen in germinating spores increased during the first 30 min (the time when optical density decreased) and then remained somewhat constant thereafter. Interestingly, the LC_{50} values were the same throughout germination and outgrowth, and they were comparable to the values obtained for sporulating cells (Fig. 1). SDS-polyacrylamide gel electrophoresis of alkali extracts of spores (not shown) revealed that both the protoxin and the toxin were present in germinating and outgrowing cells and that these proteins persisted in dividing cells (t_{120} to t_{150}).

The crystal protein contained in spores apparently is not degraded during germination and outgrowth because antigenicity, toxicity, and the amount of protoxin and toxin remain constant through first cell division (Fig. 2). The fate of the toxin has not yet been determined, but we do know that vegetative cells that have been repeatedly transferred are not toxic and do not contain detectable crystal antigen.

Solubility of the parasporal crystalline protein. A critical aspect to understanding the biological and toxicological properties of the parasporal crystalline protein of *B. thuringiensis* is the method used for purifying the crystal and rendering the protoxic and toxic components soluble in stable and functional forms. For example, solubilization of the crystal in reducing and denaturing agents (4) inhibits biological activity. Therefore, the procedure that we have determined to be best, and on which the foregoing toxicity and antigenicity experiments were based, involves titrating a suspension of wet crystals (0.4%, wt/vol) with portions of 1 N NaOH to pH 12 at 28°C. By titrating the crystal with 400 equivalents of base, maximum solubility was obtained and a protoxin subunit (apparent molecular weight, 1.34×10^5) was produced. Upon prolonged incubation of the protoxin at pH 8.6, a toxic protein (apparent molecular weight, 6.8×10^4) was generated (3). This latter protein was the smallest toxic component that we found, and any further breakdown was detrimental to toxic activity. When compared with the protoxin, the 68,000-dalton polypeptide had very similar toxicity ($LC_{50} = 2 \times 10^{-14}$ mol/ml) and, like the protoxin, was 2.5 times more insecticidal than the native crystal. Another method, reported by Lilley et al. (7), for toxin production involves proteolytic enzyme digestion of whole crystals. This procedure also generated a functional toxic molecule with an apparent molecular

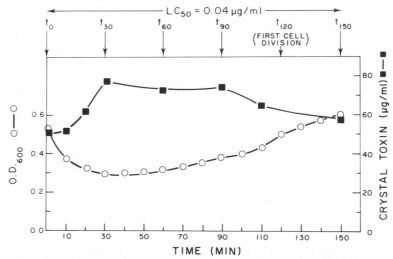

FIG. 2. *Crystal toxin content of germinated and outgrown spores of* B. thuringiensis *subsp.* kurstaki. *The spores were isolated on Renografin gradients (11), heat shocked at 80°C for 30 min, and allowed to germinate and outgrow at 28°C in 0.2% GYS medium (100 ml in 250-ml Erlenmeyer flasks shaken at 250 rpm). Alkali-extractable toxin antigen was measured by rocket immunoelectrophoresis (1). Toxicity was determined by insect bioassay (10).*

weight of 70,000. Like us, these investigators found no toxic components of lower molecular weight.

Recently, Fast and Martin (6) reported another crystal solvation system. The buffer used contained 1 M KSCN, 0.1 M *N*-morpholinopropanesulfonic acid (MOPS), and 0.05 M dithiothreitol (DTT) at pH 7.5 to 8.0. They stated that crystals treated in this buffer for 1 h at 37°C or for 24 h at 28°C were solubilized and that chromatography of such material with Sephacryl S-300 (Pharmacia Fine Chemicals, Inc.) in a buffer containing 2 M KSCN and 0.1 M MOPS separated toxic peptidic material (molecular weight, 5,000 or less).

Figure 3 is a photograph of parasporal crystals titrated with 0.1 M NaOH to pH 12 at 28°C (tube A) and of crystals subjected to 1 M KSCN, 0.1 M MOPS, and 0.05 M DTT (final pH, 7.5) and incubated for 1 h at 37°C and then for 24 h at 28°C (tube B). In contrast to alkaline titration (tube A), the strongly chaotropic buffer (tube B) did not efficiently dissolve the crystals. As can be seen, large chunks of undissolved material remained in the KSCN buffer after 24 h, whereas the alkaline buffer was uniformly clear and opalescent.

Despite the fact that the crystals were not dissolved in KSCN, we wondered whether a small toxic molecule was produced. Chromatography of supernatant material from the KSCN treatment (Fig. 4) with Bio-Gel P10 (Bio-Rad Laboratories) in 2 M KSCN and 0.1 M MOPS

FIG. 3. *Comparison of parasporal crystals of B.* thuringiensis *subsp.* kurstaki *treated with NaOH (tube A) and KSCN (tube B). Wet crystals (0.4%, wt/vol) in tube A were titrated with portions of 1 N NaOH to pH 12 and incubated for 5 h at 28°C. Crystals in tube B were treated with 1 M KSCN, 0.1 M MOPS, and 0.5 M DTT (final pH, 7.5 to 8.0) and were incubated for 1 h at 37°C and then for 24 h at 28°C.*

revealed that most of the protein appeared in the exclusion volume (V_0). The only other UV-absorbing material (DTT) was present in the inclusion volume (V_e). No peptidic components of low molecular weight were detected. Using

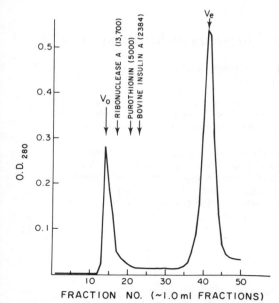

FIG. 4. *Column chromatography of supernatant material from parasporal crystals treated with KSCN (see Fig. 3); Bio-Gel P10 with elution buffer containing 2 M KSCN and 0.1 M MOPS (pH 7.5).*

gel chromatography and SDS gel electrophoresis, the smallest insecticidal component that we could find in the highly toxic KSCN buffer (KSCN is toxic at a 1:100 dilution in phosphate buffer) was approximately 68,000. Thus, the use of KSCN with parasporal crystals confirmed our earlier results (3) as well as those of Lilley et al. (7), who demonstrated that the apparent molecular weight of the insecticidal toxin of *B. thuringiensis* is 68,000 to 70,000.

ACKNOWLEDGMENT

This work was supported by grant PCM 7907591 from the National Science Foundation.

LITERATURE CITED

1. **Andrews, R. E., Jr., J. J. Iandolo, B. S. Campbell, L. I. Davidson, and L. A. Bulla, Jr.** 1980. Rocket immunoelectrophoresis of the entomocidal parasporal crystal of *Bacillus thuringiensis* subsp. *kurstaki*. Appl. Environ. Microbiol. **40**:897–900.
2. **Bechtel, D. B., and L. A. Bulla, Jr.** 1976. Electron microscope study of sporulation and parasporal crystal formation in *Bacillus thuringiensis*. J. Bacteriol. **127**:1472–1481.
3. **Bulla, L. A., Jr., L. I. Davidson, K. J. Kramer, and B. L. Jones.** 1979. Purification of the insecticidal toxin from the parasporal crystal of *Bacillus thuringiensis* subsp. *kurstaki*. Biochem. Biophys. Res. Commun. **91**:1123–1130.
4. **Bulla, L. A., Jr., K. J. Kramer, D. B. Bechtel, and L. I. Davidson.** 1976. The entomocidal proteinaceous crystal of *Bacillus thuringiensis*, p. 534–539. *In* D. Schlessinger (ed.), Microbiology—1976. American Society for Microbiology, Washington, D.C.
5. **Bulla, L. A., Jr., and A. A. Yousten.** 1979. Bacterial insecticides, p. 91–113. *In* A. H. Rose (ed.), Economic microbiology, vol. 4. Academic Press, London.
6. **Fast, P. G., and W. G. Martin.** 1980. *Bacillus thuringiensis* parasporal crystal toxin: dissassociation into toxic low molecular weight peptides. Biochem. Biophys. Res. Commun. **95**:1314–1320.
7. **Lilley, M., R. N. Ruffell, and H. J. Somerville.** 1980. Purification of the insecticidal toxin in crystals of *Bacillus thuringiensis*. J. Gen. Microbiol. **118**:1–11.
8. **Nickerson, K. J., and L. A. Bulla, Jr.** 1974. Physiology of sporeforming bacteria associated with insects: minimal nutritional requirements for growth, sporulation, and parasporal crystal formation of *Bacillus thuringiensis*. Appl. Microbiol. **28**:124–128.
9. **Schesser, J. H., and L. A. Bulla, Jr.** 1978. Toxicity of *Bacillus thuringiensis* spores to the tobacco hornworm, *Manduca sexta*. Appl. Environ. Microbiol. **35**:121–123.
10. **Schesser, J. H., K. J. Kramer, and L. A. Bulla, Jr.** 1977. Bioassay of homogeneous parasporal crystal of *Bacillus thuringiensis* using the tobacco hornworm, *Manduca sexta*. Appl. Environ. Microbiol. **33**:878–880.
11. **Sharpe, E. S., K. W. Nickerson, J. N. Aronson, and L. A. Bulla, Jr.** 1975. Separation of spores and parasporal crystals of *Bacillus thuringiensis* in gradients of certain x-ray contrasting agents. Appl. Microbiol. **30**:1052–1053.
12. **Tyrell, D. J., L. A. Bulla, Jr., R. E. Andrews, Jr., K. J. Kramer, L. I. Davidson, and P. Nordin.** 1981. Comparative biochemistry of entomocidal parasporal crystals of selected *Bacillus thuringiensis* strains. J. Bacteriol. **145**:1052–1062.

Acid-Soluble Polypeptides of *Bacillus subtilis* Spores

D. J. TIPPER, W. C. JOHNSON,[1] G. H. CHAMBLISS, I. MAHLER, M. ARNAUD,
AND H. O. HALVORSON

*Department of Molecular Genetics and Microbiology, University of Massachusetts Medical School,
Worcester, Massachusetts 01605; Department of Bacteriology, University of Wisconsin, Madison, Wisconsin
53706; and Rosenstiel Center, Brandeis University, Waltham, Massachusetts 02154*

The acid-soluble spore proteins of *Bacillus subtilis*, located in the cytoplasm of the mature spore, were synthesized between 2.5 and 6 h of sporulation, probably within the developing forespore. Antisera to the major components of acid-soluble spore proteins were used to detect specific messenger activities in sporulating cells. The time of transcription appeared to correspond to the observed time of in vivo protein synthesis, and there was no evidence of precursor formation.

About 20% of the protein of dormant spores of *Bacillus megaterium* is degraded during germination, providing free amino acids that are essential for protein synthesis in simple media, since the cell initially lacks certain amino acid-synthesizing enzymes (P. Setlow, this volume). Setlow (4, 5) found that most of these amino acids must be produced by degradation of a few acid-soluble spore proteins (ASSPs). ASSPs are so exquisitely sensitive to proteolysis that they could be recovered from mechanically ruptured spores only if spores were broken in the dry state and then extracted with 3% acetic acid. Extraction with stronger mineral acids produced additional minor species (6), and all species had a structurally determined sensitivity to a highly specific endopeptidase that is also present in mature spores (7). Setlow was able to use this dry breakage procedure to demonstrate the appearance of ASSPs during sporulation and their disappearance during germination (4, 5).

ASSPs make up a smaller fraction (7 to 10%) of the protein of mature *B. subtilis* spores (W. C. Johnson, Ph.D. thesis, University of Massachusetts, Worcester, 1980) and, together with coat proteins, comprise 60% of the total spore protein. Coat proteins and ASSPs are exclusively synthesized during sporulation and have specific functions in sporulation, maintenance, or germination; thus, they are the products of sporulation-specific genes. These genes are heavily expressed quite late in germination, after spore septum formation. Our laboratory has

been investigating these polypeptides for several years with the ultimate objective of understanding how they fulfill their special roles and how their synthesis and maturation are controlled.

We have collaborated with Peter Setlow on some of the analyses of *B. subtilis* ASSPs, including investigation of their sensitivity to the *B. megaterium* spore endopeptidase (7; Johnson, Ph.D. thesis, 1980) and cross-antigenicity with the *B. megaterium* ASSPs, and with Peter Setlow and Ed Cannon on N-terminal analysis of the γ component.

RESULTS AND DISCUSSION

Kinetics of sporulation. Most of the RNA preparations used in messenger assays (see below) were prepared from *B. subtilis* strain 168 *trp⁺*, induced to sporulate by shift of exponentially growing cells to Sterlini-Mandelstam medium (9). Under these conditions, T_0, the initiation of sporulation, is the time of shift, and the forespore septum forms after 2 h (T_2) at 37°C. In vivo experiments on coat and ASSP synthesis were performed in strain 168 *trp* which was sporulated by nutrient exhaustion in an enriched nutrient broth (Johnson, Ph.D. thesis, 1980). Under these conditions, T_0 is normally defined as the end of exponential turbidity increase and is followed at 50 min by a final synchronous symmetric cell division (3). Since, by this definition, forespore septum forms at about T_3, the termination of cell division at T_1 approximates T_0 defined under Sterlini-Mandelstam conditions. For comparative purposes, therefore, we have normalized all data so that forespore sep-

[1] Present address: Biological Laboratories, Harvard University, Cambridge, MA 02139.

tation occurs at T_2. Phase-bright forespores, with cortex and rudimentary coats, appear by T_5 and reach 50% at about T_6 under both conditions.

Complexity of *B. subtilis* ASSPs. Because ASSPs lack functions (other than protease sensitivity) assayable in vitro, the only measure of authenticity of different preparations is reproducibility of physical and chemical properties and total yield. A highly reproducible pattern and yield was produced by sodium dodecyl sulfate-polyacrylamide gel electrophoresis (SDS-PAGE) of the products of spontaneous rupture (acid "popping"; 10) of mature *B. subtilis* spores in 2 N HCl (Fig. 1, lanes 1–4). Up to 10 bands could be distinguished (Fig. 1, lane 3), although the slowest-moving component (33.7 kilodaltons [Kdaltons]) was significant only in spores iso-

lated before T_{20} (lanes 3 and 4). Its disappearance may reflect a very late stage in spore maturation, and its relationship (if any) to other ASSPs is unknown. Fractionation in nondenaturing gels at pH 4.7 (not shown) also showed 10 bands. The three major ASSPs were also the least basic and were called α, β, and γ in order of increasing mobility at pH 4.7 and pI values, respectively, of 6.58, 6.67, and 7.96 (as determined by isoelectric focusing). ASSPs α and β each comprised about 20% of the total ASSPs and ran as a single band (5.9 Kdaltons) on SDS-PAGE; γ (11 Kdaltons) comprised about 45% of the total ASSPs, and δ (12.1 Kdaltons), about another 5%. The last was found only in HCl extracts, in rather variable yield, often higher than illustrated in Fig. 1. Most minor species, including three fairly prominent bands, migrated

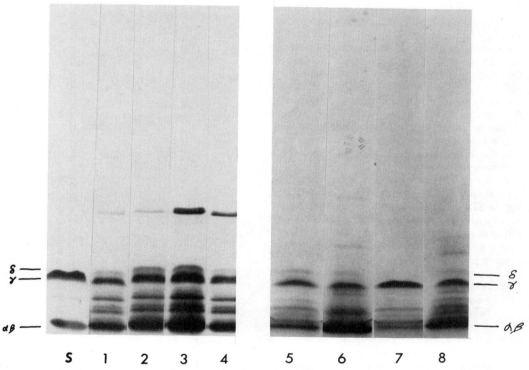

FIG. 1. *Patterns of ASSPs. Lane S: Mature spores of* B. subtilis *168 were cleaned by lysozyme hydrolysis and washing with 1 M NaCl, 0.1 M NaCl, 0.1% SDS, and water, lyophilized, broken by shaking with dry glass beads, and extracted with 3% acetic acid at 20°C (4), followed by fractionation on a column of Sephadex G-50 in 1% acetic acid. The ASSPs were fractionated by SDS-PAGE, using a 7.5 to 15% linear gradient, and stained with Coomassie blue. The upper band is* γ *ASSP and the lower is a mixture of the* α *and* β *components. Lanes 3, 4, 1, 2: Spores of strain 168 were isolated, respectively, from* T_{10}, T_{16}, T_{36}, *and* T_{48} *cultures, washed as above, lyophilized, and ruptured (acid "popped"; 10) by mixing gently in 2 N HCl at 20°C for 20 min. The lyophilized extracts were dissolved in water. Lanes 5 and 6: Mature spores of strain 168T were broken by cryoimpaction (8), and the frozen fragments were immersed in 3% acetic acid (lane 5), or 2 N HCl (lane 6). After 20 min at 20°C, the suspensions were centrifuged, the supernatants were lyophilized, and the residues were dissolved in water. Lanes 7 and 8: Mature spores of strain SMY were extracted as in lanes 5 and 6, respectively.*

between γ and $\alpha\beta$ on SDS gels (Fig. 1). They were all more basic than γ.

Extraction of dry-ruptured *B. subtilis* spores with 3% acetic acid preferentially extracted $\alpha\beta$ and γ (Fig. 1, lane S). Fractionation of this mixture on carboxymethyl cellulose separated $\alpha\beta$ (unretarded) from the more basic γ. Attempts at preparative fractionation of the $\alpha\beta$ mixture have, so far, been unsuccessful. Amino acid compositions are given in Table 1. Neither $\alpha\beta$ nor γ contains detectable tryptophan, carbohydrate, cystine, histidine, tyrosine, or proline, and only $\alpha\beta$ contains methionine. Presumably, the minor ASSP components provide these missing amino acids for outgrowth of germinated spores. The N-terminal sequence of γ is Ala-Asn-Ser-Asn-Asn-Phe-Ser-Lys-Thr-Asn-Ala-Gln-Gln-Val-Arg, and γ is cleaved twice by the *B. megaterium* spore endopeptidase (7) at the sequence Glu-Phe-Ala-Ser-Glu-Thre-(Asn or Asp) (7; Yuan, Johnson, Tipper, and Setlow, in preparation). The $\alpha\beta$ mixture is also cleaved at least once by the spore endopeptidase, so its components must contain a very similar sequence. Cleavage by this enzyme triggers the depolymerization of the ASSPs by aminopeptidases during germination (7), and this mechanism is clearly conserved in *B. megaterium* and *B. subtilis*. Wide variation in the sequences of related ASSPs, however (Setlow, this volume), argues against any function for these proteins except as a polymeric repository of amino acids.

Acid rupture occurred only with mature, refractile spores, and repeated attempts at extracting ASSPs by dry breakage of lyophilized sporulating and germinating cells of *B. subtilis* were too irreproducible to be interpretable. We found that a modification of a liquid nitrogen cryoimpaction procedure (8) produced reproducible results from all cell types. The patterns obtained from mature spores of strain 168 trp^+ (Fig. 1, lanes 5 and 6) and from strain SMY (Fig. 1, lanes 7 and 8) by this procedure illustrate the maximal strain variation seen in ASSP patterns. Spores of strain SMY had only about 75% of the total ASSP content of 168 strains, with a distinct reduction in $\alpha\beta$. Using this procedure, we have shown in both strains that ASSP accumulation is easily detectable by T_3 and that ASSPs disappear during germination, as expected.

Location of ASSPs. After removal of 70% of the coat protein from mature spores by urea-dithiothreitol-SDS extraction, the refractile spore remnant could be germinated by lysozyme, indicating that the residual layers exterior to the cortex are permeable to a protein of 14.3 Kdal-

TABLE 1. *Amino acid compositions of spore proteins*[a]

Fraction	Asx	Glx	Thr	Ser	Gly	Ala	His	Lys	Arg	Pro	Val	Leu	Ile	Phe	Tyr	Met	Cys	Trp
Total coats	8.9	10.4	7.5	6.0	15.1	6.2	3.6	7.9	5.1	5.0	4.6	4.1	3.0	4.4	5.5	0.7	2.1	ND
Insoluble coats	10	6.5	5.5	9.0	11.7	5.4	5.7	10.2	5.1	9.2	4.0	3.9	2.9	4.6	8.3	0.3	4.2	ND
12.2-Kdalton coat	1.2	2.2	0.4	1.0	19.1	1.0	0.2	0.5	13	24	0.6	0.9	0.5	6.0	32.2	0	0	ND
13-Kdalton coat	8.1	6.0	6.1	8.4	7.1	13.6	6.3	7.0	2.9	4.8	8.3	12.9	0.7	4.8	2.3	0.8	ND	ND
ASSP γ (11 Kdaltons)	14	28	6	8	9	11	0	7	4	0	5	2	1	5	0	0	0	0
ASSP $\alpha+\beta$ (5.9 Kdaltons)	14	15	6	8	12	15	0	4	3	0	5	6	7	3	0	1	0	0

[a] Data are given in moles percent; ND, not determined. The data for coat components are from Goldman and Tipper (this volume). The data on ASSPs are from Johnson (Ph.D. thesis, University of Massachusetts, Worcester, 1980).

tons. The urea-dithiothreitol-SDS–extracted spores still ruptured in 2 N HCl, producing a normal yield and pattern of ASSPs from spores of both strain SMY and strain 168 (R. C. Goldman and D. J. Tipper, this volume; Fig. 1, lanes 2 and 4). The ASSPs are clearly interior to the cortex, probably in the spore cytoplasm.

Kinetics of ASSP synthesis. Accumulation of ASSPs in cells broken by cryoimpaction was detectable by T_3, distinct by T_4, and maximal between T_5 and T_6. A variety of procedures were employed to determine the rate of ASSP synthesis. Because reproducible recoveries could be obtained only from mature spores, the best data came from pulse-chase experiments, using L-[^{3}H]valine as shown in Fig. 2. A stain of this gel (not shown) showed identical patterns, like those in Fig. 1, lane 2, in all lanes. Synthesis of $\alpha\beta$ and of ASSPs with mobility between $\alpha\beta$ and γ commenced between T_2 and $T_{2.5}$, shortly after spore septum formation. Synthesis of γ and δ commenced later, at about $T_{3.5}$, and synthesis of different ASSP components decelerated at different times, suggesting independent expression of several genes. Synthesis of $\alpha\beta$, whose messenger appears to be unusually stable (see below), peaked at T_4, and synthesis of all species was greatly diminished by T_6. This pattern is consistent with synthesis of all ASSPs within the developing forespore, which is approaching dormancy by T_6.

Detection of messenger for ASSPs. Antisera separately prepared in rabbits against purified *B. subtilis* $\alpha\beta$ and γ fractions of ASSPs showed specificity but did cross-react with each other. However, they hardly cross-reacted at all with heterologous (*B. megaterium*) ASSPs. The total L-[^{35}S]methionine-labeled translation products of T_5 RNA (Fig. 3A, lane 5) contained a prominent band with the exact mobility of $\alpha\beta$. This material was acid soluble; it was precipitated by anti-$\alpha\beta$ immunoglobulin G (IgG) (Fig. 3B, lane 12) or by anti-γ IgG (not shown), and this precipitation was specifically prevented by addition of excess $\alpha\beta$. Preliminary comparison of high-pressure liquid chromatography peptide maps of $\alpha\beta$ and of this in vitro product, labeled with a mixture of tritiated amino acids, also suggested identity, but the level of label incorporation has so far been inadequate for rigorous identification. This messenger activity was detectable in RNA from T_2 cells but not at $T_{1.5}$. It peaked at about T_4 and decreased markedly by T_6. Although methionine failed to label any other recognizable ASSP component, the tritiated amino acid mix labeled a band with the mobility of γ (which lacks methionine in its native state; Table 1). This band was also precipitated by anti-$\alpha\beta$ IgG and anti-γ IgG, unless excess competitor ASSP was added. Incorporation into this species was no more than 10% of that into $\alpha\beta$. Although the sporulating *B. sub-*

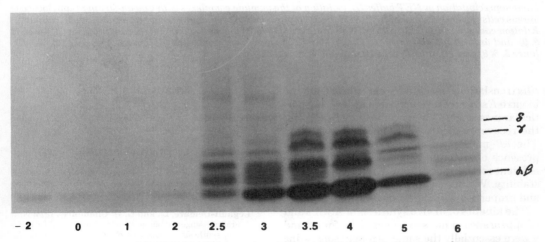

FIG. 2. *Time of synthesis of ASSPs. Cultures of strain 168 (10 ml) were labeled with L-[^{3}H]valine (5 μCi/ml) for 10 min at the indicated times; then a 500-fold excess of cold valine was added. Controls (not illustrated) demonstrated that this chase effectively prevented further uptake of label, although it did not prevent redistribution of label due to turnover of preincorporated label. Spores were isolated at T_{24}, washed, and ruptured in 2 N HCl as described in the Fig. 1 legend. This figure shows a fluorograph of equal quantities of protein from each extract fractionated by SDS-PAGE as described in the Fig. 1 legend. Bands were quantitated by densitometry, using appropriately exposed films (data not shown).*

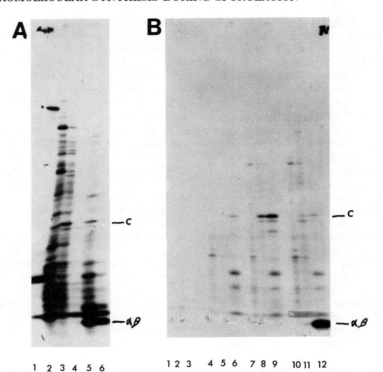

FIG. 3. *Translations of RNA from sporulating cells. Total RNA was prepared from cells as described by Legault-Demare and Chambliss (2) and translated with an E. coli S30 extract, using L-[^{35}S]methionine to label the products. These were fractionated by SDS-PAGE and detected by autoradiography. (A) Lane 1: translation products of MS2 RNA. Lanes 2, 3, and 5: translation products of RNAs from vegetative, $T_{1.5}$, and T_5 cells. Lanes 4 and 6: translation products of RNAs from, respectively, $T_{1.5}$ and T_5 cells, treated with rifampin for 10 min prior to harvesting. The translation products of these RNAs were also fractionated by immunoprecipitation in NET buffer (1), isolation of the immune complexes with Cowan-strain Staphylococcus aureus cells (1), and elution with 3% SDS. Sera employed were preimmune serum (B, lanes 1, 2, 3); anti-12.2-Kdalton coat protein IgG (B, lanes 4, 5, 6); antiserum to total low-molecular-weight coat proteins (B, lanes 7, 8, 9), and anti-αβ IgG (B, lanes 10, 11, 12). The RNAs were from vegetative (B, lanes 1, 4, 7, and 10), $T_{1.5}$ (B, lanes 2, 5, 8, and 11), and T_5 cells (B, lanes 3, 6, 9, and 12).*

tilis translational machinery may differ from the in vitro *Escherichia coli* system in favoring particular messengers (2), it seems more likely that the recovered messenger concentrations differ. The αβ messenger appears to persist in the presence of rifampin (Fig. 4A, lane 6), and its efficient recovery may simply reflect its greater stability. We intend to investigate the location and properties of this messenger.

The kinetics of in vivo synthesis of ASSPs and of appearance of messenger activities for αβ and γ were essentially the same, strongly suggesting transcriptional control, presumably of genes in the forespore genome. There is no evidence of processing, and labeling kinetics suggest distinguishable patterns of expression of several distinct genes.

ACKNOWLEDGMENTS

This work was supported by Public Health Service grants AI-10806 (to D.J.T.) and GM-18904 (to H.O.H.) from the National Institutes of Health.

LITERATURE CITED

1. **Kessler, S. W.** 1975. Rapid isolation of antigens from cells with a staphylococcal protein A-antibody absorbent: parameters of the interaction of antibody-antigen complexes with protein A. J. Immunol. **115**:1617–1624.
2. **Legault-Demare, L., and G. H. Chambliss.** 1974. Natural messenger ribonucleic acid-directed cell free protein-synthesizing system of *Bacillus subtilis.* J. Bacteriol. **120**:1300–1307.
3. **Leighton, T., G. Khachatourians, and N. Brown.** 1975. The role of semiconservative DNA replication in bacterial cell development, p. 677–687. *In* P. C. Hanawalt and M. Goulian (ed.), The mechanism and regulation of DNA synthesis. Academic Press, Inc., New York.
4. **Setlow, P.** 1975. Identification and localization of the

major proteins degraded during germination of *Bacillus megaterium* spores. J. Biol. Chem. **250**:8159–8167.

5. **Setlow, P.** 1975. Purification and properties of some unique low molecular weight basic proteins degraded during germination of *Bacillus megaterium* spores. J. Biol. Chem. **250**:8168–8173.

6. **Setlow, P.** 1978. Purification and characterization of additional low-molecular-weight basic proteins degraded during germination of *Bacillus megaterium* spores. J. Bacteriol. **136**:331–340.

7. **Setlow, P., C. Gerard, and J. Ozols.** 1980. The amino acid sequence specificity of a protease from spores of *B. megaterium*. J. Biol. Chem. **255**:3624–3628.

8. **Smucker, R. A., and R. M. Fister.** 1975. Liquid nitrogen cryo-impacting: a new concept for cell disruption. Appl. Microbiol. **30**:445–449.

9. **Sterlini, J. M., and J. Mandelstam.** 1969. Commitment to sporulation in *Bacillus subtilis* and its relationship to development of actinomycin resistance. Biochem. J. **113**:29–37.

10. **Warth, A. D.** 1979. Exploding spores. Spore Newsl. **6**(10): 4–6.

Synthesis of Putative Spore Coat Precursors in *Bacillus subtilis*

R. C. GOLDMAN AND D. J. TIPPER

Laboratory of Biochemical Pharmacology, National Institute of Arthritis and Metabolic Diseases, Bethesda, Maryland 20205, and Department of Molecular Genetics and Microbiology, University of Massachusetts Medical School, Worcester, Massachusetts 01605

Antiserum to a purified tyrosine-rich 12.2-kilodalton (Kdalton) *Bacillus subtilis* spore coat protein detected accumulation of a 32-Kdalton cross-reactive species in sporulating cells at T_4 but not at $T_{2.5}$. The same antiserum detects a species of 21 Kdaltons in $T_{4.5}$ cells pulse-labeled with methionine. This species chased rapidly, was not detectable as a stained band, and was barely detectable as a labeled species at T_3. It is presumed that the 32-Kdalton protein may be processed to a 12.2-Kdalton coat protein via a 21-Kdalton intermediate.

Spore coats comprise about 50% of *Bacillus subtilis* spore proteins, and a maximum of 70% of the protein can be solubilized from coats isolated from mature spores (1). The same fraction of coat is solubilized from intact spores by urea-dithiothreitol-sodium dodecyl sulfate (SDS) buffer, pH 9.8, at 37°C (5). Since the urea-dithiothreitol-SDS–extracted spores remain refractile, viable, and at least temporarily heat resistant and dormant, and since they retain acid-soluble spore proteins (D. J. Tipper et al., this volume; Fig. 1) and respond normally to germination inducers such as L-alanine, they must retain most noncoat functional components, indicating that the solubilized proteins are derived entirely from the coat. The 30% of the coats that cannot be solubilized nondestructively is uniquely rich in cysteine (1). The cause of the insolubility of this fraction is unclear, as it contains little if any of the known types of peptide cross-links (1, 5, 6). Much of this insoluble fraction derives from the outermost coat layer, where it presumably plays a major structural role (1).

When total solubilizable spore coat proteins of the Marburg strain SMY and of strain 168 *trp*⁺ were fractionated by SDS-polyacrylamide gel electrophoresis (SDS-PAGE) (1, 5), the type of pattern seen in Fig. 1, lanes 1 and 3, was obtained. In each, a mixture of low-molecular-weight (8- to 13-kilodalton [Kdalton]) components, including three major bands, accounts for 75% of the solubilizable protein (1, 5). When isolated, this fraction (in the absence of other coat components) is capable of reassembling into a pattern of beaded fibers resembling coat frag-

ments (1). The other 25% of solubilized coats consists of a large number of higher-molecular-weight species, some of which may be aggregates of (or partially processed precursors of) the major components. They have not been characterized.

Munoz et al. (4) selectively solubilized and purified a component of about 13 Kdaltons that almost certainly corresponds to the uppermost major band (Fig. 1, lane 1). We (1) used selective precipitation at high osmolarity and neutral pH to isolate and purify a second component, corresponding to the adjacent major 12.2-Kdalton band (Fig. 1, lane 1) and comprising about 10% of the total coat proteins. The amino acid compositions of total coat, the 13- and 12.2-Kdalton coat proteins, and the insoluble coat fraction are given in a companion paper (see Table 1 of Tipper et al., this volume). The 12.2-Kdalton polypeptide has an unusual amino acid composition with a high content of tyrosine, glycine, and proline, suggesting a collagen-like structure. This should be well suited to the structural role of this coat component, and its insolubility at neutral pH, even in the presence of SDS, may play a role in the resistance of spores to organic solvents. Antiserum to the 12.2-Kdalton polypeptide fails to react with other solubilizable coat components, suggesting that each of the major components is unique. The 13-Kdalton protein, at least, is clearly different (Table 1 of Tipper et al., this volume). Munoz et al. (4), using antiserum to this protein, detected a 25-Kdalton cross-reactive component in the soluble fraction of sporulating cells as early as T_1. (T_n is n hours after T_0, the start of sporulation. Here,

early as T_2. In this paper we describe soluble components of sporulating cells cross-reacting with the 12.2-Kdalton tyrosine-rich coat protein. They were first detectable at T_3 to T_4.

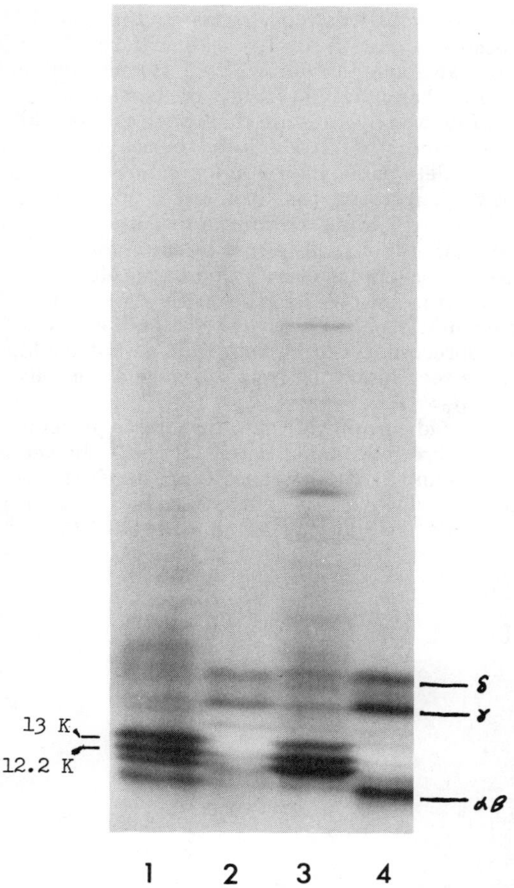

RESULTS

Accumulation of 32-Kdalton cross-reactive species. Total soluble proteins were isolated from vegetative and sporulating cells of *B. subtilis* by rupturing in a French press and ammonium sulfate precipitation. After fractionation by SDS-PAGE (2), protein bands were transferred to nitrocellulose sheets (7) and incubated with rabbit antiserum to purified 12.2-

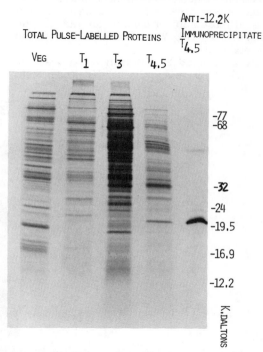

FIG. 1. *Solubilizable coat proteins and location of acid-soluble spore proteins. Extracts of spore proteins were fractionated by SDS-PAGE, using a linear gradient from 7.5 to 15% acrylamide, and stained with Coomassie blue. Spore coat proteins were extracted from mature spores by gentle shaking in urea-dithiothreitol-SDS buffer, pH 9.8, at 37°C (5). Extracted and washed spores were ruptured ("popped") by gentle shaking in 2 N HCl at 20°C (Tipper et al., this volume). After centrifugation, the supernatant was lyophilized and dissolved in water (acid-soluble spore proteins). Lane 1: urea-dithiothreitol-SDS extract of strain SMY spores. Lane 2: acid-soluble spore proteins from these extracted spores. Lane 3: urea-dithiothreitol-SDS extract of strain 168 spores. Lane 4: acid-soluble spore proteins from these extracted spores.*

as in Tipper et al. (this volume), T_0 is defined as 1 h after the end of exponential turbidity increase under our sporulation conditions [1].) Pandey and Aronson (5) used antiserum to the entire mixture of low-molecular-weight soluble coat proteins to detect an in vivo 60-Kdalton cross-reactive species which was detectable as

FIG. 2. *Putative precursor of 12.2-Kdalton coat protein. Cells were labeled for 15 min with L-[^{35}S]-methionine at the indicated times. Total soluble proteins were isolated by inducing protoplast formation with lysozyme in 0.5 M sucrose and lysis in NET buffer (3). After centrifugation, the supernatants were mixed with rabbit antiserum to the 12.2-Kdalton coat protein and the immune complexes were absorbed with Cowan-strain Staphylococcus aureus cells (3), eluted with 1% SDS, and fractionated by SDS-PAGE on a 15% gel, producing the autoradiogram shown. Lanes A, C, E, and G: anti-12.2-Kdalton immunoglobulin G. Lanes B, D, E, and F: preimmune immunoglobulin G. Lanes A and B: $T_{4.5}$ cells. Lanes C and D: T_3 cells. Lanes E and F: T_1 cells. Lanes G and H: vegetative cells. The migration positions of marker proteins and of the 12.2-Kdalton coat proteins are shown.*

Kdalton coat protein (R. C. Goldman, and D. J. Tipper, in preparation) diluted in normal goat serum. After washing, the bound rabbit antibody was detected (immunochemical staining) by using horseradish peroxidase-conjugated goat anti-rabbit serum and chromogenic substrate (7). A single band, strongly stained, was seen in extracts of T_4 cells (not shown). Its size, estimated from mobility, was 32 Kdaltons. A band of similar mobility was prominent in autoradiograms of pulse-labeled $T_{4.5}$ cell proteins fractionated by SDS-PAGE (Fig. 2) and could also be seen at T_4 as a distinct new band in gels stained with Coomassie blue (not shown). The 32-Kdalton species could not be detected immunochemically in extracts of vegetative or $T_{2.5}$ (not shown) cells and was, at most, a minor band in autoradiographically detected (Fig. 2) or Coomassie blue-stained T_3 total proteins.

Pulse-labeled 21-Kdalton cross-reactive species. Antiserum to the 12.2-Kdalton protein

detected a 21-Kdalton species in the soluble proteins isolated from $T_{4.5}$ cells that had been pulse-labeled for 15 min with L-[^{35}S]methionine (2) (Fig. 2 and 3). If $T_{4.5}$ cells were labeled for 15 min, followed by a 15-min chase with excess cold methionine, the 21-Kdalton species was no longer detectable in immunoprecipitates (not shown), suggesting that it is rapidly processed, possibly explaining the failure to detect its accumulation in T_4 cells either by immunochemical staining (above) or with Coomassie blue. The 21-Kdalton species, together with a number of other minor bands, was just detectable in immunoprecipitates of T_3 cell proteins but could not be seen in extracts of T_1 or vegetative cells (Fig. 3).

Autoradiography of the total soluble proteins from these pulse-labeled cells (Fig. 2) showed that 32- and 21-Kdalton species are major translation products at $T_{4.5}$, but not at T_3 or earlier. Our tentative hypothesis is that the 32-Kdalton

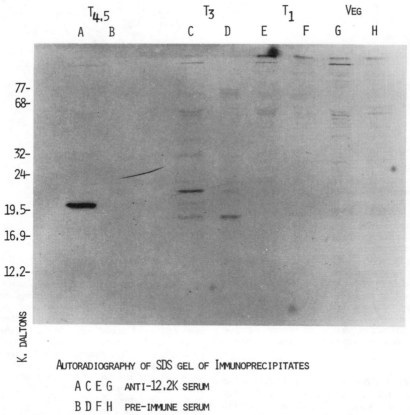

FIG. 3. *Visualization of "precursor" to coat protein in total labeled cell protein. The pulse-labeled cell extracts used in Fig. 2 were fractionated by SDS-PAGE as described in the Fig. 2 legend, but without prior immunoprecipitation. Left to right: extracts of vegetative, T_1, T_3, and $T_{4.5}$ cells and anti-12.2-Kdalton immunoglobulin G immunoprecipitate from $T_{4.5}$ cell extract.*

species is only immunochemically detectable after denaturation by SDS-PAGE and the blotting procedure (7). It may be a (primary?) precursor of the 12.2-Kdalton coat protein, whose synthesis commences at about T_3 and which is processed via an unstable 21-Kdalton intermediate to the 12.2-Kdalton final product. The latter was not seen to accumulate in these experiments because of its insolubility and because it lacks methionine (Table 1 of Tipper et al., this volume). In $T_{4.5}$ cells pulse-labeled with both L-[^{35}S]methionine and L-[^{3}H]tyrosine, the ^{3}H to ^{35}S ratio in the immunoprecipitated 21-Kdalton species was 250% of that of total soluble protein, consistent with the presence of a tyrosine-rich region in the 21-Kdalton species corresponding to the 12.2-Kdalton coat protein. Peptide sequence analysis is needed to confirm these relationships.

Detection of messenger activity for a presumed coat precursor. Using an *Escherichia coli* S30 system and L-[^{35}S]methionine, we saw a prominent band at about 30 Kdaltons in translation products of T_5 RNA precipitated by antiserum to total low-molecular-weight coat proteins (Fig. 3B, lane 9, in Tipper et al., this volume). This band (marked C) was visible in the unfractionated products of T_5 RNA translation (lane A5) and was a minor product of translation of $T_{1.5}$ RNA (lane 8). We do not at present know whether this corresponds to the 32-Kdalton in vivo "precursor" of 12.2-Kdalton coat protein. It was very poorly precipitated by anti-12.2-Kdalton serum (lane 6), and its presence at $T_{1.5}$ does not correlate with the later appearance in vivo of proteins cross-reactive with the 12.2-Kdalton species. It might be the 25-Kdalton species described by Munoz et al. (4), and following the suggestion of Pandey and Aronson (5), it may be that one or two genes give rise to polyprotein precursors, having antigenic determinants for multiple coat proteins, which are produced from it by sequential fragmentation. Again, peptide analysis should resolve this issue.

CONCLUSION

It is anticipated that the highly insoluble spore coat proteins will be synthesized as more soluble precursors, and the cross-reactive species reported here and in previous reports (4, 5) may be such precursors. The messengers for acid-soluble spore proteins, which appear at T_2 to T_3 (Tipper et al., this volume), are probably transcribed in the developing forespore. The coat messengers are probably transcribed in the mother cell for maturation and assembly of the products on the outside of the developing spore. Production of these two classes of sporulation-specific gene products, even if simultaneous, may therefore be controlled by different mechanisms. Selective induction of coat protein synthesis at T_3 to T_4 in a compartmentalized cell, as indicated by our in vivo experiments, is easier to rationalize than induction before compartmentalization at $T_{1.5}$, as suggested by in vitro assays (Fig. 4 of Tipper et al., this volume) or by others (4), but may only require a new dimension to our efforts to rationalize sporulation control (see R. Losick, this volume).

ACKNOWLEDGMENT

This work was supported in part by Public Health Service grant AI-10806 to D.J.T. from the National Institute of Allergy and Infectious Diseases.

LITERATURE CITED

1. **Goldman, R. C., and D. J. Tipper.** 1978. *Bacillus subtilis* spore coats: complexity and purification of a unique polypeptide component. J. Bacteriol. **135:**1091–1106.
2. **Goldman, R. C., and D. J. Tipper.** 1979. Morphology and patterns of protein synthesis during sporulation of *Bacillus subtilis* Ery^r Spo(Ts) mutants. J. Bacteriol. **138:**625–637.
3. **Kessler, S. W.** 1975. Rapid isolation of antigens from cells with a staphylococcal protein A-antibody absorbent: parameters of the interaction of antibody-antigen complexes with protein A. J. Immunol. **115:**1617–1624.
4. **Munoz, L. E., Y. Sadaie, and R. H. Doi.** 1978. Spore coat protein of *Bacillus subtilis.* Structure and precursor synthesis. J. Biol. Chem. **253:**6694–6701.
5. **Pandey, N. K., and A. I. Aronson.** 1979. Properties of the *Bacillus subtilis* spore coat. J. Bacteriol. **137:**1208–1218.
6. **Tipper, D. J., and J. J. Gauthier.** 1972. Structure of the bacterial endospore, p. 3–12. *In* H. O. Halvorson, R. Hanson, and L. L. Campbell (ed.), Spores V. American Society for Microbiology, Washington, D.C.
7. **Towbin, M., T. Staehelin, and J. Gordon.** 1979. Electrophoretic transfer of proteins from polyacrylamide gels to nitrocellulose sheets: procedure and some applications. Proc. Natl. Acad. Sci. U.S.A. **76:**4310–4354.

Synthesis of Spore Coat Proteins and Their Assembly During Sporulation in *Bacillus subtilis* 168

HOWARD F. JENKINSON AND WILLIAM D. SAWYER[1]

Microbiology Unit, Department of Biochemistry, University of Oxford, Oxford OX1 3QU, United Kingdom

Proteins extracted with sodium dodecyl sulfate-dithioerythritol from mature spores or their integuments were fractionated by sodium dodecyl sulfate-polyacrylamide gel electrophoresis. About 12 proteins accounted for most of the coat, and four of these made up >50% of the total protein extracted. Pulse-labeling with radioactive amino acids during sporulation showed that nine of the proteins had begun to be synthesized by t_3 (i.e., 3 h after the induction of sporulation) or soon after. The proteins on the surface of spores isolated at various times during sporulation from $t_{5.3}$ onwards were identified by in vitro labeling with ^{125}I. Those proteins on the surface in early stage V were successively covered by other proteins as the spore matured. The outermost layer of the mature spore was mainly composed of an alkali-soluble polypeptide (molecular weight, 12,000), which began to be synthesized in stage II, and another polypeptide (molecular weight, 36,000), which was synthesized much later, in stages V to VI.

Earlier observations suggested that the late events occurring after stage IV in sporulation, namely, the development of spore resistance and germination properties, result from some form of processing and self-assembly of proteins formed much earlier (2, 6). An exception to this is the development of lysozyme resistance, which requires de novo protein synthesis during stage V (6). The appearance of the spore coat layers in electron micrographs occurs concomitantly with the development of the spore properties just mentioned and may also result, at least in part, from the assembly of preformed proteins (6). This seems to follow from the fact that it is possible to detect, immunologically, spore coat antigens in sporulating cells as early as stage II (12, 14, 18). In this paper, a combination of in vivo and in vitro radioactive labeling is described which enabled us to determine the sequence of synthesis and deposition of the spore coat proteins.

RESULTS

Identification of the proteins in the spore coat. The insoluble integuments generated after breakage of spores with glass beads contain peptidoglycan, lipid, and membrane and coat pro-

[1] Present address: Department of Microbiology and Immunology, Indiana University School of Medicine, Indianapolis, IN 46223.

teins (4, 13). Treatment with DS buffer (see legend to Fig. 1) solubilized about 60% of the protein from the washed integuments. Four major proteins, 36K, 20K, 12K, and 11K (molecular weights of 36,000, 20,000, etc.), predominated in the DS-soluble coat fraction (Fig. 1, lanes A and B) and accounted for >50% of the total integrated density in the stained profile. An additional 10 to 12 polypeptides were always visible. The 12K band often ran as a doublet, and the 20K component contained varying amounts of three polypeptides, 20K, 19K, and 17.5K. Changing the temperature of extraction did not alter the profile.

Most of the polypeptides could also be extracted with DS from whole spores (Fig. 1, lane C), which remained intact and heat resistant during the extraction (5, 18). However, in this case, only a small fraction (<20%) of spore protein was solubilized, so the gel pattern is not truly representative of the polypeptides in the coat. The 12K protein was preferentially solubilized by alkali extraction of t_{20} spores (Fig. 1, lane D; see also 18).

Times of synthesis of the spore coat proteins. All the protein bands which stained with Coomassie blue could be identified on fluorograms of coat proteins from spores labeled with two amino acids (Fig. 2, lane F). The 65K, 20K, and 11K polypeptides were clearly sulfur rich (Fig. 2, lane G).

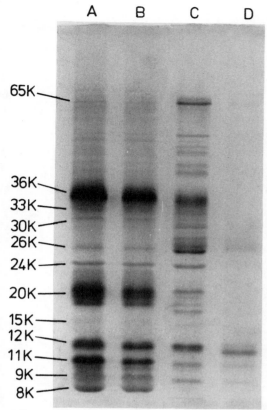

65K—

36K—

33K—

30K—

26K—

24K—

20K—

15K—

12K—

11K—

9K—

8K—

FIG. 1. *Electropherograms of coat proteins extracted from whole spores or their integuments. Sporulation was induced at 37°C by the resuspension method of Sterlini and Mandelstam (16). Spores and sporangia were harvested by centrifugation after 9 h (t_9) or 20 h (t_{20}) and passed twice through a French pressure cell at 4°C to liberate spores from sporangia. Cleaned whole spores were then incubated with DS buffer (see below) or with 0.1 M NaOH (18) or broken by shaking with glass beads. The cleaned integuments from broken spores (6a) were suspended in freshly prepared DS buffer (cyclohexylaminoethanesulfonic acid-NaOH buffer [5 mM, pH 9.8] containing sodium dodecyl sulfate [1%, wt/vol] and dithioerythritol [50 mM]) and incubated for 30 min at 70°C. Proteins in extracts were fractionated by electrophoresis as described by Laemmli and Favre (8) with a 5% (wt/vol) stacking gel and a 15% resolving gel. Gels were stained with 0.2% (wt/vol) Coomassie brilliant blue in propan-2-ol–acetic acid–water (25:10:65) and destained in propan-2-ol–acetic acid–water (10:10:80). Stained bands are numbered by their approximate molecular weights estimated by comparison of their migration distances with those of marker proteins. Lane A; DS buffer extract of washed integuments from t_9 spores (40 µg of protein applied). Lane B: as in lane A (20 µg of protein). Lane C: DS buffer extract of whole t_9 spores (30 µg of protein). Lane D: NaOH extract of whole t_{20} spores (6 µg of protein). Protein was estimated by the method of Lowry et al. (10).*

Pulse-labeling during stages II to III showed that the 33K, 30K, 26K, 24K, and 12K polypeptides were being synthesized (Fig. 2, lanes A and B). Labeling during stage IV showed that the 65K, 20K, 12K, 11K, and 8K polypeptides were being synthesized (Fig. 2, lane C). The 36K component was labeled only at later times (after t_6; Fig. 2, lanes D and E). The results show that: (i) the times of onset of synthesis of the proteins are spread over several hours, i.e., some are formed early and some only later; and (ii) the durations of their synthesis differ, e.g., whereas the 12K polypeptide is made from t_2 throughout sporulation (see also 18), the 33K, 26K, and 24K polypeptides that appear in the spore are made only between t_2 and t_4.

Order of assembly of the coat proteins. To determine which proteins of the coat are on the surface, we isolated spores at various times during sporulation and surface labeled them with ^{125}I, using a lactoperoxidase-catalyzed reaction (11, 17). At $t_{5.3}$, the 30K, 24K, and 15K polypeptides were the most prominent surface components (Fig. 3A). At $t_{6.5}$, the 20K group of polypeptides was predominantly labeled (Fig. 3B), and the labeling in the 30K, 24K, and 15K polypeptides was reduced. Later, as the 36K polypeptide was synthesized, it appeared on the surface of the spores together with the 12K, 9K, and 8K polypeptides (Fig. 3C). At t_9, these four polypeptides were the major surface components; the 30K, 24K, and 15K bands were virtually unlabeled, and the labeling in the 20K group was reduced (Fig. 3C).

When isolated spore coat fragments were taken through the same labeling procedure, all the protein bands were iodinated, i.e., they all contain tyrosine and can be labeled if they are accessible. This seems to justify the interpretation of the results: that is, if a protein is labeled it is on the surface, and if it is unlabeled it is covered. Surface iodination by a different method (3) gave similar results. Confirmation of the surface location of the 36K and 12K polypeptides on mature spores was obtained with cyanogen bromide-activated dextran used as coupling agent (7).

DISCUSSION

The beginnings and approximate durations of synthesis of the major coat proteins are shown in Fig. 4. Nine of the polypeptides began to be formed between t_2 and $t_{3.5}$, and their synthesis must be controlled by stage II or stage III operons. The 36K polypeptide was synthesized only late, and its appearance in the coat was

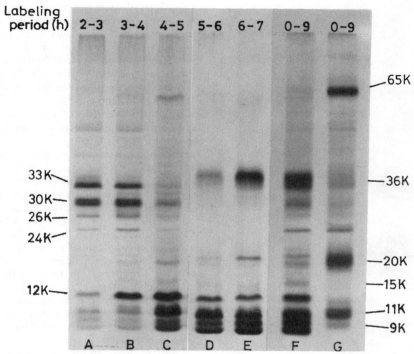

FIG. 2. *Times of synthesis of coat proteins. At intervals after resuspension (16), portions (70 ml) of sporulating culture were removed and incubated for 1 h with [^{14}C]tyrosine and [^{35}S]methionine (0.1 μCi/ml, 20 μg/ml each). The bacteria were harvested by centrifugation and suspended to their original volumes in supernatants from parallel cultures incubated without radioactive amino acids. An excess of L-tyrosine and L-methionine (500 μg/ml each) was then added, and the cultures were incubated to t_9. The cleaned spores from each culture were broken, and their integuments were extracted with DS buffer (see legend to Fig. 1). After electrophoresis the distribution of radioactivity in the stained gels was visualized by fluorography (9). Lanes A, B, and C were exposed to X-ray film for a longer period than lanes D to G, to emphasize some of the fainter bands. Lanes A, B, C, D, and E were radioactively labeled between t_2 and t_3, t_3 and t_4, t_4 and t_5, t_5 and t_6, and t_6 and t_7, respectively. Lane F was labeled with the same radioactive amino acids from t_0 to t_9 (0.1 μCi/ml, 40 μg/ml each). Lane G was labeled with [^{35}S]sulfate (0.5 μCi/ml, 75 μg/ml) from t_0 to t_9, for comparison. Each lane had 40 μg of protein applied.*

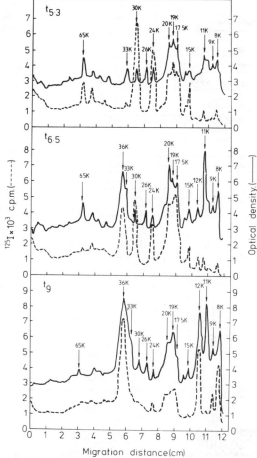

FIG. 3. *Identification of surface proteins on spores during coat formation. Spores were isolated at the times indicated by passage of sporangia through a French pressure cell at 4°C. The cleaned spores (3 to 5 mg, dry weight) were surface labeled with ^{125}I (0.15 mCi), using lactoperoxidase (11, 17). After labeling, the spores were washed and broken, and their integuments were extracted with DS buffer (see legend to Fig. 1), which solubilized >80% of the ^{125}I. After electrophoresis, the stained gels were scanned with a microdensitometer. The radioactivity associated with each protein band was determined by counting 1-mm slices in a gamma counter.*

inhibited by the addition of chloramphenicol (100 μg/ml) at any time up to t_6 (data not shown), resulting in the formation of lysozyme-sensitive spores (6).

The order of assembly of the proteins during stages V and VI is shown schematically in Fig. 4. The existence of three layers is implied, but there are probably multiple layers and the individual polypeptides may not be confined to one

layer. Some of the proteins, e.g., 33K and 26K, are not surface labeled on the spores at $t_{5.3}$ and are thus represented as being already below the surface. At this time the 30K, 24K, and 15K polypeptides are uppermost. In the mature spore most of the proteins are buried, and the outer layer consists of four proteins, 36K, 12K, 9K, and 8K (Fig. 4). The resistance and germination properties of the spores which develop during stages V and VI are also shown in Fig. 4.

The sequence of synthesis and assembly of the coat proteins described shows that some ambiguity could arise when classifying late-blocked sporulation mutants by their morphological stage reached. For example, a stage V mutant, defined as being arrested in spore coat deposition, may evidently have the primary lesion in a stage II or stage III operon. Further characterization of the coat proteins should enable the primary lesions in the late-blocked mutants to be identified.

ACKNOWLEDGMENTS

We are grateful to J. Mandelstam for his continued help and interest and to Angela Maunder for excellent technical assistance.

This work was supported by the Science Research Council.

LITERATURE CITED

1. **Dion, P., D. Kay, and J. Mandelstam.** 1978. Intrasporangial germination and outgrowth of *Bacillus subtilis* prespores. J. Gen. Microbiol. **107:**203–210.
2. **Dion, P., and J. Mandelstam.** 1980. Germination properties as marker events characterizing later stages of *Bacillus subtilis* spore formation. J. Bacteriol. **141:**786–792.
3. **Fraker, P. J., and J. C. Speck, Jr.** 1978. Protein and cell membrane iodination with sparingly soluble chloramide 1,3,4,6-tetrachloro-3a,6a-di-phenylglycoluril. Biochem. Biophys. Res. Commun. **80:**849–857.
4. **Goldman, R. C., and D. J. Tipper.** 1978. *Bacillus subtilis* spore coats: complexity and purification of a unique polypeptide component. J. Bacteriol. **135:**1091–1106.
5. **Gould, G. W., J. M. Stubbs, and W. L. King.** 1970. Structure and composition of resistant layers in bacterial spore coats. J. Gen. Microbiol. **60:**347–355.
6. **Jenkinson, H. F., D. Kay, and J. Mandelstam.** 1980. Temporal dissociation of late events in *Bacillus subtilis* sporulation from expression of genes that determine them. J. Bacteriol. **141:**793–805.
6a. **Jenkinson, H. F., W. D. Sawyer, and J. Mandelstam.** 1981. Synthesis and order of assembly of spore coat proteins in *Bacillus subtilis*. J. Gen. Microbiol. **123:**1–16.
7. **Kamio, Y., and H. Nikaido.** 1977. Outer membrane of *Salmonella typhimurium*. Identification of proteins exposed on cell surface. Biochim. Biophys. Acta **464:**589–601.
8. **Laemmli, U. K., and M. Favre.** 1973. Maturation of the head of bacteriophage T4. I. DNA packaging events. J. Mol. Biol. **80:**575–599.

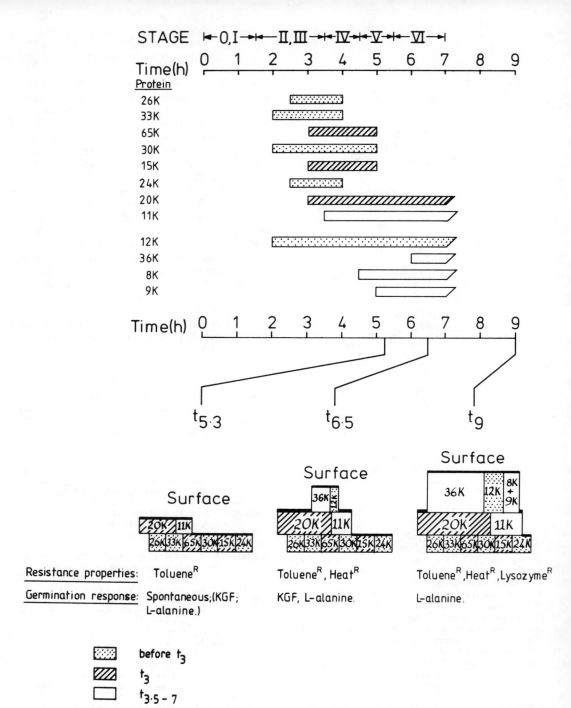

FIG. 4. *Schematic representation of the times of synthesis and order of assembly of some of the proteins in the spore outer layers (redrawn after Jenkinson et al. [6a]). The times of synthesis have been estimated mainly from the data in Fig. 2. The order of assembly has been based on the relative amounts of each protein in the coat extract at each time and on the distribution of ^{125}I among them after surface labeling (Fig. 3). Each protein is represented as a unit, the area of which approximates to the proportion of that protein in the coat extract. The outer surface is depicted uppermost. The diagram is not meant to imply either that any specific protein overlays another (e.g., at $t_{5.3}$ the 26K unit is not necessarily covered by the 20K) or that, because a protein is shown as surface exposed, all molecules of that protein are exposed. Shading indicates the approximate times at which the synthesis of the polypeptides begins: before t_3, at t_3, or after $t_{3.5}$. The stages of sporulation are as described by Ryter (15). The germination and resistance properties of the spores (2, 6) associated with the deposition of the layers are also included. "Spontaneous" refers to germination of prespores in phosphate buffer without any added germinants, after cold shock (1). KGF is a mixture of KCl, glucose, and fructose (2).*

9. **Laskey, R. A., and A. D. Mills.** 1975. Quantitative film detection of ^{3}H and ^{14}C in polyacrylamide gels by fluorography. Eur. J. Biochem. **56**:335–341.

10. **Lowry, O. H., N. J. Rosebrough, A. L. Farr, and R. J. Randall.** 1951. Protein measurement with the Folin phenol reagent. J. Biol. Chem. **193**:265–275.

11. **Marchalonis, J. J.** 1969. An enzymic method for the trace iodination of immunoglobulins and other proteins. Biochem. J. **113**:299–305.

12. **Munoz, L. E., T. Nakayama, and R. H. Doi.** 1978. Expression of spore coat protein gene, an "early sporulation gene," and its relationship to RNA polymerase modification, p. 213–219. *In* G. Chambliss and J. C. Vary (ed.), Spores VII. American Society for Microbiology, Washington, D.C.

13. **Murrell, W. G.** 1967. The biochemistry of the bacterial endospore. Adv. Microb. Physiol. **1**:133–251.

14. **Pandey, N. K., and A. I. Aronson.** 1979. Properties of the *Bacillus subtilis* spore coat. J. Bacteriol. **137**:1208–1218.

15. **Ryter, A.** 1965. Étude morphologique de la sporulation de *Bacillus subtilis*. Ann. Inst. Pasteur Paris **108**:40–60.

16. **Sterlini, J. M., and J. Mandelstam.** 1969. Commitment to sporulation in *Bacillus subtilis* and its relationship to development of actinomycin resistance. Biochem. J. **113**:29–37.

17. **Swanson, J., G. King, and B. Zeligs.** 1975. Studies on gonococcus infection. VIII. ^{125}Iodine labeling of gonococci and studies on their in vitro interactions with eukaryotic cells. Infect. Immun. **11**:453–459.

18. **Wood, D.** 1972. Sporulation in *Bacillus subtilis*. Properties and time of synthesis of alkali-soluble protein of the spore coat. Biochem. J. **130**:505–514.

Two Active Forms of Valyl-tRNA Synthetase of Sporulating *Bacillus subtilis*

K. OHYAMA, I. KANEKO,* AND S. OHKUMA

Tokyo College of Pharmacy and The Institute of Physical and Chemical Research, Wako-shi, Saitama 351, Japan

Two active forms of valyl-tRNA synthetase were found in sporulating cells of *Bacillus subtilis*. Peaks of valyl-tRNA synthetase activity occurred at t_0 and t_3 of sporulation, and no activity was detectable at t_5. The valyl-tRNA synthetase purified from the cells at various stages of sporulation was fractionated into two active forms, E-I and E-II, by hydroxyapatite column chromatography. The ratios of E-II to E-I were 28.4, 6.8, and 2.5 between $t_{-0.5}$ and t_0, $t_{0.5}$ and t_1, and at t_3, respectively. Affinities of E-I and E-II for two isoaccepting tRNA species were different and altered during sporulation. These results suggest a role for the existence and alteration of E-I and E-II of valyl-tRNA synthetase in the regulation of protein biosynthesis during sporulation of *B. subtilis*.

Multiple forms of aminoacyl-tRNA synthetase (ARSase) for a single amino acid have been reported in procaryotic cells (6, 12, 20, 27, 43, 46). Multiple forms of valyl-tRNA synthetase (ARSase^{val}) have been found in both vegetative and sporulating cells of *Bacillus subtilis* (33, 34), and their possible roles in the regulation of protein synthesis have been discussed (41).

To obtain more precise evidence of the alteration in these multiple forms of ARSase^{val} during spore formation, the chromatographic patterns of two forms of ARSase^{val}, E-I and E-II, purified from cells at various stages in sporulation were investigated, and the aminoacylating properties of these forms were examined by analysis of the valyl-tRNA products.

RESULTS AND DISCUSSION

Activity of valyl-tRNA synthetase in sporulating cells of *B. subtilis*. It was previously shown (35) that ARSase activity from t_3 cells was 30 to 50% lower than that from t_0 cells and that the ARSase^{val} from both types of cell was composed of two enzymatically active forms (33). Furthermore, the relative concentrations of two isoaccepting valyl-tRNA's in vegetative and early sporulating cells differed (7, 8). These results prompted us to purify the multiple forms of ARSase^{val} from cells at various stages during sporulation. ARSase^{val} showed peaks of total activity at t_0 and t_3 (Fig. 1). The peak at t_0 was

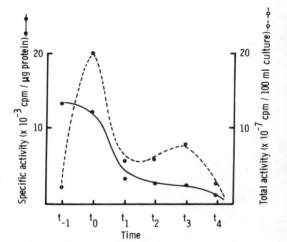

FIG. 1. *Patterns of ARSase^{val} activity during sporulation of* B. subtilis. *The crude ARSase^{val} was fractionated from* B. subtilis *cells as previously described (33), except that enzyme protein was eluted by 0.4 M NaCl in buffer solution from a DEAE-cellulose column. Protein concentration was determined by the method of Lowry et al. (24). The extraction of tRNA and assay of ARSase^{val} activity were in accordance with the methods described by Zubay (47) and Ohyama et al. (34), respectively. Cells were grown in modified Schaeffer medium (2 × SG; 21). Synchrony and the stages of sporulation were checked with a phase-contrast microscope and with an electron microscope (18).*

considerably higher than that at t_3, and specific activity, which was high at t_0, decreased sharply between $t_{0.5}$ and t_1 and was maintained at a lower level until t_4. No activity was detected at t_5 under these experimental conditions. The higher activity of ARSase^{val} at t_3 might be involved in protein synthesis regulation during spore formation, as illustrated by Linn and Losick (23) and which is consistent with the increased rate of RNA and protein synthetic activities seen at this stage of spore formation (45).

Two active forms of ARSase^{val} in cells at various stages of sporulation. ARSase^{val}, purified 100-fold from cells obtained between $t_{0.5}$ and t_0, $t_{0.5}$ and t_1, and at t_3, was fractionated into two active forms, E-I and E-II, by hydroxyapatite column chromatography (Fig. 2). E-I and E-II activities from cells at all stages of sporulation were eluted at the same molarities of potassium phosphate. When each active fraction was rechromatographed on a hydroxyapatite column, the enzymatically active fraction was eluted at the same fraction numbers as in the first chromatography (data not shown), indicating that

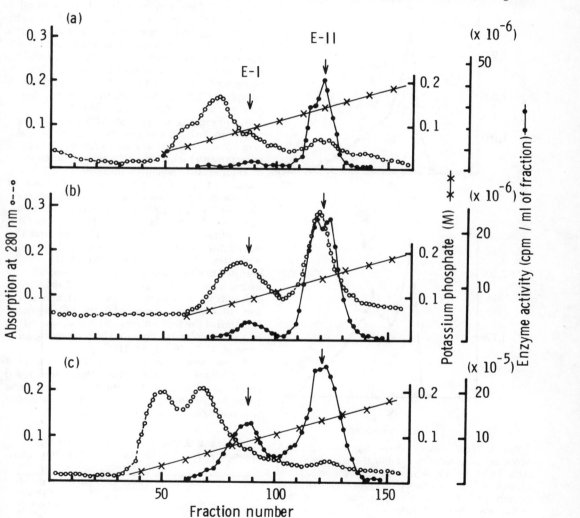

FIG. 2. *Hydroxyapatite column chromatography of ARSase^{val} purified from cells at three stages of sporulation. ARSase^{val} was purified from cells disrupted in a French press by passage through a hemoglobin-Sepharose 4B column (29), ammonium sulfate fractionation (40 to 80% saturation) (15), DEAE-cellulose column chromatography (33), and gel filtration on a Sephadex G-200 column. The active fraction was dialyzed and applied to a hydroxyapatite column (2 by 30 cm) equilibrated with 0.05 M potassium phosphate buffer, pH 7.5, and eluted with 400 ml of a linear gradient of potassium phosphate from 0.05 to 0.25 M.*

TABLE 1. *Alteration of the ratio of E-I activity to total activity during sporulation of* B. subtilis

Expt	Ratio of E-I activity (%) at given cell stage		
	$t_{-0.5-0}$	$t_{0.5-1.0}$	$t_{3.0}$
1	2.96	12.06	22.64
2	3.67	11.95	36.81
3	3.57	14.62	

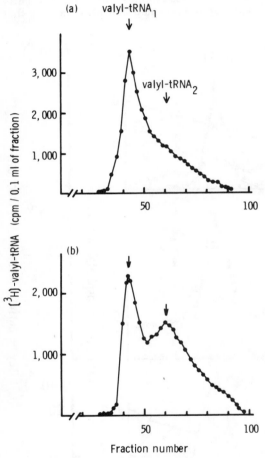

FIG. 3. *MAK column chromatography of valyl-tRNA aminoacylated by E-I and E-II derived from t_0 cells. Valyl-tRNA's, acylated by E-I (b) or E-II (a) according to the standard assay procedure (34), were directly applied to a MAK column, made as described by Kaneko and Doi (17). After the column was washed with 0.05 M phosphate buffer, pH 6.8, containing 0.1 M NaCl, tRNA was eluted by a linear gradient, using 75 ml each of 0.1 M and 0.7 M NaCl. A 0.1-ml portion of each fraction (1 ml) was mixed with 10 ml of ACS-II (aqueous counting scintillant, Amersham Corp.), and the radioactivity was measured with a Packard scintillation spectrometer.*

the two peaks were not an artifact of the hydroxyapatite technique, whose utility has been well established (1–3, 6, 19, 20, 25, 36, 42, 44, 46).

To exclude artifacts due to proteolysis during extraction and purification of the enzyme, the harvested cells were washed with buffer containing 1 M KCl, and the extract from the cells was passed through a hemoglobin-Sepharose 4B column immediately after preparation of the crude extract. These techniques have been used for the preparation of native RNA polymerase from *B. subtilis*, and protease activity was essentially completely removed from the cell extract (29).

The total activity of E-I and E-II from cells between $t_{0.5}$ and t_1 was 50% of that from cells between $t_{-0.5}$ and t_0, and the activity from t_3 cells was <10% of that from cells between $t_{0.5}$ and t_1. Whereas the E-I activity between $t_{-0.5}$ and t_0 comprised only 3 to 4% of the total activity, it comprised 12% of the total between $t_{0.5}$ and t_1 and 30% of the total t_3 cells (Table 1).

Steinberg suggested that ARSase^{lys} in vegetative-phase cells was modified during sporulation by some physiological event or structural association which altered the enzyme phenotype in temperature-sensitive mutants (39, 40). Instances of sporulation-specific modification of enzyme protein include fructose 1,6-diphosphate aldolase by P_i (37, 38), purine nucleoside phosphorylase by sporulation-specific protease (10), and catalase (32). It has been shown that the association of sporulation-specific polypeptides with the RNA polymerase core of *B. subtilis* resulted in alteration of the template specificities of holoenzymes (13, 14, 22). When *Escherichia coli* cells were infected by phage T4, a new active form of ARSase^{val} appeared in the cells by modification of the host enzyme, with the association of a small polypeptide coded by a phage-specific gene (4, 5, 25, 26, 28, 30, 31).

These facts suggest that one of the forms of the ARSase^{val} in our study might be formed from the other by certain modifications such as phosphorylation, methylation or adenylation, or the association of small polypeptide factors.

TABLE 2. *Ratio of valyl-tRNA$_1$/valyl-tRNA$_2$ aminoacylated by E-I and E-II purified from t_0 and t_3 cells^a*

Cell stage	% of valyl-tRNA$_1$ in:		Ratio of tRNA$_1$/tRNA$_2$ in:	
	E-I	E-II	E-I	E-II
Vegetative (t_0)	40.7	75.0	0.7	3.0
Sporulating (t_3)	50.2	59.5	1.0	1.5

^{a} These values were calculated from the data in Fig. 3 and 4 for the t_0 enzyme and from similar data for the t_3 enzyme.

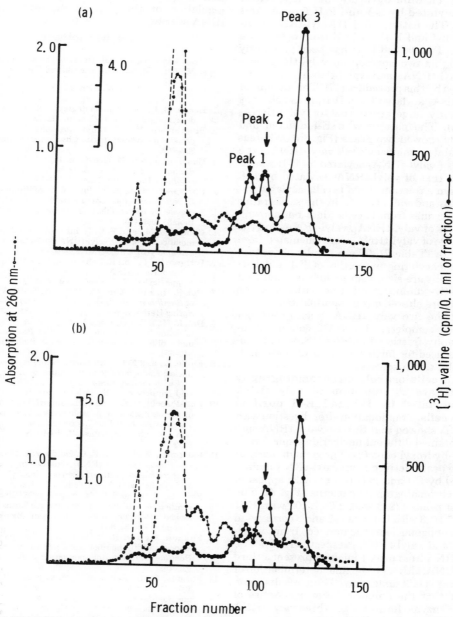

Fig. 4. *DEAE-cellulose column chromatography of RNase T_1-digested valyl-tRNA's aminoacylated by E-I and E-II purified from t_0 cells. The method of Ishida and Miura was used for RNase T_1 digestion and DEAE-cellulose column chromatography of the digest (16). [^{3}H]valyl-tRNA's aminoacylated by E-II (a) and E-I (b) were extracted from the standard assay medium by Zubay's method (47). Oligonucleotides were eluted from this column with a linear gradient from 150 ml of 0.01 M ammonium formate, pH 6.7, to 150 ml of 0.7 M ammonium formate, pH 7.1.*

MAK (methylated albumin Kieselguhr) column chromatography of valyl-tRNA aminoacylated by E-I and E-II from t_0 and t_3 cells. The patterns of $[^3H]$valyl-tRNA acylated by E-I and E-II purified from t_0 cells are shown in Fig. 3a and b. It has been previously reported that two isoaccepting valyl-tRNA species, valyl-tRNA$_1$ and valyl-tRNA$_2$, could be separated by this procedure (17). The amount of valyl-tRNA$_1$ acylated by E-II from t_0 cells was significantly larger than that of valyl-tRNA$_2$ (Fig. 3a). The pattern with E-I differed and definitely showed two peaks (Fig. 3b). Whereas E-II mainly aminoacylated valyl-tRNA$_1$, the amount of valyl-tRNA$_2$ acylated by E-II was as much as that of valyl-tRNA$_1$ acylated by E-I. The enzyme fractions from t_3 cells acylated valyl-tRNA$_1$ and valyl-tRNA$_2$ in the same ratios as the enzymes from t_0 cells (data not shown). The ratios of valyl-tRNA$_1$/valyl-tRNA$_2$ and the percentage of valyl-tRNA$_2$ were calculated from these data (Table 2). These results show that the aminoacylating properties of E-I and E-II from t_0 cells are significantly different and that the relative aminoacylating activities of these fractions are altered during sporulation.

Since the two valyl-tRNA species were not separated completely by MAK column chromatography, fractionation after RNase T$_1$ digestion was used for further analysis of these multiple valyl-tRNA species.

DEAE-cellulose column chromatography after RNase T$_1$ digestion of valyl-tRNA aminoacylated by E-I and E-II purified from t_0 cells. Fractionation after digestion with RNase T$_1$ showed that the two valyl-tRNA species contained different nucleotide sequences at their 3′-hydroxyl ends (8). Three main oligonucleotide peaks derived from valyl-tRNA aminoacylated by E-I and E-II from t_0 cells appeared in the chromatographic patterns (Fig. 4). The ratios of peaks 1/2/3 were 1.06:1:3.02 with E-II and 0.37:1:1.3 with E-I. Peaks 2 and 3 correspond to valyl-oligonucleotides derived from valyl-tRNA$_1$ and valyl-tRNA$_2$, respectively (8). The valyl-tRNA from peak 1 might represent a third valyl-tRNA, although the final determination of this point is still under way. These results again suggest that the aminoacylation properties of the two enzyme forms for the three valyl-tRNA species differ from each other.

Because E-I activity was relatively increased during sporulation and E-I and E-II from t_3 cells aminoacylated the valyl-tRNA species at equal levels, it is suggested that ARSases play a role in controlling the relative amounts of the aminoacyl-isoaccepting tRNA species during sporulation. Furthermore, this phenomenon may be involved in the control mechanisms of RNA (11) and protein (9) syntheses and in the transcription of sporulation-specific genes through the regulation of the amount of aminoacylated tRNA species.

ACKNOWLEDGMENT

This research was partially supported by a grant from the Japan Ministry of Education, Science, and Culture to I.K.

LITERATURE CITED

1. **Bernardi, G.** 1971. Chromatography of proteins on hydroxyapatite. Methods Enzymol. **22**:325–339.
2. **Bluethmann, H.** 1979. Chromatography of chromosomal proteins on hydroxyapatite. Chromatogr. Symp. Ser. **1**:297–308.
3. **Charlier, J., and H. Grosjean.** 1972. Isoleucyl-transfer ribonucleic acid synthetase from *Bacillus stearothermophilus.* Eur. J. Biochem. **25**:163–174.
4. **Chrispeels, M. J., R. F. Boyd, L. S. Williams, and F. C. Neidhardt.** 1968. Modification of valyl-tRNA synthetase by bacteriophage in *Escherichia coli.* J. Mol. Biol. **31**:463–475.
5. **Comer, M. M., and F. C. Neidhardt.** 1975. Effect of T$_4$ modification of host valyl-tRNA synthetase on enzyme action in vivo. Virology **67**:395–403.
6. **Dittgen, R. M., and R. Leberman.** 1976. Multiple active forms of lysyl-tRNA synthetase from *Escherichia coli.* Hoppe Seyler's Z. Physiol. Chem. **357**:543–551.
7. **Doi, R. H., and I. Kaneko.** 1966. Transfer RNA patterns of *Bacillus subtilis* during sporulation and growth. Cold Spring Harbor Symp. Quant. Biol. **31**:581–582.
8. **Doi, R. H., I. Kaneko, and R. T. Igarashi.** 1968. Pattern of valine transfer ribonucleic acid of *Bacillus subtilis* under different growth conditions. J. Biol. Chem. **243**:945–951.
9. **Eidlic, L., and F. C. Neidhardt.** 1965. Protein and ribonucleic acid synthesis in two mutants of *Escherichia coli* with temperature-sensitive aminoacyl-ribonucleic acid synthetases. J. Bacteriol. **89**:706–711.
10. **Engelbrecht, H. L., and H. L. Sadoff.** 1969. Properties of purine nucleoside phosphorylases from spores and vegetative cells of *Bacillus cereus* and their modification by orthophosphatase. J. Biol. chem. **244**:6228–6232.
11. **Fangman, W. L., and F. C. Neidhardt.** 1964. Protein and ribonucleic acid (RNA) synthesis in a mutant of *Escherichia coli* with an altered aminoacylribonucleic acid synthetase. J. Biol. Chem. **239**:1844–1847.
12. **Francis, T. A., and G. M. Nagel.** 1976. Glycyl-tRNA synthetase: evidence for two enzyme forms and sigmoidal saturation kinetics. Biochem. Biophys. Res. Commun. **70**:862–868.
13. **Fukuda, R., and R. H. Doi.** 1977. Two polypeptides associated with the ribonucleic acid polymerase core of *Bacillus subtilis* during sporulation. J. Bacteriol. **129**:422–432.
14. **Fukuda, R., G. Keilman, E. McVey, and R. H. Doi.** 1975. Ribonucleic acid polymerase pattern of sporulating *Bacillus subtilis* cells, p. 213–220. *In* P. Gerhardt, R. N. Costilow, and H. L. Sadoff (ed.), Spores VI. American Society for Microbiology, Washington, D.C.
15. **Green, A. A., and W. L. Huges.** 1955. Protein fractionation on the basis of solubility in aqueous solutions of salts and organic solvents. Methods Enzymol. **1**:67–90.
16. **Ishida, T., and K. Miura.** 1965. Heterogeneity in the nucleotide sequence near the amino acid accepting terminal of transfer RNA. J. Mol. Biol. **11**:341–357.
17. **Kaneko, I., and R. H. Doi.** 1966. Alteration of valyl-sRNA during sporulation of *B. subtilis.* Proc. Natl. Acad. Sci. U.S.A. **55**:564–571.
18. **Kaneko, I., and H. Matsushima.** 1975. Crystalline inclu-

sions in sporulating *Bacillus subtilis* cells, p. 580–585. *In* P. Gerhardt, R. N. Costilow, and H. L. Sadoff, (ed.), Spores VI. American Society for Microbiology, Washington, D.C.

19. **Kawasaki, T.** 1969. Study of hydroxyapatite column chromatography—on heterogeneity of tropocollagen molecules. Proteins Nucleic Acids and Enzymes **14**: 982–1067. (In Japanese.)

20. **Kisselev, L. L., and I. D. Baturina.** 1972. Two enzymatically active forms of leucyl-tRNA synthetase from *E. coli* B. FEBS Lett. **22**:231–234.

21. **Leighton, T. J., and R. H. Doi.** 1971. The stability of messenger ribonucleic acid during sporulation of *Bacillus subtilis*. J. Biol. Chem. **246**:3189–3195.

22. **Linn, T., A. L. Greenleaf, and R. Losick.** 1975. RNA polymerase from sporulating *Bacillus subtilis*. Purification and properties of a modified form of the enzyme containing two sporulation polypeptides. J. Biol. Chem. **250**:9256–9261.

23. **Linn, T., and R. Losick.** 1976. The program of protein synthesis during sporulation in *Bacillus subtilis*. Cell **8**: 103–114.

24. **Lowry, O. H., N. J. Rosebrough, A. L. Farr, and R. J. Randall.** 1951. Protein measurement with the Folin phenol reagent. J. Biol. Chem. **193**:265–275.

25. **Marchin, G. L., M. M. Comer, and F. C. Neidhardt.** 1972. Viral modification of the valyl-transfer ribonucleic acid synthetase of *Escherichia coli*. J. Biol. Chem. **247**: 5132–5145.

26. **Marchin, G. L., U. R. Müller, and G. H. Al-Khateeb.** 1974. The effect of transfer ribonucleic acid on virally modified valyl transfer ribonucleic acid synthetase of *Escherichia coli*. J. Biol. Chem. **249**:4705–4711.

27. **Marshall, R. D., and P. C. Zamecnik.** 1969. Some physical properties of lysyl- and argininyl-transfer RNA synthetase. Biochim. Biophys. Acta **181**:454–464.

28. **Müller, U. R., and G. L. Marchin.** 1977. Purification and properties of a T₄ bacteriophage factor that modified valyl-tRNA synthetase of *Escherichia coli*. J. Biol. Chem. **252**:6640–6645.

29. **Nakayama, T., L. Munoz, and R. H. Doi.** 1977. A procedure to remove protease activities from *Bacillus subtilis* sporulating cells and their crude extract. Anal. Biochem. **78**:165–170.

30. **Neidhardt, F. C., and C. F. Earhart.** 1966. Phage-induced appearance of a valyl-tRNA synthetase activity in *Escherichia coli*. Cold Spring Harbor Symp. Quant. Biol. **31**:557–563.

31. **Neidhardt, F. C., G. L. Marchin, W. H. McClain, R. F. Boyd, and C. F. Earhart.** 1969. Phage-induced modification of valyl-tRNA synthetase. J. Cell Physiol. **74**: 87–102.

32. **Norris, J. R., and A. Baillie.** 1964. Immunological specificities of spore and vegetative cell catalases of *Bacillus cereus*. J. Bacteriol. **88**:264–265.

33. **Ohyama, K., I. Kaneko, T. Yamakawa, and S. Oh-**

kuma. 1978. Valyl-tRNA synthetase of sporulating *Bacillus subtilis*, p. 232–236. *In* G. Chambliss and J. C. Vary (ed.), Spores VII. American Society for Microbiology, Washington, D.C.

34. **Ohyama, K., I. Kaneko, T. Yamakawa, and T. Watanabe.** 1977. Alteration in two enzymatically active forms of valyl-tRNA synthetase during sporulation of *Bacillus subtilis*. J. Biochem. **81**:1571–1574.

35. **Ohyama, K., T. Yamakawa, and I. Kaneko.** 1975. The protein biosynthesis and aminoacyl-tRNA synthetase during sporulation of *Bacillus subtilis*. J. Jpn. Biochem. Soc. (Seikagaku) **47**:496.

36. **Okabe, K., M. Hayakawa, and M. Koike.** 1968. Purification and comparative properties of human lactate dehydrogenase isozymes from uterus, uterine, myeloma and cervical cancer. Biochemistry **7**:79–90.

37. **Sadoff, H. L., E. Celikkol, and H. L. Engelbrecht.** 1970. Conversion of bacterial aldolase from vegetative to spore form by a sporulation-specific protease. Proc. Natl. Acad. Sci. U.S.A. **66**:844–849.

38. **Sadoff, H. L., A. D. Hitchins, and E. Celikkol.** 1969. Properties of fructose 1,6-diphosphate aldolases from spores and vegetative cells of *Bacillus cereus*. J. Bacteriol. **98**:1208–1218.

39. **Steinberg, W.** 1974. Properties and developmental roles of the lysyl- and tryptophanyl-transfer ribonucleic acid synthetases of *Bacillus subtilis*: common genetic origin of the corresponding spore and vegetative enzymes. J. Bacteriol. **118**:70–82.

40. **Steinberg, W.** 1975. Sporulation of *Bacillus subtilis* cells: in vivo evidence for phenotype modification of a thermosensitive lysyl-transfer ribonucleic acid synthetase, p. 290–300. *In* P. Gerhardt, R. N. Costilow, and H. L. Sadoff (ed.), Spores VI. American Society for Microbiology, Washington, D.C.

41. **Strehler, B., G. Hirsch, D. Gusseck, R. Jonson, and M. Bick.** 1971. Codon-restriction theory of aging and development. J. Theor. Biol. **33**:429–474.

42. **Surguchov, A. P., and I. G. Surguchova.** 1975. Two enzymatically active forms of glycyl-tRNA synthetase from *Bacillus brevis*. Eur. J. Biochem. **54**:175–184.

43. **Tu, C.-T.** 1966. Multiple active forms of leucyl-tRNA synthetase of *E. coli*. Cold Spring Harbor Symp. Quant. Biol. **31**:565–570.

44. **Yaniv, M., and F. Gross.** 1969. Studies on valyl-tRNA synthetase and tRNA^val from *Escherichia coli*. J. Mol. Biol. **44**:1–5.

45. **Yasumoto, C. S., and R. H. Doi.** 1974. Transcription from the complementary deoxyribonucleic acid strands of *Bacillus subtilis* during various stages of sporulation. J. Bacteriol. **117**:775–782.

46. **Yem, D. W., and L. S. Williams.** 1973. Evidence for existence of two arginyl-transfer ribonucleic acid synthetases. J. Bacteriol. **113**:891–894.

47. **Zubay, G.** 1962. The isolation and fractionation of soluble ribonucleic acid. J. Mol. Biol. **4**:347–356.

Membrane Development During *Bacillus thuringiensis* Sporulation[1]

DONALD B. BECHTEL AND LEE A. BULLA, JR.

U.S. Grain Marketing Research Laboratory, Agricultural Research, Science and Education Administration, U.S. Department of Agriculture, Manhattan, Kansas 66502, and Division of Biology, Kansas State University, Manhattan, Kansas 66506

The development of forespore membrane in *Bacillus thuringiensis* subsp. *kurstaki* was studied by use of freeze-fracturing and serial-sectioning techniques. Vesicular mesosomes were associated with the initiation and development of the forespore septum. Upon completion of the septum, engulfment of the incipient forespore commenced. The engulfment membranes always maintained an orientation consistent with that of the plasma membrane when viewed with freeze fracture; i.e., the C face (the intracellular membrane surface; opposite side of the A face) was always in contact with the cytoplasm, and the D face (the extracellular membrane surface; opposite side of the B face) was always in contact with the external environment. Forespore development was complete when the forespore membranes detached from the plasma membrane and transformed into the inner and outer forespore membranes.

During bacterial sporulation, the cell divides into two unequal parts via asymmetric cell division. The larger portion remains as mother-cell cytoplasm while the smaller incipient forespore cytoplasm (1) is engulfed to form a forespore. The membranous forespore septum appears to be formed by fusion of mesosomal vesicles (1). Engulfment of the incipient forespore cytoplasm begins immediately after the septum is completed and also presumably is the result of mesosome proliferation. The ultimate fore-

spore is a double membrane-bounded spheroid containing genetic material and the other components normally present in the cytosol.

Although there have been many ultrastructural investigations of bacterial sporulation (1, 3, 5, 7, 9, 10, 12), little is known about the mechanism(s) of septum formation and of the engulfment process. For example, mesosomes have been implicated in forespore membrane development by several researchers (1, 3, 7, 9, 10, 12), but a critical question regarding the authenticity of mesosomes has been posed (13). In an attempt to resolve this issue, we utilized freeze-fracturing and serial-sectioning techniques to examine forespore septum and membrane formation in *Bacillus thuringiensis* subsp. *kurstaki*.

[1] Contribution No. 81-365-j, Division of Biology, Kansas Agricultural Experiment Station, Manhattan. Cooperative investigation between Agricultural Research, Science and Education Administration, U.S. Department of Agriculture, and the Kansas Agricultural Experiment Station.

FIG. 1. *(a) Mesosome from freeze-fractured freshly frozen* B. thuringiensis *cell. All freeze-fractured samples used in this study were snap frozen in Freon 22 that was cooled in liquid nitrogen. Fracturing was carried out in a modified Denton DF-2 freeze-fracture unit at −150°C and 10⁻⁶-torr pressure. No etching was conducted. Micrographs were reproduced with dark shadows. (b) Mesosome from freeze-fractured chemically fixed cell. Cells were fixed in glutaraldehyde followed by osmium tetroxide as previously described (1). (c) A face of fresh-frozen freeze-fractured cell showing depression corresponding to mesosome. (d) B face of fresh-frozen freeze-fractured cell showing mesosomal convexity.*

FIG. 2. *(a–i) Serial cross sections through initiating forespore septum (arrows). Note oblique sections of forespore septum membrane (arrows) and associated mesosomes.*

FIG. 3. *Cross section through developing forespore septum, showing oblique view of septum (arrow) and mesosomes (M).*

FIG. 4. *Freeze-fracture micrography of freshly frozen untreated cell, showing forespore septum (arrows) and mesosome (M) at terminus of one side of septum.*

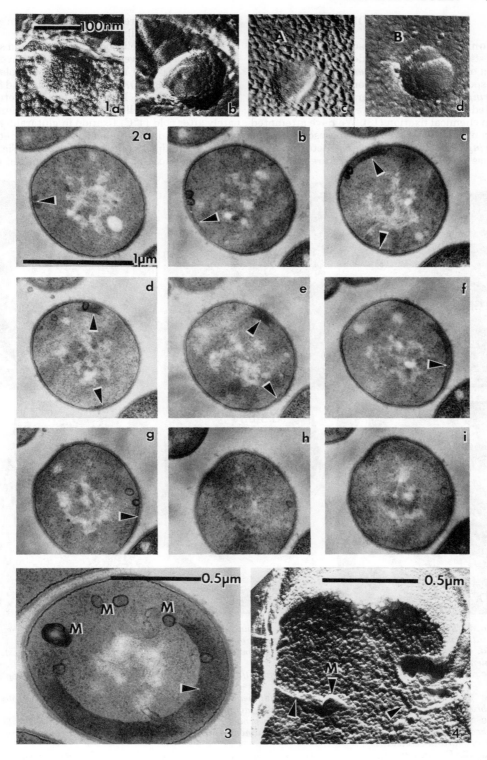

RESULTS AND DISCUSSION

Confirmation of mesosomes. Our previous investigations provided evidence for the involvement of vesicular mesosomes in the formation of forespore septa and forespore membranes in *B. thuringiensis*. Other workers, however, have reported that mesosomes are artifacts, resulting from chemical fixation (13). We have attempted to confirm the existence of mesosomes in *B. thuringiensis* by using freeze-fracturing techniques and comparing the fracture images of freshly frozen untreated cells to those of cells that were chemically fixed.

Freshly frozen cells, both vegetative and sporulating, possessed vesicular mesosomes (average diameter, 100 nm) associated with the plasma membrane (Fig. 1a), cell division septa, and forespore membranes. Prechilled (4°C) vegetative and sporulating cells that were frozen immediately after cold (4°C) fixation in glutaraldehyde or in glutaraldehyde followed by osmium tetroxide (1) contained vesicular mesosomes (Fig. 1b) similar to those of freshly frozen, unfixed cells

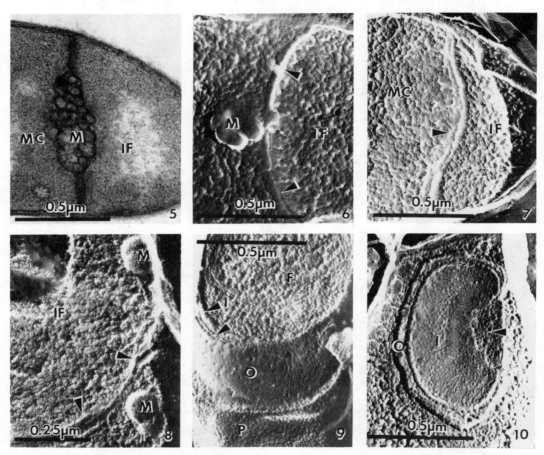

FIG. 5. *Thin section of developing forespore septum with mass of mesosomes (M) in center of septum separating mother-cell cytoplasm (MC) from incipient forespore (IF).*

FIG. 6. *Freeze-fractured cell with mass of mesosomes (M) in center of forespore septum (arrows). IF, incipient forespore.*

FIG. 7. *Completed forespore septum (arrow) separating incipient forespore (IF) from mother-cell cytoplasm (MC).*

FIG. 8. *Freeze-fractured cell undergoing engulfment, showing mesosomes (M) at junction of plasma membrane and engulfment membranes (arrows). IF, incipient forespore.*

FIG. 9. *Fracture through cell with completed forespore (F). Cell was fractured through A face of plasma membrane (P), B face of outer membrane (O), and inner membrane (I). Middle fracture (arrow) of forespore is the primordial cell wall.*

FIG. 10. *Fracture through forespore exposing B face of inner membrane (I) and A face of outer membrane (O) with numerous particles (arrow) typical of A face fracture.*

(Fig. 1a). Regions corresponding to mesosomal attachment sites in the plasma membrane were observed on both A and B fracture faces (Fig. 1c and d). These areas were the same size as mesosomes and appeared as depressions on the A fracture face (Fig. 1c) and as convexities on the B face (Fig. 1d). Prechilled (4°C) cells freeze-fractured after fixation in osmium tetroxide alone were extremely distorted and were not studied further.

We tested freshly frozen vegetative cells for viability and found that our freeze-thaw process did not prohibit cell survival. The comparison of freeze-fractured cells fixed in glutaraldehyde-osmium tetroxide with those that were freshly frozen confirmed that our standard fixation procedure (1) does not render artifacts that could be misinterpreted as mesosomes. Rather, mesosomes are real structures that can be preserved in frozen or in chemically fixed cells of *B. thuringiensis*.

Forespore septum formation and engulfment. Figure 2 is a series of nine serial sections illustrating septum initiation. Note that vesicular mesosomes were associated with both the plasma membrane and the obliquely sectioned forespore septum (arrows). Superimposition of

the nine micrographs revealed that the forespore septum was initiated at the same time all the way around the cell. Septation continued with mesosomal association as the growing membrane divided the cell into two unequal portions (Fig. 3 and 4). Prior to septum completion, a mass of mesosomes in the central portion of the septum (Fig. 5 and 6) fused to render the septum complete (Fig. 7).

Engulfment of the incipient forespore commenced once the forespore septum was complete and involved vesicular mesosomes. Figure 8 depicts the mesosomes at the junction of the forespore and plasma membranes. Engulfment was complete when the septum became detached from the plasma membrane, isolating the forespore from the mother-cell cytoplasm.

The numerous observations that we have made on mesosomes in both sectioned and freeze-fractured material have led us to believe that mesosomes represent a dynamic structure whose presence at a specific site on the membrane is transient. They are synthesized and carry out their function, i.e., add new membrane, and then new ones are synthesized. The close association of mesosomes with the plasma membrane division septa and forespore membranes

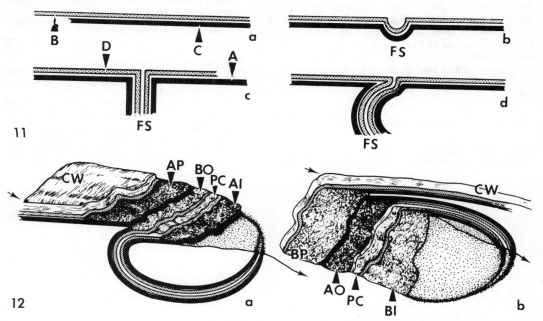

FIG. 11. *Diagrammatic scheme depicting (a) plasma membrane, (b) initiation of forespore septum (FS), and (c and d) forespore septum (FS) development where A and B are the freeze-fracture faces and C and D are the natural membrane surfaces.*

FIG. 12. *Three-dimensional artist's conception of fractures through a completed forespore within the mother cell showing the two complimentary images (a, view from above; b, view from below) obtained with freeze-fracture. CW, cell wall; AP, A face of plasma membrane; BO, B face of outer forespore membrane; PC, primordial cell wall; AI, A face of inner forespore membrane; BP, B face of plasma membrane; AO, A face of outer membrane; BI, B face of inner membrane.*

also may explain the difficulty in obtaining purified preparations of isolated mesosomes.

Forespore membrane orientation. In
freeze-fracture terminology (2, 11), the A face is
the inner membrane leaflet left frozen to the
cytoplasm and is visualized as if viewed from
outside the cell. The opposite side of the A face,
the natural surface that is in contact with the
cytoplasm, is called the C face. The B face is the
outer membrane leaflet visualized from inside
the cell. The opposite side of the B face is the
cell's external membrane surface, the D face.
The A face has more particles on its surface than
does the B face, and this criterion is used to
distinguish the two faces. Figures 9 and 10 por-
tray the two different oblique fractures obtainable by freeze-fracturing forespore membrane.
Many such freeze-fracture micrographs were
used to construct the diagrams in Fig. 11 and 12,
which represent the two types of fractures. From
such micrographs (Fig. 9 and 10) and the recon-
structions (Fig. 11 and 12), it is obvious that
proper membrane orientation is preserved
throughout forespore development; i.e., the D
face always contacts the external environment
and the C face touches the cytoplasm, be it
mother-cell or forespore cytoplasm. Thus, the
engulfment membranes maintain the same orientation as that of the plasma membrane and do
not possess a reversed polarity with respect to
the cytoplasm, as had been suggested (4, 6). This
arrangement probably is essential to further development of the forespore such as elaboration
of the primordial cell wall, cortex, and spore
coats.

LITERATURE CITED

1. **Bechtel, D. B., and L. A. Bulla, Jr.** 1976. Electron microscope study of sporulation and parasporal crystal formation in *Bacillus thuringiensis.* J. Bacteriol. **127:** 1477–1481.
2. **Bullivant, S.** 1973. Freeze-etching and freeze-fracturing, p. 67–112. *In* J. K. Koehler (ed.), Advanced techniques in biological electron microscopy. Springer-Verlag, Berlin.
3. **Decker, S., and S. Maier.** 1975. Fine structure of mesosomal involvement during *Bacillus macerans* sporulation. J. Bacteriol. **121:**363–372.
4. **Ellar, D. J., M. W. Eaton, C. Hogarth, B. J. Wilkinson, J. Deans, and J. La Nauze.** 1975. Comparative biochemistry and function of forespore and mother-cell compartments during sporulation of *Bacillus megaterium* cells, p. 425–433. *In* P. Gerhardt, R. N. Costilow, and H. L. Sadoff (ed.), Spores VI. American Society for Microbiology, Washington, D.C.
5. **Ellar, D. J., and D. G. Lundgren.** 1966. Fine structure of sporulation in *Bacillus cereus* grown in a chemically defined medium. J. Bacteriol. **92:**1748–1764.
6. **Ellar, D. J., and J. A. Postgate.** 1974. Characterization of forespores isolated from *Bacillus megaterium* at different stages of development into mature spores, p. 21–40. *In* A. N. Barker, G. W. Gould, and J. Wolf (ed.), Spore research 1973. Academic Press, London.
7. **Fitz-James, P. C.** 1960. Participation of the cytoplasmic membrane in growth and spore formation of bacilli. J. Biophys. Biochem. Cytol. **8:**507–528.
8. **Fooke-Achterrath, M., K. G. Lickfeld, V. M. Reusch, Jr., U. Aebi, U. Tschoepe, and B. Menge.** 1974. Close to life preservation of *Staphylococcus aureus* mesosomes for transmission electron microscopy. J. Ultrastruct. Res. **49:**270–285.
9. **Highton, P. J.** 1978. Changes in the structure of mesosomes and cell membrane of *Bacillus cereus* during sporulation, p. 13–18. *In* H. O. Halvorson, R. Hanson, and L. L. Campbell (ed.), Spores V. American Society for Microbiology, Washington, D.C.
10. **Holt, S. C., J. J. Gauthier, and D. J. Tipper.** 1975. Ultrastructural studies of sporulation in *Bacillus sphaericus.* J. Bacteriol. **122:**1322–1338.
11. **McNutt, N. S., and R. S. Weinstein.** 1970. The ultrastructure of the nexus. A correlated thin-section and freeze-cleave study. J. Cell Biol. **47:**666–688.
12. **Ohye, D. F., and W. G. Murrell.** 1962. Formation and structure of the spore of *Bacillus coagulans.* J. Cell Biol. **14:**111–123.
13. **Silva, M. T., J. C. F. Sousa, J. J. Polónia, M. A. E. Macedo, and A. M. Parente.** 1976. Bacterial mesosomes. Real structures or artifacts? Biochim. Biophys. Acta **443:**92–105.

Revertants of a Streptomycin-Resistant Mutant of *Bacillus subtilis*

T. M. HENKIN, K. M. CAMPBELL, AND G. H. CHAMBLISS

Department of Bacteriology and Laboratory of Genetics, University of Wisconsin, Madison, Wisconsin 53706

Revertants of a streptomycin-resistant, oligosporogenous mutant of *Bacillus subtilis* were selected for the ability to sporulate. Growth rate, sporulation frequency, streptomycin resistance, and electrophoretic pattern of ribosomal proteins were determined for these revertants and for the parent strain.

An important tool in elucidating the role of translational control in the regulation of sporulation is the isolation and characterization of sporulation-deficient mutants with altered susceptibility to antibiotics which act on the bacterial ribosome. This approach was taken in the isolation of strain SRB15, a streptomycin-resistant, oligosporogenous mutant of *Bacillus subtilis* (3). Preliminary work with this strain indicated that the two phenotypes were inseparable by transformation. They mapped to the *cysA*-linked ribosomal region of the *B. subtilis* chromosome. The ribosomes of this strain were highly resistant to streptomycin in the in vitro translation of natural mRNA and misread at reduced levels, but no ribosomal protein alteration was detected by two-dimension polyacrylamide gel electrophoresis (K. M. Campbell, Ph.D. thesis, University of Wisconsin, Madison, 1979). For the present study, derivatives of SRB15 were selected for the ability to sporulate by use of a serial heat shock selection procedure.

RESULTS AND DISCUSSION

Spo⁺ revertants isolated from Str^r Spo⁻ strain SRB15 fell into two classes. Class I revertants (TR1) became Spo⁺ by the reversion of the *strR* mutation of SRB15. Class II revertants (TR7, TR21) retained the original *strR* mutation and had acquired an additional mutation which suppressed the asporogenous phenotype of SRB15. In TR7 this suppressor mutation was 25% linked by transformation to the *strR* marker. In TR21 the suppressor was unlinked to *strR*.

Mapping studies intended to define the linkage of these suppressor mutations to *cysA* were complicated by the discovery that SRB15 and all *strR* derivatives of it contained another mutation in addition to the original *strR* mutation.

This mutation, designated *fun* as a trivial name for an as yet undefined mutation, mapped between *cysA* and *strR* and was 21% linked to *cysA* by transformation. The *fun* mutation appears to be required for the survival of strains containing the streptomycin resistance mutation of SRB15 since all *strR* derivatives obtained through a variety of genetic manipulations aimed at separating the two mutations proved to contain the *fun* marker, which gives an identifiable colony morphology. It is possible to isolate the *fun* mutation by transformation of strain K1 (*cysA purA trpC*) (4), selecting for Cys⁺. This transformant, K1*fun*, could then be compared with other derivatives of K1 containing appropriate markers.

Growth rate, sporulation frequency, and streptomycin resistance in vivo were determined for these strains (Table 1). TR1, a class I revertant, was similar to the wild type except that its resistance to streptomycin was slightly higher. Class II revertants (TR7 and TR21) were slightly less resistant than was SRB15. Sporulation in TR7 and TR21 was consistently less efficient than in the wild type, suggesting that the asporogeny caused by the *strR* marker was not completely suppressed. The *fun* mutation conferred a small increase in streptomycin resistance, but had little effect on sporulation.

The ribosomes of these strains were studied by polyacrylamide gel electrophoresis of isolated ribosomal proteins. One-dimensional sodium dodecyl sulfate-polyacrylamide gel electrophoresis (Fig. 1) allows rapid comparison of ribosomes from several strains and an estimate of the molecular weights of ribosomal proteins. All Str^r strains (lanes 2, 5, 6, 9, 11, and 13) contained the protein alteration indicated by arrow A. This alteration is apparently caused by the *fun* mutation because it was present in K1*fun* (lane 11),

TABLE 1. *Growth characteristics*[a]

Strain	Phenotype	Origin	Doubling time (min)	Sporulation frequency[b] (%)	Inhibitory streptomycin concn[c] (μg/ml)
BR151	Str^s Spo$^+$	Wild type	46	100	10
SRB15	Str^r Spo$^-$	Mutant of BR151	75	4	1,000
TR1	Str^s Spo$^+$	Revertant of SRB15	45	100	25
TR7	Str^r Spo$^+$	Revertant of SRB15	54	84	500
TR21	Str^r Spo$^+$	Revertant of SRB15	48	78	375
K1	Str^s Spo$^+$	*cysA purA trpC* mapping strain	48	50	25
K1str	Str^r Spo$^-$	Str^r transformant of K1 by DNA from SRB15	80	8	500
K1fun	Str^s Spo$^+$	Cys$^+$ transformant of K1 by DNA from TR7	55	71	50
K1–7	Str^r Spo$^+$	Cys$^+$ transductant of K1str by TR7 phage	66	100	500

[a] Growth was at 37°C in 2× Schaeffer nutrient sporulation medium (9).
[b] Sporulation frequency = heat-resistant colony-forming units/total colony forming units × 100%.
[c] Defined as the concentration of streptomycin required to reduce growth of an overnight culture to less than 50% of levels obtained in the absence of drug.

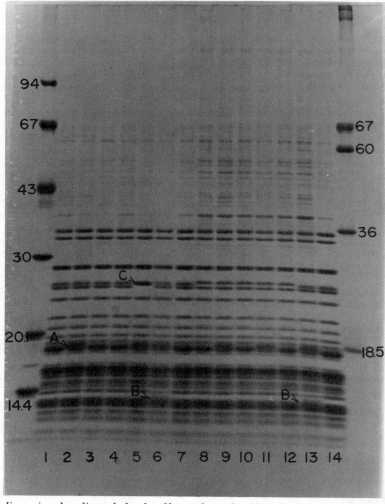

FIG. 1. *One-dimensional sodium dodecyl sulfate-polyacrylamide gel electrophoresis of 70S ribosomal proteins. Molecular weight standards are labeled. (1) BR151; (2) SRB15; (3) TR1; (4) BR151; (5) TR21; (6) TR7; (7) BR151; (8) K1; (9) K1str; (10) K1; (11) K1fun; (12) K1; (13) K1–7; (14) BR151. A 12 to 20% linear acrylamide gradient was used, with a 5% stacking gel. Sodium dodecyl sulfate concentration was 0.2% (1).*

206

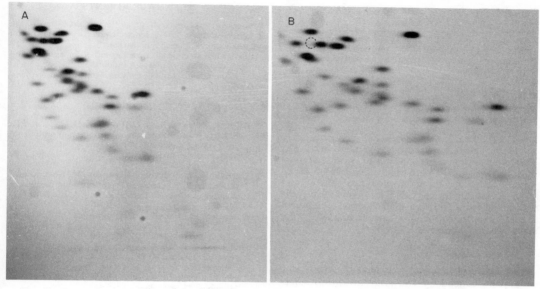

FIG. 2. *Two-dimension polyacrylamide gel electrophoresis of basic 70S ribosomal proteins. (A) BR151; (B) TR21. Proteins were acetone extracted (2), and gels were run as described by Kaltschmidt and Wittmann (7), as modified by Howard and Traut (6).*

which does not contain the *strR* mutation. TR1 (lane 3), an Str^s Spo$^+$ revertant, did not contain the *fun* alteration. TR7 (lane 6) and TR21 (lane 5) did show the *fun* alteration and exhibited in addition modifications of different ribosomal proteins (arrows B and C). K1-7 (lane 13), a transductant of K1*str* selected for Cys$^+$ and screened for Spo$^+$, showed both the *fun* and TR7 modifications; this confirms the genetic evidence that the suppressor mutation of TR7 maps to the *cysA*-linked region. Bands with molecular weights above 36,000 are believed to be contaminants, since they disappear when ribosomes are passed through an additional sucrose cushion. The identity of the protein bands in this gel system is not known.

Two-dimensional polyacrylamide gel electrophoresis (Fig. 2) revealed that TR21 contained a modification in ribosomal protein S4 (nomenclature of Geisser et al. [5]), which is the protein affecting translational ambiguity in *Escherichia coli* (10). Gels of all other strains, including SRB15, were indistinguishable from the wild type.

Resistance of isolated ribosomes to streptomycin in cell-free protein synthesis was assayed for these strains. The data in Table 2 indicate that TR1 was slightly more resistant to streptomycin in vitro than was the wild-type strain (BR151), as was observed for streptomycin resistance in vivo (Table 1). These results suggest that TR1 may not be a true wild-type revertant. The resistance of TR7 was close to that of

TABLE 2. *In vitro streptomycin resistance of isolated ribosomes*

Ribosomes	Inhibitory streptomycin concn (μg/ml)^a	% Activity with streptomycin at 50 μg/ml
BR151	13	34
SRB15	100	66
TR1	30	37
TR7	100	65
TR21	100	52
K1	3	8
K1*str*	100	72
K1*fun*	25	41
K1-7	75	56

^{a} Defined as the concentration of streptomycin required to reduce activity to less than 50% of levels obtained in the absence of drug. Incorporation of [^{14}C]phenylalanine into trichloroacetic acid-insoluble protein was measured using phage SP01 mRNA as template. Assay conditions and cell fraction preparation were as described by Legault-Demare and Chambliss (8).

SRB15, whereas TR21 was less resistant. Again, the in vitro and in vivo data agree and indicate that the ribosomal alterations suppressing the Spo$^-$ phenotype of SRB15 can also modify the expression of the *strR* mutation of SRB15. Table 2 shows that ribosomes from K1*fun* were significantly resistant to streptomycin, although not as resistant as those from K1*str*, which contains the *strR* marker of SRB15.

One possible explanation for the reduced abil-

TABLE 3. *Synthesis of proteins $\alpha\beta$ in vitro with T_8 mRNA as template[a]*

Ribosome source	Total [^{35}S]methionine incorporated (cpm)	[^{35}S]methionine bound by anti-$\alpha\beta$ serum (cpm)	% [^{35}S]methionine in $\alpha\beta$ proteins
BR151	161,575	22,563	14.0
SRB15	317,363	12,375	3.9
TR7	150,200	17,858	11.9
TR21	191,550	21,938	11.5
TR1	187,413	14,598	7.8

[a] In vitro protein synthesis was at 37°C for 30 min. Treatment with anti-$\alpha\beta$ was for 1 h. Washed *Staphylococcus aureus* cells were then added and left for 15 min on ice. *S. aureus* cells were then pelleted and washed. Antigen-antibody complex bound to the cells was eluted by boiling in sodium dodecyl sulfate and urea. Cells were removed by centrifugation and the supernatant was counted.

ity of SRB15 to sporulate is that the ribosomal alteration caused by the *strR* mutation selectively inhibits the translation of mRNA for spore products. This was assayed by comparing the ability of BR151 and SRB15 and its revertants to translate in vitro the mRNA for the acid-soluble spore proteins $\alpha\beta$ (J. M. Leventhal, W. C. Johnson, D. J. Tipper, and G. H. Chambliss, this volume). Table 3 shows that SRB15 ribosomes synthesized lower levels of proteins $\alpha\beta$ than did wild-type ribosomes. Spo$^+$ revertants synthesized levels closer to that of the wild type. The reduced synthesis by SRB15 could be caused by slower activity, since a fixed incubation period was used; however, total incorporation of [^{35}S]methionine by SRB15 ribosomes was comparable to that by the other ribosome types. This and possible artifactual explanations for these results must be tested before a firm statement about the significance of this observation can be made.

ACKNOWLEDGMENTS

This work was carried out under the support of the College of Agricultural and Life Sciences, University of Wisconsin, and the National Science Foundation.

Our thanks go to Judith M. Leventhal for assaying the synthesis of proteins $\alpha\beta$ and to A. Adoutte for the one-dimensional sodium dodecyl sulfate-polyacrylamide gel electrophoresis procedure.

LITERATURE CITED

1. **Adoutte-Panvier, A., J. E. Davies, L. R. Gritz, and B. S. Littlewood.** 1980. Studies of ribosomal proteins of yeast species and their hybrids. Gel electrophoresis and immunochemical cross-reactions. Mol. Gen. Genet. **179:** 273–282.
2. **Barritault, D., A. Expert-Bezançon, M. Guerin, and D. Hayes.** 1976. The use of acetone precipitation in the isolation of ribosomal proteins. Eur. J. Biochem. **63:** 131–135.
3. **Campbell, K. M., and G. H. Chambliss.** 1977. Streptomycin-resistant, asporogenous mutant of *Bacillus subtilis*. Mol. Gen. Genet. **158:**193–200.
4. **Dedonder, R. A., J.-A. Lepesant, J. Lepesant-Kejzlarova, A. Billault, M. Steinmetz, and F. Kunst.** 1977. Construction of a kit of reference strains for rapid genetic mapping in *Bacillus subtilis* 168. Appl. Environ. Microbiol. **33:**989–993.
5. **Geisser, M., G. W. Tischendorf, and G. Stöffler.** 1973. Comparative immunological and electrophoretic studies on ribosomal proteins of Bacillaceae. Mol. Gen. Genet. **127:**129–145.
6. **Howard, G. A., and R. R. Traut.** 1973. Separation and radioautography of microgram quantities of ribosomal proteins by two-dimensional polyacrylamide gel electrophoresis. FEBS Lett. **29:**177–180.
7. **Kaltschmidt, E., and H. Wittmann.** 1970. Ribosomal proteins. VII. Two-dimensional polyacrylamide gel electrophoresis for fingerprinting of ribosomal proteins. Anal. Biochem. **36:**401–412.
8. **Legault-Demare, L., and G. H. Chambliss.** 1974. Natural messenger ribonucleic acid-directed cell-free protein-synthesizing system of *Bacillus subtilis*. J. Bacteriol. **120:**1300–1307.
9. **Schaeffer, P., J. Millet, and J.-P. Aubert.** 1965. Catabolite repression of bacterial sporulation. Proc. Natl. Acad. Sci. U.S.A. **54:**704–711.
10. **Zimmerman, R. A., R. T. Garvin, and L. Gorini.** 1971. Alteration of a 30S ribosomal protein accompanying the *ram* mutation in *Escherichia coli*. Proc. Natl. Acad. Sci. U.S.A. **68:**2263–2267.

In Vitro Synthesis of Protein by Extracts of *Bacillus subtilis*

JUDITH M. LEVENTHAL, W. CHARLES JOHNSON, DONALD J. TIPPER, AND GLENN H. CHAMBLISS

Department of Bacteriology, University of Wisconsin, Madison, Wisconsin 53706; Biological Laboratories, Harvard University, Cambridge, Massachusetts 02138; and Department of Microbiology, University of Massachusetts Medical School, Worcester, Massachusetts 01605

Cell extracts from *Bacillus subtilis* were used in the in vitro synthesis of the phage-specific enzyme dCMP deaminase and the synthesis of the low-molecular-weight, acid-soluble spore proteins $\alpha\beta$. The in vitro synthesis of dCMP deaminase was shown to depend on the presence of RNA extracted 10 to 15 min after infection of *B. subtilis* with phage SP01. Synthesis of $\alpha\beta$ depended on the presence of RNA extracted from *B. subtilis* harvested late in sporulation.

The in vitro synthesis of protein by a system containing cell extracts of *Bacillus subtilis* was previously shown to be dependent upon the addition of exogenous DNA or RNA to the reaction mixture (5). Through sodium dodecyl sulfate-polyacrylamide gel electrophoresis (SDS-PAGE), it was shown that the molecular weight of the products synthesized in vitro depended upon the origin of the DNA template. The present study demonstrated the ability of the in vitro system to make specific proteins when the appropriate RNA was present. These proteins are the phage-specific enzyme dCMP deaminase (6) and the low-molecular-weight, acid-soluble, spore proteins $\alpha\beta$ (W. C. Johnson, Ph.D. thesis, University of Massachusetts, Amherst, 1980; D. J. Tipper, W. C. Johnson, G. H. Chambliss, I. Mahler, M. Arnaud, and H. O. Halvorson, this volume). The $\alpha\beta$ spore proteins appear to be analogous to proteins A and C synthesized by *B. megaterium*. A and C, along with another low-molecular-weight, acid-soluble protein (protein B), make up about 15% of the total proteins of *B. megaterium* spores. A, B, and C are the major proteins degraded during germination of *B. megaterium* (7).

RESULTS AND DISCUSSION

Before studying the in vitro synthesis of dCMP deaminase, its time of synthesis in vivo was determined. Most in vivo synthesis of dCMP deaminase occurred between 6 and 15 min after infection of a vegatative culture of *B. subtilis* by phage SP01 (Fig. 1).

RNA extracted from *B. subtilis* at various times after infection by SP01 stimulated incorporation of [^{35}S]methionine into protein. dCMP deaminase synthesis, however, occurred only when RNA from cells harvested 10, 15, or 20 min after infection was present in the in vitro reaction mixture (Fig. 2). The greatest amount of dCMP deaminase was synthesized with RNA extracted from cells 15 min after infection as the template.

The in vitro synthesis of the acid-soluble spore proteins $\alpha\beta$ was detected by immunoprecipitating in vitro-synthesized protein with anti-$\alpha\beta$ antiserum and *Staphylococcus aureus* Cowan I. Anti-$\alpha\beta$ was raised in rabbits injected with glutaraldehyde–cross-linked purified $\alpha\beta$. Authentic $\alpha\beta$ labeled with ^{125}I in vitro was immunoprecipitated by this antiserum. The precipitated protein was run on an SDS-polyacrylamide gel, and a single band with the molecular weight of $\alpha\beta$ was observed (Johnson, Ph.D. thesis, 1980). In Fig. 3, the dependence of in vitro $\alpha\beta$ synthesis on the presence of RNA produced during sporulation is clearly visible. RNAs from t_6 and t_8 cells both stimulated the synthesis of protein that was precipitated by anti-$\alpha\beta$ antiserum. Under the growth and sporulation conditions indicated in the legend to Fig. 3, in vivo synthesis of alkaline phosphatase (assayed with disodium *p*-nitrophenyl phosphate as substrate) and protease (assayed with Azocoll as substrate) began at t_3. Heat-resistant spores began to appear between t_{12} and t_{14}.

The fluorograms in Fig. 4 show the results of SDS-PAGE of in vitro-synthesized protein immunoprecipitated by using anti-$\alpha\beta$ (Fig. 4A). Anti-12.2K coat protein (Fig. 4B) was used as a control. Anti-12.2K coat protein antiserum was raised in rabbits injected with the 12.2K protein (protein with a molecular weight of 12,200) extracted from insoluble spore coats by treatment with SDS and reducing agents at pH 10 (1). Figure 4 shows that in vitro-synthesized protein

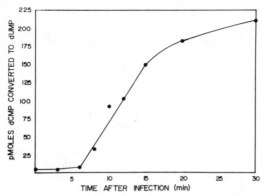

FIG. 1. *In vivo synthesis of dCMP deaminase in SP01-infected* B. subtilis. *A vegetative culture of* B. subtilis *(absorbance at 570 nm, 0.40) grown at 37°C in M medium (8) was infected with phage SP01 (multiplicity of infection, 5), and 20-ml samples were harvested at various times after infection. The cells were lysed by incubation at 37°C for 5 min in the presence of 1 ml of lysis buffer (0.1 M Tris-hydrochloride, pH 8.0, 200 μg of chloramphenicol per ml, 100 μg of lysozyme per ml, and 10 μg of DNase per ml). Cell debris was removed by centrifugation. The amount of dCMP deaminase present in 2.5 μg of each sample was determined as described by Schweiger and Gold (6), allowing the calculation of the number of picomoles of substrate (dCMP) converted to product (dUMP) in 15 min.*

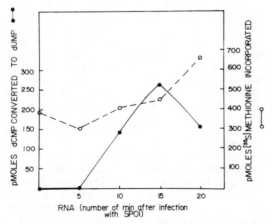

FIG. 2. *In vitro synthesis of dCMP deaminase in the presence of various RNA preparations. (●) Picomoles of dCMP converted to dUMP after a 45-min incubation of in vitro-synthesized protein in the presence of [³H]dCMP. (○) Total protein synthesized in terms of picomoles of [³⁵S]methionine incorporated into trichloroacetic acid-insoluble protein after a 30-min incubation. All incubations were carried out at 37°C. In vitro protein synthesis using a system based on cell extracts from* B. subtilis *(5) and the procedure for RNA extraction (4) were described previously.*

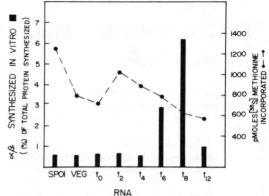

FIG. 3. *In vitro synthesis of spore proteins in the presence of various RNAs. SP01: RNA from* B. subtilis *harvested 5 min after infection with SP01. VEG: RNA from vegetative* B. subtilis. *RNA was extracted from* B. subtilis *cells grown and sporulated at 30°C in complex sporulation medium (16 g of GIBCO nutrient broth per liter, 13 mM KCl, 1 mM MgSO₄, 0.5 mM CaCl₂, 0.001 mM FeSO₄, 0.01 mM MnCl₂). Protein αβ synthesized in vitro = counts per minute immunoprecipitated by anti-αβ/total counts per minute incorporated into protein × 100%. The RNAs were added to the in vitro reaction mixture containing [³⁵S]methionine and incubated at 37°C for 30 min. After incubation, anti-αβ antiserum was added to the mixture. The antiserum had been produced in rabbits injected with low-molecular-weight proteins that were acid extracted from spores. It was allowed to react with in vitro-synthesized proteins for 1 h on ice before the addition of washed* S. aureus *Cowan I cells (2). After 5 min, the* S. aureus *cells were pelleted and washed four times to remove weakly bound radioactivity. Washing buffers contained 50 mM Tris-hydrochloride, pH 7.4, 5 mM EDTA, 0.02% sodium azide, 0.5% Triton X-100, 1 mg of ovalbumin per ml, and 5 mM L-methionine. The first wash also contained 0.5 M NaCl, the second and third washes contained 0.25 M NaCl, and the last wash contained 0.15 M NaCl. In vitro-synthesized antigen bound to the* S. aureus *was eluted by boiling in the presence of 4% (wt/vol) SDS–6 M urea. The samples were then either counted in aqueous counting scintillant or precipitated by acetone. Acetone precipitates were dried and resolubilized in 3% (vol/vol) acetic acid–2% (wt/vol) SDS. The acid-soluble protein was lyophilized and redissolved in sample buffer (0.05 M Tris-hydrochloride, pH 6.8, 2 mM EDTA, 1% β-mercaptoethanol, and 10% glycerol). Samples of 5 μl were spotted on glass fiber filters and counted.*

comigrating with in vivo-synthesized αβ is immunoprecipitated by anti-αβ but not by anti-12.2K coat protein. As shown in Fig. 3, αβ spore proteins were synthesized in vitro only in the presence of t_6 or t_8 mRNA. To ascertain that the in vitro-synthesized, immunoprecipitated protein was actually αβ, we performed a competi-

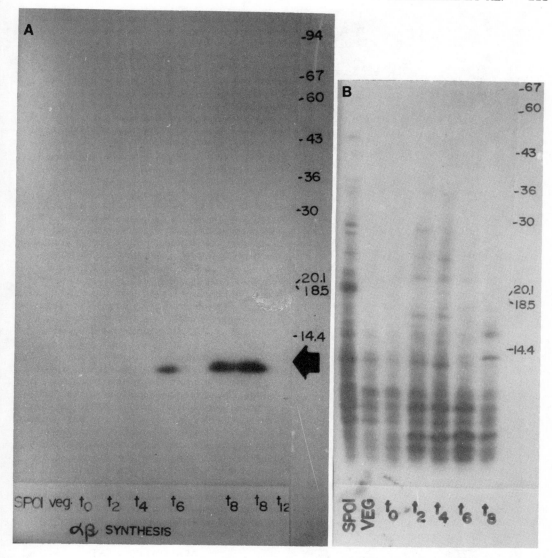

FIG. 4. (A) Fluorogram of SDS-polyacrylamide gel of in vitro-synthesized, immunoprecipitated αβ. Samples prepared as described in the legend to Fig. 3 were run on an SDS-polyacrylamide gel made up of a 1.5-cm, 5% acrylamide stacking gel and a 20-cm, 12 to 20% acrylamide linear gradient running gel by a modified Laemmli procedure (3). Molecular-weight standards from high- and low-molecular-weight calibration kits (Pharmacia Fine Chemicals) were run alongside the radioactive samples. The gel was run at 200 V for 24 h. Before treatment with EN³HANCE (New England Nuclear), the gel was cut vertically to remove the molecular-weight standards. The section of the gel containing the standards was stained with 0.1% Coomassie blue in 50% trichloroacetic acid and destained in acetic acid-methanol-water (10:30:60, vol/vol). The type of RNA present in the in vitro protein-synthesizing reaction mixture is indicated at the bottom of the figure. Molecular weights of standards (in thousands) are given at the side of the figure. Kodak XR-5 film was exposed to the gel for 3 days before being developed. The arrow indicates the position of purified αβ. (B) In vitro-synthesized protein immunoprecipitated by anti-12.2K antiserum. RNA present in the in vitro protein-synthesizing reaction mixture is given at the bottom of the figure. Molecular weights of standards (in thousands) are given at the sides. Procedures for sample preparation and SDS-polyacrylamide gel electrophoresis were the same as those described in the legends Fig. 3 and 4 with the exception that anti-12.2K was used in place of anti-αβ in the immunoprecipitation reaction.

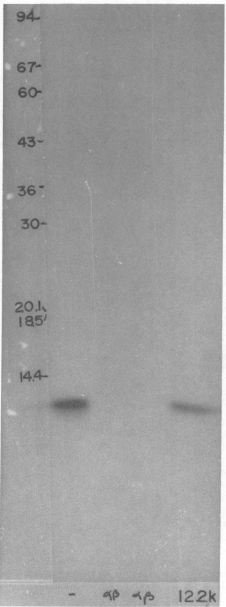

Table 1. *Competition of αβ immunoprecipitation*[a]

% Total cpm immunoprecipitated by anti-αβ antiserum	Addition before immunoprecipitation
2.9	—
0.22	Unlabeled αβ (8 μg/ml)
0.33	Unlabeled αβ (8 μg/ml)
1.5	Unlabeled 12.2K protein (17 μg/ml)

[a] In vitro protein synthesis took place in the presence of t_6 RNA.

added to separate in vitro reaction mixtures prior to the addition of anti-αβ, and the immunoprecipitation procedure was continued as described in the legend to Fig. 3. The immunoprecipitated protein was spotted on glass fiber filters and counted (Table 1). The remainder of each sample was run on SDS-polyacrylamide gels (Fig. 5). It can be seen that the unlabeled αβ prevented the binding to anti-αβ of otherwise immunoprecipitable in vitro-synthesized protein. Unlabeled 12.2K protein did not have this effect.

When a mixture of ^{3}H-labeled amino acids was used to label in vitro-synthesized products, the protein immunoprecipitated by anti-αβ had the same molecular weight as the protein immunoprecipitated after in vitro synthesis in the presence of [^{35}S]methionine (results not shown).

The in vitro protein-synthesizing system used in this study has been demonstrated not only to be dependent upon the presence of an exogenously added template, but also to translate that template accurately. This results in the synthesis of proteins antigenically and enzymatically comparable to their in vivo-synthesized counterparts.

ACKNOWLEDGMENTS

This work was carried out under the support of the College of Agriculture and Life Sciences, University of Wisconsin, and Public Health Science grant AI 16812 from the National Institutes of Health.

We thank A. Adoutte for the SDS-PAGE procedure.

FIG. 5. *Competition for binding of anti-αβ. (A) No addition before immunoprecipitation; (B, C) 8 μg of unlabeled αβ per ml added; (B) 17 μg of unlabeled 12.2K protein per ml added. The RNA used in the in vitro protein-synthesizing reaction mixture was from* B. subtilis *harvested at* t_6. *Procedures for sample preparation and SDS-polyacrylamide gel electrophoresis are given in the legends to Fig. 3 and 4.*

tion experiment using unlabeled authentic αβ. In this experiment 12.2K coat protein served as a control. These two unlabeled proteins were

LITERATURE CITED

1. **Goldman, R. C., and D. J. Tipper.** 1978. *Bacillus subtilis* spore coats: complexity and purification of a unique polypeptide component. J. Bacteriol. **135:**1091–1106.
2. **Kessler, S. W.** 1975. Rapid isolation of antigens from cells with a staphylococcal protein A- antibody adsorbent: parameters of the interaction of antibody-antigen complexes with protein A. J. Immunol. **115:**1617–1624.
3. **Laemmli, U. K.** 1970. Cleavage of structural proteins during the assembly of the head of bacteriophage T4.

Nature (London) **227**:680–685.

4. **Legault-Demare, L., and G. H. Chambliss.** 1974. Natural messenger ribonucleic acid-directed cell-free protein synthesizing system of *Bacillus subtilis.* J. Bacteriol. **120**:1300–1307.

5. **Leventhal, J. M., and G. H. Chambliss.** 1979. DNA-directed cell-free protein-synthesizing system of *Bacillus subtilis.* Biochim. Biophys. Acta **564**:162–171.

6. **Schweiger, M., and L. M. Gold.** 1970. *Escherichia coli* and *Bacillus subtilis* phage deoxyribonucleic acid-directed deoxycytidylate deaminase synthesis in *Escherichia coli* extracts. J. Biol. Chem. **245**:5022–5025.

7. **Setlow, P.** 1975. Identification and localization of the major proteins degraded during germination of *Bacillus megaterium* spores. J. Biol. Chem. **250**:8159–8167.

8. **Yehle, C. O., and R. H. Doi.** 1967. Differential expression of bacteriophage genomes in vegetative and sporulating cells of *Bacillus subtilis.* J. Virol. **1**:935–947.

In Vitro System to Study Enterotoxin Synthesis and Sporulation in *Clostridium perfringens*

JAMES L. McDONEL AND WALTER P. SMITH[1]

Department of Microbiology, Cell Biology, Biochemistry, and Biophysics, The Pennsylvania State University, University Park, Pennsylvania 16802

Polysomes were isolated from an enterotoxigenic strain of *Clostridium perfringens* during vegetative growth and at timed intervals after the onset of sporulation. Nascent chains were completed with L-[^{35}S]methionine in an *Escherichia coli* in vitro system and examined by sodium dodecyl sulfate gel electrophoresis. Gels of sporulation products contained a major band with the mobility of enterotoxin (molecular weight, 35,000) that was immunoprecipitated, together with a species of about 17,000 molecular weight by antiserum to enterotoxin.

Clostridium perfringens type A produces a sporulation-specific protein enterotoxin that is responsible for one of the most common forms of food poisoning in the United States (14). Considerable progress has been made in recent years in understanding how the enterotoxin acts in the intestine to cause the typical disease symptoms, which are diarrhea and abdominal cramps (8, 9). The enterotoxin is now also being used as a tool to study membrane structure-function relationships in intestinal epithelial cells (9). There remains, however, considerable room for progress in describing the enterotoxin molecule and its mode of action at the molecular level.

The enterotoxin is also important because of its direct relationship to the sporulation process. It is produced only during sporulation (2, 3) and, in fact, is thought to be serologically homologous to spore coat protein (4, 5). This is important because sporulation can serve as a relatively simple system for the study of cellular differentiation in terms of specificity of translation, modification of the translation apparatus, and transcription modification, all of which have been implicated in the differentiation process.

Spore coat protein is one of the few sporulation-specific gene products that have been well characterized and therefore serves well as a specific probe to analyze transcriptional regulation during sporulation. The work described in this paper deals with establishing a polypeptide-synthesizing system from vegetative and sporulating

C. perfringens cells. This system can be used to delineate the fundamentals of sporulation and, as emphasized in this paper, the synthesis of enterotoxin by this species.

MATERIALS AND METHODS

Organism and culture. *C. perfringens* type A, strain NCTC 8239 (Hobbs serotype 3), was inoculated from cooked-meat medium (Difco Laboratories) into 10 ml of fluid thioglycolate medium (BBL Microbiology Systems) and incubated overnight (16 to 18 h) at 37°C. A 2% inoculum was then made into either Duncan-Strong sporulation medium (4) for sporulation or fluid thioglycolate medium for vegetative cells. The percentage of cells in Duncan-Strong sporulation medium that entered into the sporulation process ranged from 85 to 90%.

Preparation of cell fractions. The procedures used were as described elsewhere (13). Briefly, chloramphenicol was added to exponentially growing and sporulating cultures, the cultures were poured onto excess ice, and the cells were pelleted. Cells were then suspended in 25 ml of buffer A (10 mM magnesium acetate, 50 mM KCl, 10 mM Tris-hydrochloride, pH 8.0) and lysed by sonication. DNase (5 mg/ml) was added to the lysates, and cell debris was pelleted by centrifugation at 30,000 × *g*. The supernatant was centrifuged at 160,000 × *g* to sediment the ribosomes/polysomes. Polysomes were separated from free ribosomes by Sepharose 2B chromatography (16). Preincubated S-30 fractions from *Escherichia coli* were prepared by standard methods (10). Crude initiation factors (IF), IF-free ribosomes (7), and mRNA (12) were prepared by methods described previously.

Purified *C. perfringens* polysomes (2 units of absorbance at 260 nm [A_{260} U]) were incubated with 3 A_{260} U of an *E. coli* S-100 cell extract for 45 min at 37°C in a 0.1-ml protein-synthesizing system as described previously (12), except that 5 μCi of [^{35}S]me-

[1] Present address: Dermatological Research Department, Vick Divisions Research and Development, Richardson-Merrell Inc., Mt. Vernon, NY 10553.

thionine (600 Ci/mmol) and the 19 other standard unlabeled amino acids were substituted for a mixture of ^{3}H-amino acids. Protein synthesis was terminated by diluting the entire reaction mixture in 0.9 ml of cold buffer B (0.14 M NaCl, 0.1 M NaPO$_4$, pH 7.4) with 3 mM phenylmethylsulfonyl fluoride–5 mM o-phenan-throline (pH 8.0). The mixture was clarified by centrifugation, and total protein was either fractionated by electrophoresis (15) or quantitated by trichloroacetic acid precipitation, with a subsequent counting of radioactivity.

In some experiments enterotoxin was precipitated from the reaction mixture by the addition of an excess of rabbit anti-enterotoxin serum to the 1-ml reaction mixture and incubation for 90 min at 37°C. Goat anti-rabbit serum was added and incubation was continued for 12 h at 4°C to precipitate the anti-enterotoxin. The precipitate was collected by centrifugation, washed three times with buffer B, and, after final precipitation, assayed for total radioactivity or analyzed by sodium dodecyl sulfate-polyacrylamide gel electrophoresis (10).

In vitro protein synthesis (ribosomal) and identification of products. One A$_{260}$ U of IF-free ribosomes was incubated with 0.1 A$_{280}$ U of endogenous or R17 mRNA at 37°C in a 0.1-ml protein-synthesizing system as described above.

RESULTS

Distribution and activity of ribosomes and polysomes in vegetative and sporulating cells. *C. perfringens* crude polysomal preparations isolated from cultures of vegetative and sporulating cells were incubated with an S-100 extract (*E. coli* S-100 fractions were used since they were found to be more active than *C. perfringens* extracts), and the extent of protein synthesis was monitored.

Synthesis by polysomes from vegetative cells was very active, whereas very little synthesis occurred with polysomes from cells that had been in sporulation medium for only 1 or 2 h (Table 1). Maximum protein synthesis by sporulating-cell polysomes occurred with extracts from cells that had been in sporulation medium for 6 h.

We found that rates of protein synthesis correlated well with the ratio of polysomes to monosomes in the crude polysome preparations. Whereas 67% of the ribosomes from vegetative cells were complexed with mRNA in the polysome form, only 17% from cells 1 h after inoculation into sporulation medium were in the polysome form. The polysome content increased with time after transfer into sporulation medium to a maximum of about 45% at 6 h.

Polysomes separated from the free 70S monomers by Sepharose 2B chromatography (Table 1) had identical protein synthetic activity, whether obtained from vegetative or sporulating cells. These results suggest that no apparent modifications of the polysomes (ribosomes) occurred during sporulation. Activity differences observed with crude vegetative- and sporulation-phase extracts probably reflect the ability to find available mRNA.

Translational specificity of vegetative- and sporulating-cell polysomes. To test the possibility that modified translation specificity of the protein-synthesizing components of vegetative or sporulating cells causes breakdown of polysomes upon onset of sporulation, we isolated IF, mRNA, and washed ribosomes from vegetative and sporulating cells. The ability to translate mRNA of vegetative-cell, sporulating-cell, or viral (R17) origin was identical for ribosomes from vegetative and sporulating cultures (data not shown).

TABLE 1. *Incorporation of [*^{35}*S]methionine into protein by various protein-synthesizing systems*[a]

S-100	*C. perfringens* polysomes	% Polysomes[b]	Activity[c] (cpm)	
			Crude	Purified
C. perfringens (vegetative)	Vegetative		671 × 10^3	1,007 × 10^3
E. coli	Vegetative	67	1,063 × 10^3	1,279 × 10^3
E. coli	Sporulation + 1 h	17	214 × 10^3	1,171 × 10^3
E. coli	Sporulation + 2 h	23	241 × 10^3	1,231 × 10^3
E. coli	Sporulation + 3 h	35	472 × 10^3	1,417 × 10^3
E. coli	Sporulation + 4 h	37	427 × 10^3	1,201 × 10^3
E. coli	Sporulation + 6 h	45	786 × 10^3	1,137 × 10^3
E. coli	Sporulation + 8 h	10^d	121 × 10^3	—[e]
E. coli	Sporulation + 10 h	6^d	103 × 10^3	—[e]

[a] Protein was synthesized in vitro as described in Materials and Methods.

[b] Calculated as: polysomes/(polysomes + ribosomes [70S]) × 100.

[c] Crude polysomes were prepared as described in Materials and Methods. Purified polysomes were prepared by Sepharose 2B chromatography as described. Values given are total trichloroacetic acid-precipitable counts corrected for background.

[d] Approximations due to depleted state of system.

[e] No polysomes could be isolated at these times.

Protein synthesis by polysomes from vegetative and sporulating cells. When the proteins synthesized by each polysome fraction were examined by single-dimension sodium dodecyl sulfate electrophoresis (Fig. 1), the majority of the protein bands appeared in preparations from both vegetative and sporulating cells. However, several distinct differences were noted in gel banding patterns, including the presence of a major band at a molecular weight of about 35,000 from the sporulating-cell extracts which was absent in vegetative-cell extracts.

Enterotoxin synthesis by polysomes from sporulating cells. Studies were undertaken to determine whether the major band found at a molecular weight of 35,000 in the sporulating-cell extracts had any characteristics in common with enterotoxin (molecular weight, 35,000 [12]), which had been purified by standard methods (14) from batch cultures of *C. perfringens* type A. The proteins synthesized by polysomes incubated with *E. coli* S-100 fractions were allowed to react with antisera prepared against the purified *C. perfringens* enterotoxin (Table 2). Very little enterotoxin (less than 1% of total protein) was synthesized by polysomes from vegetative cells, whereas appreciable amounts (up to 12% of total protein) were synthesized by polysomes from sporulating cells. Maximal synthesis of enterotoxin by polysomes from sporulating cells occurred at 6 h after transfer into sporulation medium. This time corresponds well with data for enterotoxin synthesis in vivo (1, 6). Figure 2 shows the results of analysis of the immunoprecipitated protein by sodium dodecyl sulfate-polyacrylamide gel electrophoresis. Three bands were observed: a major band at a molecular weight of 35,000 and two minor bands at molecular weights of 17,000 and 52,000.

DISCUSSION

We have developed a system for studying enterotoxin synthesis and sporulation in *C. perfringens* through the utilization of an in vitro protein synthesis system. Crude ribosomal fractions isolated from sporulating cells were less active in protein synthesis than those from vegetative cells. However, no difference in activity by polysomes from vegetative or sporulating cells was noted when they were separated from ribosomes and used with S-100 fractions in the in vitro synthesizing system. The difference in crude ribosomal activity appears to have been due to the degree of ribosomal association with mRNA, ribosomes, or IF. It is concluded that no significant translational specificity differences are detectable between components of protein-

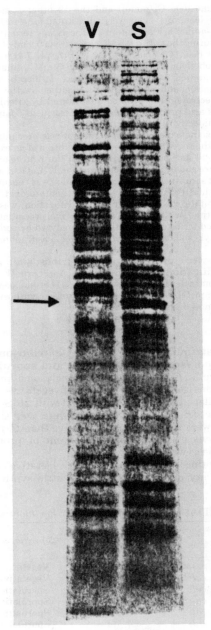

FIG. 1. *Gel electrophoresis (sodium dodecyl sulfate) of total proteins synthesized in vitro by polysomes from vegetative and 6-h sporulating cultures. Electrophoresis was performed as described in Materials and Methods. Arrow indicates location of 35,000-molecular-weight band. V, Vegetative system; S, sporulation system.*

synthesizing systems of vegetative and sporulating cells.

Although most of the gel banding patterns of

protein synthesized by vegetative- and sporulating-cell polysomes were similar, the most notable difference was the appearance in the sporulating-cell pattern of a band at a position that implies a protein with a molecular weight of about 35,000. This corresponds closely to the molecular weight reported for *C. perfringens* enterotoxin synthesized in vivo (9). Because the enterotoxin is a sporulation-specific product (2,

3), it is not surprising that only polysomes from sporulating cells produced appreciable amounts of the 35,000-molecular-weight protein believed to be the enterotoxin. This particular protein comprised 12% of the total protein synthesized by polysomes isolated from cells that had been in sporulation medium for 6 h.

The protein immunoprecipitated from the in vitro protein-synthesizing system by antisera against purified *C. perfringens* enterotoxin appeared as three bands in sodium dodecyl sulfate gels: a major band at a molecular weight of 35,000 and two minor bands with molecular weights of 17,000 and 52,000. It is possible that the 17,000-molecular-weight band represents a monomer, and the 35,000- and 52,000-molecular-weight bands could be a dimer and trimer. It is also conceivable that the immunoprecipitated enterotoxin is a complex mixture of labile polypeptides or that cross-reactive peptide fragments in *C. perfringens* cells exist. The 17,000- and 35,000-molecular-weight species may be spore coat proteins immunoprecipitated by anti-enterotoxin sera (4). We believe that the 52,000-molecular-weight product may represent the primary translation product and the 35,000- and

TABLE 2. *Synthesis of enterotoxin by vegetative and sporulation polysomes*

Polysomes	Immunoprecipitated enterotoxin (cpm)[a]	% Total protein synthesized
Vegetative	5.5×10^3	0.5×10^3
Sporulation		
1 h	45.5×10^3	4.3×10^3
2 h	66.6×10^3	6.3×10^3
6 h	131.2×10^3	12.3×10^3

[a] Enterotoxin was immunoprecipitated from $1,063 \times 10^3$ cpm of in vitro-synthesized protein from each system. Values given are corrected for background counts (5,200 cpm) precipitated when all in vitro conditions were similar but in the absence of added polysomes.

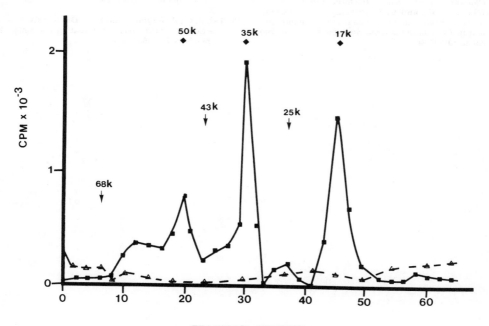

FIG. 2. *Electrophoretic profile of immunoprecipitated protein synthesized in vitro by vegetative and sporulation cultures. Anti-enterotoxin (rabbit) was allowed to react with total protein synthesized in vitro by polysomes from vegetative (△) and sporulating (■) cultures. Goat anti-rabbit serum was used to precipitate the above prior to sodium dodecyl sulfate gel electrophoresis. Counts from a sliced gel are given for each type of culture system. Arrows indicate the location of molecular-weight standards. Diamonds indicate the estimated molecular weights.*

17,000-molecular-weight proteins may represent cleavage products. The enterotoxin may be synthesized in vivo as a precursor molecule that is normally cleaved into units of 35,000 and 17,000. The 17,000 fragment may be inserted into the spore coat as an integral component. In fact, Frieben and Duncan (5) have shown that spores from enterotoxigenic and non-enterotoxigenic strains of *C. perfringens* contain proteins with molecular weights of 14,500, 23,000, and 36,000 which have serological and biological activities similar to those of purified enterotoxin. It is of central importance that further studies be performed to determine the serological and biological activities of the different molecular-weight forms synthesized in vitro.

LITERATURE CITED

1. Duncan, C. L. 1973. Time of enterotoxin formation and release during sporulation of *Clostridium perfringens* type A. J. Bacteriol. **113**:932–936.
2. Duncan, C. L., and D. H. Strong. 1969. Ileal loop fluid accumulation and production of diarrhea in rabbits by cell-free products of *Clostridium perfringens*. J. Bacteriol. **100**:86–94.
3. Duncan, C. L., D. H. Strong, and M. Sebald. 1972. Sporulation and enterotoxin production by mutants of *Clostridium perfringens*. J. Bacteriol. **110**:378–391.
4. Frieben, W. R., and C. L. Duncan. 1973. Homology between enterotoxin protein and spore structural protein in *Clostridium perfringens* type A. Eur. J. Biochem. **39**:293–401.
5. Frieben, W. R., and C. L. Duncan. 1975. Heterogeneity of enterotoxin-like protein extracted from spores of *Clostridium perfringens* type A. Eur. J. Biochem. **55**:455–463.
6. Labbe, R. G., and C. L. Duncan. 1977. Spore coat protein and enterotoxin synthesis in *Clostridium perfringens*. J. Bacteriol. **131**:713–715.
7. Landau, J. V., W. P. Smith, and D. H. Pope. 1977. Role of the 30S ribosomal subunit, initiation factors and specific ion concentration in barotolerant protein synthesis in *Pseudomonas bathycetes*. J. Bacteriol. **130**:154–159.
8. McDonel, J. L. 1980. Mechanism of action of *Clostridium perfringens* enterotoxin. Food Technol. **34**:91–95.
9. McDonel, J. L. 1980. *Clostridium perfringens* toxins (Type A, B, C, D, E). Pharm. Ther. **10**:617–655.
10. Modolell, J. 1971. The S-30 system from *Escherichia coli*, p. 1–65. *In* J. A. Last and A. I. Laskin (ed.), Protein biosynthesis in bacterial systems. Marcel Dekker, New York.
11. Skjelkvale, R., and C. L. Duncan. 1975. Characterization of enterotoxin purified from *Clostridium perfringens* type C. Infect. Immun. **11**:1061–1068.
12. Smith, W. P. 1980. Cotranslational secretion of alkaline phosphatase and diphtheria toxin in vitro: involvement of membrane protein(s). J. Bacteriol. **141**:184–189.
13. Smith, W. P., and J. L. McDonel. 1980. *Clostridium perfringens* type A: in vitro system for sporulation and enterotoxin synthesis. J. Bacteriol. **144**:306–311.
14. Stark, R. L., and C. L. Duncan. 1972. Purification and biochemical properties of *Clostridium perfringens* type A enterotoxin. Infect. Immun. **6**:662–673.
15. Studier, F. W. 1973. Analysis of bacteriophage T7. Early RNA's and proteins on slab gels. J. Mol. Biol. **79**:237–248.
16. Tai, P. C., B. Wallace, and B. D. Davis. 1974. Selective action of erythromycin on initiation ribosomes. Biochemistry **13**:4653–4656.

RNA Polymerase Forms in Vegetative and Sporulating Cells of *Bacillus subtilis*

ROY H. DOI, TOSHIAKI KUDO, AND CYNTHIA DICKEL

Department of Biochemistry and Biophysics, University of California, Davis, California 95616

At least three different forms of RNA polymerase were found to exist in vegetative cells of *Bacillus subtilis*, and four forms were found in sporulating cells. The transcription apparatus of *B. subtilis* differs from that found in *Escherichia coli* cells in terms of the large number of different forms of the enzyme, the core subunits responsible for antibiotic susceptibility, and the requirement for two polypeptide factors in *B. subtilis* instead of the one factor in *E. coli* for increased promoter selectivity.

A clear understanding of transcription specificity during growth and sporulation of *Bacillus subtilis* will require a thorough knowledge of the structures and interactions of RNA polymerase and promoters. Studies of RNA polymerase from *B. subtilis* have revealed that the enzyme can exist in several different forms (4–6, 10, 12). These forms consist of the core enzyme, with the subunit composition $\alpha_2\beta\beta'$, in association with various polypeptides. These polypeptides can be dissociated from the core, purified, and used in reconstitution experiments with the core to determine their functions. The results with *B. subtilis* RNA polymerase have shown that, besides its multistructural forms, the enzyme has some properties which are quite different from the enzymes found in the gram-negative organism *Escherichia coli* (1) and in the recently distinguished archaebacteria (17, 18). In this paper we summarize the properties of the RNA polymerase forms we have found in *B. subtilis*, describe several new features of the enzyme, and compare some of the properties of the *B. subtilis* and *E. coli* RNA polymerases.

RESULTS

Polypeptides associated with RNA polymerase core. The RNA polymerase purification scheme that we used has been described in great detail by Halling et al. (7). At the DNA-cellulose column step of purification, a sodium dodecyl sulfate-polyacrylamide gel electrophoresis pattern of the enzyme revealed the subunit composition of the various enzyme forms. The subunit composition was further substantiated by glycerol gradient centrifugation, which shows whether a polypeptide is tightly associated with the core (E). Figure 1 displays a typical example of the subunit composition of three vegetative RNA polymerase forms which were eluted sequentially from a DNA-cellulose column. The delta-containing core (Eδ) eluted in fractions 72 to 84; the sigma-containing core (Eσ), in fractions 82 to 102; and the epsilon-containing core (Eϵ), in fractions 94 to 102. The ϵ factor may be identical to the P37 polypeptide reported by Haldenwang and Losick (6). The Eδ form also contains ω^1 factor and the Eσ form also contains ω^1 and ω^2 factors. The Eσ, Eδ, and Eϵ forms comprise about 60 to 65%, 30 to 35%, and 3 to 5%, respectively, of the total enzyme. The σ and δ factors can be removed readily from the core (13), but the ϵ factor cannot be removed from the core by passage through a phosphocellulose column; therefore, this step can be used to obtain Eϵ and E enzymes free from σ and δ factors. Significantly, Eσ and Eϵ exhibit high enzymatic activity, whereas Eδ has little or no activity (15). The subunit composition and activity of these vegetative enzyme forms are summarized in Table 1.

Using our purification scheme (7), we found that sporulating cells contain an enzyme similar to the Eσ found in vegetative cells and, in addition, new enzyme forms consisting of core associated with small polypeptides which are not found in vegetative cells and which we originally designated as sporulation *delta* or *differentiation* factors (5). The Eσ form of the enzyme comprised about 80% of the total enzyme mass and about 60% of the total activity when assayed in vitro on poly(deoxyadenylate-deoxythymidylate). The Eσ form found in sporulating cells has a subunit composition similar to that of the Eσ found in vegetative cells, but the two sigma-

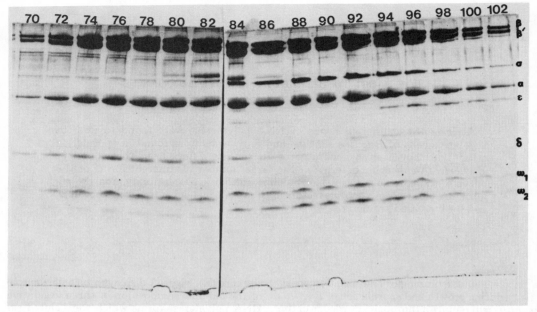

FIG. 1. *Sodium dodecyl sulfate-polyacrylamide gel electrophoretic patterns of RNA polymerase fractions from a DNA-cellulose column. See Halling et al. (7) for details of the method.*

TABLE 1. *RNA polymerase forms present in vegetative and sporulating cells of* B. subtilis

Enzyme form[a]	Mol wt of factor	Synthetic activity on poly-(deoxyadenylate-deoxy-thymidylate)
Vegetative cells		
$E\sigma$	σ 55,000	High
$E\delta$	δ 21,000	Very low
$E\epsilon$	ϵ 37,000	High
Sporulating cells		
Stage 0		
$E\sigma$	σ 55,000	High
$E\delta$	δ 21,000	Very low
Stage II		
$E\sigma$	σ 55,000	High
$E\delta^1$	δ^1 28,000	High
$E\delta^3$	δ^3 34,000	High
Stage III		
$E\sigma$	σ 55,000	High
$E\delta^1$	δ^1 28,000	High
$E\delta^2$	δ^2 20,000	High
$E\delta^3$	δ^3 34,000	High

[a] $E = \alpha_2\beta'\beta$. The molecular weight of the core subunits α, β', and β are 45,000, 130,000, and 140,000, respectively.

containing forms have different activities under identical ionic conditions, suggesting that their conformations are not identical (3).

The sporulation delta-containing enzymes were found to comprise about 20% of the mass and about 40% of the activity. The sporulation delta-containing enzymes have the following important properties: (i) they are eluted from the DNA-cellulose column at higher ionic strengths than the vegetative $E\delta$, $E\sigma$, and $E\epsilon$ forms; (ii) they have high enzymatic activity in the absence of σ factor; (iii) some of the sporulation delta factors are associated very tightly with the core and are more difficult to dissociate from the core than vegetative σ and δ factors; and (iv) the $E\delta^1$ form is more sensitive to the antibiotic Netropsin than is the vegetative $E\sigma$ enzyme. The δ^1 factor may be related to the P27 polypeptide reported by Linn et al. (10). Table 1 summarizes the properties of the forms of RNA polymerase found at different stages of sporulation, and Table 2 compares the nomenclature for the various forms of RNA polymerase as designated by Losick's and our groups.

Delta factor increases promoter selectivity of sigma-containing RNA polymerase. To understand the function of the many polypeptide factors associated with core, we examined the physical interaction of vegetative delta and sigma factors and showed that delta factor has a significantly greater affinity to the core than sigma factor (15) and that delta and sigma factors may function simultaneously or sequentially during the initiation of RNA synthesis (14, 16). Recent data have, in fact, shown that the delta factor increases the promoter selectivity of

TABLE 2. *RNA polymerase designations*

Cell form	Doi's group	Losick's group
Vegetative	$E\sigma$	$E\sigma^{55}$
	$E\delta$	$E\delta$
	$E\epsilon$	$E\sigma^{37}$
Sporulating	$E\delta^1$	$E\sigma^{29}$
	$E\delta^2$	$E + P20$
	$E\delta^3$	$E + P36$

$E\sigma$ (2). $E\sigma$ alone was found to bind to both promoter-containing and promoter-free DNA fragments of *Hind*III-restricted ϕ29 phage DNA (Fig. 2, lane 1). However, in the presence of δ factor, the $E\sigma$ enzyme formed complexes only with DNA fragments A, B, C, E, F, H, and L, which contain early ϕ29 phage promoters (Fig. 2, lane 2) (2). These results indicate that both the σ and the δ factors are required for promoter recognition specificity, in contrast to the single σ factor required for promoter specificity of the *E. coli* RNA polymerase holoenzyme (11).

Free sigma factor of *B. subtilis* RNA polymerase binds DNA. An observation by Williamson and Doi (16) indicated that sigma factor was not released during transcription by *B. subtilis* RNA polymerase, suggesting the possibility that the elongation complex consisted of $E\sigma$-DNA-RNA instead of the canonical E-DNA-RNA complex reported for *E. coli*. A further analysis of this phenomenon led to the interesting observation that free sigma factor has an affinity for DNA. In these experiments sigma factor was mixed with supercoiled plasmid pGR1-3 DNA and then analyzed for σ-DNA complex formation by use of a nondenaturing polyacrylamide gel electrophoresis system which separates free sigma from a σ-DNA complex (9). As the results in Fig. 3 indicate, sigma factor, but not delta factor, forms a complex with DNA which decreases its electrophoretic mobility. Supercoiled DNA binds more sigma than linear double- or single-stranded DNA (data not shown). At a σ/DNA mass ratio of 0.67, which is equivalent to one σ molecule per 140 base pairs of DNA, σ factor was bound completely to pGR1-3 plasmid DNA (Fig. 4). These results allow the previous data of Williamson and Doi (16) to be interpreted in the following ways: (i) the elongation complex contains $E\sigma$-DNA-RNA as originally proposed or (ii) the elongation complex contains E-DNA-RNA, and the released σ factor reassociates with DNA and therefore only appears not to be released after initiation of transcription.

Differences between *B. subtilis* and *E. coli* RNA polymerase. These and previous results with *B. subtilis* RNA polymerase indicate that

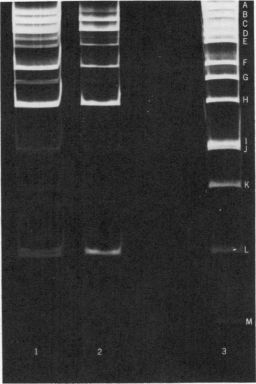

FIG. 2. *Formation of ternary initiation complexes between RNA polymerase and HindIII-restricted fragments of ϕ29 phage DNA in the presence and absence of δ factor. The reaction mixtures (100 μl) for ternary complex formation contained 10 μmol of Trishydrochloride, pH 8, 1 μmol of $MgCl_2$, 0.08 μmol of spermidine chloride, 0.1 μmol of ATP, GTP, and CTP, 10 μg of bovine serum albumin, 1.5 μl of glycerol, 0.5 pmol of HindII-restricted ϕ29 phage DNA, 50 pmol of B. subtilis RNA polymerase, and, if added, 50 pmol of δ factor (2). DNA, RNA polymerase, and δ factor were premixed on ice in 80 μl and then preincubated at 32°C for 5 min. Nucleotides were added in 20 μl to start the reaction, and the reaction was continued for 3 min. The enzyme-DNA complexes were trapped on nitrocellulose filters and treated with a 1.0 M salt wash; then the DNA was removed from the complexes and analyzed by polyacrylamide gel electrophoresis as detailed by Jones et al. (8). Lane 1, reaction without δ factor; lane 2, reaction containing δ factor; lane 3, control lane with HindIII-restricted ϕ29 phage DNA.*

some of the structural and functional properties of this enzyme are quite different from those observed with the *E. coli* enzyme. The *B. subtilis* enzyme can exist in several different forms, whereas the *E. coli* enzyme exists primarily in the single holoenzyme form. The initiation process of *B. subtilis* RNA polymerase appears to

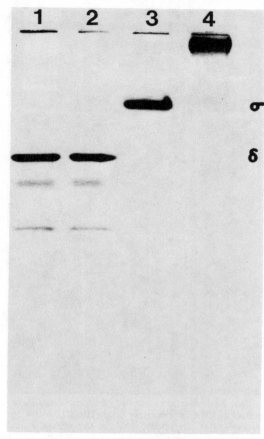

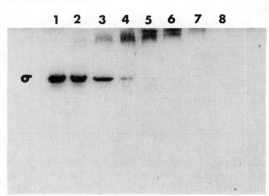

FIG. 4. Effect of increasing concentrations of pGR1-3 DNA on σ binding. The reaction mixture and assay are described in the Fig. 3 legend. Each reaction mixture contained 3 μg of σ factor. Lane 1, σ; lane 2, σ + 0.75 μg of DNA; lane 3, σ + 1.5 μg of DNA; lane 4, σ + 3 μg of DNA; lane 5, σ + 4.5 μg of DNA; lane 6, σ + 6 μg of DNA; lane 7, σ + 12 μg of DNA; lane 8, σ + 24 μg of DNA.

TABLE 3. Comparison of B. subtilis and E. coli RNA polymerases

Property	B. subtilis	E. coli
Size of β subunits	β = 140,000	β' = 160,000
	β' = 130,000	β = 150,000
Size of σ subunit	σ = 55,000	σ = 82,000
Size of α subunit	α = 45,000	α = 39,000
Size of δ subunit	δ = 21,000	None
Size of ε subunit	ε = 37,000	None
Size of ω subunits	ω^1 = 11,000	ω = 10,000
	ω^2 = 9,500	
Rifampin resistance	β	β
Streptolydigin resistance	β'	β
Specificity factors	σ, δ, ε	σ
Zinc-containing subunit	β'	β'
Sporulation factors	$\delta^1, \delta^2, \delta^3$	None

FIG. 3. Nondenaturing polyacrylamide gel electrophoresis of σ and δ factors in the presence and absence of DNA. pGR1-3 plasmid DNA (5.6 μg) was mixed with 3.2 μg of σ factor or 2 μg of δ factor in buffer A, which contained 50 mM Tris-hydrochloride, pH 7.9, 0.1 mM EDTA, 0.1 mM dithiothreitol, 100 mM KCl, 15% glycerol, 10 mM MgCl₂, and 5% sucrose. The reaction mixtures (usually 50 to 100 μl) were incubated for 5 min at 37°C and then immediately analyzed by nondenaturing polyacrylamide gel electrophoresis as described in detail by Williamson and Doi (16). Lane 1, only free δ factor; lane 2, δ factor + pGR1-3 DNA; lane 3, only free σ factor; lane 4, σ factor + pGR1-3 DNA.

require the activity of at least two protein factors, σ and δ factors, whereas the E. coli enzyme has only a single polypeptide initiation factor, σ. Interestingly, the molecular weights of B. subtilis σ and δ factors add up to 76,000, whereas the molecular weight of the E. coli σ factor is 82,000 (1). This suggests the possibility that the larger E. coli σ factor is an evolutionary product of gene fusion. The properties of these two transcription systems are compared in Table 3.

Is the relative complexity of the RNA polymerases related to the fact that B. subtilis cells can differentiate into spores? Or is the RNA polymerase from gram-positive organisms generally more complex than that found in gram-negative organisms? Further studies are required to answer these questions.

ACKNOWLEDGMENTS

This research was supported in part by National Science Foundation grant PCM 7924872 and by Public Health Service grant GM 19673 from the National Institute of General Medical Sciences.

We thank David Goldfarb for pGR1-3 DNA and Deborah Jaffe and Mary Sharp-Hayes for technical assistance.

LITERATURE CITED

1. **Burgess, R. R.** 1976. Purification and physical properties of *E. coli* RNA polymerase, p. 69–100. *In* R. Losick and M. Chamberlin (ed.), RNA polymerase. Cold Spring Harbor Laboratory, Cold Spring Harbor, N.Y.
2. **Dickel, C. D., K. C. Burtis, and R. H. Doi.** 1980. Delta factor increases promoter selectivity by *Bacillus subtilis* vegetative cell RNA polymerase. Biochem. Biophys. Res. Commun. **95:**1789–1795.
3. **Dooley, M. M., S. M. Halling, and R. H. Doi.** 1980. Template-independent poly(A)·poly(U) synthesizing activity of different forms of *Bacillus subtilis* RNA polymerase. Biochim. Biophys. Acta **610:**158–166.
4. **Fukuda, R., and R. H. Doi.** 1977. Two polypeptides associated with ribonucleic acid polymerase core of *Bacillus subtilis* during sporulation. J. Bacteriol. **129:** 422–432.
5. **Fukuda, R., G. Keilman, E. McVey, and R. H. Doi.** 1975. Ribonucleic acid polymerase pattern of sporulating *Bacillus subtilis*, p. 213–220. *In* P. Gerhardt, R. N. Costilow, and H. L. Sadoff (ed.), Spores VI. American Society for Microbiology, Washington, D.C.
6. **Haldenwang, W. G., and R. Losick.** 1979. A modified RNA polymerase transcribes a cloned gene under sporulation control in *Bacillus subtilis*. Nature (London) **282:**256–260.
7. **Halling, S. M., K. C. Burtis, and R. H. Doi.** 1977. Reconstitution studies show that rifampicin resistance is determined by the largest polypeptide of *Bacillus subtilis* RNA polymerase. J. Biol. Chem. **252:**9024–9031.
8. **Jones, B. B., H. Chan, S. Rothstein, R. D. Wells, and W. S. Reznikoff.** 1977. RNA polymerase binding sites in λplac5 DNA. Proc. Natl. Acad. Sci. U.S.A. **74:**4914–4918.
9. **Kudo, T., D. Jaffe, and R. H. Doi.** 1981. Free sigma subunit of *Bacillus subtilis* RNA polymerase binds to DNA. Mol. Gen. Genet. **181:**63–68.
10. **Linn, T., A. L. Greenleaf, and R. Losick.** 1975. RNA polymerase from sporulating *Bacillus subtilis*. Purification and properties of a modified form of enzyme containing two sporulation polypeptides. J. Biol. Chem. **250:**9256–9261.
11. **Lowe, P. A., D. A. Hager, and R. R. Burgess.** 1979. Purification and properties of the σ subunit of *Escherichia coli* DNA-dependent RNA polymerase. Biochemistry **18:**1344–1352.
12. **Nakayama, T., V. Williamson, K. Burtis, and R. H. Doi.** 1978. Purification and properties of two RNA polymerases from sporulating cells of *Bacillus subtilis*. Eur. J. Biochem. **88:**155–164.
13. **Pero, J., J. Nelson, and T. D. Fox.** 1975. Highly asymmetric transcription by RNA polymerase containing phage SPO1-induced polypeptides and a new host protein. Proc. Natl. Acad. Sci. U.S.A. **72:**1589–1593.
14. **Spiegelman, G. B., and H. R. Whiteley.** 1979. Subunit composition of *Bacillus subtilis* RNA polymerase during transcription. Biochem. Biophys. Res. Commun. **87:** 811–817.
15. **Williamson, V. M., and R. H. Doi.** 1978. Delta factor can displace sigma factor from *Bacillus subtilis* RNA polymerase holoenzyme and regulate its initiation activity. Mol. Gen. Genet. **161:**135–141.
16. **Williamson, V. M., and R. H. Doi.** 1979. Sigma factor is not released during transcription in *Bacillus subtilis*. Mol. Gen. Genet. **174:**47–52.
17. **Zillig, W., K. Stetter, and D. Janekovic.** 1979. DNA-dependent RNA polymerase from the Archaebacterium *Sulfolobus acidocaldarius*. Eur. J. Biochem. **96:**597–604.
18. **Zillig, W., K. O. Stetter, and M. Tobein.** 1978. DNA-dependent RNA polymerase from *Halobacterium halobium*. Eur. J. Biochem. **91:**193–199.

D. Dormancy, Germination, and Outgrowth

Alterations in *Bacillus megaterium* QM B1551 Spore Membranes with Acetic Anhydride and L-Proline

F. M. RACINE, J. F. SKOMURSKI, AND J. C. VARY

Department of Biochemistry, University of Illinois Medical Center, Chicago, Illinois 60612

Studies on L-proline–triggered germination in *Bacillus megaterium* QM B1551 spores have implicated a size class of membrane proteins (band 18 as defined by electrophoresis) by labeling with acetic anhydride. By fluorescence depolarization studies, L-proline, but not D-proline, was seen to cause an in vitro biophysical change in isolated spore membranes. However, isolated spore membranes, first acetylated to inhibit L-proline triggering, did not respond to L-proline. Therefore, L-proline–triggered germination may be a membrane-associated event controlled by membrane proteins.

Heat-activated spores of *Bacillus megaterium* QM B1551, exposed to L-proline or D-glucose, rapidly trigger germination, apparently not requiring metabolism of either proline (14) or glucose (11), and we have proposed that triggering is a membrane-associated event (21). We have reported that a proline affinity analog, proline chloromethyl ketone, triggered germination and covalently labeled spores (15). When techniques to purify spore inner membranes were used (12), [^{3}H]proline chloromethyl ketone labeled proteins in the spore inner membrane (D. P. Rossignol and J. C. Vary, Fed. Proc. **39**:1614, 1980). In addition, acetic anhydride (Ac$_2$O) specifically blocked proline-triggered germination, and [^{3}H]Ac$_2$O labeled spore inner membrane proteins (Rossignol and Vary, Fed. Proc. **39**:1614, 1980).

We report here additional studies on the membrane proteins that are labeled with [^{3}H]Ac$_2$O and the effect of acetylation on the properties of the inner membrane as measured by fluorescence depolarization with 1,6-diphenyl-1,3,5-hexatriene (DPH).

RESULTS AND DISCUSSION

Spores of *B. megaterium* QM B1551 were prepared and extracted with sodium dodecyl sulfate-dithiothreitol (SDS-DTT) at pH 10 as previously described (16, 20). References to spore weights are on a dry weight basis. Spore

inner membranes were prepared from SDS-DTT–extracted spores by method A (12).

Ac$_2$O treatment. It was previously shown that treatment of spores with Ac$_2$O inhibited proline- but not glucose-triggered germination, and treatment with 1 M NH$_2$OH at pH 9, but not at pH 7, reversed the inhibition (15). With SDS-DTT–extracted spores, different concentrations of Ac$_2$O were tested. The degree of Ac$_2$O inhibition depended on which parameter of germination was measured (Table 1). Based on absorbance loss, 0.03% Ac$_2$O caused 90% inhibition, but dipicolinic acid (DPA) loss was inhibited only 23%. However, 0.1% Ac$_2$O inhibited DPA loss by almost 90%. Loss in heat resistance was also blocked by at least 90% with 0.1% Ac$_2$O (data not shown).

Reversal of the inhibition caused by Ac$_2$O required 0.1 M NH$_2$OH at pH 9; at pH 7 there was no reversal of inhibition (Table 1). Of particular importance was the differential effect of NH$_2$OH at pH 7 and pH 9. NH$_2$OH at pH 9, but not at pH 7, specifically reversed Ac$_2$O inhibition as measured by DPA release. Considering the sequence of events defined for this spore strain (3, 8), we think that these results are probably the result of Ac$_2$O blocking not only triggering of germination as judged by DPA and heat resistance loss but also a later event involved with absorbance loss. It is interesting to recall that mutants with altered spore coats (1, 18) or altered proteases (2) in *B. cereus* were somewhat

TABLE 1. *Effect of Ac₂O and NH₂OH on SDS-DTT-extracted spores*

Ac₂O (%)[a]	NH₂OH[b] (M)	Percent inhibition[c]	
		Absorbance loss	DPA loss
0	0	0	0
0.01	0	30	5
0.03	0	90	23
0.10	0	100	87
0.1	0, pH 9	96	100
0.1	0.01, pH 9	96	37
0.1	0.10, pH 9	80	0
0.1	0.10, pH 7	100	100

[a] Acetylation of SDS-DTT–extracted spores was by a modification of the technique for nonextracted spores (D. P. Rossignol and J. C. Vary, submitted for publication). The SDS-DTT–extracted spores were heat activated, centrifuged at $5,000 \times g$ for 10 min, and then suspended in 10 mM sodium borate buffer (pH 9) at 5 mg/ml. Acetic anhydride was added at the indicated concentrations (vol/vol) at 25°C, and the pH was maintained between 8.8 and 9.0 by the addition of 1 M NaOH. After 10 min, the spores were centrifuged and washed with water.

[b] Deacetylation with NH₂OH was for 1 h at 0°C at either pH 7 or 9, followed by washing twice with water.

[c] Percent inhibition was calculated from the untreated control, which showed a 60% decrease in absorbance at 660 nm and a loss of 100 μg of DPA per mg of spores within 30 min at 30°C. Percent decrease in absorbance and DPA losses were measured as previously described (3).

defective in germination if judged by absorbance loss, a late event. Certain chemicals such as HgCl₂ (13) have also been reported to separate the early reactions of DPA loss from later ones of absorbance loss.

Labeling of spores with Ac₂O. SDS-DTT–extracted spores were labeled with [³H]Ac₂O, and the suspension was divided and treated with 0.1 M NH₂OH at pH 7 for one half and at pH 9 for the other. The spores were disrupted and membranes were prepared by method A (12). By SDS-polyacrylamide gel electrophoresis (Fig. 1) a peak corresponding to the position of membrane band 18 (12) was found to contain more radioactivity from the samples treated at pH 7 than from those treated at pH 9. Thus, it appears that the differential labeling by [³H]-Ac₂O and NH₂OH that might account for the in vivo differential effect of Ac₂O and NH₂OH on triggering is in band 18 of the spore inner membrane. Support for this possibility comes from the labeling of this same band by [³H]proline chloromethyl ketone (C. Ugolini, unpublished data).

Fluorescence   depolarization   studies. Spore inner membranes were isolated from heat-activated spores, purified, and labeled with DPH. Anisotropy values were then determined, and the results were plotted by the method of Shinitzky and Barenholz (17). Because of current contradictions on the interpretations of thermotropic transition temperatures (M. Glaser, personal communication), no attempt has been made to emphasize breaks in Arrhenius

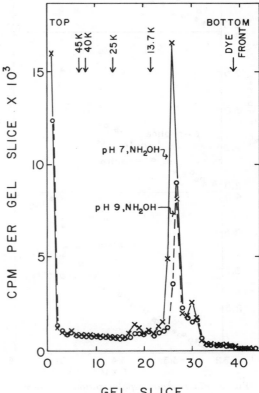

FIG. 1. *[³H]Ac₂O-labeled components in the membrane fractions. A portion corresponding to about 150 μg of protein of each membrane preparation, as in Table 1 and references 12 and 15, was boiled in 1% SDS and then analyzed by SDS-polyacrylamide gel electrophoresis (7) with 15% acrylamide and 0.98% cross-linker. Gels were stained with Coomassie blue or cut into 2-mm slices and digested in 0.4 ml of a 9:1 (vol/vol) mixture of NCS solubilizer (Amersham Corp.) and water at 50°C for 2 h. After the addition of 4.2 ml of toluene-based scintillation fluid (16), the samples were counted in a Beckman scintillation counter. Published methods (9) were used for protein determination. The arrows indicate the positions of molecular-weight standards (45K = 45,000).*

APPARENT MICROVISCOSITY OF SPORE MEMBRANES

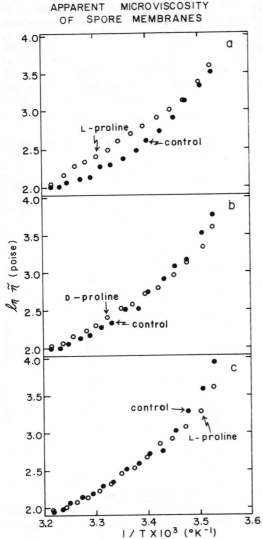

plots, which is why lines were not drawn between the points in Fig. 2. There was a decrease in apparent microviscosity with increasing temperatures (Fig. 2a). This is interpreted to mean that DPH, which is located within the hydrophobic portion of the lipid bilayer (19), reports an increased mobility with increased temperatures. When L-proline was added, there was a shift up in this curve, showing that L-proline altered the properties of these membranes so that DPH reported less freedom of movement. Apparent microviscosity changes of the same order of magnitude as shown in Fig. 2a have been reported for liver plasma cell membranes after the binding of insulin (10). Interpretation of these types of data as a decrease in membrane "fluidity" should be taken with a great deal of caution so as not to promulgate any misconceptions by comparing complex biological membranes and model membrane systems. The significant point is that L-proline altered the environment around DPH in the membranes, and this alteration occurred in the temperature range (14 to 38°C) where L-proline triggers germination in vivo (14). However, in the presence of D-proline, which is not a trigger reagent, a much smaller change in the Arrhenius plot occurred than with L-proline (Fig. 2b). It is not certain why D-proline should have any effect, but similar small changes caused by analogs relative to controls have also been observed by DPH fluorescence using insulin analogs on liver cell membranes (10) and by com-

FIG. 2. *Fluorescence depolarization of spore membranes. Fluorescence depolarization was measured with a Perkin-Elmer fluorescence spectrophotometer model MPF-44B equipped with polarization accessories and a temperature-controlled cuvette holder. DPH was dissolved in CHCl₃, and the CHCl₃ was evaporated under a stream of N₂ in a conical 10-ml test tube. Inner membranes were isolated from heat-activated spores as previously described (12). To 20 nmol of dried DPH, 0.6 to 1.0 ml of purified membrane (10 to 12 mg of protein) was added, and the mixture was gently sonicated at room temperature for 15 min as previously described (6). The sample was then diluted to 25 ml with 10 mM N-2-hydroxethylpiperazine-N'-2-ethanesulfonic acid (HEPES) buffer (pH 7.5), and the membranes were sedimented at 304,000 × g for 105 min to remove unincorporated DPH. The membranes were resuspended in 10 mM HEPES (pH*

7.5) at a protein concentration of 15 to 20 mg/ml and stored on ice. Prior to each run, membranes were diluted into the above buffer to give a final concentration of 0.5 mg/ml, and any particulate matter was removed by centrifugation at 5,000 × g for 5 min. Equal volumes were placed in matched quartz cuvettes and equilibrated to about 10°C. Light scattering was less than 1 to 2%, and therefore it was not necessary to correct routinely for scattering. The excitation and emission wavelengths were 360 nm and 430 nm, respectively, with a 390-nm emission cutoff filter. Corrected spectra were measured with a Perkin-Elmer DCSU-2 microprocessor, and then anisotropy values were determined by published methods (22) in duplicate or triplicate for each temperature as the temperature was varied in 2 to 3°C increments with 12 min of equilibration between each temperature. When additions were made to one cuvette, the same volume of buffer was added to the control cuvette to correct for dilution. The data were analyzed by calculating apparent microviscosities from anisotropy values, r, using r₀ = 0.362, by the method of Shinitzky and Barenholz (17). Symbols are (a) membranes with (○) and without (●) L-proline; (b) membranes with (○) and without (●) D-proline; (c) membranes from heat-activated spores treated in vivo with Ac₂O before isolation and purification run with (○) and without (●) L-proline.

paring the effects of D- and L-alanine on membrane conductance in pancreatic cells (4). Also, a similar effect of D-proline, compared with L-proline, was observed with spore membranes in experiments using electron spin resonance spectroscopy (5). We doubt that the results could be explained by contaminating L-proline, since the same solution of D-proline had no effect on triggering of germination in vivo. Also, it should be noted that these changes are a property of the intact membranes, since the isolated phospholipids from these membranes did not respond to L-proline (data not shown).

Ac$_2$O inhibited proline-triggered germination, and [^{3}H]Ac$_2$O differentially labeled band 18 of the spore inner membrane. If acetylation exerted its effect by altering membrane components, such an alteration might be reflected by DPH. To test this possibility, we acetylated heat-activated spores and prepared inner membranes (legend to Fig. 1). These membranes exhibited similar Arrhenius plots as membranes from nonacetylated spores, with some possible slight deviations at low temperatures. However, in the presence of L-proline (or D-proline, data not shown), no change in the Arrhenius plots was observed (Fig. 2c). Therefore, there is a direct correlation between the in vivo properties of acetylated spores and the in vitro properties of purified spore inner membranes with respect to the effect of L-proline. Our next experiment will center around a similar analysis of membranes from acetylated spores that have been treated with NH$_2$OH at pH 7 or pH 9.

ACKNOWLEDGMENTS

We thank M. Glaser, D. Soucek, and J. A. F. Op den Kamp for helpful discussions and Kathy Joseph for technical assistance.

This work was supported by Public Health Service grant AI 12678 from the National Institutes of Health.

LITERATURE CITED

1. **Aronson, A. I., and P. C. Fitz-James.** 1975. Properties of *Bacillus cereus* spore coat mutants. J. Bacteriol. **123:**354–365.
2. **Cheng, Y. E., P. C. Fitz-James, and A. I. Aronson.** 1978. Characterization of *Bacillus cereus* protease mutant defective in an early stage of spore germination. J. Bacteriol. **133:**336–344.
3. **Hsieh, L. K., and J. C. Vary.** 1975. Peptidoglycan hydrolysis during initiation of spore germination in *Bacillus megaterium*, p. 465–471. *In* P. Gerhardt, H. L. Sadoff, and R. N. Costilow (ed.), Spores VI. American Society for Microbiology, Washington, D.C.
4. **Iwatsuki, N., and O. H. Peterson.** 1980. Amino acids evoke short-latency membrane conductance increase in pancreatic acinar cell. Nature (London) **283:**492–494.
5. **Janoff, A. S., R. T. Coughlin, F. M. Racine, E. J. McGroarty, and J. C. Vary.** 1979. Use of electron spin resonance to study *Bacillus megaterium* spore membranes. Biochem. Biophys. Res. Commun. **89:**565–570.
6. **Janoff, A. S., A. Haug, and E. J. McGroarty.** 1979. Relationship of growth temperature and thermotropic lipid phase changes in cytoplasmic and outer membranes from *Escherichia coli*. Biochim. Biophys. Acta **555:**56–66.
7. **Laemmli, V. K.** 1970. Cleavage of structural proteins during assembly of the head of bacteriophage T4. Nature (London) **227:**680–685.
8. **Levinson, H. S., and M. T. Hyatt.** 1966. Sequence of events during *Bacillus megaterium* spore germination. J. Bacteriol. **91:**1811–1818.
9. **Lowry, O. H., N. J. Rosebrough, A. L. Farr, and R. J. Randall.** 1951. Protein measurement with the Folin phenol reagent. J. Biol. Chem. **193:**265–275.
10. **Luly, P., and M. Shinitzky.** 1979. Gross structural changes in isolated liver cell plasma membranes upon binding of insulin. Biochemistry **18:**445–450.
11. **Racine, F. M., S. S. Dills, and J. C. Vary.** 1979. Glucose uptake during triggering of germination of *Bacillus megaterium* spores. J. Bacteriol. **138:**442–445.
12. **Racine, F. M., and J. C. Vary.** 1980. Isolation and properties of membranes from *Bacillus megaterium* spores. J. Bacteriol. **143:**1208–1214.
13. **Rossignol, D., and J. C. Vary.** 1977. A unique method for studying the initial reactions of *Bacillus megaterium* spore germination. Biochem. Biophys. Res. Commun. **79:**1098–1103.
14. **Rossignol, D., and J. C. Vary.** 1979. Biochemistry of L-proline-triggered germination in spores of *Bacillus megaterium*. J. Bacteriol. **138:**431–441.
15. **Rossignol, D. P., and J. C. Vary.** 1979. L-proline site for triggering *Bacillus megaterium* spore germination. Biochem. Biophys. Res. Commun. **89:**547–551.
16. **Shay, L. K., and J. C. Vary.** 1978. Biochemical studies on glucose initiated germination in *Bacillus megaterium*. Biochim. Biophys. Acta **538:**284–292.
17. **Shinitzky, M., and Y. Barenholz.** 1978. Fluidity parameters of lipid regions determined by fluorescence polarization. Biochim. Biophys. Acta **515:**367–394.
18. **Stelma, G. N., A. I. Aronson, and P. C. Fitz-James.** Properties of *Bacillus cereus* temperature-sensitive mutants altered in spore coat formation. J. Bacteriol. **134:**1157–1170.
19. **Stubbs, G. W., B. J. Litman, and Y. Barenholz.** 1976. Microviscosity of the hydrocarbon region of the bovine retinal rod outer segment disk membrane determined by fluorescent probe measurements. Biochemistry **15:**2766–2772.
20. **Vary, J. C.** 1973. Germination of *Bacillus megaterium* spores after various extraction procedures. J. Bacteriol. **116:**797–802.
21. **Vary, J. C.** 1978. Glucose-initiated germination in *Bacillus megaterium* spores, p. 104–108. *In* G. Chambliss and J. C. Vary (ed.), Spores VII. American Society for Microbiology, Washington, D.C.
22. **Weidekamm, E., C. Schudt, and D. Brdiczka.** 1976. Physical properties of muscle cell membranes during fusion. Biochim. Biophys. Acta **443:**169–180.

Reversion of Phase-Dark Germinated Spores of *Clostridium perfringens* Type A to Refractility

HILLEL S. LEVINSON AND FLORENCE E. FEEHERRY

Science and Advanced Technology Laboratory, U.S. Army Natick Research and Development Laboratories, Natick, Massachusetts 01760

The increase in optical density of suspensions of germinated spores of *Clostridium perfringens* NCTC 8238 and *C. perfringens* NCTC 8798, aerobically incubated in an outgrowth medium, was not attributable to growth; the phase-dark germinated spores had become phase-bright refractile bodies, similar in appearance to resting spores. Optical density increase and refractile body formation under aerobic conditions did not occur in buffer without the organic constituents of the outgrowth medium. At high pH levels, a lower temperature was required for optical density increase and for refractile body formation than at lower pH levels.

Microcycle sporogenesis (1, 3, 6) and other types of cytodifferentiation, as described by Vinter and co-workers (5, 7), involve the production of sporelike refractile bodies from germinated spores in the absence of an intervening stage of growth and multiplication. The present paper is, we believe, the first report of this phenomenon with germinated spores of an anaerobic species. We emphasize that these are preliminary results which we believe are important and unusual enough to be called to the attention of spore scientists; their interpretation and significance remain to be assessed. We shall document rather simple conditions for the reversion of phase-dark germinated spores to refractile bodies, and we hope that others will join us in pursuing further research on this fascinating phenomenon.

As part of a study of nutritional and other requirements for outgrowth, spores of two strains of *Clostridium perfringens* type A (NCTC 8238 and 8798) were produced as described previously (2). The spores were germinated in 10 mM $KH_2-K_2HPO_4$ buffer, pH 6.5, at 35°C for 1 h, by which time suspensions had lost 40 to 50% of their initial optical density (OD) of ca. 1.0, and ca. 95% of the spores had lost refractility, heat resistance, and dipicolinic acid and had become dark under phase-contrast optics (Fig. 1). Similar results were obtained with both strains, although only data obtained with strain NCTC 8238 are presented here. The germinated spores (Fig. 1B) were incubated in an outgrowth medium at a final ratio of germinated spore suspension to outgrowth medium of 1:4. The initial OD of the germinated spore suspension

was 0.125 to 0.150. The medium contained (final concentration): 1.0% salt-free casein hydrolysate (Hy-Case SF, Humko-Sheffield), 0.25% yeast extract (Difco), 25 mM glucose, 25 mM NH_4NO_3, 1.0 mM K_2SO_4, and 50 mM $KH_2-K_2HPO_4$ buffer. Incubation was at pH levels from 5.4 to 7.45 at temperatures ranging from 30 to 60°C. Parallel anaerobic and aerobic experiments were run, the latter serving as controls. Outgrowth was followed by change in OD as measured with a Spectronic 21 (Bausch & Lomb) spectrophotometer at 560 nm.

Results of the anaerobic experiments will be reported elsewhere. Briefly, outgrowth of germinated spores of *C. perfringens* was optimal at pH 6.75 and, at the optimal pH, was maximal at 45°C. The aerobic controls, however, offer the more interesting results, and it is on these that the remainder of this brief report will focus.

When the aerobic controls accompanying anaerobic cultures were incubated in outgrowth medium at pH 7.35 or 7.45, OD increased markedly, especially at the higher temperatures (Fig. 2). The minimum temperature at which the increase in OD occurred was lower at pH 7.45 (35°C) than at pH 7.35 (40°C). The increasing temperature requirement with decreasing pH was also evident at pH 7.1 (Fig. 3), where the OD of aerobically incubated suspensions did not increase unless the temperature was 50°C or higher, and at pH 6.75 (Fig. 3), where the minimum temperature for a detectable OD increase was ca. 60°C (no OD increase after 5 h of incubation at pH 6.75, 50°C).

C. perfringens is an anaerobe, and increasing

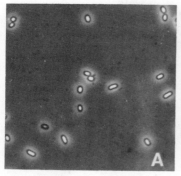
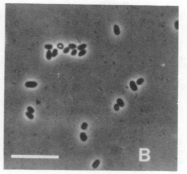

FIG. 1. *Phase-contrast photomicrographs of heat-activated (A) and germinated (B) spores of* C. perfringens *NCTC 8238. Spores, activated at 65°C for 15 min, were germinated for 1 h at 35°C in 10 mM KH₂-K₂HPO₄, pH 6.5. Preparations of germinated spores, similar to those in (B), were used as inoculum in subsequent experiments. Marker represents 10 μm.*

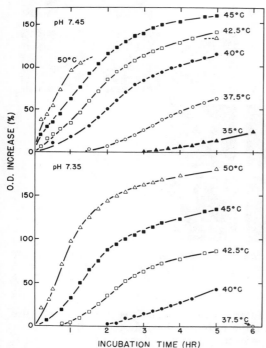

FIG. 2. *Increase in OD of suspensions of germinated spores of* C. perfringens *NCTC 8238 incubated aerobically at various temperatures in outgrowth medium at pH 7.45 and pH 7.35. Percentage increase in OD was calculated as* $(OD_t - OD_i)/OD_i \times 100$, *where* OD_i *and* OD_t *were the initial OD and OD after* t *min, respectively.*

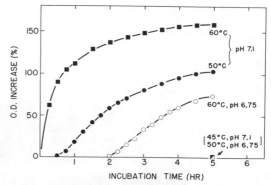

FIG. 3. *Increase in OD of suspensions of germinated spores of* C. perfringens *NCTC 8238, incubated aerobically at 50 and 60°C in outgrowth medium at pH 6.75 and 7.1. Percentage increase in OD was calculated as in Fig. 2.*

OD under aerobic conditions was unexpected. Microscopic observations under phase-contrast optics convinced us that the increase in OD of aerobically incubated suspensions was not attributable to growth, but rather to transformation of some of the phase-dark germinated spores (similar to Fig. 1B), constituting the inoculum, into phase-bright refractile bodies (Fig. 4), at least superficially resembling resting spores (similar to those in Fig. 1A). As the pH of aerobic incubation increased, the temperature requirement for the transformation from phase-dark to phase-bright (sporelike) bodies decreased. At pH 5.7 or lower, the transformation to phase-bright bodies did not occur even at 60°C; at pH 7.45, refractile bodies were seen after aerobic incubation at temperatures ranging upward from 37.5°C; the minimal temperatures for refractile body formation appeared to be ca. 40, 50, and 60°C at pH 7.35, 7.1, and 6.75, respectively. This appearance of refractile bodies was reflected as an increasing OD (Fig. 2 and 3) during aerobic incubation in outgrowth medium. Refractile bodies were also formed under anaerobic conditions at suitable pH and temperature levels, but their presence was often obscured by the proliferation of vegetative cells. Under anaerobic conditions, growth, with production of

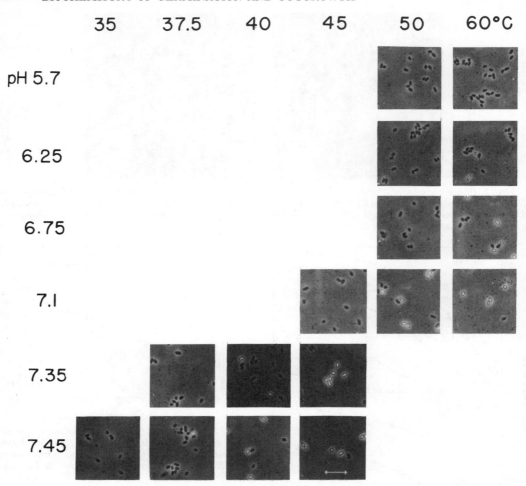

FIG. 4. *Formation of refractile bodies from phase-dark germinated spores of* C. perfringens *NCTC 8238, incubated aerobically in outgrowth medium at various temperatures and pH levels. Incubation was for 5 h, except for the 50°C preparations at pH 5.7, 6.25, and 6.75, which were incubated for 7.5, 8.0, and 7.75 h, respectively. Marker represents 10 μm.*

numerous vegetative cells, was evident at temperatures up to 45°C throughout the experimental pH range. Indeed, Smith and Holdeman (4) indicated that *C. perfringens* grows readily at temperatures from 20 to 50°C and at pH levels from 5.5 to 8.0. We have seen some growth at 50°C at pH 6.25 and a lesser amount at pH 6.75, but not at lower or higher pH levels; no growth was observed, at any pH, when the incubation temperature was 60°C or higher.

The OD increase and acquisition of refractility under conditions precluding growth, i.e., aerobic conditions, may have a nutritional basis, since they do not occur in phosphate buffer in the absence of the other (including organic) constituents of the outgrowth medium. The relationship of this phenomenon to other types of cyto-

differentiation (1, 3, 5–7) is, at present, obscure. In the system we have described, the conditions permitting germinated spores of the anaerobe *C. perfringens* to regain refractility without intervening growth and multiplication are notably simple, requiring only manipulation of pH and temperature together with a medium suitable for outgrowth. This system, unlike microcycle sporogenesis, does not appear to be a variant of normal *endo*sporulation. For production of refractile bodies, free from vegetative cells, prevention of outgrowth was desirable and was achieved by maintenance of aerobic conditions.

We make no pretense that this constitutes a completed study. We plan a detailed investigation of the refractile bodies, of the conditions necessary for their formation, of their properties of dormancy and resistance, and of their capability to germinate and grow. If these refractile, sporelike bodies are indeed hypometabolic, resistant forms with a normal complement of dipicolinic acid and capable of germination, this may be of the utmost fundamental and practical significance.

LITERATURE CITED

1. **Freer, J. H., and H. S. Levinson.** 1967. Fine structure of *Bacillus megaterium* during microcycle sporogenesis. J. Bacteriol. **94:**441–457.
2. **Levinson, H. S., and F. E. Feeherry.** 1978. Kinetics of heat activation of spores of *Clostridium perfringens*, p. 34–40. *In* G. Chambliss and J. C. Vary (ed.), Spores VII. American Society for Microbiology, Washington, D.C.
3. **MacKechnie, I., and R. S. Hanson.** 1968. Microcycle sporogenesis of *Bacillus cereus* in a chemically defined medium. J. Bacteriol. **95:**355–359.
4. **Smith, L. DS., and L. V. Holdeman.** 1968. The pathogenic anaerobic bacteria. Charles C Thomas, Publisher, Springfield, Ill.
5. **Vinter, V., J. Chaloupka, J. Šťastná, and J. Čáslavská.** 1972. Possibilities of cellular differentiation of bacilli into different hypometabolic forms, p. 390–397. *In* H. O. Halvorson, R. Hanson, and L. L. Campbell (ed.), Spores V. American Society for Microbiology, Washington D.C.
6. **Vinter, V., and R. A. Slepecky.** 1965. Direct transition of outgrowing bacterial spores to new sporangia without intermediate cell division. J. Bacteriol. **90:**803–807.
7. **Vinter, V., and J. Šťastná.** 1967. Spores of microorganisms. XXI. Conversion of outgrowing spores of *Bacillus cereus* to refractile forms by basic peptides and proteins. Folia Microbiol. (Prague) **12:**301–307.

Rate and Lag of KNO$_3$-Induced Germination of Spores of the Food-Poisoning Anaerobe *Clostridium perfringens* Type A, NCTC 8238

HILLEL S. LEVINSON AND FLORENCE E. FEEHERRY

Science and Advanced Technology Laboratory, U.S. Army Natick Research and Development Laboratories, Natick, Massachusetts 01760

Temperature, pH, KNO$_3$ concentration, sodium phosphate buffer concentration and spore concentration affected the germination rate and germination lag of heat-activated spores of *Clostridium perfringens* type A, NCTC 8238. Optimal conditions for germination rate were pH 6.1 to 6.3, 32.5 to 35°C, 0.325 mg of spores per ml, 100 mM KNO$_3$, and 2 to 5 mM sodium phosphate. Both lag and rate of KNO$_3$-induced germination were the same under anaerobic or aerobic conditions.

The increasing recognition of the significance of the anaerobic sporeformer *Clostridium perfringens* type A and of its sporulation-associated enterotoxin (6) in food-poisoning outbreaks, particularly in institutional settings (4), has done much to stimulate research on the physiology of the spores of this organism.

Spores of certain food-poisoning strains of *C. perfringens* (1, 2), as well as of *Bacillus megaterium* QM B1551 (10, 14), are able to germinate in inorganic solutions, without organic adjuvants, K$^+$ being much more effective than Na$^+$ in supporting ionic germination (2, 10). In this paper, which constitutes part of our continuing program (11) to establish a data base on the activation, germination, and outgrowth of *C. perfringens* spores, comparable to that on *Bacillus*, we report on our investigations of optimal conditions for the KNO$_3$-induced germination of *C. perfringens* spores.

RESULTS AND DISCUSSION

In the following series of experiments, special attention was paid to a comparison of germination rate and germination lag under various conditions, these parameters of germination being determined as in Fig. 1.

pH. Germination rate increased from near undetectable at pH 5.4 to a peak of ca. 1.25 ± 0.2% optical density (OD) loss per min at pH 6.1 to 6.3, and then declined at higher pH (Fig. 2). Germination lag was ca. 13 min under standard conditions (pH 6.1), 27 min at pH 5.4, and 9.0 min at pH 6.5 to 7.5 (Fig. 2).

The extent of early germination, i.e., within the first 30 min (data not shown), was dependent on both rate and lag and had a pH optimum at ca. pH 6.2. However, spores continued germinating slowly and spore suspensions continued decreasing in OD, even under less than optimal conditions. Even spore suspensions with slow or delayed germination (at pH 5.7, for example) eventually attained an extent of germination equal to that achieved under the best conditions. This resulted in a broadening of the apparent optimal pH range, so that by 5 h, germination at pH levels from 5.7 to 6.3 had attained equivalent values, i.e., near 50% loss in OD. This pH range (5.7 to 6.3) is approximately that of meat and poultry, important vehicles (3–5) of *C. perfringens* food poisoning.

Temperature. Spores were incubated at various temperatures in the standard KNO$_3$ germination mixture (Fig. 1 legend) and, in some cases, were shifted after 1 h to a secondary incubation at 30°C (Table 1). The rate and extent (KNO$_3$-induced percent OD loss at 2 h) of germination appeared to be optimal at 32.5 to 35°C. Germination lag showed no clearly defined temperature optimum. Rather, with increasing temperature, there was a progressive decrease in the lag from >45 min at 15°C to <2 min at 45°C (Table 1). At the temperature where germination rate was maximal, the lag was ca. 8 to 10 min. The energy of activation (μ) for germination, calculated from an Arrhenius plot of reciprocal lags between 20 and 35°C was ca. 17.5 kcal (7.3×10^5 J), a value close to that for germination of *B. megaterium* spores in KNO$_3$ (10) or in glucose (13) and of a magnitude consistent with an enzymatic basis for KNO$_3$-induced germination of *C. perfringens* spores.

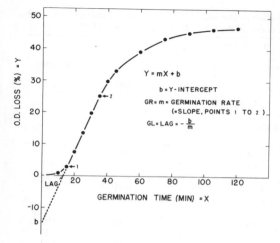

FIG. 1. *KNO₃-induced germination of heat-activated (65°C, 10 min) spores of* C. perfringens *NCTC 8238 under standard conditions.* C. perfringens *type A, NCTC 8238, used throughout these investigations, was isolated by Hobbs et al. (7) from salt beef and is classified as serotype 2 (8). Spore production, heat activation, glassware preparation, and chemicals were as previously described (11). Standard conditions for germination were 5 mM KNO_3, 10 mM sodium phosphate buffer, pH 6.1, 0.325 mg of spores per ml (OD, ca. 1.0; 1.1×10^8 spores per ml), and 30°C. No attempt was made to maintain anaerobiosis. Sodium phosphate was used because spores germinated readily in potassium phosphate buffer. OD (560 nm; disposable 13 by 100 mm Pyrex glass tubes; Bausch & Lomb Spectronic 21 colorimeter) was followed during incubation (11, 12). Each experiment examining a variable (pH, temperature, and spore, KNO_3, and buffer concentrations) was accompanied by two controls: one in the appropriate sodium phosphate buffer (without KNO_3) and the other under standard conditions. KNO_3-induced OD loss was corrected by subtracting OD loss in the buffer control. OD loss in buffer controls (10 mM sodium phosphate, pH 6.1 to 6.3; no KNO_3), at various temperatures, is shown in Table 1. The slope (m) of the least-squares regression line between points 1 and 2 (the rectilinear portion of the plot) represents germination rate (GR) and is expressed as percent OD lost per minute. The intercept of the regression line with the x-axis (no OD loss) represents germination lag (GL), which can be calculated from the regression line equation as $-b/m$, where b is the y-intercept of the regression line.*

The rate of OD loss (germination rate) was slightly higher at 40°C than at 23.5°C (Table 1). After 30 to 40 min of incubation, however, the rate at 40°C had begun to decrease, although the rate at 23.5 or 30°C was still maximal. As a result, the extent of germination at 2 h was substantially lower at 40°C than at 23.5°C. At 15°C, the rate of KNO_3-induced germination was low, but there was appreciable OD loss after 2 h, and *total* germination, as determined by

phase-contrast microscopy, had reached about 20% (including those spores which had germinated in buffer without KNO_3). Even at 0°C, there were some 10% phase-dark spores after 3 h of incubation, suggesting that *C. perfringens* spores may germinate in foods even under refrigeration.

Spores, first incubated with KNO_3 at 0°C for 1 h, had lost virtually none of their ability to germinate at 30°C; their rate of OD loss after the shift to 30°C was only slightly less than that of spores kept continuously at 30°C (Table 1). Spores incubated with KNO_3 at 45 or 50°C had almost identical germination rates after shifting to 30°C (Table 1); this rate was lower than that of spores initially at 0°C or at 60°C for 1 h (Table 1). Why did spores, incubated in KNO_3 for 1 h at 60°C, germinate at 30°C at a higher rate than spores which were in KNO_3 for 1 h at 50°C before being shifted to 30°C? The following interpretation seems plausible. In 1 h at 50°C, 22% of the spores had germinated, i.e., became

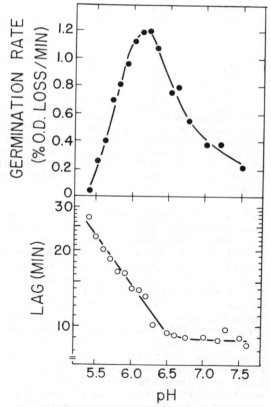

FIG. 2. *Effect of pH on KNO_3-induced germination of spores of* C. perfringens *NCTC 8238. Germination rate and lag of spores, incubated at various pH levels at 30°C in the standard germination mixture, were calculated from plots similar to that in Fig. 1.*

TABLE 1. *Germination of spores of* C. perfringens *NCTC 8238 in the standard KNO$_3$ germination mixture and the effect of this primary incubation for 1 h on subsequent secondary germination at 30° C[a]*

Temp (°C)	Primary				GL	GR	Secondary GR
	OD loss (%)						
	1 h		2 h				
	Buffer control	KNO$_3$-induced	Buffer control	KNO$_3$-induced			
0	1.9	0.4	2.6	2.4	0.0	0.0	1.13
15	3.4	1.0	4.2	4.8	47.9	0.04	ND
17.5	4.1	2.8	5.6	9.6	37.6	0.14	ND
20	4.8	4.6	6.5	14.1	32.4	0.19	ND
22	4.5	9.3	6.4	22.5	24.8	0.29	0.91
23.5	4.6	22.0	6.2	35.6	22.8	0.63	0.91
25	5.2	26.8	6.7	37.1	19.4	0.76	ND
27.5	6.0	34.2	8.0	42.3	16.4	1.08	ND
30	6.2	40.5	8.4	45.5	12.8	1.25	—
32.5	5.0	41.9	6.5	46.6	9.8	1.41	ND
35	7.1	44.3	8.6	47.5	8.3	1.40	ND
37.5	5.9	32.0	7.2	36.0	7.4	0.98	ND
40	5.5	21.5	7.5	26.0	5.5	0.71	0.36
42.5	5.4	14.9	6.6	22.3	2.5	0.42	ND
45	5.7	7.3	8.1	13.0	1.8	0.18	0.43
47	6.1	4.7	8.5	5.6	1.3	0.08	ND
50	4.3	2.4	6.8	3.1	ND	0.03	0.48
60	4.3	1.9	6.6	3.1	ND	0.03	0.67
65	6.2	1.4	ND	ND	ND	ND	1.64
70	10.3	0.5	ND	ND	ND	ND	1.68

[a] Primary incubation was in standard sodium phosphate-KNO$_3$ germination mixture (legend to Fig. 1). Buffer control (sodium phosphate buffer, pH 6.1; no KNO$_3$) and KNO$_3$-induced (corrected for control) OD losses after 1 and 2 h of incubation at the primary incubation temperatures are shown. Incubation at the primary temperature for 1 h was followed, in some cases, by a secondary incubation at 30°C. Germination lag (GL), in minutes, calculated as in Fig. 1, was determined during primary incubation. Germination rate (GR), as percent OD lost per minute, was determined during both primary and secondary incubation periods. ND, no data obtained.

dark under phase optics. However, at this temperature, these germinated spores, being more heat sensitive than dormant spores, were inactivated or injured before a substantial decrease in OD had occurred, acquisition of heat sensitivity preceding loss of OD during germination (12). Indeed, even at 40°C, the rate of OD loss had declined abruptly before the 1-h primary incubation had been completed. Further, at 50°C, the ungerminated spores would not be additionally heat activated for germination during the secondary incubation (11). At 60°C, in contrast, fewer spores (14%) were germinated in buffered KNO$_3$, and although these were rapidly inactivated or injured, there *was* some additional heat activation of the ungerminated spores. After primary incubation at 60°C, there were more ungerminated spores poised to germinate than after primary incubation at 50°C. Similarly,

after primary incubation for 1 h at 65 or 70°C, where only 6% germination had occurred, the germination rate during secondary incubation at 30°C was significantly higher than that of spores held at 60°C, or even at 0°C, during primary incubation (Table 1). Even less germination-inactivation-injury and even more heat activation had occurred at 65 or 70°C than at 60°C. We speculate that spores maintained at 60°C on a serving table in a medium suitable for germination (e.g., in foods) can be heat activated (11) and a certain percentage of them will germinate, but these germinated spores will be rapidly inactivated. When transferred to a lower, more suitable germination temperature, the remaining ungerminated spores will germinate rapidly and, under favorable conditions, may grow and be a health hazard.

Spore concentration. Germination rate was

optimal at about the standard spore concentration of 0.325 mg of spores per ml and decreased to ca. 70% of the standard rate when the spore concentration was decreased to 20% of the standard or increased to 5 times the standard. With increase in spore concentration from 20% to 40% of the standard, germination lag decreased from ca. 1.5 times that under standard conditions to a lag equal to that obtained under standard conditions and remained equal to the standard lag with further increase in spore concentration to 5 times the standard.

KNO$_3$ concentration. Germination rate was maximal (ca. 4 times that with the standard 5 mM KNO$_3$ concentration) at 100 mM KNO$_3$ (Fig. 3). With yet higher KNO$_3$ concentrations, the rate decreased, declining to approximately standard germination rate with 500 mM KNO$_3$. The effect of KNO$_3$ concentration on germination lag was less marked. Increasing the KNO$_3$ concentration from 0.5 mM to 50 mM was accompanied by a modest decrease in lag, whereas at concentrations exceeding 100 to 150 mM the lag increased slightly.

Sodium phosphate concentration. There was a broad peak of germination rate with buffer concentrations between 0.5 mM and the standard 10 mM (Fig. 4). With 100 mM sodium phosphate, the rate was only one-tenth of that under standard conditions. Lags with from 2 to 30 mM sodium phosphate buffer were approximately equal. The increased lag with lesser sodium phosphate concentrations may be attributable to an inability to control pH at very low buffer concentration, rather than to sodium phosphate concentration per se.

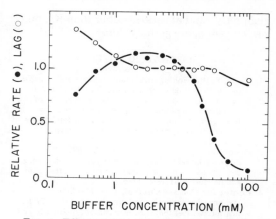

FIG. 4. *Effect of concentration of sodium phosphate buffer on the KNO$_3$-induced germination of spores of* C. perfringens *NCTC 8238. Germination rate and lag, calculated from plots similar to that in Fig. 1, are shown relative to these parameters under standard conditions (10 mM sodium phosphate, ◑).*

Spore investigators have frequently used lag as an index of the capacity of spores to germinate. It is of some interest to note that, in none of the conditions examined—pH, temperature, spore concentration, KNO$_3$ concentration, and sodium phosphate concentration—were the rate and lag responses equivalent as indices of germination. We believe that, if one must select a single criterion of germinability, germination rate would be the most useful and valid such criterion.

With our system, then, the conditions for maximum rate of KNO$_3$-induced germination were pH 6.1 to 6.3, 32.5 to 35°C, 0.325 mg of spores per ml, 100 mM KNO$_3$, and 2 to 5 mM sodium phosphate. We now find that spores of *C. perfringens* NCTC 8238 germinate equally well under air and under N$_2$. The ratio of germination rate under aerobic conditions to that under anaerobic conditions (flushed with N$_2$) was ca. 1.0. Similarly, the rate, lag, and extent of germination of spores of *C. sporogenes* (putrefactive anaerobe PA 3679h) were virtually equal under anaerobic or aerobic conditions (15). This, we believe, is analogous to the capability of the spores of the strict aerobe *B. megaterium* to germinate under anaerobic conditions (9). The duality of lack of necessity for aerobiosis in *B. megaterium* spore germination and lack of a requirement for anaerobiosis for *C. perfringens* spore germination suggests the generality of the thesis that germination and outgrowth are processes with separate and definable nutritional requirements and metabolic pathways. In addition to exploring this thesis, we plan to investigate more fully the effect of anions, other

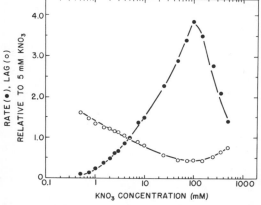

FIG. 3. *Effect of KNO$_3$ concentration on KNO$_3$-induced germination of spores of* C. perfringens *NCTC 8238. Germination rate and lag, calculated from plots similar to that in Fig. 1, are shown relative to these parameters under standard conditions (5 mM KNO$_3$, ◑).*

than NO_3^-, and of cations, other than K^+, on *C. perfringens* spore germination.

LITERATURE CITED

1. **Ando, Y.** 1974. Ionic germination of spores of *Clostridium perfringens* type A. Jpn. J. Microbiol. **18**:433–439.
2. **Ando, Y.** 1975. Studies on germination of spores of clostridial species capable of causing food poisoning (VII). Ionic germination of spores of heat-resistant strains of *Clostridium perfringens* type A. J. Food Hyg. Soc. Jpn. **16**:25–29.
3. **Bryan, F. L., and E. G. Kilpatrick.** 1971. *Clostridium perfringens* related to roast beef cooking, storage, and contamination in a fast food service restaurant. Am. J. Public Health **61**:1869–1885.
4. **Center for Disease Control.** 1979. Foodborne and waterborne disease outbreaks, annual summary 1977. Publication no. 79-8185. U.S. Department of Health, Education, and Welfare, Washington, D.C.
5. **Duncan, C. L.** 1970. *Clostridium perfringens* food poisoning. J. Milk Food Technol. **33**:35–41.
6. **Duncan, C. L., D. H. Strong, and M. Sebald.** 1972. Sporulation and enterotoxin production by mutants of *Clostridium perfringens*. J. Bacteriol. **110**:378–391.
7. **Hobbs, B. C., M. E. Smith, C. L. Oakley, G. H. Warrack, and J. C. Cruickshank.** 1953. *Clostridium welchii* food poisoning. J. Hyg. **51**:75–101.
8. **Hughes, J. A., P. C. B. Turnbull, and M. F. Stringer.** 1976. A serotyping system for *Clostridium welchii* (*C. perfringens*) type A, and studies on the type-specific antigens. J. Med. Microbiol. **9**:475–485.
9. **Hyatt, M. T., and H. S. Levinson.** 1959. Utilization of phosphates in the postgerminative development of spores of *Bacillus megaterium*. J. Bacteriol. **77**:487–496.
10. **Levinson, H. S., and F. E. Feeherry.** 1975. Influence of cations on nitrate-induced germination of *Bacillus megaterium* QM B1551 spores, p. 495–505. *In* P. Gerhardt, R. N. Costilow, and H. L. Sadoff (ed.), Spores VI. American Society for Microbiology, Washington, D.C.
11. **Levinson, H. S., and F. E. Feeherry.** 1978. Kinetics of heat activation of spores of *Clostridium perfringens*, p. 34–40. *In* G. Chambliss and J. C. Vary (ed.), Spores VII. American Society for Microbiology, Washington, D.C.
12. **Levinson, H. S., and M. T. Hyatt.** 1966. Sequence of events during *Bacillus megaterium* spore germination. J. Bacteriol. **91**:1811–1818.
13. **Levinson, H. S., and M. T. Hyatt.** 1970. Activation energy for glucose-induced germination of *Bacillus megaterium* spores. J. Bacteriol. **103**:270–271.
14. **Rode, L. J., and J. W. Foster.** 1962. Ionic germination of spores of *Bacillus megaterium* QM B1551. Arch. Mikrobiol. **43**:183–200.
15. **Uehara, M., and H. A. Frank.** 1965. Factors affecting alanine-induced germination of clostridial spores, p. 38–46. *In* L. L. Campbell and H. O. Halvorson (ed.), Spores III. American Society for Microbiology, Ann Arbor, Mich.

Role of Charged Groups of the Spore Surface in Germination

TSUTOMU NISHIHARA, TOMIO ICHIKAWA, AND MASAOMI KONDO

Faculty of Pharmaceutical Sciences, Osaka University, Suita, Osaka, Japan 565

Strong acidic groups (such as phosphate) in the spore coat of *Bacillus megaterium* QM B1551 may block the action of ionic germinants and, thus, act as a barrier against initiation of ionic germination. Weak acidic groups (such as carboxylate) may be involved later in germination.

The spore coat, composed of proteins with phosphate, is the outermost layer of *Bacillus megaterium* spores and is the contact between the spores and germinants. Many reactive groups of the coat are exposed on the spore surface, including such components as sulfhydryl, hydroxyl, carboxylic, phosphoric, and amino groups. We previously reported that sulfhydryl and hydroxyl groups play a role in the recognition of a germinant for initiation of germination (6). We have now studied the role of negatively charged groups by using colloid titration.

Colloid titration was carried out as follows. The spores (2.5 to 10 mg/2 ml) were incubated at room temperature for 5 min with an excess amount of a positive colloid (N/200, 2 ml), and after centrifugation to remove the spores with the colloid, the remaining positive colloid in the supernatant was back-titrated with a negative colloid, potassium polyvinyl sulfate (N/800, Wako Jun-Yaku), using toluidine blue as an indicator. Colloid titration of the spores using protamine as a positive colloid was reported by Watanabe and Takesue (8), but we used glycol chitosan (GCh; Wako Jun-Yaku) and methylglycol chitosan (MGCh; Wako Jun-Yaku) as positive colloids because of their reaction specificity; GCh reacts only with strong acidic groups such as phosphate and sulfate, whereas MGCh reacts with all negatively charged groups including carboxylate and phenol (5). The values obtained by titrating the spores with these colloids indicate the surface charge of the spores because the colloids, having an average molecular weight of about 100,000 (5), cannot penetrate into the spores (4).

Spores of *B. megaterium* QM B1551, prepared on glutamate sporulation agar plates as described previously (1), were titrated (Fig. 1). The surface charge was negative; about -0.2 μeq/mg with GCh and -0.4 μeq/mg with MGCh. On the basis of the reaction specificity of the colloids (5) and the spore coat composition (2), these values suggested that about half of the charge might be due to phosphate and half to carboxylate.

When spores of this strain (10 mg/ml) were treated with 0.5% sodium dodecyl sulfate (SDS; Nakarai Chemicals) plus 10 mM dithiothreitol (DTT; Sigma Chemical Co.) plus 0.1 M NaCl at pH 10, some parts of the coat components were extracted, and the treated spores became sensitive to lysozyme (3, 7), implying a change in permeability of the spore surface layers. The charge of the SDS-DTT–treated spores with GCh was similar to that of the intact spores (Fig. 2), but the charge of the treated spores with MGCh was higher, suggesting that the inside of the coat which had been exposed by the SDS-DTT treatment was rich in carboxylate-like residues.

During germination, the negative charge was nearly doubled, perhaps as a result of the release of positively charged compounds or the change in permeability during germination. When the

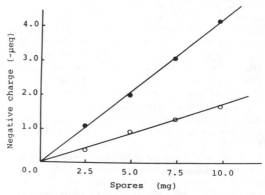

FIG. 1. *Colloid titration of* B. *megaterium* QM *B1551 spores. The spores were titrated with GCh* (O) *and MGCh* (●).

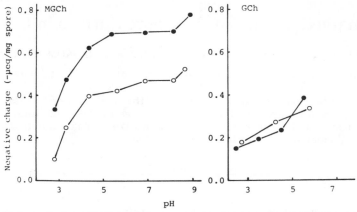

FIG. 2. *Colloidal charge versus pH curves of the intact and the SDS-DTT–treated spores. The intact (○) and the SDS-DTT–treated (●) spores were titrated at various pH's.*

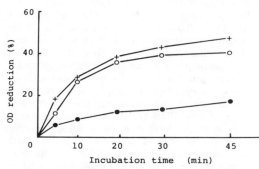

FIG. 3. *Germination of the spores in nutrient broth with a positive colloid. The heat-activated spores were incubated in nutrient broth alone (+) or with GCh (○) or MGCh (●).*

spores (0.2 mg/ml) were incubated in nutrient broth at 37°C in the presence of the positive colloids (1 mN), GCh had little effect on germination, but MGCh hindered the germination, measured as optical density reduction (Fig. 3). The spores with GCh became phase dark and heat sensitive, released dipicolinic acid, and grew after a longer lag time than the spores without the colloid. On the other hand, the spores with MGCh stayed at the stage of phase gray although they lost dipicolinic acid; they did not initiate outgrowth. This suggests that MGCh-reactive weak acidic groups may be associated with a later process in germination.

Phosphate is the most likely candidate for the strong acidic group. High-phosphate spores, produced in a medium containing 43 mM phosphate, had a coat with 56 μg of phosphate per mg of coat, whereas low-phosphate spores, produced in a medium containing 0.3 mM phosphate, had coats with lower phosphate content

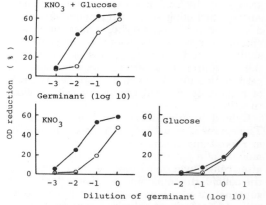

FIG. 4. *Germinability of high-phosphate and low-phosphate spores. High-phosphate (○) and low-phosphate (●) spores, after heat activation, were incubated in a dilute solution of 40 mM KNO_3, 1 mM glucose, or both at 37°C for 60 min, and the percent reduction in optical density at 610 nm was plotted.*

(25 μg/mg of coat), and the surface charge was about -0.1 μeq/mg of spore with GCh or MGCh. The germinabilities of the high- and the low-phosphate spores were compared by incubating them in a dilute solution of the germinants glucose and KNO_3, singly or in combination. (Fig. 4). A mixture of 4 mM KNO_3 and 0.1 mM glucose resulted in a 40% reduction in optical density of the high-phosphate spores, but a greater than 10-fold diluted solution resulted in almost same optical density reduction of the low-phosphate spores, indicating an increase in sensitivity to the germinants. This increase was observed with KNO_3, but not with glucose, as a germinant, suggesting that the main strong acidic group of

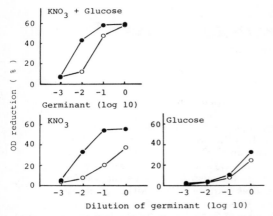

FIG. 5. *Germinability of calcium-enriched spores. The germinabilities of spores treated with calcium acetate (●) and of native spores (○) were examined as for Fig. 4.*

blocking effect of the phosphate residue on germination.

the spore surface is phosphate residue in the coat and that the phosphate residue plays a role as a barrier against initiation of germination by the ionic germinant.

Enrichment with calcium (spores plus 10 mM calcium acetate, 60°C, 60 min) increased the germinability of the native spores in KNO_3 to that of the low-phosphate spores (Fig. 5), suggesting that positively charged compounds such as calcium in the coat might compensate for the

ACKNOWLEDGMENT

We are grateful to Ikuko Yoshimoto for assistance in the colloid titration experiments.

LITERATURE CITED

1. **Kawasaki, C., M. Kondo, and T. Nishihara.** 1970. Studies on the bacterial spore coat. 1. Morphological structure of the spore coat of *Bacillus megaterium.* Nippon Saikingaku Zasshi **25:**209–214.
2. **Kondo, M., and T. Nishihara.** 1970. Studies on the bacterial spore coat. 2. Chemical structures of fractions isolated from the dormant spore coat of *Bacillus megaterium.* Nippon Saikingaku Zasshi **25:**215–221.
3. **Nishihara, T., T. Yutsudo, T. Ichikawa, and M. Kondo.** 1981. Studies on the bacterial spore coat. 8. On the SDS-DTT extract from *Bacillus megaterium* spore. Microbiol. Immunol. **25:**327–331.
4. **Scherrer, R., T. C. Beaman, and P. Gerhardt.** 1971. Macromolecular sieving by the dormant spore of *Bacillus cereus.* J. Bacteriol. **108:**868–873.
5. **Senju, R.** 1969. On the colloid titration. Nankodo, Tokyo.
6. **Ueno, A., S. Kametaka, T. Nishihara, T. Ichikawa, and M. Kondo.** 1978. Changes in the specificity of germinants for *Bacillus megaterium* spores by *p*-chloromercuribenzoate treatment, p. 109–112. *In* G. Chambliss and J. C. Vary (ed.), Spores VII. American Society for Microbiology, Washington, D.C.
7. **Vary, J. C.** 1973. Germination of *Bacillus megaterium* spores after various extraction procedures. J. Bacteriol. **116:**797–802.
8. **Watanabe, K., and S. Takesue.** 1976. Colloid titration for determining the surface charge of bacterial spores. J. Gen. Microbiol. **96:**221–223.

Germination of *Clostridium perfringens* Spores: a Proposed Role for Ions

YOSHIAKI ANDO

Hokkaido Institute of Public Health, Sapporo 060, Japan

Unactivated spores of *Clostridium perfringens* S40 were able to germinate in combinations of L-alanine, inosine, and $CaCl_2$ in the presence of CO_2. In contrast, heat-activated spores were able to germinate in $CaCl_2$ alone in the presence of CO_2. Ca^{2+} was replaceable not only by divalent cations other than Mg^{2+}, which was a competitive inhibitor for germination, but also by various monovalent cations. The germination of hydrogen spores required Ca^{2+}, although it was greatly stimulated by monovalent cations. The germination of calcium spores, which proceeded to some extent in the absence of Ca^{2+}, was more efficient in the presence of Ca^{2+} as well as monovalent cations. When heat-activated spores were germinated in $CaCl_2$ plus CO_2, monovalent cations could bypass the requirement for CO_2.

The significance of ions in spore germination has been emphasized by several investigators (4, 5) since Rode and Foster (6) proposed the hypothesis of the primary germinative capacity of ions. However, the exact role of ions in germination remains unclear.

Previously, I studied (1) the ionic germination of spores of *C. perfringens* NCTC 8238 whose germination was initiated by either organic or inorganic salts alone. Studies have now been extended to another strain of the same species, *C. perfringens* S40, which is unique since unactivated spores germinate in combinations of certain organic germinants and inorganic salts in the presence of CO_2, whereas heat-activated spores germinate in inorganic salts alone in the presence or absence of CO_2.

These spore germination properties prompted me to examine the effect of ions on germination, particularly as regards the augmentative action of organic germinants in ionic germination.

RESULTS AND DISCUSSION

Effect of CO_2. Preliminary observation showed that spores of *C. perfringens* S40 were dependent on bicarbonate or CO_2 for satisfactory germination in a certain complex medium. In a synthetic medium consisting of L-alanine, inosine, $CaCl_2$, and Tris buffer (pH 7.5), germination of unactivated spores did not proceed until CO_2 was introduced. The pH of the medium fell rapidly from 7.5 to 6.0 during passage of CO_2. A

pH of 6.0 favored germination of the spores, but there was no appreciable germination in medium adjusted to this pH if CO_2 was omitted (data not shown).

Germination requirements. Spores germinated with or without heat activation in various combinations of L-alanine, inosine, and $CaCl_2$ in the presence of CO_2 (Fig. 1). Of particular interest are the following observations: (i) with unactivated spores, maximal germination occurred in a combination of L-alanine, inosine, and $CaCl_2$ in the presence of CO_2, although germination proceeded to some extent in the absence of added $CaCl_2$; (ii) with heat-activated spores, sufficient germination occurred in $CaCl_2$ alone in the presence of CO_2, although the rate of germination was increased by the addition of L-alanine but not of inosine; and (iii) the germination of unactivated spores was more extensive in a combination of L-alanine and inosine than in either alone. L-Alanine plus inosine plus $CaCl_2$ plus CO_2 was the most effective; inosine plus $CaCl_2$ plus CO_2 was next most effective; and L-alanine plus $CaCl_2$ plus CO_2 was the least effective.

Effect of $CaCl_2$ concentration. The germinative response of unactivated spores to inosine plus $CaCl_2$ plus CO_2 displayed apparently sigmoidal kinetics against $CaCl_2$ concentrations. When L-alanine was added to the mixture, the kinetic curve was transformed into a hyperbolic one (Fig. 2A). Similar experiments carried out with heat-activated spores showed that the sigmoidal curves obtained from the respective responses to inosine plus $CaCl_2$ plus CO_2 and $CaCl_2$ plus CO_2 are also transformed into hyperbolic ones after the addition of L-alanine to these mixtures (Fig. 2B).

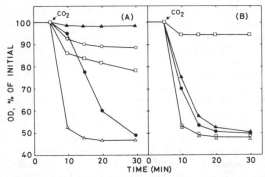

FIG. 1. *Germination of* C. *perfringens S40 spores in combinations of L-alanine (1 mM), inosine (1 mM), and* $CaCl_2$ *(40 mM) in the presence of* CO_2. *Spores were incubated at* $40°C$ *in Tris buffer (50 mM, pH 7.5) containing the germinants, and germination was initiated by introducing* CO_2 *into the medium.* CO_2 *was passed at ca. 20 ml/min through a capillary tube into the germination mixtures in Klett tubes for the initial 5 min of incubation, and then the tubes were closed with rubber stoppers. Germination was estimated by measuring the decrease in optical density (OD) of spore suspensions in germination media with a Klett-Summerson photoelectric colorimeter (no. 56 filter). (A) Unactivated spores. (B) Heat-activated (80°C for 10 min) spores. Symbols:* $\bigcirc$, *L-alanine* + $CaCl_2$ + CO_2; $\bullet$, *inosine* + $CaCl_2$ + CO_2; $\triangle$, *L-alanine* + *inosine* + $CaCl_2$ + CO_2; $\blacktriangle$, $CaCl_2$ + CO_2; $\square$, *L-alanine* + *inosine* + CO_2.

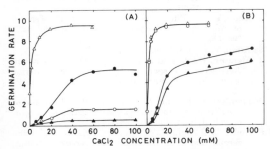

FIG. 2. *Kinetic curves of germination rates of* C. *perfringens S40 spores against* $CaCl_2$ *concentrations. Germination was performed as described in the legend of Fig. 1. Germination rate was expressed as the maximum percent decrease in optical density per minute, as measured over a 5-min period during the first 10 or 20 min of germination. The concentration of L-alanine or inosine was 1 mM. (A) Unactivated spores. (B) Heat-activated (80°C for 10 min) spores. Symbols:* $\bigcirc$, *L-alanine* + $CaCl_2$ + CO_2; $\bullet$, *inosine* + $CaCl_2$ + CO_2; $\triangle$, *L-alanine* + *inosine* + $CaCl_2$ + CO_2; $\blacktriangle$, $CaCl_2$ + CO_2.

TABLE 1. *Effect of various inorganic salts on germination of* C. perfringens *S40 spores[a]*

Salt	% Decrease in optical density/30 min[b]	
	Unactivated spores	Heat-activated spores
$CaCl_2$	54	53
$SrCl_2$	53	49
$BaCl_2$	50	38
$MgCl_2$	6	5
$MnCl_2$	48	15
NaCl	52	43
KCl	54	49
CsCl	52	49
RbCl	51	50
LiCl	48	33
NH_4Cl	51	48
None	24	4

[a] The method for preparing spores is described elsewhere (3a).

[b] Unactivated spores were germinated in L-alanine (1 mM) plus inosine (1 mM) plus one of the salts (divalent, 40 mM; monovalent, 80 mM) plus CO_2. Heat-activated (80°C for 10 min) spores were germinated in one of the salts (divalent, 100 mM; monovalent, 200 mM) plus CO_2. Germination was performed as described in the legend of Fig. 1.

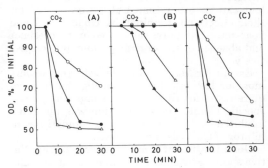

FIG. 3. *Germination of native spores, hydrogen spores, and calcium spores of* C. *perfringens S40 in combinations of L-alanine (1 mM), inosine (1 mM),* $CaCl_2$ *(40 mM), and NaCl (80 mM). The methods for preparing hydrogen spores and calcium spores were similar to those previously described with spores of* C. *perfringens S45 (3). Unactivated spores of these ionic forms were germinated as described in the legend of Fig. 1. (A) Native spores. (B) Hydrogen spores. (C) Calcium spores. Symbols:* $\bigcirc$, *L-alanine* + *inosine* + CO_2; $\bullet$, *L-alanine* + *inosine* + *NaCl* + CO_2; $\triangle$, *L-alanine* + *inosine* + $CaCl_2$ + CO_2; $\blacktriangle$, *L-alanine* + *inosine* + $CaCl_2$ + *NaCl*.

It is evident from these results that the increase in germination rates is of an allosteric nature, and hence L-alanine singly or in combination with inosine induces a cooperative interaction between the Ca^{2+} binding sites within a spore.

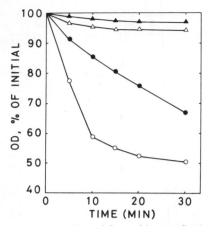

FIG. 4. *Germination of* C. perfringens *S40 spores in combinations of various salts in the absence of* CO_2. *Heat-activated (80°C for 10 min) spores were germinated at 40°C in aqueous solutions (pH not adjusted) of the salts that had been boiled and cooled to expel* CO_2. *Symbols:* ○, $CaCl_2$ *(50 mM)* + NH_4Cl *(100 mM)*; ●, $CaCl_2$ *(50 mM)* + *NaCl (100 mM)*; △, *NaCl (100 mM)* + NH_4Cl *(100 mM)*; ▲, $CaCl_2$ *(50 mM)*.

Effect of other inorganic ions. Inorganic chloride salts of various divalent and monovalent cations other than $CaCl_2$ were tested for their effects on germination. Unactivated and heat-activated spores responded in the presence and absence of the organic germinants, respectively.

Except for $MgCl_2$, all of the salts tested were more or less effective, indicating that the requirements for these ions are apparently nonspecific (Table 1). Germination was markedly inhibited by Mg^{2+}, which was a competitive inhibitor of Ca^{2+} (data not shown).

Germination of ionic forms of spores. Native spores germinated to some extent in L-alanine plus inosine plus CO_2 in the absence of $CaCl_2$, but did so more extensively in the presence of $CaCl_2$ or of NaCl (Fig. 3A). Calcium spores showed essentially the same germination pattern as native spores (Fig. 3C). By contrast, hydrogen spores (Fig. 3B) did not germinate in L-alanine plus inosine plus CO_2 in the absence of $CaCl_2$, or in the presence of NaCl. Appreciable germination occurred only when $CaCl_2$ was pres-

ent in the medium. Moreover, germination of hydrogen spores was greatly enhanced by the simultaneous addition of $CaCl_2$ and NaCl to the medium (Fig. 3B).

Bypassing of CO_2 requirement. In a previous paper (2), I reported that spores of *C. perfringens* S40 germinated in various combinations of divalent and monovalent salts when heat activated at 80°C for 10 min. This implies that CO_2 is not absolutely required for germination of the spores. In fact, such heat-activated spores germinated in combinations of $CaCl_2$, NaCl, and NH_4Cl, any of which alone was ineffective in inducing germination in the absence of CO_2 (Fig. 4). It seems possible that heat may act on the spores by causing the bypassing of the CO_2 requirement by monovalent salts.

Based on the results reported here, the following possible roles of ions, organic germinants, and CO_2 are proposed. (i) Heat or CO_2 modifies the membrane lipid bilayers so as to facilitate the permeability of the spore membrane to Ca^{2+}. (ii) L-Alanine and inosine stimulate Ca^{2+} transport by functioning as allosteric effectors. (iii) Monovalent cations nonspecifically stimulate Ca^{2+} transport. Further studies on Ca^{2+} transport during germination and identification of Ca^{2+} binding site(s) are needed to verify the above hypothesis.

LITERATURE CITED

1. **Ando, Y.** 1974. Ionic germination of spores of *Clostridium perfringens* type A. Jpn. J. Microbiol. **18**:433–439.
2. **Ando, Y.** 1974. Studies on germination of spores of clostridial species capable of causing food poisoning. V. Ionic germination of spores of some heat-sensitive strains of *Clostridium perfringens* type A. J. Food Hyg. Soc. Jpn. **15**:373–376.
3. **Ando, Y.** 1975. Effect of lysozyme on ionic forms of spores of *Clostridium perfringens* type A. J. Bacteriol. **122**:794–795.
3a.**Ando, Y.** 1980. J. Appl. Bacteriol. **49**:527–535.
4. **Fleming, H. P., and Z. J. Ordal.** 1964. Responses of *Bacillus subtilis* spores to ionic environments during sporulation and germination. J. Bacteriol. **88**:1529–1537.
5. **Foerster, H. F., and J. W. Foster.** 1966. Response of *Bacillus* spores to combinations of germinative compounds. J. Bacteriol. **91**:1168–1177.
6. **Rode, L. J., and J. W. Foster.** 1962. Ions and the germination of spores of *Bacillus cereus* T. Nature (London) **194**:1300–1301.

Role of Gramicidin S in the Producer Organism, *Bacillus brevis*

JACQUELINE M. PIRET AND ARNOLD L. DEMAIN

Massachusetts Institute of Technology, Cambridge, Massachusetts 02139

A comparison was made of the properties of spores from the gramicidin S-producing *Bacillus brevis* strain Nagano and its gramicidin S-negative mutant BI-7. Both cultures grew and sporulated similarly. Resistance and germination of the spores were the same. Outgrowth of parental spores was delayed in comparison with BI-7; exogenous gramicidin S mimicked this effect. Extraction of gramicidin S from parental spores reduced the outgrowth delay. Outgrowth delay was dose dependent, and spores were sensitive to gramicidin S for only a distinct time span. Gramicidin S appears to act on spores from the outside.

Most sporulating microorganisms produce antibiotics. Since they do so just prior to and during sporulation, the idea of a role for these compounds in the sporulation cycle has been popular (18). However, results have often been only correlative and confusing as a result of the use of pleiotropic mutants. There is apparently no one universal role for all antibiotics (1). For example, in the genus *Bacillus* the literature suggests a number of cell functions which might be controlled by peptide antibiotics: spore quantity (14–17), spore quality (9, 10), or spore outgrowth (1, 9, 11). A direct cause-and-effect relationship between antibiotic synthesis and the extent of sporulation in *Bacillus* is apparently ruled out because there exist antibiotic-negative mutants which still sporulate at parental levels (2, 5, 9, 11, 13). That antibiotic synthesis might affect the quality of spores was suggested by Mukherjee and Paulus (10). Alternatively, antibiotic synthesis and sporulation may be two independent processes regulated by a common control mechanism. Concurrent antibiotic synthesis and sporulation would ensure that antibiotic and spores are in close physical contact so that they can interact at some later time, for example, when the antibiotic might activate or inhibit germination or outgrowth (2). Another possibility is that concomitant antibiotic synthesis and spore formation protect the spore during germination or outgrowth, or both, when it is most vulnerable to competition.

The system we are using is gramicidin S (GS) production by *B. brevis* Nagano as compared with its GS-negative mutant BI-7, isolated by Saito's group (7). This mutant lacks the phenyl-alanine activation activity of the light GS synthetase (Table 1).

RESULTS AND DISCUSSION

The parent and mutant were first compared with respect to sporulation. Sporulation rate and extent and sporulation-associated events such as intracellular protease synthesis and acquisition of heat, UV, and solvent resistances followed similar time courses in parent and mutant. Media supporting high (45 to 110 µg/ml), moderate (15 to 20 µg/ml), and low (2 to 4 µg/ml) GS production by the parent were used to produce spores in these experiments.

Next, the qualities of the mature spores of the two strains were compared. The inactivation by heat, UV irradiation, and solvents (ethanol, octanol, and chloroform) of parental and mutant spores was found to be virtually identical (3). The literature on the heat resistance of various GS-negative mutants of *B. brevis* is in some disagreement. Although the data of Nandi and Seddon (11), who used two other mutants of the Nagano strain (BI-9 and E-1), agree with ours, Marahiel et al. (9) claimed that GS-negative mutants of the ATCC 9999 strain (EB16, MIV, Sm170, and R5) produce heat-sensitive spores. As in the earlier case of the linear gramicidin mutants of *B. brevis* ATCC 8185 (10), these mutants were partially "cured" when allowed to sporulate in the presence of the antibiotic. The discrepancy may reflect strain differences or the use of pleiotropic mutants (Table 1) by Marahiel et al, i.e., mutants altered in both heavy and light GS synthetases.

TABLE 1. *Enzymatic and fermentative characteristics of GS-negative mutants*

| Strain | Isolate | Reference | "Light" enzyme | "Heavy" enzyme | | | | Formation of: | |
			D-Phenylalanine	L-Proline	L-Valine	L-Ornithine	L-Leucine	DKP[a]	GS
Nagano	Parent		+	+	+	+	+	+	+
	BI-7	3	−	+	+	+	+	−	−
	BI-9	11	+	+	+	+	−	+	−
	E-1	11	+	+	+	+	+	−	−
ATCC 9999	Parent		+	+	+	+	+	ND[b]	+
	EB16	9	−	−	−	−	−	ND	−
	MIV	9	−	−	−	−	−	ND	−
	Sm170	9	±	−	−	−	−	ND	−
	R5	9	−	−	+	−	±	ND	−

[a] D-Phenylalanyl-L-prolyl diketopiperazine.
[b] Not determined.

Spore germination and outgrowth were next examined. Germination, as measured by rate and extent of spore darkening, was similar in complex and defined media for both strains. However, in outgrowth, the behaviors of parent and mutant spores differed. Whereas the mutant was observed to germinate, outgrow, and begin vegetative growth in 1 to 2 h (in nutrient broth), the parent took 6 to 10 h, remaining in outgrowth for a prolonged period of time.

The delay in outgrowth was seen with spores made in a medium supporting high GS synthesis by the parent. Spores harvested from a medium yielding low GS production did not exhibit outgrowth delay; outgrowth of both parent and mutant spores was rapid. When spores produced in a medium supporting a moderate level of GS synthesis were used, outgrowth inhibition was observed only if the parental spore concentration was sufficiently high. These data indicate that outgrowth delay occurs only when a certain volumetric concentration of GS is present. This can be realized by a low concentration of spores obtained from a good GS production medium or a high concentration of spores produced in a medium supporting moderate GS production.

When parental spores were extracted, their outgrowth was rapid, like that of mutant spores (Fig. 1). When this extract or exogenous GS was added to mutant spores, their outgrowth was delayed in a dose-dependent manner. Outgrowth delay caused by GS has now been reported by four groups (1, 4, 9, 11). We have also found that parental spores treated with a proteolytic enzyme (Nagarse [EC 3.4.4.16]) capable of hydrolyzing GS (20) outgrow quickly. Furthermore, when parent and mutant spores are mixed, the outgrowth of the mixture is delayed. These two

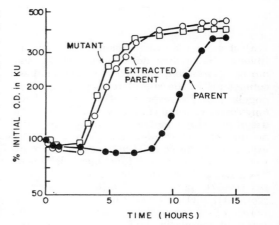

FIG. 1. *Effect of extraction of GS on outgrowth. Parental spores were extracted with 95% ethanol–0.2 N HCl (9:1) at 80°C for 5 min. Extracted parental spores, untreated parental spores, and untreated mutant spores were resuspended (all at the same inoculum concentration) in 2.5× Difco nutrient broth and shaken at 37°C. Treated mutant spores behaved like untreated mutant spores.*

results suggest that GS can act from outside the spore. GS, a hydrophobic compound, may be adsorbed to spore surfaces after its excretion during sporulation or germination. In view of its known activity on bacterial membranes (8, 12), we favor the possibility that GS affects the transport properties of the membrane of germinating spores.

Outgrowth delay caused by GS is dose-dependent and also time dependent. Addition of GS to mutant spores at or after 125 min no longer delayed outgrowth. The antibiotic did not inhibit vegetative growth, even at 80 times the concentration which delayed outgrowth.

We believe that the inhibition of outgrowth by GS is permanent, despite the fact that parental spore suspensions eventually began to grow, as measured by absorbance increase. However, microscopic observation showed that, whereas the mutant spore population darkens, swells, and begins to grow in a homogeneous manner, the behavior of parental spores is heterogeneous. Most parental spores remain as dark spores, and only a few are converted into rods. Thus, in the growing parental culture a large number of dark spores are still present. Those rare parental spores which are able to grow may somehow be resistant to GS. For the vast majority of the parental spores, inhibition by the antibiotic is apparently permanent as long as a sufficient concentration of GS is present.

That certain secondary metabolites are produced in fungi which inhibit spore germination in a crowded environment is well known (19). This phenomenon may also occur in streptomycetes since Hirsch and Ensign (6) have described an antibiotic which is present in *Streptomyces viridochromogenes* spores and inhibits germination. We propose that GS has a similar function in *B. brevis*, that of inhibiting spore outgrowth under crowded conditions.

ACKNOWLEDGMENT

This work was supported by U.S. Army Research Office grant DAAG29-78-C-0015.

LITERATURE CITED

1. **Demain, A. L.** 1980. Do antibiotics function in nature? Search **11**:148–151.
2. **Demain, A. L., and J. M. Piret.** 1978. Speculations on the role(s) of antibiotics in spore-forming microorganisms, p. 16–17. *In* G. Chambliss and J. C. Vary (ed.), Spores VII. American Society for Microbiology, Washington, D.C.
3. **Demain, A. L., and J. M. Piret.** 1979. Relationship between antibiotic synthesis and sporulation, p. 183–188. *In* M. Luckner and K. Schreiber (ed.), Regulation of secondary product and plant hormone metabolism. Pergamon Press, New York.
4. **Egorov, N. S., A. N. Vyipiyach, G. G. Zharikova, and V. N. Maksimov.** 1975. Effect of various factors on spore germination in S and P⁻ variants of *Bacillus brevis*. Mikrobiologiya **44**:237–240.
5. **Haavik, H. I., and S. Thomassen.** 1973. A bacitracin-negative mutant of *Bacillus licheniformis* which is able to sporulate. J. Gen. Microbiol. **76**:451–454.
6. **Hirsch, C. F., and G. C. Ensign.** 1978. Some properties of *Streptomyces viridochromogenes* spores. J. Bacteriol. **134**:1056–1063.
7. **Iwaki, M., K. Shimura, M. Kanda, E. Kaji, and Y. Saito.** 1972. Some mutants of *Bacillus brevis* deficient in gramicidin S formation. Biochem. Biophys. Res. Commun. **48**:113–118.
8. **Kaprel'yants, A. S., V. V. Nikiforov, A. I. Mironikov, L. G. Snezkova, V. A. Eremin, and D. N. Ostrovskii.** 1977. Bacterial membranes and mechanism of action of gramicidin S. Biokhimiya **42**:329–337.
9. **Marahiel, M. A., W. Danders, M. Krause, and H. Kleinkauf.** 1979. Biological role of gramicidin S in spore functions. Studies on gramicidin S-negative mutants of *Bacillus brevis* ATCC 9999. Eur. J. Biochem. **99**:49–55.
10. **Mukherjee, P. K., and H. Paulus.** 1977. Biological function of gramicidin: studies on gramicidin-negative mutants. Proc. Natl. Acad. Sci. U.S.A. **74**:780–784.
11. **Nandi, S., and B. Seddon.** 1978. Evidence for gramicidin S functioning as a bacterial hormone specifically regulating spore outgrowth in *Bacillus brevis* Nagano. Biochem. Soc. Trans. **6**:409–411.
12. **Ostrovskii, D. N., V. G. Bulgarova, I. G. Zhukova, A. S. Kaprel'yants, E. G. Rozantzev, and I. M. Simakova.** 1976. Changes of lipid-protein interactions in bacterial membranes under the effect of gramicidin S. Biokhimiya **41**:175–182.
13. **Ray, B., and S. K. Bose.** 1971. Polypeptide antibiotic-negative sporeformer mutants of *Bacillus subtilis*. J. Gen. Appl. Microbiol. **17**:491–498.
14. **Ristow, H., W. Pschorn, J. Hansen, and U. Winkel.** 1979. Induction of sporulation in *Bacillus brevis* by peptide antibiotics. Nature (London) **280**:165–166.
15. **Ristow, H., B. Schazschneider, and H. Kleinkauf.** 1975. Effects of the peptide antibiotics tyrocidine and the linear gramicidin on RNA synthesis and sporulation. Biochim. Biophys. Acta **63**:1085–1092.
16. **Sarkar, N., D. Langley, and H. Paulus.** 1977. Biological function of gramicidin: selective inhibition of RNA polymerase. Proc. Natl. Acad. Sci. U.S.A. **74**:1478–1482.
17. **Sarkar, N., and H. Paulus.** 1972. Function of peptide antibiotics in sporulation. Nature (London) New Biol. **239**:228–230.
18. **Schaeffer, P.** 1969. Sporulation and the production of antibiotics, exoenzymes, and exotoxins. Bacteriol. Rev. **33**:48–71.
19. **Sussman, A. S., and H. A. Douthit.** 1973. Dormancy in microbial spores. Annu. Rev. Plant Physiol. **24**:311–352.
20. **Yukioka, M., Y. Saito, and O. Otani.** 1966. Enzymatic hydrolysis of gramicidin S. J. Biochem. **60**:295–302.

Two-Dimensional Polyacrylamide Gel Analysis of Protein Synthesized During Outgrowth of *Bacillus subtilis* 168

HELEN B. MULLIN AND J. NORMAN HANSEN

Division of Biochemistry, Department of Chemistry, University of Maryland, College Park, Maryland 20742

Two-dimensional polyacrylamide gels of the O'Farrell type were used to examine protein populations synthesized during outgrowth of *Bacillus subtilis* 168. Several classes of proteins in which synthesis was modulated were identified, including some unique to outgrowth and others unique to vegetative growth.

The outgrowth stage of the life cycle of sporulating bacilli is a developmental process with an ordered sequence of morphological and biochemical events. The usefulness of this system as a model for the study of mechanisms of gene expression is limited by a lack of information about gene products which participate in the developmental steps. We have undertaken to examine the population of proteins produced during various stages of outgrowth of *Bacillus subtilis*, using two-dimensional polyacrylamide gel analysis. The purpose of this was to identify specific proteins whose synthesis is modulated. Of particular interest would be the identification of classes of proteins which appear to be under coordinate control, such as a class which is synthesized in outgrowth but not in vegetative growth or one which is synthesized in vegetative growth but not in outgrowth. It is possible that various studies of these protein classes, including nucleic acid sequencing studies, will reveal control elements.

RESULTS AND DISCUSSION

The experimental approach was to germinate spores in defined minimal medium (3) and, at selected times during outgrowth, transfer samples of spores to flasks containing [^{35}S]methionine. After continued incubation for 20 min in the presence of label, unlabeled methionine was added, followed by an additional 10 min of incubation to allow the completion of synthesis of protein chains initiated during exposure to label. The spores were washed and broken in a glass bead shaker, and total detergent-soluble supernatant was analyzed by two-dimensional gels of the O'Farrell type (4).

The germination stage of *B. subtilis* extended over a period of about 30 min, and cell division

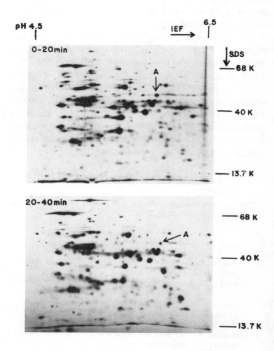

FIG. 1. *A 2-mg amount of spores was germinated in 10 ml of minimal medium (3), using 1 µg of unlabeled methionine and 60 µCi (top) or 10 µCi (bottom) of labeled methionine per ml. After labeling for 20 min, 100 µg of methionine per ml was added, and incubation at 37°C was continued for 10 min. Spores were disrupted by glass beads in a dental amalgamator. The lysate was examined by the O'Farrell procedure, with isoelectric focusing in the first dimension and sodium dodecyl sulfate (SDS) in the second (4). The gels were soaked in Enhance (New England Nuclear) and fluorographed on Kodak XR-5 X-ray film. Labeled methionine was added at zero time (top) or at 20 min (bottom).*

246

began at about 100 min. Uptake of labeled methionine was detectable during the first 20 min and accelerated greatly at 60 min (data not shown). Although little methionine was taken up during the first 20 min, a wide range of proteins was synthesized (Fig. 1). Indeed, the major burst of synthesis at 60 min produced a population of proteins with a similar composition, but with a few significant differences. These can be compared with successive outgrowth time points (Fig. 2 and 3) and with vegetative cells (Fig. 4). The protein patterns from all the outgrowth time points and the vegetative cells are much the same. Therefore, it does not appear that there is a problem with protease action modifying the proteins and introducing artifactual peptides. This was further supported by a control experiment in which labeled vegetative cells were mixed with unlabeled outgrowing spores, and the total mixture was fractionated on the two-dimensional system. No differences were observed with this mixed population in

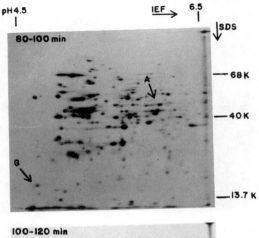

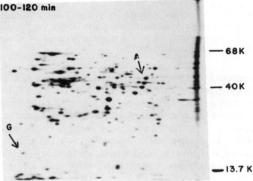

FIG. 3. Spores labeled at 80 min (top) or at 100 min (bottom) with 10 μCi of methionine per ml (see Fig. 1 legend).

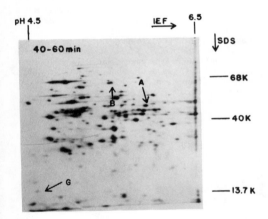

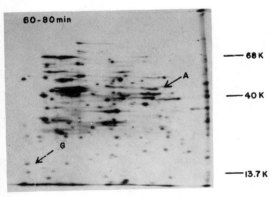

FIG. 2. Spores labeled at 40 min (top) or at 60 min (bottom) by addition of 10 μCi of methionine per ml (see Fig. 1 legend).

comparison with vegetative cells alone (not shown).

Although the protein populations appeared quite similar throughout outgrowth and into vegetative growth, careful examination by overlaying the fluorograms permitted the identification of at least 12 different classes of proteins, each class containing one or more members. Among these is class V, synthesized by vegetative cells, but not by outgrowing spores (for example, Fig. 4, V). Class A was synthesized from 0 to 100 min (for example, Fig. 1–3, A). Class B was synthesized during the period from 40 to 60 min only (Fig. 2, B). The synthesis of class G began at 40 min and became more substantial in vegetative cells (Fig. 2–4, G).

This partial list illustrates that some proteins are synthesized only during outgrowth, some only during vegetative growth, and still others only during a brief period of outgrowth. Periodicity of protein synthesis during outgrowth is well known (3, 5, 6). Also, mutants which are

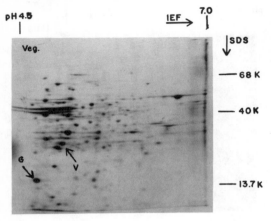

pH 4.5   IEF →   7.0

FIG. 4. *Vegetative cells at an absorbance at 600 nm of 0.2 labeled by addition of 10 μCi of methionine per ml followed by unlabeled methionine after 20 min. Cells were isolated and disrupted after an additional 10 min (see Fig. 1 legend).*

temperature sensitive in outgrowth but not in vegetative growth have been isolated (1, 2), indicating that there are proteins required for outgrowth that are not required for vegetative growth. Our results are therefore consistent with what is known about gene expression during spore outgrowth, and these data have made possible the direct identification of specific pro-

teins unique to specific stages of the life cycle. The factors influencing the differential levels of synthesis of these proteins are central to the question of developmental control in *B. subtilis*. Isolation of some of these proteins is in progress.

LITERATURE CITED

1. **Galizzi, A., A. M. Albertini, M. L. Baldi, E. Farrari, E. Isnenghi, and M. T. Zambelli.** 1978. Genetic studies of spore germination and outgrowth in *Bacillus subtilis*, p. 150–157. *In* G. Chambliss and J. C. Vary (ed.), Spores VII. American Society for Microbiology, Washington, D.C.
2. **Gallizzi, A., F. Gorrini, A. Rollier, and M. Polsinelli.** 1973. Mutants of *Bacillus subtilis* temperature sensitive in the outgrowth phase of spore germination. J. Bacteriol. **113**:1482–1490.
3. **Kennett, R. H., and N. Sueoka.** 1971. Gene expression during outgrowth of *Bacillus subtilis* spores. The relationship between gene order on the chromosome and temporal sequence of enzyme synthesis. J. Mol. Biol. **60**:153–160.
4. **O'Farrell, P. H.** 1975. High resolution two-dimensional electrophoresis of proteins. J. Biol. Chem. **250**:4007–4021.
5. **Steinberg, W., and H. O. Halvorson.** 1968. Timing of enzyme synthesis during outgrowth of spores of *Bacillus cereus*. I. Ordered enzyme synthesis. J. Bacteriol. **95**:469–478.
6. **Yeh, Y. C., and W. Steinberg.** 1978. Gene expression during outgrowth of *Bacillus subtilis* spores: influence of gene position, gene dosage, and DNA repair functions, p. 164–170. *In* G. Chambliss and J. C. Vary (ed.), Spores VII. American Society for Microbiology, Washington, D.C.

E. Resistance of Spores

Stabilization of Spore Enzymes to Heat by Reduction in Water Activity

A. D. WARTH

Commonwealth Scientific and Industrial Research Organisation Division of Food Research, North Ryde, New South Wales, 2113, Australia

Glucose-6-phosphate dehydrogenase and other spore enzymes were greatly stabilized to heat by reduction in water content. Dipicolinic acid and other low-molecular-weight spore solutes had little stabilizing effect. A water activity of 0.73 gave a stability in vitro similar to that in intact spores.

Spore enzymes are not in general intrinsically heat resistant, but are stabilized in the spore. In *Bacillus cereus* a number of enzymes were on average 39°C more stable in intact spores than in extracts (11). Two phenomena could account for the stabilization: reduction in water activity (a_w) and the presence of stabilizing substances. The effects may be interdependent, as reduction in a_w increases the concentration and probably the effectiveness of stabilizing factors. Dry proteins are very heat stable, but there are few quantitative data on the relationship between protein stability and water at intermediate levels. To evaluate the relative importance of partial dehydration and of specific protective factors in spore heat resistance, we need data on the heat stability of spore enzymes in the presence and absence of stabilizing compounds, as a function of the a_w and water content.

Determination of the a_w and water content necessary to give the same stabilization of enzymes as in the intact spore will give an indication of the water content of the core in vivo and the pressure needed to maintain it. Heat stability of enzymes at defined a_w during various stages in their purification should indicate the importance of protective (or damaging) factors present.

Glucose-6-phosphate dehydrogenase was chosen for this study because in *B. cereus* it is relatively heat sensitive and does not have any unusual properties (11). Leucine dehydrogenase, an intrinsically heat-stable enzyme (11), and glucose dehydrogenase were also studied.

RESULTS

Preparation and heating of spore extracts. An aqueous suspension of *B. cereus* spores was shaken with glass beads and quickly centrifuged. The supernatant (spore contents) contained spore enzymes, ribosomes, DNA, and calcium dipicolinate (CaDPA) and other solutes. For most experiments, freeze-dried contents were dissolved in a little water, spread in the heating tubes, and dried. X-ray diffraction showed no sign of CaDPA crystal formation. Tubes were heated in an apparatus in which the partial pressure of water vapor during equilibration and heating was controlled by the temperature of water in the side arm (Fig. 1).

Heat stability of glucose-6-phosphate dehydrogenase in solution and at high a_w. It is advantageous in studying the general effects of water content and spore stabilizing factors for the test enzyme to be relatively sensitive to heat and insensitive to changes in pH and ionic composition. Glucose-6-phosphate dehydrogenase (EC 1.1.1.49) meets these criteria. The enzyme, purified 10-fold from spores and heated in Tris buffer, showed only small (<40%) differences in inactivation rates at pH 7.0 to 7.8 and in Ca^{2+}, Mg^{2+}, EDTA, and DPA at concentrations up to 1 mM. Phosphate, at pH 6.3 to 7.3, reduced the inactivation rate by 60%. In Tris buffer there was no difference in stability between the crude extract and the partly purified enzyme, indicating that any stabilizing factors in spore contents were ineffective in dilute solution. Dried contents equilibrated to 0.99 a_w were more stable (Table 1). However, the stabilization of 1 to 2°C was small compared with the 39°C increase in heat resistance that this enzyme shows in vivo (11). During heating, the freeze-dried material remained solid but lost some of its open texture. Hence, even under these much more concentrated conditions, spore contents did not show a marked stabilizing effect.

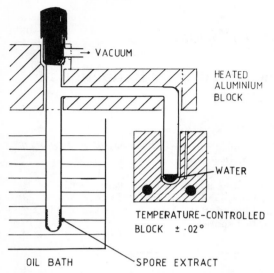

FIG. 1. *Apparatus for heating at controlled a_w. Samples were placed or dried in the main arm of side-arm tubes fitted with a removable Teflon stopcock. Water (200 μl) was placed in the side arm, which was then cooled, and the tubes were evacuated, sealed, and equilibrated at or below the required a_w at ambient temperature. For heating, the tubes were placed directly in the heating apparatus and removed at intervals over 3 h. The apparatus consisted of an oil bath to heat the samples, an aluminum block to control the temperature of the water in the side arm, and a milled aluminum block which largely enclosed and supported the tubes and kept all their remaining parts above the temperature of the side arm. The entire apparatus was enclosed and insulated with balsa wood coated with silicone rubber. Proportional controllers using thermistor sensors maintained sample and water temperatures to ±0.05°C. Temperature readings were made by thermocouples fixed to various parts of the apparatus.*

Effect of a_w on enzyme stability. At 0.80 a_w, glucose-6-phosphate dehydrogenase was inactivated more slowly at 65°C than it was at 45°C at 0.99 a_w (Table 1), which is equivalent to a decrease in inactivation rate of more than 1,000-fold. Lower a_w's gave additional large increases in stability (Table 1). To express this wide range of stability, inactivation rates were measured at two or more temperatures for each a_w, and the temperature at which the rate constant k was 0.010 min^{-1} was determined from a plot of log k against K^{-1}. There was a steady, almost linear increase in inactivation temperature as the a_w was lowered (Fig. 2).

Two other spore enzymes were investigated. Leucine dehydrogenase was very heat stable both in the spore and in Tris buffer (11). Glucose dehydrogenase was of intermediate stability in

spores and in Tris at pH 7.3 (11). Its heat stability however, is very dependent on pH, nucleotides, and salts (12). At 0.99 a_w, glucose dehydrogenase was much more stable than in Tris at pH 7.3 (Table 2), perhaps as a consequence of the lower pH of spore extract (5.8 after rehydration)

TABLE 1. *Heat inactivation of glucose-6-phosphate dehydrogenase in solution and at controlled a_w[a]*

a_w	Temp (°C)	Inactivation rate constant (min^{-1})
1.00[b]	45	0.037
0.99	45	0.027
1.00[b]	50	0.26
0.99	50	0.12
0.80	65	0.014
0.70	80	0.024
0.50	110	0.021
0.30	125	0.009
0.10	145	0.036

[a] *B. cereus* T spores were disrupted at 0°C by shaking for 10 min with 0.1-mm glass beads in water containing phenylmethylfluorosulfonate as protease inhibitor and were centrifuged at 20,000 × g for 15 min. The supernatant (spore contents fraction) was freeze-dried. Freeze-dried contents (50 mg) were dissolved in 1 ml of water, and 50 μl was immediately spread in "h" tubes (Fig. 1), dried in vacuo, and then equilibrated to an a_w approximately 20% less than that in the heating apparatus (Fig. 1). After heating, the contents were dissolved in Tris buffer and passed through Sephadex minicolumns, and the enzymes were assayed (11).

[b] Heated in 0.05 M Tris, pH 7.3.

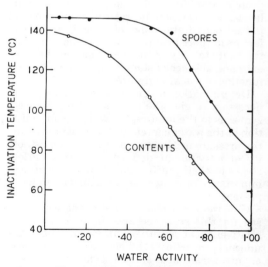

FIG. 2. *Heat stability of glucose-6-phosphate dehydrogenase at different a_w's. Symbols: ○, spore contents; ●, intact spores. Inactivation temperature was the temperature at which the inactivation rate constant k was 0.01 min^{-1}.*

TABLE 2. *Inactivation of* B. cereus *spore enzymes at reduced* a_w

Temp (°C)	a_w	Inactivation rate constant (min^{-1})		
		Glucose-6-phosphate dehydrogenase	Glucose dehydrogenase	Leucine dehydrogenase
55	—^{a}		0.07	
70	—^{a}			0.07
70	0.99	>1	0.006	0.027
70	0.80	>0.7	0.003	0.008
90	0.70	>0.7	0.004	0.022
110	0.50	0.021	0.011	0.016
135	0.30	0.073	0.068	0.083
135	0.10	0.009	0.011	0.015
145	0.10	0.036	0.030	0.040
145	0.05	0.027	0.021	0.034

^{a} Heated in 0.05 M Tris buffer, pH 7.3.

and the cations present. On the other hand, leucine dehydrogenase was less stable. The outstanding feature of the results at lower humidities is that the very large differences in stability of the three enzymes, evident in solution and at high a_w, largely disappeared at 0.5 a_w and lower (Table 2).

Inactivation of enzymes in spores. Spores were heated and the enzymes were subsequently extracted. All three enzymes showed greatly increased stability as the a_w was lowered to 0.50 and little further change at lower a_w. As with the in vitro experiments, differences in stability among the enzymes at low a_w were small. The dry heat stabilities were a little greater in the spore than in the extract. Inactivation rate constants at 145°C and 0.05 a_w were 0.011, 0.022, and 0.018 min^{-1} for glucose-6-phosphate, glucose, and leucine dehydrogenases, respectively.

A plot of inactivation temperatures ($k = 0.01$ min^{-1}) for glucose-6-phosphate dehydrogenase in spores (Fig. 2) paralleled that of the extract down to 0.50 a_w and was displaced by about 0.30 a_w unit. At 1.00 a_w (Tris buffer), the enzyme was 37°C more stable in vivo than in the spore contents preparation.

Estimation of the a_w and water content equivalent to the in vitro stability. In freshly prepared spore contents, dried directly in the heating tubes, glucose-6-phosphate dehydrogenase had the same stability at 0.73 a_w as it had in intact spores in water. The water content at 0.73 a_w and 85°C was 20% of the dry weight and at 25°C it was 28%.

Factors affecting enzyme heat stability at 0.7 a_w. Freeze-dried contents which had been redissolved in water and dried had lower stability than freshly prepared extract. Unexpectedly, suspensions of disrupted spores which had not

been centrifuged to remove spore integuments had greater heat stability. The contribution that CaDPA and other low-molecular-weight solutes make to stabilization under conditions of limited water content was investigated by removal of low-molecular-weight solutes by gel filtration on Sephadex G-25. The inactivation rate was reduced by 50%, which is insignificant compared with the approximately 10^4-fold stabilization due to water limitation.

Purification of the enzyme 10-fold by use of $(NH_4)_2SO_4$ precipitation and DEAE-cellulose chromatography reduced heat stability, suggesting that some macromolecular compounds in the spore extract might have a stabilizing role. Addition of ovalbumin restored heat stability. Several low-molecular-weight compounds including glycine, proline, potassium glutamate, sucrose, and the calcium salts of phosphoglycerate, sulfolactate, and glutamate were tested at pH 7.3. Only calcium glutamate restored heat stability to a level similar to that in unfractionated contents.

DISCUSSION

These results demonstrate directly that reduction in a_w greatly stabilizes spore enzymes and show the quantitative relationship between a_w and stability for a relatively heat-sensitive spore enzyme in a chemical environment approximating that of the spore core. The dryness necessary to achieve the same stability as in intact spores was ~0.73 a_w, corresponding to 20% water content, well within the estimates of spore core water content made from refractive index (6) or from dry density and buoyant density data (3, 10), which indicate water contents generally less than 20%.

Some factors could affect the accuracy of this result. Despite rapid disruption and drying, changes in the composition and structure of the spore contents, perhaps due to dilution, proteolysis, and surface effects during drying, would seem inevitable. Generation of peptides and amino acids by proteolysis would increase the water content for a given a_w. Insofar as these might reduce enzyme stability, then the in vivo a_w estimate will be low. However, the effect of a_w on stability was so great that large differences in inactivation rate had small effects on the equivalent a_w. The stabilizing effect of spore coats and cortex was not understood. Since the coats and cortex do not contact intracellular enzymes in vivo, the phenomenon was not considered of physiological significance. Likewise, the reduced stability of the partly purified enzyme (at 0.75 a_w) should be interpreted with caution, as, experimentally, very small quantities of protein were being handled and these may well suffer from surface effects. Ovalbumin, for example, was protective.

Reduction in a_w is capable of giving much greater stabilization than was achieved naturally by *B. cereus* spores. Stability of the enzymes in vivo increased uniformly with reduction in a_w, and the spore behaved as if it lowered the a_w by about 0.3 unit. This result is inconsistent with the swollen or osmoregulatory cortex model for the spore dehydration mechanism (2), as reduction in the external a_w would eliminate the coupling between cortical swelling pressure and the pressure on the core. Additionally, the a_w-lowering potential predicted for this model (2) is about one-tenth of the 0.25 a_w which appears necessary for enzyme stabilization. The pressure necessary to maintain the lowered water content of the core must result from tension in the cortex itself (7, 10). However, the cortex does appear expanded in the radial direction and is therefore anisotrophic in its mechanical properties (10). The expansion must depend considerably on electrostatic repulsion between the fixed anions in the cortex, which, in comparison with polymer swelling pressure and the osmotic effect of counterions, would not be greatly reduced by a small reduction in a_w.

At very low a_w, spores are less heat resistant than at an optimal a_w near 0.3 (1). Enzyme stability did not show this effect, suggesting that spore death under very dry conditions is not due to enzyme inactivation.

The apparent water content of 20% corresponds on the water absorption isotherm to the loss of most, but not all, of the free water (5). This is a biologically plausible result, for up to this point water is readily removable by reverse osmosis, but much higher pressures are required to remove much more water. Spores which have more effective heat-protective mechanisms (9) probably have lower internal water contents.

In contrast to the consistent and very large stabilization caused by a_w reduction, stabilizing factors in the spore clearly do not have the major role. This is so not only in dilute solution, where partly purified enzyme had the same stability as the crude extract, but also in conditions much more closely resembling the inside of the spore, e.g., at 0.99 a_w and 0.7 to 0.8 a_w, where removal of CaDPA and other low-molecular-weight solutes did not significantly reduce heat stability.

That is not to say that CaDPA, etc., do not interact with spore enzymes. As a result of their high concentration, they undoubtedly do (10). However, the heat stability of glucose-6-phosphate dehydrogenase in an environment containing a high concentration of CaDPA and other low-molecular-weight solutes is not very different from that seen when the molecular environment is other cell molecules. The mechanism of irreversible denaturation of proteins involves transitions mediated by a fluid solvent (8). Denial of this fluidity by replacing protein-water with protein-protein and other interactions prevents this mechanism and hence increases stability in a fashion similar to enzyme immobilization (4).

ACKNOWLEDGMENTS

I thank S. K. Meldrum for skilled technical assistance, I. C. Watt for the water absorption isotherm measurements, and J. Clousten for the X-ray diffraction.

LITERATURE CITED

1. **Alderton, G., and N. Snell.** 1970. Chemical states of bacterial spores: heat resistance and its kinetics at intermediate water activity. Appl. Microbiol. **17**:745–749.
2. **Gould, G. W., and G. J. Dring.** 1975. Heat resistance of bacterial endospores and concept of an expanded osmoregulatory cortex. Nature (London) **258**:402–405.
3. **Hsieh, L. K., and J. C. Vary.** 1975. Germination and peptidoglycan solubilization in *Bacillus megaterium* spores. J. Bacteriol. **123**:463–470.
4. **Klibanov, A. M.** 1979. Enzyme stabilization by immobilization. Anal. Biochem. **93**:1–25.
5. **Kuntz, I. D., and W. Kauzmann.** 1974. Hydration of proteins and polypeptides. Adv. Protein Chem. **28**:239–345.
6. **Leman, A.** 1973. Interference microscopical determination of bacterial dry weight during germination and sporulation. Jena Rev. 263–270.
7. **Lewis, J. C., N. S. Snell, and H. K. Burr.** 1960. Water permeability of bacterial spores and the concept of a contractile cortex. Science **132**:544–545.
8. **Privalov, P. L.** 1979. Stability of proteins. Adv. Protein Chem. **33**:167–241.
9. **Warth, A. D.** 1978. Relationship between the heat resistance of spores and the optimum and maximum growth temperatures of *Bacillus* species. J. Bacteriol. **134**:699–705.
10. **Warth, A. D.** 1978. Molecular structure of the bacterial spore. Adv. Microb. Physiol. **17**:1–45.
11. **Warth, A. D.** 1980. Heat stability of *Bacillus cereus* enzymes within spores and in extracts. J. Bacteriol. **143**:27–34.
12. **Yokota, A., K. Sasajima, and M. Yoneda.** 1979. Reactivation of inactivated D-glucose dehydrogenase of a *Bacillus* species by pyridine and adenine nucleotides. Agric. Biol. Chem. **43**:271–278.

Water Vapor Adsorption by *Bacillus stearothermophilus* Endospores

IAN C. WATT

Commonwealth Scientific and Industrial Research Organisation Division of Textile Physics, Ryde, New South Wales, 2112, Australia

Water vapor sorption isotherms were determined by a gravimetric technique on intact spores and fractions obtained from *Bacillus stearothermophilus* spores. The extracted fractions were, essentially, coat material, a mixture of coat plus cortex, and protoplast material. At low humidities the water content of the intact spores was the sum of the water contents of the fractions, but the coat played a determinant role in restricting the swelling of the spore near saturation.

Bacterial spores have a significant affinity for water, the amount sorbed varying with spore species. The relationship between water content and the relative humidity of the environs at a constant temperature, given by the adsorption isotherm, is usually not a unique relationship for spores and varies with the preparative treatments, the sorption history, and the conditions of adsorption (4). The hydration of bacterial spores is of particular complexity because of the variety of hydrophilic sites and differing physical structures and solubilities of the morphological components. The purpose of this study was to determine the distribution of water among the components of *Bacillus stearothermophilus* spores. These spores were chosen for study because of their high heat resistance characteristics.

Spores were prepared as previously described (3). Sorption isotherms were determined at different temperatures by a gravimetric technique; 10-mg samples of spore material were pressed into disks and suspended from a calibrated quartz helical spring balance enclosed in a sorption chamber connected to a water vapor reservoir and a wide-bore mercury manometer.

RESULTS AND DISCUSSION

Adsorption isotherms of intact endospores. The adsorption and desorption isotherms of intact *B. stearothermophilus* spores at 35°C are shown in Fig. 1. An increased water content when some of the outer spore coats were ruptured is evident above 75% relative humidity, and the divergence becomes greater approaching saturation. The water content at low humidities

was less for the pressed disks of spore material than for freeze-dried samples of the same material. It is probable that the pressing eliminates voids in the structure.

The effect of temperature on the water content of *B. stearothermophilus* spores over the range of 20 to 100°C is shown by the data in Table 1. There was a decrease in "dry" weight with increasing temperature, and subsidiary experiments indicated a weight loss of 1.2% at zero humidity on increasing the temperature from 20°C to 100°C. Consequently, the "dry" spore sample at 20°C has residual water compared with the same sample at higher temperatures.

The water content of the spores decreased with increasing temperature over the entire humidity range; the higher water contents at lower temperatures confirm that water vapor adsorption by bacterial spores is an exothermic reaction. Analysis by the D'Arcy and Watt equation (1) indicated that the sorbed water can be represented as water associated with high binding energy sites such as the basic and acidic side-chain polar groups, water sorbed on specific sites of lower binding energy, and multilayer formation through water molecules sorbed on the primary sorbed water, predominantly at high humidities. Limitations imposed by interfacial tension effects and mechanical constraints to swelling determine the ability of the water to form a multilayer. The more water which is condensed into the multilayer, the closer it becomes to liquid water in its properties.

Water contents of spore fractions. Three spore fractions composed essentially of protoplast, a coat plus cortex preparation, and, by removal of the cortical material, an outer coat

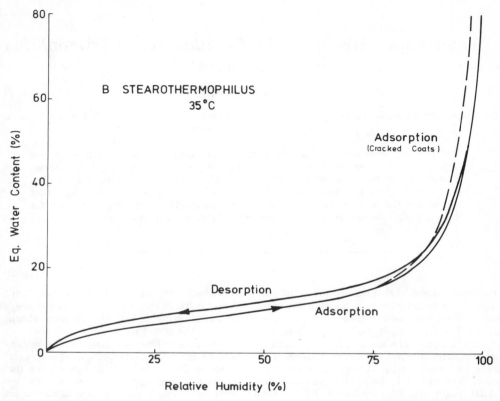

FIG. 1. *Water vapor adsorption and desorption isotherms at 35°C for* B. stearothermophilus *endospores pressed into a disk at 60 MPa of pressure. The isotherms were unchanged for pressures in the range of 30 to 120 MPa. The dashed curve represents the course of adsorption after a pressure of 600 MPa was applied to form the disk. Microscopic examination showed that many of the outer coats of the spores were cracked by this pressure.*

TABLE 1. *Water contents of* B. stearothermophilus *spores as a function of temperature*

Relative humidity (%)	Percent water content[a] at	
	20°C	100°C
5	3.1	2.3
20	6.3	4.7
35	8.4	6.5
50	10.8	8.6
65	13.7	11.1
80	18.5	14.1
90	28	21.5
100	85	53

[a] Equilibrium water content expressed as the weight of water as a percentage of the sample weight at zero humidity at the temperature of the isotherm determination.

fraction were obtained from *B. stearothermophilus* endospores. The sorptivity of each fraction was measured over the range of vapor pressures applicable for that fraction. The water contents, expressed as a percentage of the dry weight of the spore component, are shown in Table 2, with corresponding data for the intact spores.

At high relative humidities there was considerable divergence in sorptivity for the fractions, and the relation between the water content of the intact spores and the water contents of the fractions changed dramatically. This was most pronounced for the protoplast material, which showed a strong upward swing in the adsorption isotherm at relative humidities above 65% and went into solution before saturation. Similar behavior was exhibited by the coat plus cortex fraction, and values of water content above 80 to 90% relative humidity are not reliable for either fraction. The variability between fractions at low humidities, which was almost a factor of 2:1 for protoplast/coat material, is significant and can be related directly to the larger number of strongly hydrophilic sites in the protoplast fraction.

Because the water content of intact spores lies between the extremes of water content of the fractions at low and intermediate humidities, it

TABLE 2. *Percentage water contents of intact spores and fractions[a] of* B. stearothermophilus *at 35°C*

Relative humidity (%)	Intact spore	Spore fractions			
		Coat	Coat + cortex	Proto-plast	Cortex[b]
5	2.9	2.1	2.6	3.3	3.3
20	6.0	4.8	5.4	7.3	6.3
35	8.1	6.8	7.6	10.1	8.8
50	10.2	9.2	9.9	13.1	11.0
65	12.9	11.9	13.1	16.8	14.8
80	17.5	16.1	19.6	26.8	24.9
90	27.0	21.9	34	54	52
100	80	42	—	—	—

[a] The three fractions and the intact spore sample were prepared from spores of the same crop by the method of Warth et al. (5). Spores were disrupted by shaking with glass beads in water. The insoluble portion was a coat plus cortex fraction from which coat material was obtained by treatment with lysozyme. The soluble protoplast fraction was freeze-dried to obtain a sample for sorption measurements.

[b] Calculated values based on the amounts of cystine plus cysteine present in the coat and coat plus cortex fractions.

appears that the water content of the intact spores can be directly attributed to the sorptivities of the fractions and is in accord with the number of hydrophilic groups present. The lowest value of saturation water content was obtained for the coat material. This fraction has the highest concentration of the amino acid residue cystine, which provides disulfide cross-links between adjacent molecular chains. One of the most prevalent components in the protoplast of spores is calcium dipicolinic acid. However, this material made little contribution to water sorption at low and intermediate humidities, but a rapid increase in uptake approaching saturation reflects the lack of constraints to the swelling of the material in the free state.

It is reasonable to suppose that the level of swelling of the coat material is also a limiting factor with respect to the swelling of the interior of the intact spores. This deduction is supported by the fact that pressing of the spores, which ruptures the outer coats, enabled more water to be taken up at high humidities. The fact that there was no change in water content due to pressing at low and intermediate humidities implies that the coat has no mechanical influence on adsorption in these regions, an argument which is borne out by the relativity between

water contents and chemical composition of the spore material at relative humidities below 50%.

Sorptivity of the cortex. Although no cortex fraction was isolated, it is possible to deduce the sorption behavior from that of the two fractions—coat and coat plus cortex—if the composition of the mixture can be established. The hypothesis (2) that all of the cystine residues of bacterial spores are concentrated in the coat material has been adopted. Amino acid analyses of the coat and coat plus cortex fractions showed the cystine plus cysteine content of the coat plus cortex fraction to be six-tenths that of the coat material. A calculation made on this basis yielded the water contents listed in Table 2. There is considerable uncertainty in these values because the calculations are based on the differences between two experimental isotherm determinations and between two amino acid analyses.

The data suggest that the sorptivity of the cortex is only a little less than that of the protoplast. The cortex is richer than coat material in strongly hydrophilic primary sites for water sorption, as expected from the higher glutamic acid content, and exhibits a greater uptake at high humidities, reflecting the low number of covalent cross-links present. Clearly, the extracted fraction has a different sorption behavior when removed from the environment of the spore, reinforcing previous deductions as to the role of the coat in restricting the swelling of the spore at saturation.

ACKNOWLEDGMENTS

I gratefully acknowledge J. E. Algie and my colleagues at CSIRO Division of Food Research for the provision of the experimental samples.

LITERATURE CITED

1. **D'Arcy, R. L., and I. C. Watt.** 1970. Analysis of sorption isotherms of non-homogeneous sorbents. Trans. Faraday Soc. **66:**1236–1245.
2. **Kadota, H., K. Iijima, and A. Uchida.** 1965. The presence of keratin-like substance in spore coat of *Bacillus subtilis*. Agric. Biol. Chem. **29:**870–875.
3. **Marshall, B. J., and W. G. Murrell.** 1970. Biophysical analysis of the spore. J. Appl. Bacteriol. **33:**103–129.
4. **Neihof, R., J. K. Thompson, and V. R. Deitz.** 1967. Sorption of water vapor and nitrogen gas by bacterial spores. Nature (London) **216:**1304–1306.
5. **Warth, A. D., D. F. Ohye, and W. G. Murrell.** 1963. The composition and structure of bacterial spores. J. Cell Biol. **16:**579–592.

Biological Significance of Radiation Studies in Bacterial Spores

Laboratory of Radiation Biology, University of Texas, Austin, Texas 78712

Studies on physical and physicochemical mechanisms of radiation-induced damage in cells are possible in bacterial spores because of their resistance to environments ordinarily lethal to vegetative cells. This brief review describes some of these studies that demonstrate the existence of a number of independent radiation effects, the roles of water in modifying these, the participation of the hydroxyl radical in some, the importance of DNA as a target molecule, the multicomponent nature of the "O_2 effect," and the kinds of the associations of water with the critical target.

Most of the good information concerning the very reactive (and, therefore, very short-lived) free-radical and excited-state chemistry induced by X rays and gamma rays has been generated over the last 20 years with the use of pulsed-radiation systems. This information has been applied to the known body of information concerning steady-state irradiation of chemical systems with much success. When the biologist attempts to apply this mass of information to the cell, however, he finds that the ways in which the experiments are done in the two general systems (physicochemical and biological) make direct application of one to the other very difficult. Table 1 shows the differences in the experimental conditions which must be reconciled before accurate application of knowledge from one area to the other can be accomplished. In the radiation chemistry experiment, the conditions are very well defined, and an approach to "single variable" investigations is made, whereas in the irradiated cell, the physics and chemistry occur under very different conditions. It might seem that application of one area to the other is not possible.

Fortunately, several investigators have ignored the difficulties and have proceeded with some success. The use of microorganisms, and of the bacterial spore particularly, has allowed recognition of some of the fundamental actions of radiations in cells. The inertness of the spore and its capability to behave as a complex biological organism make it almost ideal for analysis of early events in the development of radiation injury in cells. I shall review here some of the radiation studies with the bacterial spore performed in several laboratories.

COMPARTMENTS OF DAMAGE

We began (over 30 years ago) believing that radiation effects are multitudinous in cells and that the many elements must be separated before analysis of the "effect" of radiation is possible (10). Our system exploits the fact that spore lethality induced by X rays is first order. As in Fig. 1, exponential behavior is observed in O_2, in N_2, and in N_2 followed by heat or radical scavenger treatment prior to exposure to O_2. These relations can be expressed as

$$N/N_0 = e^{-(k_I + k_{II} + k_{III})D}$$

in which three kinds of radiation damage are described: (i) that measured by k_I, the anoxic component of damage; (ii) that measured by k_{III}, an oxygen-dependent process(es) that involves reaction betwen a radiation-induced free radical and O_2, subsequent to radical formation; and (iii) that measured by k_{II}, which is a "fast" O_2 process, seen only when O_2 is present during irradiation. Note well that these are independent of one another. Each component has a separate, unique set of properties.

In all experiments it is required that the behavior of each component be recognized and marked. We separate a mélange into components that can be varied independently. There might, then, be hope of understanding the physics and chemistry basic to these separate components, all adding up to a general statement of radiation sensitivity.

SITE OF ACTION IN THE SPORE

The fact that these first-order relations strongly imply one-hit kinetics means that a

TABLE 1. *Summary of parameters of experiments in radiation chemistry contrasted with those in radiation biology*

Radiation chemistry	Radiation biology
1. Homogeneous	1. Heterogeneous and compartmentalized
2. Dilute solution	2. Concentrated solution
3. Aqueous	3. Aqueous and nonaqueous
4. Single variable	4. Many unknown variables
5. Single shot	5. Steady state
6. High dose rate	6. Low dose rate

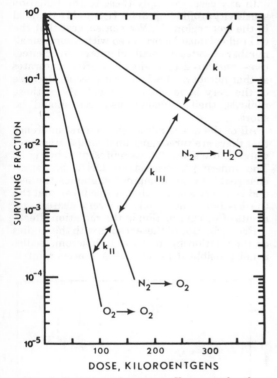

FIG. 1. *Response of spores to X rays under three experimental conditions. The vertical lines indicate the increase in the slope of the survival curve as the experimental conditions change (7).*

molecule of critical importance to the cell is affected; DNA is such a molecule, and there is direct evidence that lethality induced by radiation in the spore is truly due to damage to DNA.

1. The thymine substitute, 5-bromo-2'-deoxyuridine, incorporated into *thy⁻* spores of *Bacillus subtilis* by Al-Shaickly and Tallentire (1) changes radiation resistance. In these dried spores, class I damage was increased 45%, class II, 32%, and class III, 29% over the values seen in the normal spores. Further, whereas enhancement of class I by water appears to be independ-

ent of the 5-bromo-2'-deoxyuridine, the single O_2-dependent damage observed in unsubstituted cells is divisible into two on this basis, one affected and the other unaffected. The introduction of the "fake" thymidine changes the complex radiation response.

2. Studies by Tanooka and Sakakibara (17, 20) on transforming DNA support the notion that the physical configuration of DNA in the spore determines radiation sensitivity. Spores containing transforming DNA were irradiated with 6-MeV electrons or with UV. The DNA resistance was high compared with that in vegetative cells, in vitro, or in osmotically ruptured spheroplasts. This behavior parallels the resistance of spores vis à vis vegetative cells (12).

3. Tanooka (18, 19) demonstrated a reversible action of postirradiation heat on dried irradiated spores by measuring radiation-induced His⁺ revertants, and this reversal paralleled spore survival studies. Class III events participate in these two effects. Tanooka showed (19) that the killing/mutation ratio is the same in the wild type and the *polA* strains of *B. subtilis* spores, whether O_2 is present or absent during irradiation. This measures class II in our terms. It is the same for mutation and lethality. Or, mutagenesis and lethality are based on the same molecular mechanisms. Therefore, we deal with DNA changes when we measure lethality in these experiments. Inferences concerning molecular behavior of DNA are allowable from the lethality studies.

THE DRIED SPORE

One important candidate as an intermediary between radiation and biological damage is water, and many studies of water and its modification of radiation effects have been made in spores over the past 30 years. Some of these are briefly described here to show what inferences can be made concerning the role(s) of water in radiation damage.

Taking advantage of the resistance of the spore to hard vacuum, we can remove water and then restore it bit by bit to measure the importance of the *amount* of water in the spore. In Fig. 2 (16) note that, for the anoxic component, water changes sensitivity only at high water contents; for the two O_2 components, water is *protective*, reducing III to operative zero at saturation and reducing the overall O_2 effect by a factor of 3. The behavior of each of these components is peculiar to itself as water content varies. These relations have been observed in bacterial vegetative cells (21) and in the spore of the eucaryote *Aspergillus* (22). There is no one answer to the question: "What is the effect of

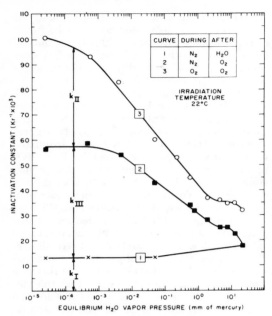

FIG. 2. *Responses of the various classes of radiation damage to increasing amounts of water in the spore (16).*

water on radiation sensitivity of the bacterial spore?" I believe the same is true for other organisms.

Another way of investigating the role of water is to change the *kind* of water, that is, substitute D_2O for H_2O (6). There are some differences in the physics of these two types of water and a little difference in the chemistry. If we add D_2O rather than H_2O to dry spores and measure the behavior of k_{II} and k_{III}, the results are about the same, that is, D_2O as a protective agent operates approximately like H_2O over most of the range. So, for the annealment of free radicals and the reduction of II, H_2O and D_2O added to the spore have about the same efficiency. The behavior of k_I, however, is affected by the substituton at high water contents, the region depicted by the slightly sloping upward line on Fig. 2.

At a critical point of water content, there is a sharp increase in the magnitude of the anoxic (k_I) component of damage, and this increases to a maximum that is approximately 1.3 times the value of the base line (5). But, in raising sensitivity, D_2O operates quantitatively differently from H_2O; the cells to which D_2O is added are always more sensitive, and the plateau is at a higher value (Fig. 3).

One difference between the chemsitry of H_2O and that of D_2O is that the production of $^{\cdot}OD$ is approximately 10% higher than that of $^{\cdot}OH$ for the same radiant energy absorbed (4). If this is

significant, then the rise in anoxic sensitivity at high water content is an expression of the reaction of this oxidative radical with the DNA target. Our adsorption studies (13) show that at monolayer formation there are 1.2 molecules of H_2O per phosphate of DNA, agreeing well with 2 H_2O per phosphate for in vitro DNA. But at saturation vapor pressure, we see 4.5 H_2O and 5.3 D_2O molecules per phosphate, contrasted to 20 for in vitro DNA. This may be a reflection of the restraints imposed by the physical state of DNA in the spore.

In any case, we apply these to the radiation sensitivity profile (Fig. 3) for H_2O and D_2O added in the wet region (5). We can suggest that the OH radical must be produced within some small number of water molecule diameters to be effective—perhaps 2 to 4, since the effect saturates at that number. This is not surprising. Because of the very large chemical reactivity of these radicals, their diffusion distances should be short.

All of the above allows the general inference that there are three functional compartments of water in the spore (9), as represented in Fig. 4. The innermost, nearest to DNA, is nonexchangeable by our pumping techniques; the second is a very small exchangeable layer; and the third is bulk water, only the very nearest elements of which function in the radiation effect.

Reconciliation of these results with the studies on the relationship of water to macromolecules is not possible at present. It is, however, inter-

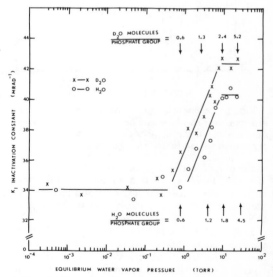

FIG. 3. *Response of anoxic sensitivity to addition of H_2O and D_2O. The arrows denote the estimated number of water molecules per phosphate DNA at the indicated equilibrium vapor pressure (5).*

esting that our compartmentalization of water into three kinds, based on responses to radiation in the spore, agrees formally with the ideas of several others (2), which include three kinds of water recognized experimentally in other systems by other means—bulk water, the water of "nonspecific" hydration, and that of "specific" hydration.

THE WET SPORE

A large series of experiments on spores in aqueous suspension gives other evidence concerning radiation mechanisms. The very early products of water radiolysis are

$$H_2O \rightarrow H^{\cdot}, \; {}^{\cdot}OH, \; e_{aq}^-, \; H_2, \; H_2O_2$$

It is possible, through manipulation of the rela-

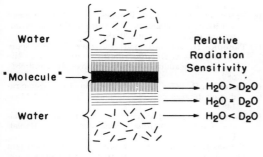

FIG. 4. *Scheme of kinds of water associated with target molecule.* $H_2O > D_2O$ *means that the spore is more sensitive to X rays when that compartment is occupied by* H_2O *than it is when occupied by* D_2O *(9).*

tive concentrations of the individual species, to assess their individual contributions to radiation damage. For instance, addition of N_2O gives

$$N_2O + H_2O + e_{aq}^- \rightarrow {}^{\cdot}OH + {}^-OH + N_2$$

resulting in doubling of ${}^{\cdot}OH$ yield while e_{aq}^- goes to zero. Addition of alcohols that react with ${}^{\cdot}OH$ reduces ${}^{\cdot}OH$ to zero without affecting e_{aq}^-.

Addition of N_2O to the suspending water markedly increases the radiation sensitivity of spores, implicating ${}^{\cdot}OH$, and the fact that addition of alcohol to this system prevents the sensitization confirms the implication (8). Thus, N_2O sensitizes by increasing [${}^{\cdot}OH$].

The radiation sensitizer of most importance in radiation biology is O_2. In the dried spore system, the evidence that O_2 acts through two mechanisms, one a free-radical mechanism, is clear, but evidence for two actions of O_2 in aqueous suspension has been found only recently. Tallentire et al. (14) showed that increases in [O_2] cause increases in sensitivity to the maximum, with a plateau at intermediate [O_2]. This finding suggested two kinds of action, one at low and one at high [O_2]. Subsequently, we (3) confirmed this (Fig. 5) and showed that alcohols partially reverse the sensitization observed at intermediate [O_2]. There may now be three O_2 effects: an ${}^{\cdot}OH$-associated and an ${}^{\cdot}OH$-independent effect at intermediate [O_2] and an effect at high [O_2].

Other experiments suggest two O_2 effects in the spore in aqueous suspension. With stopped-flow mixing and pulsed-irradiation techniques O_2 can be added in short times after the radiation pulse (11). There are two decay kinetics—one

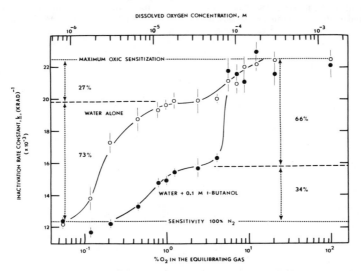

FIG. 5. *Radiation sensitivity of spores in water suspension related to [O_2]. Symbols:* $\bigcirc$, *in water alone;* $\bullet$, *in 0.1 M* t-butanol *(3).*

with a half-time of 9 s and the other with one of 120 s. In addition to suggesting two O_2 effects, these results strongly imply (15) that the spore suspended in water is not fully hydrated. In the dry spore, the lifetimes of O_2-sensitive radicals are very long; in vegetative cells, the lifetimes are on the order of 0.5 ms; and in the fully wetted spore, they are intermediate. The degree of wetness in the spore must, then, be intermediate.

LITERATURE CITED

1. **Al-Shaickly, M. A. S., and A. Tallentire.** 1974. 5-Bromouracil and oxygen effects in bacterial spores. Radiat. Res. **59**:249.

2. **Eagland, D.** 1975. Nucleic acids, peptides and proteins, p. 305–518. *In* F. Franks (ed.), Water: a comprehensive treatise. Aqueous solutions of amphiphiles and macromolecules, vol. 4. Plenum Press, New York.

3. **Ewing, D., and E. L. Powers.** 1976. Irradiation of bacterial spores in water: three classes of oxygen-dependent damage. Science **194**:1049–1051.

4. **Fielden, E. M., and E. J. Hart.** 1968. Primary radical yields and some rate constants in heavy water. Radiat. Res. **33**:426–436.

5. **Iwasaki, T., A. Tallentire, B. F. Kimler, and E. L. Powers.** 1974. The influence of added H_2O and D_2O on anoxic radiation sensitivity in bacterial spores. Radiat. Res. **57**:306–310.

6. **Powers, E. L.** 1965. Some physicochemical bases of radiation sensitivity in cells, p. 286–304. *In* Eighteenth annual symposium on fundamental cancer research, Cellular radiation biology. The Williams & Wilkins Co., Baltimore.

7. **Powers, E. L.** 1978. Water as a modulator of radiation damage to microorganisms, p. 97–118. *In* E. R. L. Gaughran and A. J. Goudie (ed.), Sterilization by ionizing radiation, vol 2. Multiscience Publications, Ltd., Montreal.

8. **Powers, E. L., and M. Cross.** 1970. Nitrous oxide as a sensitizer of bacterial spores to x-rays. Int. J. Radiat. Biol. **17**:501–514.

9. **Powers, E. L., and A. Tallentire.** 1968. The roles of water in the cellular effects of ionizing radiations, p. 3–

67. *In* M. Haissinsky (ed.), Actions chimiques et biologiques des radiations, vol. 12. Mason, Paris.

10. **Powers, E. L., R. B. Webb, and C. F. Ehret.** 1960. Storage, transfer and utilization of energy from x-rays in dry bacterial spores. Radiat. Res. **2**:94–121.

11. **Stratford, I. J., R. L. Maughan, B. D. Michael, and A. Tallentire.** 1977. The decay of potentially lethal oxygen-dependent damage in fully hydrated *Bacillus megaterium* spores exposed to pulsed electron irradiation. Int. J. Radiat. Biol. **32**:447–455.

12. **Tallentire, A.** 1970. Radiation resistance of spores. J. Appl. Bacteriol. **33**:141–146.

13. **Tallentire, A., J. R. Hayes, B. F. Kimler, and E. L. Powers.** 1974. H_2O and D_2O sorption studies on spores of *Bacillus megaterium*. Radiat. Res. **57**:300–305.

14. **Tallentire, A., A. B. Jones, and G. P. Jacobs.** 1972. The radiosensitizing actions of ketonic agents and oxygen in bacterial spores suspended in aqueous and nonaqueous milieux. Isr. J. Chem. **10**:1185–1197.

15. **Tallentire, A., R. L. Maughan, B. D. Michael, and I. J. Stratford.** 1977. Radiobiological evidence for the existence of a dehydrated core in bacterial spores, p. 649 ff. *In* A. N. Barker, D. J. Dring, D. J. Ellar, G. W. Gould, and J. Wolf (ed.), Spore research, vol. 2. Academic Press, London.

16. **Tallentire, A., and E. L. Powers.** 1963. Modification of sensitivity of x-irradiation by water in *Bacillus megaterium*. Radiat. Res. **20**:270–287.

17. **Tanooka, H.** 1968. Ultraviolet resistance of DNA in spore spheroplasts of *Bacillus subtilis* as measured by the transforming ability. Biochim. Biophys. Acta **166**:581–583.

18. **Tanooka, H.** 1978. Thermorestoration of mutagenic radiation damage in bacterial spores. Science **200**:1493–1494.

19. **Tanooka, H.** 1980. Oxygen effects and mutagenic radiation damage in *Bacillus subtilis* spores. Int. J. Radiat. Biol. **37**:556.

20. **Tanooka, H., and Y. Sakakibara.** 1968. Radioresistant nature of the transforming activity of DNA in bacterial spores. Biochim. Biophys. Acta **155**:130–142.

21. **Webb, R. B.** 1964. Physical components of radiation damage in cells, p. 267–285. *In* Physical processes in radiation biology. Academic Press, Inc., New York.

22. **Wilson, J. D., and E. L. Powers.** 1970. X-ray sensitivity and modifying effects of water in conidia of *Aspergillus nidulans*. Radiat. Res. **43**:698–710.

Injury and Resuscitation of Germination and Outgrowth of Bacterial Spores[1]

F. F. BUSTA, P. M. FOEGEDING, AND D. M. ADAMS

Department of Food Science and Nutrition, University of Minnesota, St. Paul, Minnesota 55108, and Campbell Soup Company, Camden, New Jersey 08101

Injury has long been recognized as a factor in the evaluation of apparent survival of bacterial spores after treatments to control these resistant entities. Heat, irradiation, and chemical treatments damage spores of anaerobic or aerobic bacteria. Injury has been manifested by increased sensitivity to selective or antimicrobial agents or by increased requirements for germination and growth and has been associated with germination or specific steps in outgrowth, or both. Damage of enzymes, DNA, membranes, or other systems has been implied by resuscitation studies. There is a continuing need for fundamental research on bacterial spore injury.

There has been an increasing awareness of injury in cells and spores after exposure to one or more environmental stresses. Recognition of sublethally damaged microorganisms is imperative for accurate interpretation of microbiological data. The concepts of injury in vegetative cells of bacteria, bacterial spores, yeasts, and molds and the cellular sites involved have been indicated or reviewed in several articles (3, 7, 9, 10, 14, 27, 32, 39, 41–47, 53, 55).

CURRENT SITUATION

Table 1 is a tabulation of reports of injury and resuscitation of spores of aerobic and anaerobic bacteria, showing the current status in the area of injury of spores of *Bacillus* and *Clostridium* species.

FUTURE NEEDS

Obviously, the reports of injury and resuscitation in Table 1 are not all-inclusive. Recent observations in our laboratory by L. S. Donnelly indicated that *Desulfotomaculum nigrificans* spores responded to multiple heat activation treatments in a manner that resembled injury. Also, K. M. Johnson in our laboratory has demonstrated that *Bacillus cereus* spores exposed to a sublethal heat treatment were recovered more readily with the addition of phosphate to a nonselective plating medium. Other laboratories must have similar unpublished data.

It is readily evident that there is need for more research in the area of spore injury and resuscitation. Use of the current information and new techniques to elucidate specific damage sites and

TABLE 1. *Summary of influence of various treatments or environmental stresses on spores of aerobic or anaerobic bacteria and the manifestations that indicate damage*

Stress	Spore of organism	Manifestation(s), proposed mechanism, or description	Reference
Heat			
50–70°C	*B. cereus*	Sucrose, glycerol, or NaCl required for stage V forespores to form colonies. The solute effect appeared to be related to establishing a suitable medium osmolality or water activity.	12
75–90°C	*B. cereus*	Microlag and microgermination times delayed. A germination system was injured; the cortex lytic enzyme activity was lost. Inactivation was due to interference with water imbibition into core during germination.	31
96–117°C	*B. pumilus*	Sensitivity to NaCl in recovery medium varied with stress dose.	8

[1] Paper No. 11430, Scientific Journal Series, Minnesota Agricultural Experiment Station, St. Paul.

TABLE 1. *Continued*

Stress	Spore of organism	Manifestation(s), proposed mechanism, or description	Reference
96–117°C, 155°C	*B. stearothermophilus*	Sensitivity to NaCl in recovery medium.	8, 17
121°C	*B. stearothermophilus*	Starch, charcoal, or sterile *B. stearothermophilis* culture supernatant fluid required in plating media. Culture supernatant fluid action was probably at the outgrowth or cell division stage. Lysozyme was ineffective in promoting recovery.	33
—	*B. stearothermophilus*	Sensitivity to various lots of commercial soybean casein digest medium varied with stress dose.	38
140–170°C	*B. subtilis*	Amino acids, particularly combinations of valine with isoleucine, or arginine, or glutamine, were required in minimal growth agar. Valine plus isoleucine provided an alternative germination system when the glucose and alanine system was inactivated.	30
90°C	*B. subtilis*	Glycine, homoserine, or threonine required in minimal agar medium. Amino acid requirements were genetically inherited. DNA damage was caused by heat. Heat did not cause single-strand DNA breaks.	54
81.8–95.1°C	*B. subtilis*	Complex reaction model for heat inactivation characterized by three states of decreasing activity of spores, followed by death, was proposed.	52
110–132°C	*B. subtilis*	$CaCl_2$ and disodium dipicolinate required in fortified nutrient agar. Altered glucose, NaCl, L-alanine, sodium phosphate germination system.	11, 22
115–125°C	*B. subtilis*	L-Alanine-induced germination was eliminated; germination below 35°C was enhanced. Thermodynamic values and influence of heat are consistent with protein denaturation mechanism.	4
96–117°C	*B. subtilis, B. subtilis* var. *niger*	Sensitivity to NaCl in recovery medium varied with stress dose.	8
110–115°C	*Bacillus* spp.	Sensitivity of germination to media components. Inhibitors were absorbed on starch, charcoal, or serum albumin.	34
—	*C. botulinum*	Sensitivity to media components. Inhibitors were absorbed on starch.	35
—	*C. botulinum*	Sensitivity to inhibitors in media, probably unsaturated fatty acids. Starch probably acted by absorbing inhibitors. Injured spores were more dependent on thioglycolate, pH, and incubation temperature.	36
105°C	*C. perfringens*	Sensitivity to antibiotics. Inactivation of the normal lytic system for cortical degradation during germination or inactivation of the mechanism for lytic enzyme release.	6
70–100°C	*C. perfringens*	Sensitivity to curing salts and antibiotics (surface-active agents) indicated spore membrane damage.	16
105–120°C	*C. perfringens*	Sensitivity to various plating media and antibiotics.	5
105°C	*C. perfringens*	Sensitivity to polymyxin and neomycin. Probably spore plasma or cortical membrane was damaged.	23
—	*C. perfringens*	Lysozyme or initiation protein required for germination. Lytic system apparently damaged.	13
105°C	*C. perfringens*	Lysozyme required for germination. Enumeration of all spores required sensitization to lysozyme.	2
90°C	*C. welchii*	Sensitivity to pH and incubation temperature. Longer incubation was needed for maximum count.	24
90°C	*C. welchii*	Sensitivity to certain gas atmosphere compositions. Inclusion of O_2 or CO_2 altered recovery. More heated spores were recovered in pure N_2 or H_2 than in a mixture.	25
98°C	*C. sporogenes*	Sensitivity to the presence of curing agents (NaCl, $NaNO_3$, $NaNO_2$). Injured organisms were more sensitive at pH 6 than at pH 7.	19

TABLE 1. *Continued*

Stress	Spore of organism	Manifestation(s), proposed mechanism, or description	Reference
110, 121°C	*C. sporogenes*	Sensitivity to various plating media.	37
—	*Clostridium* spp.	Sensitivity to inhibitors in media, probably unsaturated fatty acids. Starch probably acted by absorbing inhibitors. Injured spores were more dependent on thioglycolate, pH, and incubation temperature.	36
Cold			
7–8°C	*C. botulinum*	DNA breakage decreased spore recovery upon storage in several menstrua or lyophilized.	—[a]
Irradiation			
Ionizing	*B. megaterium*	Thermorestoration occurred. Viability was increased by heat if anoxic conditions were used for irradiation and heating.	40, 57
Ionizing	*B. pumilus*	Sensitivity to NaCl in recovery medium varied with stress dose.	8
UV	*B. pumilus*	Sensitivity to components of plating media; none in broth.	1
Ionizing	*B. stearothermophilus*	Sensitivity to NaCl in recovery medium varied with stress dose.	8
Ionizing	*B. subtilis*	Sensitivity to NaCl in recovery medium varied with stress dose.	8
—	*B. subtilis*	Single-strand breaks in DNA not repaired prior to germination.	28
Ionizing	*B. subtilis*	Single-strand breaks in DNA. Repair did not occur unless germination was induced, but occurred before normal DNA replication.	50
Ionizing	*B. subtilis*	Single-strand breaks in DNA. Rejoining of single-strand breaks occurred without detectable DNA synthesis and with little DNA degradation. Repair occurred in the initial germination stage, before DNA replication.	51
Ionizing	*B. subtilis*	Single-strand breaks in DNA. Radiation-resistant strains may have repaired single-strand breaks before germination and may have contained more DNA repair enzyme activity than sensitive strains.	21
Ionizing	*B. subtilis*	Single-strand breaks in DNA. Spore DNA was more resistant than cell DNA. DNA structural integrity was important to resistance.	49
Ionizing	*B. subtilis*	Heat and anoxic conditions permitted thermorestoration. DNA damage was the origin of thermorestoration.	48
Ionizing	*B. subtilis* var. *niger*	Sensitivity to NaCl in recovery medium varied with stress dose.	8
Ionizing	*C. botulinum*	Single-strand breaks in DNA decreased survivor curve shoulder and repair ability. DNA ligase rejoined DNA in radiation-dormant 33A during dormancy. Excision or other repair may have occurred after germination. No relationship of the number of genomes and radiation resistance.	29
—	*C. botulinum*	Single-strand breaks in DNA. *C. botulinum* may have repaired single-strand breaks.	28
Ionizing	*C. botulinum*	Sensitivity to incubation temperature, NaCl, and dilution. Repair appeared to involve protein synthesis, was more operative at 40 than at 30°C, and occurred during outgrowth.	15
Ionizing	*C. botulinum*	Single-strand breaks in DNA. Resistant strains may have repaired single-strand breaks before germination and may have contained more DNA repair enzyme activity than sensitive strains.	21
—	*C. botulinum*	Single-strand breaks in DNA. Repair dependent on Mg^{2+}, perhaps ligase.	—[b]

TABLE 1. *Continued*

Stress	Spore of organism	Manifestation(s), proposed mechanism, or description	Reference
Ionizing	C. perfringens	Sensitivity to sublethal heat treatment related to loss of ability to maintain the core in a dehydrated state. Presence of sucrose or glycerol reversed the effect.	26
Ionizing	C. welchii	Sensitivity to pH and incubation temperature. Longer incubation was needed for maximum count.	24
Ionizing	C. welchii	Sensitivity to certain gas atmospheric conditions. Inclusion of O_2 or CO_2 altered recovery.	25
Chemical			
EtO[c]	B. stearothermophilus	Sensitivity to various recovery media.	18
EtO	B. subtilis	Sensitivity to various recovery media.	18
EtO	C. welchii	Sensitivity to pH and incubation temperature. Longer incubation was needed for maximum count.	24
EtO	C. welchii	Sensitivity to certain gas atmospheric conditions. Inclusion of O_2 or CO_2 altered recovery.	25
H_2O_2	B. subtilis var. niger	Sensitivity to various plating media. Yeast extract, glucose, and vitamin-free Casamino Acids were important for recovery; optimal pH was 5.8 to 7.0; ferrous sulfate and manganous sulfate improved recovery.	56
NaOH	C. perfringens	Lysozyme required for germination. May be similar to heat injury requirement for lysozyme or initiation protein.	20

[a] A. S. Ntamere and N. Grecz, Abstr. Annu. Meet. Am. Soc. Microbiol. 1979, P4, p. 211.
[b] C. Waitr, Abstr. Annu. Meet. Am. Soc. Microbiol. 1974, E96, p. 17.
[c] EtO, Ethylene oxide.

mechanisms should permit greater insight into the complexities of spore resistance, increase understanding of spore germination and outgrowth, and improve recovery of injured spores.

ACKNOWLEDGMENTS

This research was supported in part by University of Minnesota Agricultural Experiment Station Project 18-59. P.M.F. was supported in part by a Ralston Purina Food Science Fellowship.

LITERATURE CITED

1. **Abshire, R. L., B. Bain, and T. Williams.** 1980. Resistance and recovery studies on ultraviolet-irradiated spores of *Bacillus pumilus*. Appl. Environ. Microbiol. **39:**695–701.
2. **Adams, D. M.** 1974. Requirement for and sensitivity to lysozyme by *Clostridium perfringens* spores heated at ultrahigh temperatures. Appl. Microbiol. **27:**797–801.
3. **Adams, D. M.** 1978. Heat injury in bacterial spores. Adv. Appl. Microbiol. **23:**245–261.
4. **Adams, D. M., and F. F. Busta.** 1972. Heat injury as the selective inactivation of a *Bacillus subtilis* spore germination system, p. 368–377. *In* H. O. Halvorson, R. Hanson, and L. L. Campbell (ed.), Spores V. American Society for Microbiology, Washington, D.C.
5. **Barach, J. T., D. M. Adams, and M. L. Speck.** 1974. Recovery of heated *Clostridium perfringens* type A spores on selective media. Appl. Microbiol. **28:**793–797.
6. **Barach, J. T., R. S. Flowers, and D. M. Adams.** 1975. Repair of heat-injured *Clostridium perfringens* spores during outgrowth. Appl. Microbiol. **30:**873–875.
7. **Beuchat, L. R.** 1978. Injury and repair of gram-negative bacteria, with special considerations of the involvement of the cytoplasmic membrane. Adv. Appl. Microbiol. **23:**219–243.
8. **Briggs, A., and S. Yazdany.** 1970. Effect of sodium chloride on the heat and radiation resistance and on the recovery of heated and irradiated spores of the genus *Bacillus*. J. Appl. Bacteriol. **33:**621–632.
9. **Busta, F. F.** 1976. Practical implications of injured microorganisms in food. J. Milk Food Technol. **39:**138–145.
10. **Busta, F. F.** 1978. Introduction to injury and repair of microbial cells. Adv. Appl. Microbiol. **23:**195–201.
11. **Busta, F. F., and D. M. Adams.** 1972. Identification of a germination system involved in the heat injury of *Bacillus subtilis* spores. Appl. Microbiol. **24:**412–417.
12. **Busta, F. F., E. Baillie, and W. G. Murrell.** 1977. Heat-induced sublethal damage of *Bacillus cereus* forespores, p. 431–450. *In* A. N. Barker, L. J. Wolf, D. J. Ellar, G. J. Dring, and G. W. Gould (ed.), Spore research 1976, vol. 2. Academic Press, London.
13. **Cassier, M., and M. Sebald.** 1969. Germination lysozyme-dépendante des spores de *Clostridium perfringens* ATCC 3624 après traitement thermique. Ann. Inst. Pasteur Paris **117:**312–324.
14. **Cerf, O.** 1977. Tailing of survival curves of bacterial spores. J. Appl. Bacteriol. **42:**1–19.
15. **Chowdhury, M. S. U., D. B. Rowley, A. Anellis, and H. S. Levinson.** 1976. Influence of postirradiation incubation temperature on recovery of radiation-injured *Clostridium botulinum* 62A spores. Appl. Environ. Microbiol. **32:**172–178.
16. **Chumney, R. K., and D. M. Adams.** 1980. Relationship between the increased sensitivity of heat injured *Clostridium perfringens* spores to surface active antibiotics and to sodium chloride and sodium nitrite. J. Appl. Bacteriol. **49:**55–63.
17. **Cook, A. M., and R. J. Gilbert.** 1969. The effect of sodium chloride on heat resistance and recovery of heated spores of *Bacillus stearothermophilus*. J. Appl. Bacteriol. **32:**96–102.
18. **Davis, S. B., R. A. Carls, and J. R. Gillis.** 1978. Recov-

ery of sublethal sterilization damaged *Bacillus* spores in various culture media. Dev. Ind. Microbiol. **1978**: 427–438.

19. **Duncan, C. L., and E. M. Foster.** 1968. Role of curing agents in the preservation of shelf-stable canned meat products. Appl. Microbiol. **16**:401–405.

20. **Duncan, C. L., R. G. Labbe, and R. R. Reich.** 1972. Germination of heat- and alkali-altered spores of *Clostridium perfringens* type A by lysozyme and an initiation protein. J. Bacteriol. **109**:550–559.

21. **Durban, E., N. Grecz, and J. Farkas.** 1974. Direct enzymatic repair of deoxyribonucleic acid single-strand breaks in dormant spores. J. Bacteriol. **118**:129–138.

22. **Edwards, J. L., Jr., F. F. Busta, and M. L. Speck.** 1965. Heat injury of *Bacillus subtilis* spores at ultrahigh temperatures. Appl. Microbiol. **13**:858–864.

23. **Flowers, R. S., and D. M. Adams.** 1976. Spore membrane(s) as the site of damage within heated *Clostridium perfringens* spores. J. Bacteriol. **125**:429–434.

24. **Futter, B. V., and G. Richardson.** 1970. Viability of clostridial spores and the requirements of damaged organisms. I. Method of colony count, period and temperature of incubation, and pH value of the medium. J. Appl. Bacteriol. **33**:321–330.

25. **Futter, B. V., and G. Richardson.** 1970. Viability of clostridial spores and the requirements of damaged organisms. II. Gaseous environment and redox potentials. J. Appl. Bacteriol. **33**:331–341.

26. **Gomez, R. F., D. E. Gombas, and A. Herrero.** 1980. Reversal of radiation-dependent heat sensitization of *Clostridium perfringens* spores. Appl. Environ. Microbiol. **39**:525–529.

27. **Gray, T. R. G., and J. R. Postgate.** 1976. The survival of vegetative microbes. Cambridge University Press, London.

28. **Grecz, N., and J. Grice.** 1978. The use of alkaline sucrose gradient sedimentation of DNA to study injury and repair of bacterial spores. Spore Newsl. **6**(Special Issue, March 1978):61–62.

29. **Grecz, N., C. Wiatr, E. Durban, T. Kang, and J. Farkas.** 1978. Bacterial spores: biophysical aspects of recovery from radiation injury. J. Food Process. Preserv. **2**:315–337.

30. **Gurney, T. R., and L. B. Quesnel.** 1979. Dry-heat induced and sub-lethal damage in *Bacillus subtilis* spores. Spore Newsl. **6**(10):16–17.

31. **Hashimoto, T., W. R. Frieben, and S. F. Conti.** 1972. Kinetics of germination of heat-injured *Bacillus cereus* spores, p. 409–415. *In* H. O. Halvorson, R. Hanson, and L. L. Campbell (ed.), Spores V. American Society for Microbiology, Washington, D.C.

32. **Hurst, A.** 1977. Bacterial injury: a review. Can. J. Microbiol. **23**:935–944.

33. **Labbe, R. G.** 1979. Recovery of spores of *Bacillus stearothermophilus* from thermal injury. J. Appl. Bacteriol. **47**:457–462.

34. **Murrell, W. G., A. M. Olsen, and W. J. Scott.** 1950. The enumeration of heated bacterial spores. II. Experiments with *Bacillus* species. Aust. J. Sci. Res. Ser. B **3**:234–244.

35. **Olsen, A. M., and W. J. Scott.** 1946. Influence of starch in media used for detection of heated bacterial spores. Nature (London) **157**:337.

36. **Olsen, A. M., and W. J. Scott.** 1950. The enumeration of heated bacterial spores. I. Experiments with *Clostridium botulinum* and other species of *Clostridium*. Aust. J. Sci. Res. Ser. B **3**:219–233.

37. **Pflug, I. J., M. Scheyer, G. M. Smith, and M. Kopel-**

man. 1979. Evaluation of recovery media for heated *Clostridium sporogenes* spores. J. Food Protect. **42**: 946–947.

38. **Pflug, I. J., G. M. Smith, and R. Christensen.** 1979. Effect of the lot of soybean casein digest medium on the number of *B. stearothermophilus* spores recovered. Spore Newsl. **6**(10):15–16.

39. **Pierson, M. D., R. F. Gomez, and S. E. Martin.** 1978. The involvement of nucleic acids in bacterial injury. Adv. Appl. Microbiol. **23**:263–285.

40. **Powers, E. L., R. B. Webb, and C. F. Ehret.** 1960. Storage, transfer, and utilization of energy from x-rays in dry bacterial spores. Radiat. Res. Suppl. **2**:94–121.

41. **Ray, B., and M. L. Speck.** 1973. Freeze-injury in bacteria. Crit. Rev. Clin. Lab. Sci. **4**:161–213.

42. **Roberts, T. A.** 1970. Recovering spores damaged by heat, ionizing radiations or ethylene oxide. J. Appl. Bacteriol. **33**:74–94.

43. **Schmidt, C. F.** 1955. The resistance of bacterial spores with reference to spore germination and its inhibition. Annu. Rev. Microbiol. **9**:387–400.

44. **Skinner, F. A., and W. B. Hugo (ed.).** 1976. Inhibition and inactivation of vegetative microbes. Academic Press, Inc., New York.

45. **Speck, M. L., and B. Ray.** 1977. Effects of freezing and storage on microorganisms in frozen foods: a review. J. Food Protect. **40**:333–336.

46. **Stevenson, K. E., and T. R. Graumlich.** 1978. Injury and recovery of yeasts and mold. Adv. Appl. Microbiol. **23**:203–217.

47. **Strange, R. E.** 1976. Microbial response to mild stress. Meadowfield Press Ltd., Durham, England.

48. **Tanooka, H.** 1978. Thermorestoration of mutagenic radiation damage in bacterial spores. Science **200**:1493–1494.

49. **Tanooka, H., and H. Terano.** 1970. Resistance of DNA against radiation-induced strand breakage in bacterial spores. Radiat. Res. **43**:613–626.

50. **Terano, H., H. Tanooka, and H. Kadota.** 1969. Germination-induced repair of single-strand breaks of DNA in irradiated *Bacillus subtilis* spores. Biochem. Biophys. Res. Commun. **37**:66–71.

51. **Terano, H., H. Tanooka, and H. Kadota.** 1971. Repair of radiation damage to deoxyribonucleic acid in germinating spores of *Bacillus subtilis*. J. Bacteriol. **106**:925–930.

52. **Toda, K.** 1970. Studies on heat sterilization. VIII. Complex reaction model for heat inactivation of bacterial spores. J. Ferment. Technol. **48**:811–818.

53. **Tomlins, R. I., and Z. J. Ordal.** 1976. Thermal injury and inactivation of vegetative bacteria, p. 153–190. *In* F. A. Skinner and W. B. Hugo (ed.), Inhibition and inactivation of vegetative microbes. Academic Press, London.

54. **Uchida, A., and H. Kadota.** 1979. DNA injury in *Bacillus subtilis* cells induced by heat treatment. Spore Newsl. **6**(8):222.

55. **van Schothorst, M.** 1976. Resuscitation of injured bacteria in foods, p. 317–328. *In* F. A. Skinner and W. B. Hugo (ed.), Inhibition and inactivation of vegetative microbes. Academic Press, London.

56. **Wallen, S. E., and H. W. Walker.** 1979. Influence of media and media constituents on recovery of bacterial spores exposed to hydrogen peroxide. J. Food Sci. **44**: 560–563.

57. **Webb, R. B., E. L. Powers, and C. F. Ehret.** 1960. Thermorestoration of radiation damage in dry bacterial spores. Radiat. Res. **12**:682–693.

Preparation and Characterization of Various Salt Forms of *Bacillus megaterium* Spores

R. E. MARQUIS, E. L. CARSTENSEN, S. Z. CHILD, AND G. R. BENDER

Departments of Microbiology and Electrical Engineering, The University of Rochester, Rochester, New York 14642

Spores of *Bacillus megaterium* ATCC 19213 were subjected to an ion-exchange regimen that resulted in nearly complete exchange of cations from the core and enveloping structures without major viability losses. The extreme immobilization of electrolytes characteristic of bacterial spores was found not to be related specifically to calcification. The state of hydration of the spores in aqueous media was not affected by conversion of one salt form to another, but heat resistance was. The order of resistance among the salt forms was Mn > native > Ca, Mg > K > Na, H.

Previous dielectric studies (2, 3) of bacterial spores have shown that cell electrolytes, especially those in the core, have extremely low mobilities. Speculation on the bases for this striking immobilization led us to prepare various salt forms of spores so that we could determine whether immobilization is due to chemical peculiarities of calcium, the major spore mineral.

Of the procedures described in the literature for altering spore mineral contents, ion exchange (1) seemed the most versatile. Initial attempts to carry out extensive ion exchange with spores of *Bacillus cereus* subsp. *terminalis* failed because acidification of suspensions to pH values lower than 4 resulted in loss of refractility and death. However, we were successful with spores

of *B. megaterium* formed in the medium of Slepecky and Foster (5). The procedure developed involved use of aqueous suspensions containing 2 g (wet weight) of spores per 50 ml. These were titrated to a pH of 2, over 2 to 3 h, with intermittent addition of 0.02 N HCl solution and then were incubated at 60°C for up to 18 h to obtain protonated or H spores. As shown in Table 1, H spores were found to be nearly devoid of calcium and other minerals but viable. Also shown are the extents of replacement or remineralization possible when H spores were titrated with appropriate bases and incubated at 60°C for up to 72 h. All of the desired salt forms could be prepared, although all were contaminated to a degree with Na, presumably from the vessels

TABLE 1. *Mineral contents of native and salt forms of spores of* B. megaterium[a]

Spore type	Mineral content (μmol/mg of spore dry weight)					Total cationic equivalents (μeq/mg)
	Ca^{2+}	Mg^{2+}	Mn^{2+}	K^+	Na^+	
Native	0.65	0.17	0.18	0.12	0.23	2.35
Hydrogen	0.02	0.01	0.02	0.00	0.17	0.27
Calcium	0.49	0.02	0.03	0.01	0.24	1.33
Magnesium	0.12	0.33	0.05	0.02	0.21	1.23
Manganese	0.03	0.02	0.52	0.01	0.14	1.29
Potassium	0.08	0.05	0.01	0.46	0.28	1.02
Sodium	0.02	0.03	0.01	0.03	0.34	0.49

[a] Suspensions of native spores were titrated with HCl to pH 2 and heated at 60°C for approximately 18 h to produce the hydrogen form. The suspensions were then back-titrated to pH 8 with mixtures of NH_4OH plus 0.5 M $CaCl_2$, NH_4OH plus 0.5 M $MgCl_2$, NH_4OH plus 0.5 M $MgCl_2$, NH_4OH plus 0.5 M $MnCl_2$, KOH plus 0.5 M KCl, or NaOH plus 0.5 M NaCl and incubated at 60°C to produce the various salt forms. Mineral assays were carried out by means of atomic absorption spectrophotometry of extracts of spores which were autoclaved and acid treated. Total counts of the hydrogen spore suspensions made with a Petroff-Hausser chamber indicated 9.3×10^9 spores per ml; viable counts on tryptic soy agar indicated 7.5×10^9 spores per ml. These two numbers were not different statistically based on estimation of the counting errors at the 95% confidence limit.

TABLE 2. *Dielectric properties of various salt forms of spores of* B. megaterium[a]

Spore type	Frequency (MHz)	Spore conductivity, σ_2 (mho/m)		Spore dielectric constant, k_2	
		$\sigma_1 =$ 0.05	$\sigma_1 =$ 0.50	$\sigma_1 =$ 0.05	$\sigma_1 =$ 0.50
Native	1	0.05	0.17	202	260
	50	0.09	0.25	50	46
H	1	0.04	0.20	170	300
	50	0.05	0.30	42	40
Li	1	0.04	0.25	280	350
	50	0.10	0.37	44	52
Na	1	0.04	0.21	320	340
	50	0.10	0.34	49	48
K	1	0.09	0.20	320	260
	50	0.20	0.33	50	45
Ca	1	0.04	0.10	330	350
	50	0.10	0.22	50	50
Mn	1	0.02	0.17	150	250
	50	0.03	0.25	50	60

[a] The values of σ_1 are the conductivity of the suspending NaCl solution in mhos per meter.

and reagents used. The extents of exchange here are amazing and must involve core electrolytes as well as those in enveloping layers. Full remineralization in terms of total cationic equivalents was not attained with any of the salt forms, possibly because of loss of dipicolinate. For example, the dipicolinate level in native spores, determined by the method of Scott and Ellar (4), was 0.91 µmol/mg of spore dry weight compared with 0.61 for H spores after 18 h at pH 2.

In Table 2, some of the cardinal dielectric values for selected salt forms are presented. Overall, the data indicate no large differences among the various forms, although Mn spores appeared to have lowest inherent conductivity and also greatest heat resistance. It is clear that electrolyte immobilization in spores cannot be related specifically to calcification.

We found in some of our early experiments that viability declined during exchange. This loss could be very much reduced by reducing the heating times to minima needed for exchange. As shown by the data presented in Fig. 1, H spores are heat sensitive (but less so than are vegetative cells). The relative heat resistances of the forms tested were Mn > native > Mg, Ca > K > Na, H.

The data presented in Table 3 show that variation in heat resistance among the various forms was not related to changes in degrees of hydration, indicated here by dextran-imperme-

able volumes per gram (dry weight) of spores. All salt forms appeared to be hydrated in aqueous suspension to the same extent, which was much less than that of vegetative cells or outgrowing cells and less than that of spores germinated with alanine and inosine. Decoated

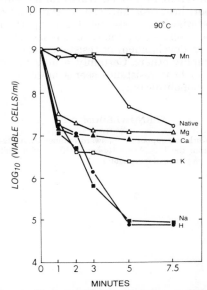

FIG. 1. *Heat sensitivities of various salt forms of* B. megaterium. *Samples (0.1-ml volumes) of spore suspensions were added to 0.9-ml volumes of water at 90°C in tubes in a heating block. After thorough mixing, 0.1-ml samples were taken at intervals and transferred to dilution blanks of 1% Difco peptone broth at room temperature. After further dilution, 0.1-ml samples were plated on the surfaces of tryptic soy agar plates. The plates were incubated at 37°C for at least 48 h before counting.*

TABLE 3. *Hydration of various forms of spores of* B. megaterium *in dextran solutions*

Cell type	Avg dextran-impermeable volume[a] (ml/g of cell dry wt)
Vegetative	7.29
Native spore	2.59
Germinated spore	5.36
Outgrowing spore	7.39
Decoated spore	1.79
H spore	2.62
Na spore	2.48
K spore	2.70
Li spore	2.58
Ca spore	2.61
Mn spore	2.42

[a] Dextran-impermeable volumes were determined by use of high-molecular-weight dextrans as described in detail previously (2).

spores of the sort we used previously (2) were even less hydrated than were native spores.

The data presented here are compatible with a view of heat resistance based on a series of mechanisms, including initial dehydration early in sporogenesis, maturation of the spore, and mineralization. Moreover, Warth (6) has shown that spore heat resistance can be related directly to vegetative heat resistance for any particular species. Apparently, many processes are involved in development of resistance, and some may amplify others. Certainly, a multifactorial theory of heat resistance seems more tenable than a monolithic one.

ACKNOWLEDGMENT

This work was supported by award number DAAG29-80-C-0051 from the U.S. Army Research Office, with Philipp Gerhardt as principal investigator.

LITERATURE CITED

1. **Alderton, G., and N. Snell.** 1963. Base exchange and heat resistances in bacterial spores. Biochem. Biophys. Res. Commun. **10**:139–143.
2. **Carstensen, E. L., R. E. Marquis, S. Z. Child, and G. R. Bender.** 1979. Dielectric properties of native and decoated spores of *Bacillus megaterium*. J. Bacteriol. **140**:917–928.
3. **Carstensen, E. L., R. E. Marquis, and P. Gerhardt.** 1971. Dielectric study of the physical state of electrolytes and water within *Bacillus cereus* spores. J. Bacteriol. **107**:106–113.
4. **Scott, I. R., and D. J. Ellar.** 1978. Study of calcium dipicolinate release during bacterial spore germination by using a new, sensitive assay for dipicolinate. J. Bacteriol. **135**:133–137.
5. **Slepecky, R. A., and J. W. Foster.** 1979. Alterations in metal content of spores of *Bacillus megaterium* and the effects on some spore properties. J. Bacteriol. **78**:117–123.
6. **Warth, A. D.** 1978. Relationship between the heat resistance of spores and the optimum and maximum growth temperatures of *Bacillus* species. J. Bacteriol. **134**:699–705.

Interaction of Nitrosothiol and Iodoacetate with Membrane Sulfhydryls of Outgrowing *Bacillus cereus* Spores

SHELDON L. MORRIS AND J. NORMAN HANSEN

Division of Biochemistry, Department of Chemistry, University of Maryland, College Park, Maryland 20742

Nitrosothiols, iodoacetate, and a number of other inhibitors derived from nitrite were all found to exert inhibitory effects on *Bacillus cereus* spores by interfering with membrane sulfhydryl groups. This type of membrane sulfhydryl group may be responsible for the inhibitory effects of currently used antimicrobial agents such as nitrite.

Nitrosothiols with the general structure $RSN{=}O$ inhibit germination and outgrowth in *Bacillus cereus* spores (5, 8). Nitrosothiols can be considered as structural analogs of nitrous acid ($HON{=}O$) in which the hydroxyl group is replaced by an RS group. It is possible that nitrosothiols and nitrite or its conjugate acid may undergo similar types of reactions. Because of this similarity, nitrite and nitrosothiols may exert their inhibitory functions in analogous ways. Indeed, nitrosothiols as well as other substances derived by reaction with nitrite have been implicated as possible components of nitrite action. These include the Roussin black salts (2, 6), "Perigo factor" (3, 5, 9), lactoferrin, and transferrin (M. C. Custer and J. N. Hansen, Abstr. Annu. Meet. Am. Soc. Microbiol. 1980, P18, p. 192). Because of their relatively simple known structures and ease of preparation (1, 5, 7, 8), nitrosothiols are suitable for detailed mechanistic studies. We have undertaken a study of the chemical mechanism by which nitrosothiols interact with a spore component(s) and of spore sites which are of critical importance to the inhibitory effect. We do this in the hope that knowledge of the sites of interaction which cause outgrowth inhibition by nitrosothiols will lead to understanding of fundamental outgrowth processes and permit the rational design of antimicrobial agents of practical value.

RESULTS AND DISCUSSION

We previously observed that the effectiveness of nitrosothiols as inhibitors of *B. cereus* was determined by the polarity of the R group, the effectiveness being greater for electron-withdrawing R groups (8). Conversely, hydrophobicity and charge had no effect. Since inhibitory effectiveness is not a function of hydrophobicity or charge, it is likely that the inhibitory event does not require transport of nitrosothiol across the spore membrane. This is further supported by results shown in Table 1. Radioactive coenzyme A was nontransportable in either germinated spores or vegetative cells; nevertheless, the nitrosothiol of coenzyme A inhibited spore outgrowth. Furthermore, the effectiveness of the nitrosothiols prepared from 8-mercaptopurine riboside and from 8-mercaptopurine riboside-5'-phosphate was essentially the same (data not shown). Nucleosides are generally much more easily transported than their corresponding nucleotides, so it appears from these results that the site of inhibitory action must be accessible to inhibitor molecules which have not traversed the membrane.

An important aspect of nitrosothiol action is its reversibility. Spores which are inhibited by nitrosothiol recover immediately after a water wash (5). This reversibility makes identification of the site of interaction difficult. Identification of the binding site was therefore attempted by competition experiments involving inhibitor mixtures. Several common protein-modifying reagents were screened for their ability to inhibit spore outgrowth and to interfere with inhibition by nitrosothiol. Iodoacetate, a sulfhydryl group modifier, irreversibly inhibited outgrowth at the same morphological stage as nitrosothiols. Moreover, addition of nitrosothiol immediately prior to addition of iodoacetate prevented iodoacetate inhibition. Labeled iodoacetate was allowed to react with germinated spores in the presence of various concentrations of nitrosothiol (Fig. 1). The double-reciprocal plots in Fig. 1 are linear, intersecting in the upper-left quadrant. The kinetics thus conform to the "mixed

TABLE 1. *Comparison of the uptake of 0.1 mM tritiated methionine (8 μCi/ml) with uptake of 0.1 mM tritiated S-nitroso-coenzyme A (2 μCi/ml) into vegetative cells and germinating spores[a]*

Compound	Prepn	Total cpm/sample	Amt (cpm) in cells	% Uptake
Methionine	Vegetative cells	553,185	55,158	10.5
	Germinating spores	527,640	3,932	0.7
S-Nitroso-coenzyme A	Vegetative cells	111,274	1,607	0
	Germinating spores	51,527	0	0

[a] Spores were germinated for 20 min (phase dark) at a concentration of 100 μg/ml and for 120 min (vegetative) at a concentration of 40 μg/ml. Chloramphenicol (100 μg/ml) was added (to prevent incorporation of methionine into protein) followed by the addition of label after 5 min. Twenty minutes later, 0.2-ml samples were collected on 0.45-μm membrane filters, washed rapidly with phosphate-buffered saline, dried, and counted. Other samples removed at the same time were precipitated with trichloroacetic acid and counted. Subtraction of the trichloroacetic acid value from the samples which were washed with buffer only gave a measure of the label transported into the cell cytoplasm, but not incorporated into macromolecules. The results show rapid uptake of methionine into the cytoplasm of 120-min (vegetative) cells, but not into germinated spores. The nitrosothiol was not taken up at either time, so its inhibitory effect must not require transport. The nitrosothiol of coenzyme A was prepared by acidification of an equimolar (0.1 M) solution of coenzyme A and nitrite to pH 2 with HCl.

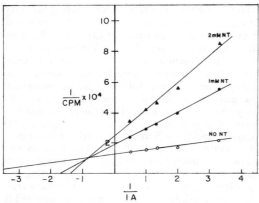

FIG. 1. *Double-reciprocal plot of nitrosothiol inhibition of labeled iodoacetate binding to germinated spores. For each point, 3 mg of spores was suspended in 5 ml of a solution containing 1% tryptone (Difco) and various amounts of S-nitrosomercaptoethanol. Two minutes later, tritiated iodoacetate (2 μCi/mmol) was added. After 20 min of further incubation at 30°C, the spores were washed and counted in 5 ml of Aquafluor. The nitrosothiol of mercaptoethanol was prepared as in Table 1. The concentrations of nitrosothiol and iodoacetate were varied for the different data points and plotted as double reciprocals of iodoacetate concentrations (1/IA mM) and the amount of radioacetate bound as counts per minute in the total sample (1/CPM), in the presence of various inhibiting levels of nitrosothiol.*

competitor" type (4, 10), in which nitrosothiol affects both the rate and the extent of iodoacetate binding. Since mixed systems often reflect multiple sites or binding modes, iodoacetate may be interacting with more than one type of site. Other inhibitors derived from nitrite ("Perigo factor" and a nitrite derivative of transferrin) were also examined for their ability to interfere

with iodoacetate incorporation (Fig. 2). Although the data show greater scatter than for nitrosothiol, the kinetics appear to be essentially the same. These results imply that these substances interfere with binding of iodoacetate to spores by the same mechanism. Additional experiments (not shown) also established that interference with irreversible binding of iodoacetate label is accompanied by interference with the ability of iodoacetate to inhibit spore outgrowth. Since iodoacetate is primarily a sulfhydryl-modifying agent, an explanation of these results is that iodoacetate inhibition of spore outgrowth is the result of covalent modification of critical sulfhydryl groups. Nitrosothiols and the other substances tested above somehow protect these groups from modification.

Experiments were next done to determine the functional significance of radioactive iodoacetate binding. Inhibition of outgrowth correlated well with uptake of label (Fig. 3). Experiments were carried out at two different temperatures. Although qualitatively similar, the 20°C experiment established that the spores neither are sensitive toward iodoacetate nor incorporate label during the very early stages of germination. The sensitive sulfhydryl groups are apparently unavailable for modification until germination is well under way. It also appears that much of the uptake of label is related to the inhibitory event and that the critical sulfhydryl groups represent the bulk of the sulfhydryl groups on the spore surface that are capable of reacting with iodoacetate at this stage of growth.

The above results suggest that iodoacetate is reacting with sulfhydryl groups in the spore membrane. With the purpose of further localizing the reactive groups, iodoacetate-labeled germinated spores were disrupted in a glass-bead shaker and centrifuged to remove debris; the

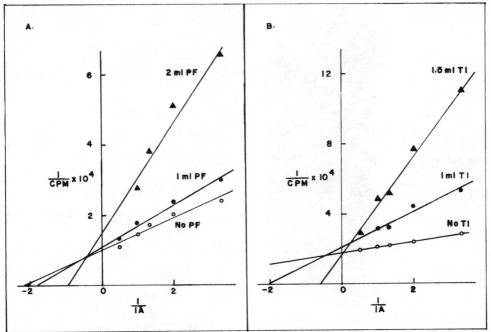

FIG. 2. *Double-reciprocal plots of the inhibition of iodoacetate binding to germinated spores by "Perigo factor" (A) and by nitrite derivative of bovine transferrin (B). The data were obtained as in Fig. 1, with 3 mg of spores for each point, addition of inhibitor, and, after 2 min of exposure, addition of labeled iodoacetate. After 20 min more, the spores were washed with unlabeled iodoacetate, suspended in Aquafluor, and counted. "Perigo factor" (PF) was prepared as previously described (1). Preparation of the nitrite derivative of transferrin (TI) will be described elsewhere.*

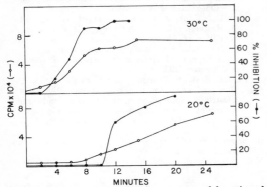

FIG. 3. *Comparison of the time course of functional sensitivity to iodoacetate with the time course of radioactive iodoacetate labeling. Time of functional inhibition was determined by first simultaneously adding 325 μg of spores and 1 mM iodoacetate to 5 ml of 1% tryptone (time zero). After the indicated time period had elapsed, the cells were washed and resuspended in outgrowth medium. Percent inhibition was evaluated by phase-contrast microscopy 3 h later. The time course of radioactive incorporation was obtained by mixing 3 mg of spores and 125 μCi of iodoacetate in 5 ml of 1% tryptone at 20°C or 30°C (time zero). At the appropriate times, the samples were washed with unlabeled 1 mM iodoacetate, resuspended in Aquafluor, and counted.*

supernatant was then centrifuged at 100,000 × g for 1 h. Approximately 20% of the label remained in the 100,000 × g supernatant; most of the label was incorporated into the insoluble fraction. Of the insoluble fraction, approximately 90% was solubilized by sodium dodecyl sulfate. This result is rationalized in terms of the label being incorporated into membrane.

At least 11 bands of incorporated label were found to be present in the sodium dodecyl sulfate-solubilized material when it was electrophoresed on a sodium dodecyl sulfate-polyacrylamide slab gel and autofluorographed (Fig. 4, lane 1). Three of these bands, with molecular weights of approximately 13,000, 28,000, and 29,000, were intense. The intensities of the bands labeled in the presence of nitrosothiol were greatly reduced, the faint bands no longer being visible (Fig. 4, lane 2). We can make no conclusions as to which labeled components are responsible for inhibition or spore outgrowth.

The data support the idea that nitrosothiols are sulfhydryl agents [probably reacting to give RSN(OH)—SX, where —SX is derived from spore membrane sulfhydryl] which protect HSX against reaction with iodoacetate. When HSX is covalently attached to either nitrosothiol or iodoacetate, the spore cannot proceed through

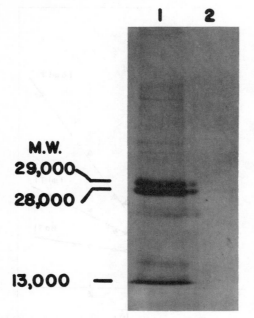

M.W.
29,000
28,000
13,000

FIG. 4. *Autofluorogram showing the pattern of io-doacetate incorporation into spore extract proteins. A 3-mg amount of spores was labeled with 1 mCi of iodoacetate for 20 min in the presence and absence of 10 mM S-nitrosomercaptoethanol. After several washings with unlabeled 1 mM iodoacetate, the cells were disrupted in a glass-bead shaker. Soluble components were electrophoresed on sodium dodecyl sulfate-polyacrylamide gels. The gels were then soaked in Enhance (New England Nuclear) and autofluorographed on Kodak XR-5 X-ray film. Lane 1 represents spore material labeled in the absence of nitrosothiol. Lane 2 represents spore proteins labeled in the presence of nitrosothiol.*

outgrowth. If the group is removed and the sulfhydryl is regenerated, outgrowth can proceed. It appears that *B. cereus* spores have highly reactive sulfhydryl groups in their membranes which react with numerous sulfhydryl agents, and these groups are critical for main-

taining the outgrowth process. Given the chemical analogies between nitrosothiols and nitrite, it is possible that the antimicrobial activity of nitrite may result from interaction at this type of sensitive membrane sulfhydryl site, a site which may be of general occurrence in sporulating and other bacteria.

ACKNOWLEDGMENT

This research was supported by the Agricultural Research Service under grant 12-14-1001-1230-WRU-801-1090-20841-4111, administered by the Eastern Regional Research Center, Philadelphia, Pa.

LITERATURE CITED

1. **Ashworth, B. W., and R. E. Keller.** 1967. Ultraviolet determination of tertiary mercaptans as thionitrites. Anal. Chem. **39**:373–374.
2. **Ashworth, J., A. Didcock, L. A. Hargreaves, B. Jarvis, and C. L. Walters.** 1974. Chemical and microbiological comparisons of inhibitors derived thermally from nitrite with an iron thionitrosyl (Roussin black salt). J. Gen. Microbiol. **84**:403–408.
3. **Ashworth, J., and R. Spencer.** 1972. The Perigo effect in pork. J. Food Technol. **7**:111–124.
4. **Cornish-Bowden, A.** 1979. Fundamentals of enzyme kinetics. Butterworths, London.
5. **Hansen, J. N., and R. A. Levin.** 1975. Effects of some inhibitors derived from nitrite on macromolecular synthesis in *Bacillus cereus*. Appl. Microbiol. **30**:862–869.
6. **Huhtanen, C. N., and A. E. Wasserman.** 1975. Effect of added iron on the formation of clostridial inhibitors. Appl. Microbiol. **30**:768–770.
7. **Incye, K., J. Farkas, V. Milhalys, and E. Zickal.** 1974. Antibacterial effect of cysteine-nitrosothiol and possible precursors thereof. Appl. Microbiol. **27**:202–205.
8. **Morris, S. L., R. A. Levin, C. Wright-Wilson, and J. N. Hansen.** 1978. Effect of S-nitrosothiol structure on inhibition of germination and outgrowth of *Bacillus cereus* spores, p. 85–89. *In* G. Chambliss and J. C. Vary (ed.), Spores VII. American Society for Microbiology, Washington, D.C.
9. **Perigo, J. A., E. Whiting, and T. E. Bashford.** 1967. Observations on the inhibition of vegetative cells of Clostridium sporogenes by nitrite which has been autoclaved in a laboratory medium discussed in the context of sublethally processed cured meats. J. Food Technol. **2**:377–397.
10. **Segel, I. H.** 1975. Enzyme kinetics. John Wiley & Sons, Inc., New York.

Mutations in *Bacillus subtilis* var. *niger (Bacillus globigii)* Spores Induced by Ethylene Oxide

L. A. JONES AND D. M. ADAMS[1]

Becton, Dickinson and Company Research Center, Research Triangle Park, North Carolina 27709, and Department of Food Science, North Carolina State University, Raleigh, North Carolina 27607

Spore preparations of *Bacillus subtilis* var. *niger*, frequently named *Bacillus globigii*, were 99.7% one colony type prior to ethylene oxide treatment, but exhibited up to 50% atypical types among the survivors of ethylene oxide treatment. Of approximately 10,000 colonies grown from spores treated with ethylene oxide, more than 100 phenotypic variants were isolated. Characterization of these variants with regard to phenotypic stability, ethylene oxide resistance characteristics, sporeforming ability, cellular morphology, biochemical characteristics, optimal growth temperature, and nutrient requirements suggested that ethylene oxide induced mutations in the surviving spores.

Ethylene oxide (ETO) is an extremely reactive alkylating agent. Its toxicity to vegetative cells and spores and its ability to permeate certain packaging materials make ETO a very effective sterilant for products that would be damaged by other sterilization methods. In ETO sterilization processes, spores of the organism *Bacillus subtilis* var. *niger (B. globigii)* commonly are used as biological indicators to monitor the process effectiveness.

On Trypticase soy agar (TSA; BBL Microbiology Systems) the organism forms colonies with a very distinctive orange pigment and colony morphology, and a distinctive orange pellicle is formed in broth culture. These characteristics are used to identify growth that develops from subcultured ETO-treated biological indicator strips as surviving *B. globigii* or as accidental contaminants. ETO is mutagenic for vegetative bacteria, fungi, plants, and animal systems (1, 3, 6, 7). The objective of this study was to determine whether ETO induced mutations in *B. globigii* spores, resulting in changes in the growth characteristics used for identification.

RESULTS AND DISCUSSION

Non-ETO-treated spores from spore crops and biological indicator spore strips were surface plated onto TSA, incubated at 35°C for 48 h, and examined for the presence of atypical colo-

nies. A typical *B. globigii* colony has a medium orange pigmentation, a translucent to opaque appearance, a dry surface with ridges radiating from the center, and an irregular shape with a diameter of approximately 0.5 cm. Of 10,000 colonies arising from untreated spores, only 26 exhibited atypical colony morphology. These were grouped into five phenotypes on the basis of degree of pigmentation, opaqueness, and surface texture characteristics (rough or smooth, wet or dry, raised or flat, presence or absence of ridges, and relative size). Of these five phenotypes, only two differed markedly from the normal colony; one had a smooth wet surface, and one was white, lacking orange pigment.

Exposure of the spores to ETO increased the percentage of atypical colonies and the variety of phenotypes dramatically (Fig. 1). During the time required to charge the Cryotherm with ETO (10 to 15 s), the proportion of surviving spores producing atypical colonies increased from 0.26% to between 10 and 20%; the number of spores yielding atypical colonies increased from an estimated 2.6×10^2 to $>1 \times 10^4$ spores per strip. Further exposure of the spores to ETO increased the proportion of atypical colonies to a constant level of approximately 50% of the total surviving population. After the first minute of ETO exposure, the inactivation curves for the total and atypical populations were parallel.

Of the spores surviving ETO treatment, 10,000 colonies were screened and 125 phenotypes were isolated. This was a 25-fold increase in the num-

[1] Present address: Campbell Institute of Research and Technology, Campbell Soup Co., Camden, NJ 08101.

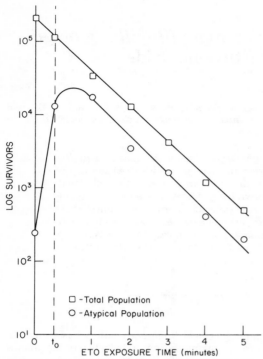

LOG SURVIVORS

□ -Total Population
○ -Atypical Population

ETO EXPOSURE TIME (minutes)

FIG. 1. *ETO resistance of* B. globigii: *comparison of the total and atypical populations. Two spore crops were prepared on a sporulation agar medium containing 0.5% Trypticase, 0.3% Phytone (BBL Microbiology Systems), 0.5% beef extract, 0.015% MnSO₄·H₂O, 0.001% CaCl₂, and 2% agar. The agar medium was inoculated from a 16-h broth culture of vegetative cells. The broth contained 1% Trypticase, 0.5% yeast extract, and 0.2% K₂HPO₄. The inoculated agar plates were incubated at 35°C until ≥90% sporulation occurred; the growth was harvested, washed by differential centrifugation methods until the spores were ≥99% free from vegetative debris as determined from spore stains, and stored in sterile distilled water at 4°C. Strips of Whatman no. 1 chromatography paper, 0.4 by 25.4 mm in size, were inoculated with approximately 2 × 10⁵ spores per strip. The colony characteristics of both spore crops and organisms recovered from paper strips were determined by use of a surface plating technique. The TSA plates were incubated at 35°C for 48 h, and 10,000 colonies were screened for phenotypic characteristics as previously described. Spore strips were exposed to ETO in a 1.6-ft³ (45.31-liter) Amsco Cryotherm. The gas mixture used was Penngas, a 12:88 ETO-Freon mixture by weight, manufactured by Pennsylvania Engineering Co. Exposure parameters used throughout these studies were 540 to 580 mg of ETO per liter, 130 to 140°F (54 to 60°C), and 40 to 60% relative humidity, with exposures of increasing time increments up to 5 min. The strips were surface plated, and colonies were differentially counted after incubation at 35°C as previously described. On the graph, time zero shows results prior to any exposure to ETO; t₀ refers to exposure to the time required to charge the Cryotherm with ETO and*

ber of identifiable phenotypes over the number seen in non-ETO-treated spore populations. There were many phenotypes that represented very small percentages of the total population. The most frequently observed single phenotype represented 12% of the total population. Three others next most frequently seen comprised 2 to 6% of the population. Least frequently seen phenotypes comprised only 0.05 to 0.5% of the population. The 125 well-defined phenotypes were grouped into general colony descriptions to make six very general classifications (Table 1). Segmented colonies, i.e., colonies consisting of two or more morphological types, were about 15% of the total population. These colonies were characterized only by the presence of segments which might have any of the criteria used in colony descriptions. When segments were isolated and streaked onto TSA, the resulting colonies were phenotypically similar to the segment from which they were isolated.

Forty representative phenotypes of the 125 isolated were selected for further study. To confirm that the changes in colony morphology were the result of mutations and to determine the stability of these mutations, we transferred each of the 40 atypical colony types five times in succession either in Trypticase soy broth or by streaking on TSA. The cultures were incubated at 35°C for 48 to 60 h. Of the 40 isolates, 11 reverted to the typical appearance, 12 changed to other atypical appearances, and 17 remained stable. In broth culture, 5 of the 40 isolates did not produce a pellicle at all, and five produced a pellicle that was white in color. These characteristics in broth remained stable through five successive transfers.

The 40 isolates were examined microscopically for changes in cell size and shape and for Gram reaction. Of the 40 isolates, 22 were rods of typical size, 0.7 to 0.8 μm by 2 to 3 μm. Seven isolates were significantly longer, up to 8 to 9 μm in length; four were shorter, approximately 1.5 μm in length. Cell widths ranged from 0.5 to 1.0 μm. Seven isolates were irregularly shaped, curved, or bent. Thirty-six isolates had the normal gram-positive reaction; four were gram variable.

Sporeforming characteristics of the 40 isolates also were determined by using TSA, soil extract agar, and the sporulation agar described in the legend to Fig. 1. After incubation for 7 days at 35°C, growth from the seeded plates was spore stained and examined microscopically for the

the time required to evacuate the gas. All other exposure times given refer to timed intervals after the Cryotherm was fully charged with ETO and before gas evacuation.

TABLE 1. *General classification of atypical colony types and frequency of occurrence in ETO-exposed populations*[a]

General classification	No. of phenotypes	% of total population
Orange pigment, rough	35	14
Orange pigment, wet	28	13
White to gray, rough	5	3
White to gray, wet	11	2
Colony center and outer area, different characteristics	46	4
Segmented colonies	—	15
Typical colonies	—	49

[a] A total of 10,000 colonies were observed after exposure to increasing time increments of ETO and were characterized in detail phenotypically. On the basis of these descriptions, some types differed in only one or two details. A study of these led to grouping the phenotypes in a more general manner. The simplest classifications obtained are shown, with the number of phenotypes and the percentage of each classification in the total ETO-treated population.

presence of spores. Thirteen isolates did not form spores on TSA; eight would not sporulate on either TSA or soil extract agar, and five isolates did not form spores on any of the screened media. Two isolates which grew on TSA and soil extract agar (but did not sporulate) did not grow on sporulation agar. To quantitate percent sporulation, spore crops were prepared as described in the legend to Fig. 1. Percent sporulation varied greatly among the seven isolates. Three exhibited sporulation of ≥90% by the second day of incubation; sporulation of the other four ranged from 1 to 60% after 5 days.

Biochemical reactions are needed to confirm the identity of *B. globigii* cultures propagated in the laboratory. The biochemical characteristics have been defined by Gordon, Haynes, and Pang (5). Their media and methods were used to characterize biochemically atypical *B. globigii* isolates. Thirty-eight percent of the isolates tested no longer produced black pigment from tyrosine. The inability to utilize citrate was the next most frequently seen change (18%), and negative arabinose fermentation, acetylmethylcarbinol production, and mannitol reactions were each observed only once. All other characteristics were normal in all 40 isolates. Four isolates differed from the normal organism in two biochemical reactions, and one isolate was atypical for three. The five phenotypes isolated prior to ETO treatment had typical reactions in all biochemical tests.

In other studies of the recovery of ETO-treated spores (2,4), it was concluded that optimal recovery temperatures are lower than for untreated spores. We also found that recovery of ETO-treated spores was up to fourfold greater at 25°C than at 35°C, whereas no difference was seen for nontreated spores. However, when 50 colonies from ETO-treated spores enumerated at 25°C were streaked onto TSA and incubated at 35°C, all formed colonies. These results indicate that greater recovery at 25°C was due to repairable cell damage (injury) rather than induction of a temperature-sensitive mutation.

The dramatic increase in atypical populations with ETO treatment, the stability of a large proportion of the isolates, and the similarity of the ETO resistances of the total and atypical populations (Fig. 1) indicated that the atypical types are phenotypic mutations induced by the ETO treatment and not extremely ETO-resistant, atypical portions of the initial spore population. The differences among the isolates in microscopic appearance, sporeforming ability, colony morphology, and some biochemical reactions indicated that ETO induces a variety of mutations in *B. globigii* spores. An organism that appears during subculturing of ETO-treated spores should not be classified as an accidental contaminant simply because it lacks certain characteristics typical of *B. globigii*.

ACKNOWLEDGMENTS

We are grateful to John L. Holland, Carolyn S. Kilpatrick, and Kuo-Sun Su Chung for technical assistance.

LITERATURE CITED

1. **Committee to Coordinate Toxicology and Related Processes.** 1977. Report of the Subcommittee on the Benefits and Risks of Ethylene Oxide for Sterilization. Department of Health, Education, and Welfare, Washington, D.C.
2. **Davis, S. B., R. A. Carls, and J. R. Gillis.** 1976. Recovery of sublethal sterilization damaged *Bacillus* spores in various culture media. Dev. Ind. Microbiol. 18:427–438.
3. **Department of Health, Education, and Welfare. Food and Drug Administration.** 1978. Ethylene oxide, ethylene chlorhydrin and ethylene glycol. Proposed maximum residue limits and maximum levels of exposure. Federal Register, 23 June 1978, Part V.
4. **Futter, B. V., and G. Richardson.** 1970. Viability of clostridial spores and the requirements of damaged organisms. I. Method of colony count, period and temperature of incubation, and pH value of the medium. J. Appl. Bacteriol. 33:321–330.
5. **Gordon, R. E., W. C. Haynes, and C. H. Pang.** 1973. The genus *Bacillus*. Agriculture Handbook no. 427, p. 3–14, 36–41, 182–183. Agricultural Research Service, U.S. Department of Agriculture, Washington, D.C.
6. **Health Industries Manufacturers Association.** 1978. HIMA report 78-3. Ethylene oxide technical report. Health Industries Manufacturers Association, Washington, D.C.
7. **Miller, E. C., and J. A. Miller.** 1971. The mutagenicity of chemical carcinogens: correlations, problems, and interpretations, p. 83–119. *In* A. Hallaender (ed.), Chemical mutagens: principles and methods for their detection, vol. 1. Plenum Publishing Corp., New York.

F. Resting Forms of Microorganisms Other than Bacillus and Clostridium

Development-Specific Localization of Myxobacterial Hemagglutinin During Fruiting Body Formation of *Myxococcus xanthus*

DAVID R. NELSON, MICHAEL CUMSKY,[1] AND DAVID R. ZUSMAN

Department of Microbiology and Immunology, University of California, Berkeley, California 94720

During fruiting body formation in *Myxococcus xanthus*, a major new protein that has lectin-like activity is synthesized. Myxobacterial hemagglutinin is apparently localized on the cell surface or in the periplasmic space, since as much as 90% of the hemagglutinating activity of developing cells can be released by washing the cells with cold buffer followed by osmotic shock. The amino-terminal residues of purified myxobacterial hemagglutinin were found to be strongly hydrophobic, suggesting that this region may be important in the transport of the protein through the cytoplasmic membrane. Fluorescence-labeled antibody probes were used to show that myxobacterial hemagglutinin is bound to the cell surface in patches. Significantly, the labeled probes were primarily localized at the cell poles, suggesting that myxobacterial hemagglutinin may function in end-to-end cellular interactions during development.

Myxococcus xanthus is a gram-negative, rod-shaped bacterium which in nature commonly grows in soils on decaying organic material or by preying upon other microorganisms (10, 18). The myxobacteria generally hunt by gliding over solid surfaces in large groups called swarms while synthesizing extracellular enzymes and antibiotics. The swarm provides the critical cell mass needed by the myxobacteria to attack and digest other microorganisms or to degrade complex organic matter. Under conditions of starvation on a solid surface, the developmental program is triggered. Cells reverse their outward movement and glide toward aggregation centers, forming raised mounds of cells. The individual rod-shaped cells then shorten and convert to ovoid, environmentally resistant resting cells called myxospores. In contrast to endospore formation in *Bacillus* species, the whole cell is converted to a myxospore. In *M. xanthus*, mounds of myxospores are termed fruiting bodies.

We have recently analyzed the developmental program of *M. xanthus* with the isolation of mutants (12, 17) and with the study of the pattern of protein synthesis (8). The mutant studies showed that the two developmental functions of aggregation and sporulation, although temporally related, behave as largely independent pathways, since most of the mutants blocked in aggregation showed normal levels of sporulation, and most mutants blocked in sporulation showed normal aggregation. Events necessary for both functions occur at early times, however, since both aggregation and sporulation mutants were found which had early temperature-sensitive periods.

One major protein which appears to be associated with sporulation was purified and called protein S (9). It is a small protein (molecular weight of 23,000) which accumulates in the soluble fraction of cells at early times in development and is later transported to the cell surface, where it assembles to form a spore surface coat. The protein is development specific since antiserum to protein S does not cross-react with vegetative extracts. Purified protein S will spontaneously self-assemble onto protein S-deficient spores in the presence of 10 mM Ca^{2+}. Electron

[1] Present address: Department of Molecular, Cellular, and Developmental Biology, University of Colorado, Boulder, CO 80307.

microscopy of the reconstituted spores reveals the assembly of new material on the spore surface.

A second major development-specific protein, called myxobacterial hemagglutinin (MBHA), was isolated. It may be involved in cell-cell interactions (5). The hemagglutinating activity has been purified and found to be a protein with an apparent molecular weight of 30,000. MBHA is abundant (1 to 2% of soluble protein in developmental extracts) and can be clearly identified in soluble extracts of developing cells. MBHA is a lectin since its hemagglutinating activity is inhibited by a trisaccharide glycopeptide (N-acetylneuraminic acid–galactose-N-acetyl-galactosamine–serine polypeptide) common to fetuin, glycophorin, and rabbit immunoglobulin G (IgG). Destruction of the penultimate galactose residue in the glycopeptide eliminated the inhibitory activity. In this paper we explore the localization of MBHA. This information is important in the continuing study of the function of this protein during development of *M. xanthus*.

MATERIALS AND METHODS

Cells and growth conditions. *M. xanthus* strain DZ-2 (3) was used for all experiments. Vegetative cultures were grown in Casitone-yeast extract (CYE) medium (2). Development was induced by spotting a concentrated cell suspension on clone fruiting agar (7) as described (5).

Hemagglutination assays of cell extracts. *M. xanthus* cells were harvested, sonicated, and assayed for hemagglutinating activity as described previously (5).

Ouchterlony assays. Ouchterlony double-diffusion assays were performed on glass microscope slides coated with 4.5 ml of agar containing 15 mM phosphate-buffered saline (PBS) (5), 1% Triton X-100, and 0.02% sodium azide. The outside wells were filled with 20 μl of antigen; the center well was filled with 20 μl of antiserum. Precipitin patterns were scored after overnight incubation at 28°C.

Cell wash and osmotic shock procedures. Developmental cells were harvested from clone fruiting agar by adding 1 ml of cold T buffer (1 mM Tris buffer, pH 7.6) per plate, disrupting the cell aggregates with a glass rod, and then pipetting the cell suspension into a centrifuge tube. The cells were collected by centrifugation (10,000 × *g* for 10 min at 4°C), and the supernatant (wash fluid) and pellets were saved. Vegetative cells were harvested as above except that the cells were pelleted from CYE broth before being washed with the buffer.

The sedimented cells were subjected to osmotic shock as follows. Cells were suspended in cold 20% sucrose–T buffer and incubated at 0°C for 20 min with gentle shaking. The cells were collected by centrifugation (10,000 × *g* for 10 min at 4°C) and then shocked with cold T buffer for 20 min at 0°C. The cells were collected by centrifugation, and the supernatant (shock fluid) and cell pellet were saved. EDTA was omitted from the shock procedure because it causes *M. xanthus* to lyse. Fewer than 10% of the cells lyse under this procedure. Osmotically shocked cells were disrupted by sonication and separated into membrane and soluble (S100) fractions by ultracentrifugation (100,000 × *g* for 60 min) as described previously (5).

Polyacrylamide gel electrophoresis. Samples were prepared for polyacrylamide gel electrophoresis by partially purifying the wash fluid, osmotic shock fluid, and S100 fractions on DEAE-cellulose (DE52, Whatman) as described (5) to remove material which seriously interferes with the resolution of protein bands of *M. xanthus*. The flow-through fractions, which contained all the hemagglutinating activity, were concentrated as described (5) and analyzed on 10% polyacrylamide gels with the buffer system of Laemmli (11).

Immunofluorescence staining of cells. Cells were harvested from CYE broth or from clone fruiting agar as described above except that 150 mM PBS buffer was used (5). The cells were washed twice with cold PBS buffer after centrifugation (10,000 × *g* for 10 min at 4°C). The final pellet was suspended in 25 μl of an appropriate dilution of immune or preimmune IgG in PB buffer (150 mM PBS plus 5 mg of bovine serum albumin per ml) and incubated on ice for 30 min. The cells were then pelleted in a microfuge (3 min at 4°C) and washed three times with cold PB buffer. The final pellet was suspended in 50 μl of a 1:50 dilution of fluorescein-labeled goat anti-rabbit IgG and incubated for 30 min at 0°C. The cells were then harvested and washed as described above. The pellet was suspended in 150 μl of PB buffer and then was examined and photographed with a Zeiss microscope fitted with standard fluorescence optics.

Amino acid sequence analysis. MBHA was analyzed by use of a Beckman 890C automatic sequencer. Individual residues were identified by at least two of the following methods: gas-liquid chromatography, thin-layer chromatography, or high-pressure liquid chromatography.

RESULTS

Synthesis of MBHA during development of *M. xanthus*. Cell surface carbohydrate binding or "lectin-like" proteins have been implicated in cell-cell interactions in the cellular slime molds *Dictyostelium discoideum* (13, 15) and *Polysphondylium pallidum* (16), as well as in other eucaryotic systems (1, 4). We therefore examined extracts of *M. xanthus* for a hemagglutinating activity. Whereas no activity was present in vegetative cultures or in cells starved in a liquid medium, as much as 7,500 U/mg of protein was found in cells plated on a solid fruiting medium. In contrast, cells induced to sporulate in liquid culture by the addition of glycerol (0.5 M) (6) did not contain the hemagglutinating activity. Thus, the activity was observed only in extracts of cells under fruiting conditions, in which cell-cell contact and aggregation occur. Figure 1 shows the kinetics of appearance of the hemagglutinating activity in soluble extracts of *M. xanthus* strain DZ-2 and a nonfruiting rifampin-resistant mutant. The activity in the wild-type strain appears in early developmental extracts and peaks at the time of tight cell-cell aggregation. The nonfruiting mutant produces very low amounts of the hemagglutinin (about 5% of the wild-type control) but shows similar kinetics. The hemagglutinating activity has been purified and shown to be a single polypeptide with an apparent molecular weight of 30,000 (M. Cumsky and D. Zusman, in preparation). We call this protein myxobacterial hemagglutinin (MBHA). Antiserum to MBHA was raised in rabbits. Figure 2 shows that the antiserum reacts with developmental extracts of *M. xanthus* but not with vegetative extracts. Thus, MBHA is a development-specific protein.

Localization of MBHA during development. The hemagglutinating activity was routinely measured in the soluble fraction of sonicated cells, but it was also present in the envelope fraction. We therefore decided to study the localization of the activity more carefully. Cells were harvested from fruiting agar, washed with

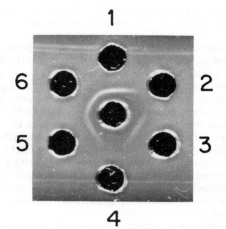

FIG. 2. *Ouchterlony double-diffusion assays. Vegetative (well 4) and developmental (wells 1, 3, and 6) extracts were analyzed against anti-MBHA sera (center well). Pure MBHA (well 5) and pure protein S (well 2) are also presented.*

TABLE 1. *Relative hemagglutinating activity in the wash, osmotic shock, and cytoplasmic fractions of developing cells of* M. xanthus

Time of development (h)	Hemagglutinating activity[a] (U)		
	Wash fluid	Osmotic shock fluid	Cytoplasmic fraction (S100)
0	0 (0%)[b]	0 (0%)	0 (0%)
24	3,012 (72%)	753 (18%)	422 (10%)
48	2,688 (76%)	640 (18%)	205 (6%)
72	1,472 (83%)	213 (12%)	81 (5%)
96	422 (83%)	53 (10%)	34 (7%)

[a] Cells of *M. xanthus* strain DZ-2 were spotted on clone fruiting agar and incubated at 28°C. At timed intervals, the cells were harvested, washed at 4°C, and then subjected to osmotic shock, as described in Materials and Methods. The cytoplasmic fraction was prepared by sonic disruption of cells previously washed and subjected to osmotic shock. The hemagglutinating activity in the various fractions was determined with sheep erythrocytes.

[b] Percentage of total activity.

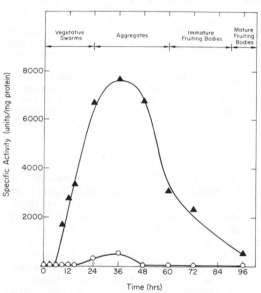

FIG. 1. *Kinetics of appearance of hemagglutinating activity in extracts of* M. xanthus. *Cells were spotted on clone fruiting agar and harvested at various times during development. Extracts were prepared, and the soluble (S100) fractions were assayed for hemagglutinating activity with formalinized sheep erythrocytes. Symbols: ▲, strain DZ-2 (wild type); ○, mutant KR9 (rifampin-resistant, nonfruiting mutant of DZ-2 isolated by Ken Rudd).*

cold buffer, and then subjected to osmotic shock. The relative amounts of hemagglutinating activity in the various fractions are presented in Table 1. As much as 90% of the total activity was found in the wash and shock fractions. In fact, the cytoplasmic fraction contained only a minor amount of total activity. The presence of MBHA in the wash and shock fractions was confirmed by sodium dodecyl sulfate-polyacrylamide gel electrophoresis of these fractions (Fig. 3). Control experiments showed very low levels of cell lysis during these treatments (e.g., cytoplasmic enzymes were not detectable in the shock fluid). Thus, MBHA is a periplasmic protein or a protein loosely bound to the surface of the cell.

MBHA can be detected on the surface of cells immunologically (Fig. 4). Cells were washed to remove loosely bound MBHA and culture fluid and then treated, first with rabbit anti-MBHA

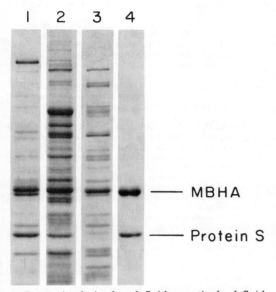

FIG. 3. *Analysis of wash fluid, osmotic shock fluid, and cytoplasmic proteins by sodium dodecyl sulfate-polyacrylamide gel electrophoresis. Developmental cells (48 h) were prepared and analyzed as described in Materials and Methods. Lane 1, wash fluid; lane 2, osmotic shock fluid; lane 3, S100 fraction; lane 4, MBHA (molecular weight, 30,000) and protein S (molecular weight, 23,000).*

IgG and then with fluorescein-conjugated goat anti-rabbit IgG. Vegetative cells were not fluorescently labeled by this sandwich technique. However, developmental cells (24 to 72 h) showed strong fluorescence that increased with the time of development. Significantly, the fluorescence was localized in patches at the cell poles in about 65% of the labeled cells. In about 35% of the cells, patches of fluorescence were observed in the center of the cells as well. Developmental cells stained with preimmune IgG showed no visible fluorescence. Thus, MBHA appears to be located on the cell surface during development.

MBHA extracted from the cytoplasmic fraction and MBHA extracted from the shock fractions appear to have identical molecular weights by sodium dodecyl sulfate-polyacrylamide gel electrophoresis. Thus, the transport of MBHA through the cytoplasmic membrane is probably not associated with the cleavage of a "signal" peptide. We therefore analyzed the amino-terminal residues of MBHA to determine whether it had a hydrophobic region. Figure 5 shows the amino-terminal sequence of MBHA. As predicted, it is extremely hydrophobic. The first 27 residues do not contain a single charged residue. These results suggest an important role for this region in the localization of MBHA during development.

DISCUSSION

During fruiting body formation in *M. xanthus*, several new proteins are synthesized (8). One of the abundant new proteins is MBHA, a hemagglutinin which first appears at 6 to 8 h of development (5). In this study, we found that MBHA is localized on the cell surface and in the peri-

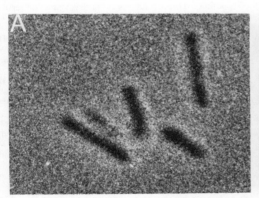

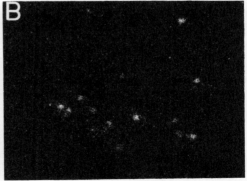

FIG. 4. *Immunofluorescence localization of MBHA on the surface of developmental cells. Developmental cells (24 h) were harvested from clone fruiting agar plates and stained immunologically for MBHA as described in Materials and Methods. (A) Photomicrograph of developmental cells under phase contrast. (B) Photomicrograph of the same field under fluorescence optics. The cells are about 4 to 6 μm in length.*

NH$_2$ — ala — ala — tyr — leu — val — gln — asn — gln — trp — gly —
                            5   10

gly — ser — gln — ala — thr — trp — asn — pro — gly — gly —
                    15   20

leu — trp — leu — ile — met — ala — cys — *asp* — *lys* —
                    25

FIG. 5. *Amino acid sequence of the N-terminal region of MBHA. The first 27 residues consist of hydrophobic (bold type) and neutral, nonpolar amino acids. The amino acids at positions 28 and 29 are polar (italicized).*

plasmic space since as much as 90% of the total hemagglutinating activity can be removed and recovered by washing the cells with cold buffer and subjecting them to osmotic shock. Analysis of the wash and shock fluids by sodium dodecyl sulfate - polyacrylamide gel electrophoresis showed that these fractions differ from the cytoplasmic fractions (S100) and are greatly enriched in MBHA. It is of interest to note that protein S, the spore surface coat protein, is also enriched in these fractions at 48 h. Apparently, both of these proteins accumulate in the periplasmic space and on the cell surface, where they must have their primary functions.

MBHA was detected on the surface of cells immunologically after extensive cell washing. The fluorescence was localized in patches, usually at the cell poles. This shows that MBHA is firmly bound to specific sites on the cell surface, presumably at the lectin receptors. The observation that MBHA is firmly bound at the cell poles suggests site-specific developmental changes in *M. xanthus* which may function in end-to-end cellular interactions during development. Lectins have been found in extracts of aggregating slime molds, e.g., *D. discoideum* and *P. pallidum* (15, 16). These lectins were hypothesized to be cohesiveness factors which act as specific molecular hooks during developmental aggregation (14). The role for MBHA in development of *M. xanthus* has yet to be determined.

ACKNOWLEDGMENTS

We thank Allan J. Smith for performing the N-terminal sequence analyses.

D.R.N. was the recipient of an American Cancer Society Postdoctoral Fellowship (PF-1718). M.C. was a National Institutes of Health Predoctoral Trainee (on grant 5-T32-GM-07232). This work was supported by Public Health Service grant GM 20509 from the National Institutes of Health and National Science Foundation grant PCM-7922249.

LITERATURE CITED

1. Burger, M. M., and J. Jumblatt. 1977. Membrane involvement in cell-cell interactions: a two-component model system for cellular recognition that does not require live cells, p. 155–172. *In* J. W. Lash and M. M. Burger (ed.), Cell and tissue interactions. Raven Press, New York.

2. Campos, J. M., J. Geisselsoder, and D. R. Zusman. 1978. Isolation of bacteriophage MX4, a generalized transducing phage for *Myxococcus xanthus.* J. Mol. Biol. 119:167–178.

3. Campos, J. M., and D. R. Zusman. 1975. Regulation of development in *Myxococcus xanthus*: effect of 3':5'-cyclic AMP, ADP, and nutrition. Proc. Natl. Acad. Sci. U.S.A. 72:518–522.

4. Crandall, M. 1977. Mating-type interactions in microorganisms. Recept. Recognition Ser. A 3:47–100.

5. Cumsky, M., and D. R. Zusman. 1979. Myxobacterial hemagglutinin: a development-specific lectin of *Myxococcus xanthus.* Proc. Natl. Acad. Sci. U.S.A. 76:5505–5509.

6. Dworkin, M., and S. M. Gibson. 1964. A system for studying microbial morphogenesis: rapid formation of microcysts in *Myxococcus xanthus.* Science 146:243–444.

7. Hagen, D. C., A. P. Bretscher, and D. Kaiser. 1978. Synergism between morphogenetic mutants of *Myxococcus xanthus.* Dev. Biol. 64:284–296.

8. Inouye, M., S. Inouye, and D. R. Zusman. 1979. Gene expression during development of *Myxococcus xanthus*: pattern of protein synthesis. Dev. Biol. 68:579–591.

9. Inouye, M., S. Inouye, and D. R. Zusman. 1979. Biosynthesis and self-assembly of protein S, a development-specific protein of *Myxococcus xanthus.* Proc. Natl. Acad. Sci. U.S.A. 76:209–213.

10. Kaiser, D., C. Manoil, and M. Dworkin. 1979. Myxobacteria: cell interactions, genetics, and development. Annu. Rev. Microbiol. 33:595–636.

11. Laemmli, U. K. 1970. Cleavage of structural proteins during assembly of the head of bacteriophage T4. Nature (London) 227:680–682.

12. Morrison, C. E., and D. R. Zusman. 1979. *Myxococcus xanthus* mutants with temperature-sensitive,stage-specific defects: evidence for independent pathways in development. J. Bacteriol. 140:1036–1042.

13. Ray, J., T. Shinnick, and R. Lerner. 1979. A mutation altering the function of a carbohydrate binding protein blocks cell-cell cohesion in developing *Dictyostelium discoideum.* Nature (London) 279:215–221.

14. Reitherman, R. W., S. D. Rosen, W. A. Frazier, and S. H. Barondes. 1975. Cell surface species-specific high affinity receptors for discoidin: developmental regulation in *Dictyostelium discoideum.* Proc. Natl. Acad. Sci. U.S.A. 72:3541–3545.

15. Rosen, S. D., J. A. Kafka, D. L. Simpson, and S. H. Barondes. 1973. Developmentally regulated carbohydrate-binding protein in *Dictyostelium discoideum.* Proc. Natl. Acad. Sci. U.S.A. 70:2554–2557.

16. Rosen, S. D., D. L. Simpson, J. E. Rose, and S. H. Barondes 1974. Carbohydrate-binding protein from *Polysphondylium pallidum* implicated in intercellular adhesion. Nature (London) 252:149–151.

17. Rudd, K., and D. R. Zusman. 1979. Rifampin-resistant mutants of *Myxococcus xanthus* defective in development. J. Bacteriol. 137:295–300.

18. Zusman, D. R. 1980. Genetic approaches to the study of development in the myxobacteria, p. 41–78. *In* T. Leighton and W. Loomis (ed.), The molecular genetics of development. Academic Press, Inc., New York.

Novel Lipids of *Azotobacter vinelandii* Cysts and Their Possible Role[1]

R. N. REUSCH, C.-J. SU, AND H. L. SADOFF

Department of Microbiology and Public Health, Michigan State University, East Lansing, Michigan 48824

We found that, during encystment, *Azotobacter vinelandii* synthesized five unique 5-*n*-alkylresorcinols and two 6-*n*-alkyl-4-hydroxy-pyran-2-ones. Each compound occurred as both C_{21} and C_{23} homologs. These lipids accounted for 80% of the readily extractable cyst lipid and were found in the exine and the central body membrane.

Azotobacter vinelandii is a large (2 to 5 μm), gram-negative N-fixing bacterium which forms cysts. These are dormant cells analogous to bacterial endospores (12); they are not heat resistant, but they do withstand desiccation and various deleterious physical and chemical agents which readily kill vegetative cells (16). Cysts consist of spherical cells (central body) bounded by a thin, layered outer shell (exine) and a thicker inner layer (intine) (6, 9). Encystment of glucose-grown cultures of this organism is very poor (0.1 to 0.5%) and occurs over a 5- to 10-day period (8). Higher levels of encystment of *A. vinelandii* can be obtained by growing cells on agar plates containing *n*-butanol (15). The alcohol is somewhat toxic, and vegetative growth on the plates is slow. A less cumbersome technique for producing quantities of cysts and studying the biochemistry of morphogenesis consists of growing cells on glucose in liquid culture and then shifting them during exponential phase to β-hydroxybutyrate (BHB) as the carbon and energy source (8). Yields of 95% encystment can be achieved. The sequelae of this specific metabolic shift include a turning off of N fixation, a marked reduction in phospholipid (P-lipid) synthesis, a final cell division, and a temporal sequence of biochemical and morphogenetic events leading to the conversion of each division product into a cyst (6, 7). A shift-down of exponential-phase cells to acetate only promotes further vegetative growth.

Cells of *A. vinelandii* contain about 10% readily extractable (2) lipids (10), whereas cysts contain twice that amount (10, 11), even though the P-lipid content of cysts is markedly less than

that of cells. In considering the role of BHB (or a closely related metabolite) as the specific inducer of encystment, we speculated that the shift to BHB metabolism could result in the synthesis of unique lipids to replace those P-lipids which were being lost. We have found such lipids and thus far have identified five species of 5-*n*-alkylresorcinols and two species of 6-*n*-alkyl-4-hydroxy-pyran-2-ones (pyrones).

RESULTS

Phospholipids. *A. vinelandii* membranes contain myristic (3%), palmitic (35%), palmitoleic (40%), and *cis*-vaccenic (22%) acids in their P-lipids (17). During the course of encystment, more than 90% of the palmitoleic and *cis*-vaccenic acyl groups add a methyl group across their double bond, yielding the corresponding C_{17}- and C_{19}-cyclopropane fatty acids. This change probably has little effect on the fluidity of the cyst membrane, but it does reduce the possibility of peroxidation of membrane lipids, a desirable trait in a dormant cell.

The rate of synthesis of the two major membrane lipids, phosphatidylglycerol and phosphatidylethanolamine, decreased exponentially upon initiation of encystment and was 15 to 20% of the initial rate by 20 h. Studies of the turnover of phosphatidylglycerol and phosphatidylethanolamine revealed that the half-lives of these lipids were 12 to 14 h over a period of 25 h (Fig. 1). The combination of the two phenomena, sharply reduced rates of synthesis and relatively rapid turnover, resulted in cyst membranes of low P-lipid content.

BHB metabolism and lipid synthesis. BHB is oxidized by an NAD-linked dehydrogenase to acetoacetic acid (13), and the coenzyme A

[1] Article no. 9677 of the Michigan Agricultural Experiment Station.

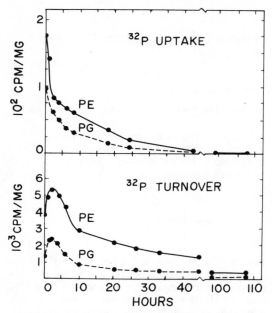

FIG. 1. *Phospholipid synthesis (^{32}P-uptake) and turnover during encystment of* A. vinelandii. *For ^{32}P uptake, a 30-min pulse of 7.1 µCi/ml was added to a 30-ml sample of the encysting culture. The P-lipids in the cells were then extracted (2), resolved on thin-layer chromatography, and counted. Turnover of P-lipids was studied by first labeling the cells for two generations (6 h) with 0.5 µCi of ^{32}PO$_4$ per ml, inducing cells to encystment with BHB with a medium containing ^{31}PO$_4$, and monitoring P-lipids as above. Time of sampling was measured from the initiation of encystment with 0.2% BHB.*

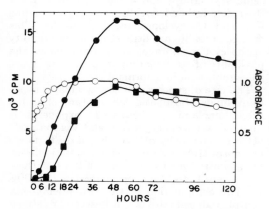

FIG. 2. *Culture turbidity (○), the uptake of [3-^{14}C]BHB into cells (●), and [3-^{14}C]BHB incorporated into extractable lipids (■) during encystment of* A. vinelandii. *[^{14}C]BHB (9 µCi) was added to a 300-ml encysting culture. Activity is shown as counts per minute per milliliter of culture.*

(CoA) ester is then formed. Acetoacetyl-CoA can be cleaved into two acetyl-CoA, which can be oxidized or utilized for synthesis. Alternatively, acetoacetyl-CoA can be reduced to BHB CoA and polymerized into poly-β-hydroxybutyrate (PHB) (14), a storage product which constitutes 8% of the cyst's weight (11). Both pathways are operative during encystment.

The incorporation of BHB carbon into cells began almost immediately upon its addition to a culture of *A. vinelandii* and continued for 55 h postinduction (Fig. 2). At that time, cells contained as much as 16% PHB, of which about 50% was depolymerized and oxidized during the latter half of the encystment process. Of the total BHB carbon incorporated into cysts, 70% was in lipids whose synthesis began 6 h postinduction and was complete 42 h later (Fig. 2). Thus, PHB and lipid synthesis in the presence of BHB are related but independent events. However, when BHB, used to induce encystment, was removed from the culture at 8 h (at the initiation of lipid synthesis), cyst lipids were formed at the expense of PHB.

Identification of cyst lipids. Crude lipids

A.vinelandii Resorcinols and Pyrones

Structure	Lab name	R	ratio	Per cent of lipid
	AR$_1$	$C_{21}H_{43}$ $C_{23}H_{47}$	$\frac{87}{13}$	29
	AR$_2$	$C_{21}H_{43}$ $C_{23}H_{47}$	$\frac{87}{13}$	18
	AR$_3$	$C_{21}H_{43}$ $C_{23}H_{47}$	$\frac{1}{1}$	1
	AR$_4$	$C_{19}H_{39}$ $C_{21}H_{43}$	$\frac{3}{4}$	3
	AR$_5$	$C_{21}H_{43}$ $C_{23}H_{47}$	$\frac{6}{4}$	2
	AP$_1$	$C_{21}H_{43}$ $C_{23}H_{47}$	$\frac{2}{5}$	18
	AP$_2$	$C_{19}H_{39}$ $C_{21}H_{43}$	$\frac{1}{1}$	9

FIG. 3. *Structures of unique cyst lipids, ratios of homologs, and relative amounts of each molecular species.*

TABLE 1. *Lipid composition of major cyst components*[a]

| Component | Dry wt (mg) | Total lipid (mg) | Unique lipids (mg) | | | $AR_1 + AR_2$ $+ AP_1$ (% of total lipid) |
			AR_1	AR_2	AP_1	
Exine	6.4	1.7 (27%)	0.6	0.5	0.1	70
Intine	3.8	1.0 (26%)	0.1	0.1	0	20
Central body	12.3	1.9 (16%)	0.7	0.5	0.5	90

[a] Mature cysts were fractionated as described by Lin and Sadoff (9).

were extracted from cysts (2) and fractionated into neutral, glyco-, and P-lipids by elution from silicic acid columns with chloroform, acetone, and methanol, respectively. When [3-^{14}C]BHB was used as inducer, about 85% of the radioactivity was found in the fraction eluting with acetone. This fraction was resolved by thin-layer chromatography into nine components, all of which reacted with diazotized sulfanilamide and five of which reacted with vanillin reagent, suggesting that these five were resorcinols. Seven of the lipid constituents were purified to homogeneity by gradient elution from silicic acid columns, preparative thin-layer chromatography, and high-pressure liquid chromatography, and their structures were determined by chemical tests and infrared, UV, nuclear magnetic resonance, and mass spectrometry. Those vanillin positive were all 5-*n*-alkylresorcinols (AR, Fig. 3) with two chain-length homologs of each species. The other two lipid species were 6-*n*-alkyl-4-hydroxy-pyran-2-ones (pyrones) with two chain-length homologs of each (AP, Fig. 3). Those components designated AR_1, AR_2, and AP_1 constituted 65% of the cyst lipid.

Synthesis of unique lipids. Orsellinic acid (5-methyl-4-carboxyresorcinol) and triacetic acid lactone (6-methyl-4-hydroxy-pyran-2-one) are homologs of the two classes of unique cyst lipids. The synthetic pathways of these and other similar compounds have been published (1, 4, 5, 18). They all arise from an aldol condensation or lactonization of polyketides resulting from the condensation of acetyl-CoA and malonyl-CoA in the absence of NADPH. The catalyst is a multienzyme system similar or identical to fatty acid synthetase (3, 18).

The fatty acid synthetase of vegetative *A. vinelandii* is a soluble ezyme which catalyzes the synthesis of fatty acids of 14 to 18 carbons. We suggest that, during encystment, a synthetase modified in chain-length control or an "encystment synthetase" catalyzes the formation of C_{28} to C_{30} acids with keto groups at the 3, 5, and 7 positions. Such a compound could cyclize to form AR_1 or AP_2 directly. Minor modifications of the substrate or product could yield all the unique lipids thus far isolated.

Distribution and function of unique lipids. The distribution of lipids in cyst components is presented in Table 1. The unique cyst lipids were found primarily in the exine and central body, both of which contain membrane or bilayer structures. The phenolic hydroxyl groups of AR_1 and AR_2 could form cross-linking glycosides with carbohydrates of the exine, thus contributing to its structure.

We have purified central body membranes and found that they do indeed contain high levels of resorcinols. Alkyl resorcinols and pyrones are remarkably well suited as substituents of membranes in possessing hydrophilic "heads" and long hydrophobic "tails," much like P-lipids. We suggest that they replace a portion of the P-lipids in membranes of encysting cells and in so doing generate a structure which may "direct" the differentiation process or contribute to the dormancy and resistance of the mature cysts.

ACKNOWLEDGMENT

This investigation was supported by Public Health Service research grant AI-01863 from the National Institute of Allergy and Infectious Diseases.

LITERATURE CITED

1. **Birch, A. J.** 1967. Biosynthesis of polyketides and related compounds. Science **156**:202–206.
2. **Bligh, E. G., and W. J. Dyer.** 1959. A rapid method of total lipid extraction and purification. Can. J. Biochem. Physiol. **37**:911–917.
3. **Dimroth, P., E. Ringelmann, and F. Lynen.** 1976. 6-Methyl-salicylic acid synthetase from *Penicillium patulum*. Eur. J. Biochem. **68**:591–596.
4. **Dimroth, P., H. Walter, and F. Lynen.** 1970. Biosynthesis von 6-Methyl-salicylsaure. Eur. J. Biochem. **13**:98–110.
5. **Gatenbeck, S., and K. Mossbach.** 1959. Acetate carboxyl oxygen (^{18}O) as donor for phenolic hydroxy groups in orsellinic acid produced by fungi. Acta Chem. Scand. **13**:1561–1564.
6. **Hitchins, V. M., and H. L. Sadoff.** 1970. Morphogenesis of cysts in *Azotobacter vinelandii*. J. Bacteriol. **104**:492–498.
7. **Hitchins, V. M., and H. L. Sadoff.** 1973. Sequential metabolic events during encystment of *Azotobacter vinelandii*. J. Bacteriol. **113**:1273–1279.
8. **Lin, L. P., and H. L. Sadoff.** 1968. Encystment and polymer production by *Azotobacter vinelandii* in the presence of β-hydroxybutyrate. J. Bacteriol. **95**:2336–2343.
9. **Lin, L. P., and H. L. Sadoff.** 1969. Preparation and

ultrastructure of the outer coats of *Azotobacter vinelandii* cysts. J. Bacteriol. **98**:1335–1341.

10. **Lin, L. P., and H. L. Sadoff.** 1969. Chemical composition of *Azotobacter vinelandii* cysts. J. Bacteriol. **100**:480–486.

11. **Reusch, R. N., and H. L. Sadoff.** 1979. 5-*n*-Alkyl resorcinols from encysting *Azotobacter vinelandii*: isolation and characterization. J. Bacteriol. **139**:448–453.

12. **Sadoff, H. L.** 1973. Comparative aspects of morphogenesis in three prokaryotic genera. Annu. Rev. Microbiol. **27**:133–153.

13. **Sadoff, H. L., E. Berke, and B. Loperfido.** 1971. Physiological studies of encystment in *Azotobacter vinelandii*. J. Bacteriol. **105**:185–189.

14. **Senior, P. J., and E. A. Dawes.** 1971. Poly-β-hydroxy-butyrate biosynthesis and the regulation of glucose metabolism in *Azotobacter beijerinckii*. Biochem. J. **125**:55–66.

15. **Socolofsky, M. D., and O. Wyss.** 1961. Cysts of *Azotobacter*. J. Bacteriol. **81**:946–954.

16. **Socolofsky, M. D., and O. Wyss.** 1962. Resistance of the *Azotobacter* cyst. J. Bacteriol. **84**:119–124.

17. **Su, C.-J., R. Reusch, and H. L. Sadoff.** 1979. Fatty acids in phospholipids of cells, cysts, and germinating cysts of *Azotobacter vinelandii*. J. Bacteriol. **137**:1434–1436.

18. **Yalpani, M., K. Willecke, and F. Lynen.** 1969. Triacetic acid lactone, a derailment product of fatty acid synthesis. Eur. J. Biochem. **8**:495–502.

Ultrastructure of *Streptomyces bambergiensis* Aerial Spore Envelope

RICHARD A. SMUCKER AND SUSANNE L. SIMON

Chesapeake Biological Laboratory, University of Maryland, Solomons, Maryland 20688

The *Streptomyces bambergiensis* aerial spore envelope was examined by electron microscopy. Information from transmission electron micrographs of negatively stained, thin-sectioned, and freeze-etched spores was used to develop a spore envelope model. The spore envelope outside the plasma membrane consists of centripetally consecutive layers: inner spore wall, outer spore wall, rodlet mosaic, granular mosaic, and fibrous sheath. The hairlike projections are composed of a basal core of spore wall material covered by the rodlet fibers and a granular matrix and the fibrous sheath.

Surface ornamentation of *Streptomyces* spores ranges from smooth spore surface in some species, e.g., *S. coelicolor* (2), to spore surfaces having extensive projections, e.g., hairlike structures in *S. bambergiensis* (R. A. Smucker and S. L. Simon, Abstr. Annu. Meet. Am. Soc. Microbiol. 1980, J15, p. 83). Spores of all *Streptomyces* species examined by freeze-etch replication have been shown to have a spore wall-associated array of rodlets (1, 12). We have shown that the rodlets in *S. coelicolor* are a complex arrangement of individual chitin fibrils (9). These rodlets were previously thought of as the outermost spore component which imparted the hydrophobic character to *Streptomyces* spores (1). Collectively, the rodlets were labeled as the "rodlet sheath" or "fibrous sheath" (1, 3). The rodlet mosaic observed in carbon replicas (1) and in freeze-etch replicas (3) has been equated with the fibrous sheath observed in thin sections. We have demonstrated in *S. coelicolor* that the rodlet mosaic is covered by two layers: a granular matrix and an outermost layer (9). The purpose of the present work is to develop a working model of the *S. bambergiensis* hairy spore envelope.

RESULTS AND DISCUSSION

S. bambergiensis ISP 5590 was obtained from E. B. Shirling and was maintained on a chitin agar medium (11). *S. bambergiensis* was grown on glycerol-asparagine agar (8) for ultrastructural studies. Cultures were harvested after 7 to 9 days of growth at 27 to 28°C. Impressions of 7- to 9-day aerial sporulated growth were coated with Au-Pd in a Technics Hummer V sputter-coater. Micrographs were taken with a JEOL SEM-U3 scanning electron microscope using an accelerating voltage of 25 kV. A modification of the Luft method (5) using ruthenium red was used for fixing the cells. Cells were prefixed in 1.2% glutaraldehyde-ruthenium red for 2 h and in 1.7% osmium tetroxide-ruthenium red for 4 h at 23°C followed by 12 to 14 h at 5°C. The samples were dehydrated in an ethanol series and embedded in Epon (4). Thin sections were poststained with 2% lead citrate (7). Grid impressions of aerial growth were negatively stained with 2% phosphotungstic acid. Freeze-etch replicas were made of deep-etched fractures using a modified Denton freeze-etch device (10). Figure 1 is a scanning electron micrograph showing secondary mycelia (M) with no spore hairs. The mature spores have the longest hairs. Figures 2 and 3 are transmission electron micrographs of thin-sectioned spore wall and associated hairs. Figure 3 shows particularly well the outer wall material in the hair base. Figures 4 and 5 are transmission electron micrographs of 2% phosphotungstic acid negatively stained hairs. The hairs are fragile and often break off (arrows) near the hair base during air drying.

The scanning electron micrograph in Fig. 1 demonstrated: (i) the relatively smooth secondary mycelial surface, (ii) the developing hairlike projections on the sporophore surfaces, and (iii) the elongated hairs on mature spores. Remaining figures refer to the hair-producing spores.

Thin sections in Fig. 2 and 3 show two wall layers typical of *Streptomyces* spores (1, 2). The osmium-ruthenium red fixation permitted differ-

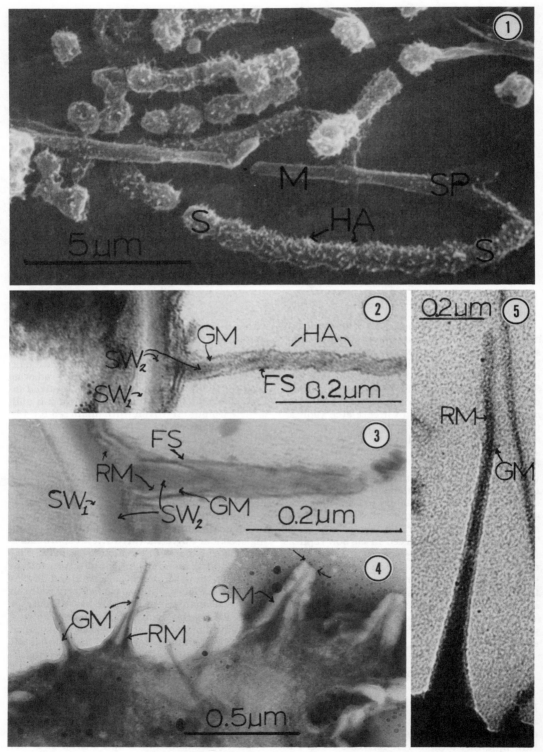

Fig. 1–5. *Electron micrographs of* S. bambergiensis. *Secondary mycelia (M), hair (HA), sporophore (SP), spore (S), inner spore wall (SW₁), outer spore wall (SW₂), rodlet mosaic (RM), granular matrix (GM), fibrous sheath (FS).*

entiation of the inner (electron-lucent) and the outer (electron-dense) spore wall. This is similar to the staining pattern which occurs in *S. coelicolor* spore walls (9).

Negatively stained hairs in Fig. 4 and 5, as well as the longitudinally sectioned hairs in Fig. 2 and 3, establish the hairs as integral components of the mature spore envelope. The negatively stained hairs in Fig. 4 and 5 show that the fibrils which originate from the rodlet mosaic layer extend throughout the length of the hair. Freeze-etch replicas (Fig. 6 and 8) and thin sections (Fig. 2 and 3) both show the twisting of rodlet mosaic fibers through the length of the hair. The fiber bundles taper to a relatively uniform diameter of 20 nm, along the length of the hair (Fig. 4 and 5).

The negative stain in Fig. 5 shows an amorphous layer(s) of material (GM) surrounding the length of the hair and surrounding the tip of the fiber bundle. The freeze-etch replica in Fig. 8 reveals one layer outside the fibril component layer (GM). This outer layer sometimes can be observed, but usually the hairs are fractured, exposing the rodlet fibril component (Fig. 7 and 8).

The model of *S. bambergiensis* envelope (Fig. 9) is presented as a summary of the evidence presented in Fig. 1–8 and of the hundreds of other micrographs examined. No single method of sample preparation has proved adequate to elucidate the complex architecture of the *Streptomyces* spore. The preparation which most clearly and singly supports the model is the longitudinal section (Fig. 3 and 4).

The assumption in earlier work by Hopwood and Glauert (3) for the smooth-spored *S. violaceoruber* (*S. coelicolor*) and Wildermuth (12) for the warty-spored *S. viridochromogenes* and hairy-spored *S. glaucescens* is that the rodlet mosaic is the external layer. In contrast, Matselyukh (6) used negative staining to show clearly that *S. olivaceus* hairs had a predominant layer of amorphous material surrounding the fibrils. This is consistent with our findings (Fig. 4 and 5).

To maintain continuity with the established literature, including the model presented for *S. coelicolor* (9), we have retained the term "fibrous sheath" to refer to the outermost layer of *Streptomyces* spores. "Rodlet mosaic" is likewise used for the composite of rodlet fibrils which

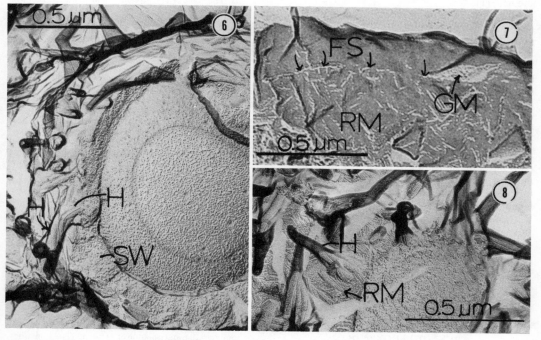

FIG. 6–8. *Electron micrographs of freeze-etch replicas of* S. bambergiensis. *Figure 6 shows fibrous bundles exposed by fracturing. The wall fractured so as to reveal the placement of rodlet mosaic and hairs adjacent to the wall. In Fig. 7, the granular matrix (GM) and the fibrous sheath are visible. Arrows delineate the fracture zone between the fibrous sheath and the GM dotted rodlet mosaic. Figure 8 shows that the RM fibrils formed partial spirals around the base of several hairs.*

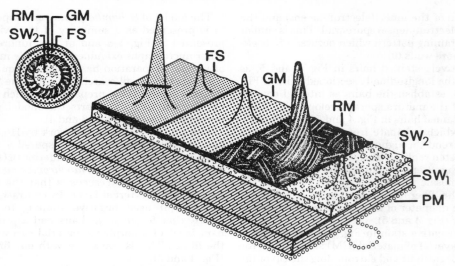

FIG. 9. *Composite model of* S. bambergiensis *envelope.*

run along the spore subsurface in a patchwork fashion in developing spores and in the mature spores are synthesized to form fibrils inside the elongated hairs (Fig. 6–8). Previously, the "rodlet mosaic" was essentially equated with "fibrous sheath" (1).

The spore envelope model in Fig. 9 is a summary of present evidence. Thin sections of osmium-ruthenium red–stained walls (Fig. 2 and 3) reveal two spore wall layers (SW$_1$ and SW$_2$). The outer wall (SW$_2$) material extends into the base of the hairs (Fig. 3 and 4). Rodlet mosaic fibers helically wrap around the outer wall layer (SW$_2$) and the wall extensions (Fig. 8). In negative stain (Fig. 5) and occasionally in freeze-etch replicas (Fig. 7), the outer amorphous material (GM) is visible. Thin sections of osmium-ruthenium red–fixed hairs indicate a second layer (FS) outside the rodlet fibrils in the hair (Fig. 2 and 3). In keeping with the spore envelope models of *S. coelicolor* (9), the outermost layer is labeled FS (1, 12).

Information currently available indicates that *Streptomyces* spore envelopes are more complex than previously realized. Biochemical characterization of these surface projections is not yet complete except for indications that the fibrils are chitin (Smucker and Simon, Abstr. Annu. Meet. Am. Soc. Microbiol. 1980, J15, p. 83), as are the rodlet mosaic fibrils in *S. coelicolor* (9).

ACKNOWLEDGMENTS

We gratefully acknowledge Russell Steere, Eric Erbe, and Mike Moseley (U.S. Department of Agriculture, Beltsville, Md.) for technical and interpretive assistance for the freeze-etch replication.

LITERATURE CITED

1. **Bradley, S. G., and D. Ritzi.** 1968. Composition and ultrastructure of *Streptomyces venezuelae.* J. Bacteriol. **95:**2358–2364.
2. **Glauert, A. M., and D. A. Hopwood.** 1961. The fine structure of *Streptomyces violaceoruber* (*S. coelicolor*). II. The walls of the mycelium and spores. J. Biophys. Biochem. Cytol. **10:**505–516.
3. **Hopwood, D. A., and A. M. Glauert.** 1961. Electron microscope observations on the surface structures of *Streptomyces violaceoruber.* J. Gen. Microbiol. **26:**325–330.
4. **Luft, J. H.** 1961. Improvements in epoxy resin embedding methods. J. Biophys. Biochem. Cytol. **9:**409–414.
5. **Luft, J. H.** 1971. Ruthenium red and violet. I. Chemistry, purification methods of use for electron microscopy and mechanism of action. Anat. Rec. **171:**347–367.
6. **Matselyukh, B. P.** 1978. Ultrastructure of hairs on the spore surface of *Streptomyces olivaceus* VKX, p. 235–240. *In* M. Mordarski, W. Kurylowicz, and J. Jeljaszewicz (ed.), Nocardia and Streptomyces, Proceedings of the International Symposium on *Nocardia* and *Streptomyces*, Warsaw, 4–8 October 1976. Gustav Fischer Verlag, New York.
7. **Reynolds, D. M.** 1963. The use of lead citrate at high pH as an electron-opaque stain in electron microscopy. J. Cell Biol. **17:**208–212.
8. **Shirling, E. B., and D. Gottlieb.** 1966. Methods for characterization of *Streptomyces* species. Int. J. Syst. Bacteriol. **16:**313–340.
9. **Smucker, R. A., and R. M. Pfister.** 1968. Characteristics of *Streptomyces coelicolor* A3(2) aerial spore rodlet mosaic. Can J. Microbiol. **24:**397–408.
10. **Steere, R. L., E. F. Erbe, and J. M. Moseley.** 1979. Controlled contamination of freeze-fractured specimens, p. 99–109. *In* J. E. Rash and C. S. Hudson (ed.), Freeze fracture; methods, artifacts, and interpretations. Raven Press, New York.
11. **Walker, J. D., and R. R. Colwell.** 1975. Factors affecting enumeration and isolation of Actinomycetes from Chesapeake Bay and Southeastern Atlantic Ocean sediments. Mar. Biol. **30:**193–201.
12. **Wildermuth, H.** 1970. Surface structure of streptomycete spores as revealed by negative staining and freeze-etching. J. Bacteriol. **101:**318–322.

L-Arginine Inhibition of the Initiation of Germination in Spores of *Thermoactinomyces thalpophilus*

HAROLD F. FOERSTER

Department of Life Sciences, Sam Houston State University, Huntsville, Texas 77341

Included among the most effective initiators for the germination of bacterial spores of thermophilic actinomycetes are L-alanine, L-leucine, glycine, L-phenylalanine, and L-serine. Of 20 common amino acids, only L-arginine blocked the germination of spores in solutions containing each of the most effective amino acid initiators singly or as mixtures. D-Arginine and agmatine were ineffective.

In determining the germination requirements for different strains of thermophilic actinomycetes, it was found that spore strains could be variously germinated in solutions containing a single inorganic salt. That is, spore populations of some strains were completely germinated in solutions containing only sodium phosphate, whereas some strains germinated moderately well and still other strains germinated poorly or not at all. The rate and extent of the germination of spore populations of each of the strains were enhanced by the addition of one or more selected organic initiators to the sodium phosphate solutions. Compared with simple sugars, purines, pyrimidines, and a nucleoside (inosine), amino acids were the most effective organic initiators for germinating the spores of thermophilic actinomycetes. However, amino acids produced a wide range of germination responses. Amino acids found to be particularly effective as initiators for spore germination included L-alanine, glycine, L-leucine, L-phenylalanine, and L-serine. Some amino acids produced intermediate (fractional) responses, and some produced no apparent effect, i.e., did not affect the sodium phosphate germination (1, 3). A few amino acids, L-arginine in particular, caused reductions in the germination responses obtained with the sodium phosphate solutions (1; H. F. Foerster, Abstr. Annu. Meet. Am. Soc. Microbiol. 1978, I114, p. 100).

L-Arginine is a powerful and highly specific inhibitor of the germination of spores of thermophilic actinomycetes. This paper describes preliminary studies on the effects of L-arginine on the germination of spores of *Thermoactinomyces thalpophilus* in solutions containing different amino acid initiators.

RESULTS

Effect of L-arginine on amino acid-initiated spore germination. In addition to blocking the germination of spores of thermophilic actinomycetes in solutions containing sodium phosphate, L-arginine also inhibited spore germination in solutions containing amino acid initiators (Table 1). Each test solution included sodium phosphate (10 mM, pH 8.0) and the most effective amino acid initiator for the strain tested, i.e., L-phenylalanine for Ha19 and L-leucine for Si01 and Ss02. Several amino acids, notably L-arginine, L-lysine, L-methionine, and L-valine, caused reductions in the germination of spore suspensions of Si01 when added individually to the germinant solutions. However, of the 14 amino acids added individually only L-arginine prevented the germination of all or most of the spores of each of the three strains of *T. thalpophilus*.

Effects of L- and D-arginine and agmatine on spore germination. The results in Table 2 show the effects of L- and D-arginine and agmatine on the germination of spores initiated by the most effective amino acid initiators for three strains of *T. thalpophilus*. L-Arginine, but not D-arginine, inhibited the germination of spores of *T. thalpophilus*. The results in Table 2 were obtained with the hydrochloride forms of L- and D-arginine. However, the free base of L-arginine was equally effective as an inhibitor of germination, indicating the stereospecific nature of the inhibition. This was also indicated by the observation that agmatine did not block spore germination. Agmatine, a compound lacking the carboxyl group found in arginine and thus without its asymmetric carbon atom, was ineffective as an inhibitor of spore germination. Furthermore, L-citrulline, in which the guanido group of L-arginine is replaced by a carbamido group, also failed to inhibit spore germination in *T. thalpophilus* (unpublished data).

TABLE 1. *Effect of individual amino acids on spore germination[a]*

Amino acid[b] added (1.0 mM)	Organism[c]		
	Ha19	Si01	Ss02
None	62	59	58
Arginine	11	0	0
Asparagine	60	59	60
Aspartic acid	60	59	60
Cysteine	59	60	57
Cystine	62	57	60
Glutamine	61	58	60
Histidine	60	60	59
Lysine	59	52	62
Methionine	59	29	58
Ornithine	59	60	61
Threonine	59	58	60
Tryptophan	61	60	60
Tyrosine	63	61	60
Valine	63	48	58

[a] Germination reported as the percent reduction in absorbance at 540 nm after 30 min of incubation. Sodium phosphate in each test: 10 mM, pH 8.0.
[b] Each amino acid was used as the L-enantiomorph.
[c] Spores of *T. thalpophilus* Ha19 were initiated by L-phenylalanine; Si01 and Ss02 were initiated by L-leucine. Strains of *T. thalpophilus* were isolated from cured baled hay (Ha19) and soil (Si02, Ss02). The growth and sporulation medium and the procedures used in the preparation, storage, and germination of spores have been described (1, 3). All spore preparations were low-temperature activated (2).

Spore populations of some strains of *T. thalpophilus* (e.g., F102) can be fully germinated by only a limited number of amino acids, whereas the spores of other strains (e.g., Ha19, Si01) can be effectively germinated by a variety of amino acids in solutions of a suitable salt. A comparison of the most potent amino acid initiators for three strains of *T. thalpophilus,* including L-alanine, glycine, L-leucine, L-phenylalanine, and L-serine, showed that each of these was inhibited by L-arginine (Table 2). However, amino acid initiators differed in their sensitivity to the inhibitory action of L-arginine; e.g., L-alanine and L-leucine were less sensitive to inhibition than were glycine, L-phenylalanine, and L-serine.

Efficacy of L-arginine as an inhibitor for the germination of spores initiated by different amino acids. To enhance the sensitivity of germination to inhibition by L-arginine, the minimum concentrations of sodium phosphate and each of the amino acid initiators needed to produce a maximum (or near maximum) germination response were determined. This was done by using reciprocal titrations for the sodium phosphate and each of the amino acid initiators; i.e., the concentration of one component (e.g., sodium phosphate or amino acid) was kept constant while the concentration of the component to be titrated (e.g., amino acid or sodium phosphate) was varied. In the titration of each component, a range of concentrations was used which included levels too low to produce detectable germination responses through levels which produced maximum obtainable responses (Table 2).

The sodium phosphate titration was carried

TABLE 2. *Effects of L-arginine, D-arginine, and agmatine on the germination[a] of spores of* T. thalpophilus[a]

Organism	Initiator[b] amino acid	Concn (mM)	Addition[c]			
			None	L-Arginine	D-Arginine	Agmatine
F102	Leucine	1.0	57	33	55	57
		0.1	57	4	55	57
Ha19	Alanine	1.0	60	51	60	61
		0.1	60	30	60	57
		0.01	60	0	57	56
	Glycine	1.0	56	0	55	57
	Phenylalanine	1.0	60	0	60	57
	Serine	1.0	56	0	53	53
Si01	Alanine	0.01	57	2	55	54
	Glycine	1.0	56	8	58	56
	Leucine	1.0	59	48	58	59
		0.1	59	8	58	56
	Phenylalanine	1.0	58	0	58	56
	Serine	1.0	58	4	53	57

[a] Germination is reported as the percent reduction in absorbance at 540 nm after 30 min of incubation at 55°C. Sodium phosphate in all tests: 10 mM, pH 8.0.
[b] Excepting glycine, each amino acid initiator was used as the L-enantiomorph.
[c] L-Arginine, D-arginine, and agmatine, 1.0 mM each.

out in 1.0 mM solutions of each of the amino acids. The minimum concentration of sodium phosphate required to support an optimal germination response in spore suspensions of Ha19 was 0.6 mM and was similar in solutions of each of the amino acid initiators tested (Table 2).

The minimum concentration of each of the amino acid initiators required for an optimal germination response, in 0.6 mM (pH 8.0) solutions of sodium phosphate, differed for the five amino acids tested. Essentially three levels of germinative potency were observed and, in decreasing order of potency, included: L-alanine, 0.01 mM; L-phenylalanine, glycine, and L-leucine, 0.1 mM; and L-serine, 1.0 mM (Table 2). At the indicated concentrations, each of the amino acids, with the exception of L-leucine, produced rapid and complete germination of spore populations as measured by reductions in absorbance at 540 nm and by loss of refractility of spores examined with a phase-contrast microscope. The incomplete germination responses

obtained with 0.1 mM L-leucine were, however, the best obtainable in solutions of sodium phosphate since higher concentrations of L-leucine gave similar (fractional) responses.

L-Arginine effectively blocked the germination initiated by each of the amino acids (Fig. 1). However, the five amino acids tested individually in solutions of sodium phosphate differed somewhat in their sensitivity to inhibition by L-arginine. The amount of L-arginine needed to reduce the germination obtained with each of the amino acids by 90% or more, as evidenced by a 5% or less reduction in absorbance, ranged from above to below stoichiometric levels. L-Phenylalanine, one of the most effective amino acids for initiating the spores of Ha19, was also one of the most sensitive to inhibition by L-arginine; essentially complete inhibition of germination was obtained with stoichiometric levels. Glycine and L-serine, two potent initiators, were also effectively blocked in their capacity to germinate spores at stoichiometric and

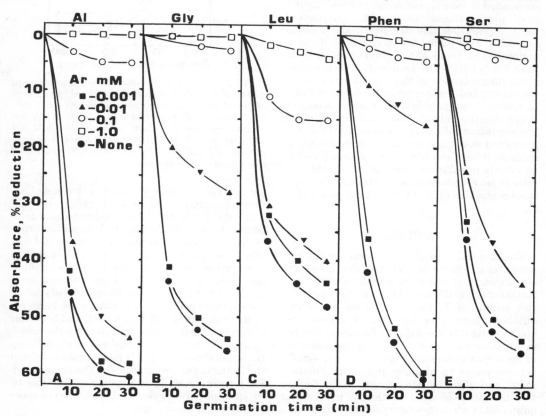

FIG. 1. *Effect of L-arginine on the germination of spores of* T. thalpophilus *Ha19 initiated by individual amino acids. Germination was measured as the percent reduction in absorbance at 540 nm. Sodium phosphate in each test: 0.6 mM, pH 8.0. Amino acids added to tests in frames A–E, respectively: L-alanine, 0.01 mM; glycine, 0.1 mM; L-leucine, 0.1 mM; L-phenylalanine, 0.1 mM; L-serine, 1.0 mM. L-Arginine (Ar) was added to tests in frames A–E as indicated.*

less than stoichiometric levels of L-arginine, respectively. On the other hand, L-alanine and L-leucine were less sensitive to inhibition, and each required more than stoichiometric levels of L-arginine to produce a 90% reduction in germination. Paradoxically, L-alanine and L-leucine were the most effective and least effective initiators, respectively, for the germination of spores of Ha19. However, for some strains of *T. thalpophilus,* L-leucine was a potent initiator of spore germination but was nevertheless relatively resistant to inhibition by L-arginine. Indeed, irrespective of their potencies as initiators for the germination of spores of different strains of *T. thalpophilus,* L-alanine and L-leucine have been found to be less sensitive to inhibition by L-arginine than other amino acid initiators (unpublished data).

L-Arginine was also an effective inhibitor of germination of spores suspended in solutions of sodium phosphate containing mixtures of the four most potent amino acid initiators. Spores suspended in solutions of sodium phosphate (0.6 mM) plus L-alanine (0.01 mM), glycine (0.1 mM), L-phenylalanine (0.1 mM), and L-serine (1.0 mM) were tested for their sensitivity to L-arginine. The levels of L-arginine needed to inhibit the initiation of germination in solutions containing mixtures of the four amino acid initiators equaled or approached the levels required to produce comparable inhibitions in solutions containing the individual amino acids. Whereas concentrations of L-arginine of 0.01 mM or less permitted essentially maximum germination responses, more than 50% of the spores failed to germinate in solutions containing 0.1 mM, and little to no germination was obtained with solutions containing 1.0 and 10.0 mM L-arginine (Fig. 2).

DISCUSSION

The unusual properties of arginine in the germination of spores of thermophilic actinomycetes were first noted by Kirillova, Agre, and Kalakoutskii (5). They observed that spores of two strains of *T. vulgaris,* grown on several different media, were not initiated or were initiated very weakly in solutions of arginine. Studies in our laboratory have confirmed the resistance of spores of numerous strains of thermophilic actinomycetes to arginine initiation. Furthermore, and of particular interest, is the finding that L-arginine blocks initiation by many other amino acids when these are present individually or as mixtures. The inhibitory action of L-arginine on the spores of thermophilic actinomycetes is of particular interest since inhibition of

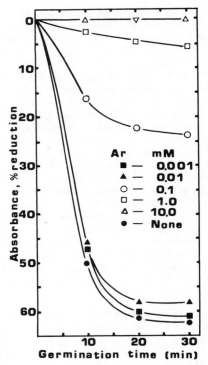

FIG. 2. *Effect of L-arginine on the germination of spores of* T. thalpophilus *Ha19 initiated by mixtures of amino acids. Germination was measured as percent reduction in absorbance at 540 nm. Each test solution contained: sodium phosphate, 0.6 mM, pH 8.0; L-alanine, 0.01 mM; glycine, 0.1 mM; L-phenylalanine, 0.1 mM; L-serine, 1.0 mM. L-Arginine (Ar) was added to tests as indicated.*

germination by amino acids in solutions of sodium phosphate was rarely observed.

In addition to the lack of L-amino acid inhibition observed (Table 1), the D-enantiomorphs of leucine, phenylalanine, and serine did not affect the germination responses of spores in solutions of sodium phosphate with or without equimolar concentrations of the corresponding L form. That is, the D-enantiomorphs of each of these amino acids did not initiate the germination of spores of F102, Ha19, Si01, and Ss02, nor did any of them inhibit initiation by the corresponding L forms (unpublished data). D-Enantiomorphs have frequently been shown to inhibit the initiation of germination by the corresponding L-amino acids in the spores of *Bacillus* (4, 6), and this may suggest a difference in the spore properties of these organisms.

That activated spores of Ha19 and Si01 can be germinated by L-leucine, L-phenylalanine, or L-serine but not by the corresponding D-enantiomorphs indicates that the initiation of germination in these spores is stereospecific for each of the amino acid initiators. That the germination of these spores can be initiated with essentially equal facility by three (or more) different L-amino acids raises the question of whether the germination of these spores is the result of different (multiple) mechanisms, each initiated by a stereospecific initiator, or the result of a single mechanism that can be set in motion by any one of a number of stereospecific initiators.

Considering the stereospecific nature of initiation by each of the amino acids, it seems unlikely that structurally different amino acid initiators would bind to or affect the same initiator sites in the spore. The apparent need for different initiator sites might be taken to support multiple mechanisms of germination, each dependent upon a specific initiator. On the other hand, the fact that L-arginine can inhibit germination by each of the amino acid initiators at close to or less than stoichiometric levels may indicate that the initiation of germination in these spores is by a single mechanism, that is, a mechanism of germination that can be initiated by a number of different amino acids but, irrespective of the initiator, is inhibitable by L-arginine. The observation that 0.1 mM L-arginine inhibited the germination of more than 50% of the spores of Ha19 in solutions containing mixtures of four different amino acid initiators at a combined concentration of 1.21 mM (Fig. 2) agrees with the latter interpretation.

ACKNOWLEDGMENTS

The assistance of Diana Cook is gratefully acknowledged. This work was supported by the University Research Fund.

LITERATURE CITED

1. **Foerster, H. F.** 1975. Germination characteristics of some thermophilic actinomycete spores, p. 36–43. *In* P. Gerhardt, R. N. Costilow, and H. L. Sadoff (ed.), Spores VI. American Society for Microbiology, Washington, D.C.
2. **Foerster, H. F.** 1978. Effect of temperature on the spores of thermophilic actinomycetes. Arch. Microbiol. **118:** 257–264.
3. **Foerster, H. F.** 1978. Properties of spores of thermophilic actinomycetes, p. 321–326. *In* G. Chambliss and J. C. Vary (ed.), Spores VII. American Society for Microbiology, Washington, D.C.
4. **Gould, G. W.** 1969. Germination, p. 397–444. *In* G. W. Gould and A. Hurst (ed.), The bacterial spore. Academic Press, Inc., New York.
5. **Kirillova, I. P., N. S. Agre, and L. V. Kalakoutskii.** 1975. Germination of spores of *Thermoactinomyces vulgaris*: significance of the composition of the medium, the role of Mg^{2+} and Ca^{2+}. Mikrobiologiya **44**:1034–1040.
6. **Sussman, A. S., and H. O. Halvorson.** 1966. Spores: their dormancy and germination. Harper and Row, New York.

Trehalase from Dormant Spores of *Dictyostelium discoideum* Isolated by Percoll Density Gradient Centrifugation

KATHLEEN A. KILLICK

Department of Developmental Biology, Boston Biomedical Research Institute, Boston, Massachusetts 02114

Purification on a preparative scale and resolution of *Dictyostelium discoideum* spores from contaminating nonspore material was achieved by Percoll density gradient centrifugation. This made possible the isolation of a sufficiently large amount of biological material which permitted detection of spore-specific trehalase. Partial purification and characterization of this enzyme from dormant spores demonstrated that it was very similar to trehalase isolated from *Dictyostelium* myxamoebae.

The life cycle of the cellular slime mold *Dictyostelium discoideum* consists of two distinct phases: growth and multicellular differentiation. Exhaustion of an exogeneous food source initiates the developmental cycle, and this event is followed by collection of vegetative myxamoebae together to form multicellular aggregates. Each aggregate subsequently differentiates over a 24- to 28-h period (at 22°C) into a mature fruiting body (or sorocarp) which consists of a spore mass supported by a slender cellulose-ensheathed stalk (1). Activation of the dormant *Dictyostelium* spores results in germination and the release of single-celled myxamoebae, which, under appropriate nutritional conditions, undergo mitosis and reenter the growth phase of the life cycle (5).

Trehalose is a major carbohydrate reserve in the dormant spores of *Dictyostelium*. This nonreducing disaccharide serves as the principal energy and carbon source during the emergence stage of germination, at which time its hydrolysis to glucose is catalyzed by the hydrolytic enzyme trehalase (α,α'-trehalose 1-D-glucohydrolase, EC 3.2.1.28) (3–5, 11). Although the enzyme from vegetative *D. discoideum* cells has been partially purified and characterized (2), limited attention has been paid to trehalase from dormant spores. As part of an ongoing series of studies on the regulation of enzyme latency during cellular differentiation and aging in *Dictyostelium*, methods were developed which permitted the purification on a preparative scale and the efficient resolution of spore cells from contaminating, nonspore material by the use of Percoll density gradient centrifugation (K. Killick, Anal. Biochem., in press). Use of these preparative procedures, as described in the present report, made possible the isolation of a sufficiently large amount of biological material to permit the detection of spore-specific trehalase. The above methodology also allowed a partial purification and characterization of this enzyme to be undertaken.

RESULTS AND DISCUSSION

Spore preparation. To ensure full spore maturation, spores were routinely prepared from sorocarps which were 2 to 3 days old. Since trehalase is secreted by *Dictyostelium* during development (3, 10), harvested sorocarps were washed at least three times with 10 mM phosphate buffer (pH 6.5) to remove all traces of the exogeneous enzyme prior to spore preparation. Washed fruiting bodies were homogenized and then subjected to a 5-min treatment on a Vortex mixer to dislodge spores from the stalks. Short time periods of fruit agitation resulted in poor recoveries of dormant spores, and excessively long periods damaged a high proportion of the spore cells and caused coincident loss of trehalase activity. After these treatments, the dispersed cell suspension was determined, by phase contrast microscopy (×40), to consist of spores, intact stalks, semi-intact stalks, stalk cells, and a few residual sorocarps.

Partial removal of spores from copurifying material was achieved by filtering the slime mold suspension under gravity through two layers of nylon mesh. After collection of cells from the filtrate by low-speed centrifugation (3,000 × *g*, 10 min), the spores were washed with 10 mM phosphate buffer (pH 6.5) and again filtered through clean nylon. Spores were again collected by centrifugation as described above, resuspended in 10 mM phosphate buffer (pH 6.5), and subjected to preparative density gradient centrifugation in Percoll.

Early studies using Percoll for purification of *Dictyostelium* spores used preformed gradients. Slime mold spores did not sediment or did so very slowly through the gradient material at unit gravity. Likewise, use of rate zonal centrifugation with preformed Percoll gradients proved to be an unsatisfactory method when applied to the *Dictyostelium* spores (Killick, Anal. Biochem., in press). Generation of Percoll gradients in situ, however, with subsequent banding of the cells at their isopycnic densities proved to be an excellent method for preparation of spores retaining morphological and biochemical integrity.

Gradients were generated in situ by subjecting the isotonic Percoll solution to centrifugation at $20,000 \times g$ for 15 to 90 min (Fig. 1). The gradients were calibrated with color-coded dextran density marker beads (Pharmacia Fine Chemicals), each of the latter having a specific reproducible buoyant density in Percoll gradients. Calibration of the gradients was obtained by plotting in an external marker tube the distance from the top

of the meniscus to each band versus the density of each band. Resultant curves corresponded extremely well with those published by Pharmacia. For isolation of *Dictyostelium* spores, a running time of 15 or 30 min proved satisfactory. After centrifugation, upper and lower bands were examined in a phase-contrast microscope. The upper band consisted mainly of stalk cells and spores that showed morphological damage, this having occurred primarily during the dislodging treatment of the sorocarps on the Vortex mixer. On the other hand, the lower band, based on morphological criteria, consisted entirely of spore cells which retained their morphological integrity at the completion of all preparatory steps in their isolation. Estimation of the densities indicated values of about 1.03 and 1.12 g/cm^3 for the upper and lower cell populations, respectively.

In view of the fact that the major purpose underlying studies on preparative spore procedures was (i) to determine whether the dormant *Dictyostelium* spores contain trehalase activity and then (ii) to compare the properties of this enzyme with those of the enzyme from vegetative cells after purification of both enzymes, the two cell layers were analyzed for trehalase activity. Enzymatic activity was detectable before and after French pressure cell treatment of material from the lower band, whereas trehalase activity was not measurable in extracts prepared from upper cell band material. Thus, the latter cell layer was routinely removed by aspiration and discarded. The lower spore layer was collected with an 18-gauge syringe and diluted fivefold with 10 mM phosphate (pH 6.5) buffer. After the cells had been collected by low-speed centrifugation, they were suspended in 50 mM morpholinoethanesulfonic acid-NaOH (pH 6.5) buffer to serve as the source of trehalase for subsequent studies.

Detection of trehalase activity in spore extracts. To effect maximal release of soluble enzymatic activity, spores were subjected to a single freeze-thaw cycle followed by passage of the crude homogenate through a French pressure cell (20,000 lb/in^2). The French pressure cell treatment increased recoverable trehalase activity about 100-fold over that achieved with a freeze-thaw cycle alone. With these extraction procedures, at least 95% of the trehalase activity in the crude homogenate was recoverable in the $39,000 \times g$ supernatant fraction. With this source of enzyme, product formation was linear with time for 90 to 120 min of incubation after correction for endogenous glucose production in the absence of trehalose addition to the assay system. The initial reaction rate was directly proportional to the amount of protein assayed.

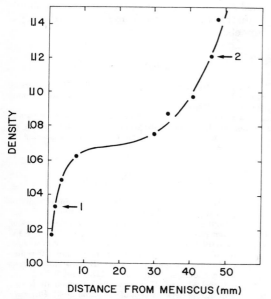

FIG. 1. *Calibration of Percoll density gradients. Gradients were generated in situ and calibrated as described in the text. The arrows designate the positions of: (1) damaged spores, stalk material, and partially ruptured sorocarps and (2) dormant spores retaining both trehalase activity and morphological integrity. Enzymatic activity was assayed as previously described (7–9).*

Although these results confirm earlier observations by Ceccarini (3) and Cotter and Raper (5) on the presence of trehalase activity in *Dictyostelium* spore extracts, they are at variance with those reported by Jefferson and Rutherford (6). The latter investigators, employing ultramicrodissection and fluorometric enzymatic recycling techniques, obtained evidence that trehalase activity was specifically localized in the stalk cells of the slime mold and that no enzymatic activity was detectable at any stage of spore cell development. It can be speculated that the inability of Rutherford and Jefferson (6) to detect trehalase activity stems from: (i) the small amounts of biological material used, (ii) the low enzymatic activity which resulted from the in situ assay methods employed, and (iii) the possible interferences with the trehalase assay by competing side reactions, e.g., the transglycosylase reaction catalyzed by β-glucosidase (unpublished data).

Comparison of the values for specific activity of enzyme from myxamoebae, sorocarps, and dormant spores indicated that the value for the spore enzyme, assayed under standard conditions (3, 5, 10), was 10% of that observed for enzyme preparations from myxamoebae. Since the low specific activity observed for the enzyme from spores relative to that from myxamoebae may have resulted from the presence of soluble trehalase inhibitors in the spore extract or the presence of soluble trehalase activators in the amoebal preparation, or from both, samples of myxamoebal and spore enzyme were assayed singly and after they had been mixed together. The results indicated that the values for enzymatic activity for the two cell types were additive and thus suggested that neither soluble inhibitors nor activators were involved with the observed trehalase activities associated with enzyme preparations from the two cell types.

General properties of spore-specific trehalase. Studies on the stability of spore trehalase at low temperature demonstrated that: (i) enzymatic activity was stable to several repeated freeze-thaw cycles, (ii) trehalase activity was stable at $-12°C$ for at least 2 months, and (iii) enzymatic activity was stable for at least 4 days at 6°C in a cold room. The stability of the enzyme to temperatures over the range from 25 to 60°C was investigated by subjecting samples of the partially purified enzyme in 25 mM Tris-maleate (pH 6.5) buffer to a 5-min incubation at each of several designated temperatures. After heat treatment, enzyme samples were stored on ice for 30 min, and residual trehalase activity was measured. Full enzymatic activity was recovered after incubation of trehalase at temper-

atures up to about 40 to 50°C. At higher temperatures, trehalase instability was apparent. For example, after 5 min at 55°C, 75% of the original activity was recovered.

The kinetic properties of trehalase were examined at 35°C in 50 mM potassium citrate (pH 5.5) buffer. Measurement of initial reaction rates as a function of trehalose level indicated a typical hyperbolic response and an apparent Michaelis constant equal to 1.2 mM trehalose. Analysis of initial velocity data as a function of temperature showed maximal activity to occur at about 46 to 50°C. Examination of kinetic data with Arrhenius plots demonstrated that the Q_{10} was 1.95 (between 30 and 40°C) and that the energy of activation, as calculated from the slope of the straight line resulting from a plot of log V_{max} versus $1/T°K$ was 12.0 ± 0.5 kcal (50 ± 2 kJ)/mol.

Parallel studies with a purified preparation of myxamoebal trehalase (unpublished data) indicated that: (i) enzymatic activity was stable to repeated freeze-thaw cycles, storage at 6°C, and 5-min incubations at temperatures up to 50°C, and (ii) the Q_{10} was 1.9 to 2.0 (between 30 and 40°C) and the energy of activation was 12.0 to 13 kcal (50 to 54 kJ)/mol. Thus, the above characterization studies show that enzyme purified from dormant spores was indistinguishable from trehalase isolated from *Dictyostelium* myxamoebae.

ACKNOWLEDGMENTS

This investigation was supported by Public Health Service research grants AG 00922 and GM 25534 from the National Institutes of Health.

I express my sincere appreciation to Judy Appleby and Catherine Thistle for their excellent artwork and typing, respectively, and lastly to Dung Do Dai Thanh for growing the *Dictyostelium* cells.

LITERATURE CITED

1. **Bonner, J. J.** 1971. Aggregation and differentiation in the cellular slime molds. Annu. Rev. Microbiol. **25**:78–92.
2. **Ceccarini, C.** 1966. Trehalase from *Dictyostelium discoideum*: purification and properties. Science **151**:454–456.
3. **Ceccarini, C.** 1967. The biochemical relationship between trehalase and trehalose during growth and differentiation in the cellular slime mold, *Dictyostelium discoideum*. Biochim. Biophys. Acta **148**:114–124.
4. **Ceccarini, C., and M. Filosa.** 1965. Carbohydrate content during development of the slime mold, *Dictyostelium discoideum*. J. Cell. Comp. Physiol. **66**:135–140.
5. **Cotter, D. A., and K. B. Raper.** 1970. Spore germination in *Dictyostelium discoideum*. Trehalase and the requirement for protein synthesis. Dev. Biol. **22**:112–128.
6. **Jefferson, B. L., and C. L. Rutherford.** 1976. A stalk specific localization of trehalase activity in *Dictyostelium discoideum*. Exp. Cell Res. **103**:127–134.
7. **Killick, K. A.** 1979. Trehalase from *Dictyostelium discoi-*

deum: evaluation of precision and sensitivity of continuous, coupled assays. Curr. Microbiol. 2:99–102.

8. Killick, K. A. 1979. A continuous coupled polarographic assay for trehalase. Anal. Biochem. 94:360–365.

9. Killick, K. A. 1980. Coupled, continuous and discontinuous fluorometric assays for trehalase activity. Anal. Biochem. 105:291–298.

10. Killick, K. A., and B. E. Wright. 1972. Trehalose synthesis during differentiation in Dictyostelium discoideum. IV. Secretion of trehalase and in the in vitro expression of trehalose 6-P synthetase activity. Biochem. Biophys. Res. Commun. 48:1476–1481.

11. Killick, K. A., and B. E. Wright. 1974. Regulation of enzyme activity during differentiation in Dictyostelium discoideum. Annu. Rev. Microbiol. 28:139–166.

Effect of Environment and Genotype on the Synthesis of Abundant Proteins During Sporulation of *Saccharomyces cerevisiae*

ELLEN KRAIG AND JAMES E. HABER

Department of Biology and Rosenstiel Basic Medical Sciences Research Center, Brandeis University, Waltham, Massachusetts 02254

We used two-dimensional gel electrophoresis to examine changes in the pattern of the abundant proteins of *Saccharomyces cerevisiae* and to identify those proteins whose synthesis is changed by the transition from growth to sporulation. Of 44 proteins that responded to nitrogen starvation, about half did so regardless of carbon source; the others appeared to respond only when acetate was the carbon source. All of six temperature-sensitive sporulation mutations were extremely pleiotropic and altered the synthesis not only of some of the spots identified as sporulation specific but also of a number of other proteins as well. There does not appear to be a clear stage-specific regulation of the synthesis of various new proteins during sporulation, because many of the proteins that were altered by "late" sporulation mutations were not affected by two "early" mutations.

When diploid cells of the yeast *Saccharomyces cerevisiae* are transferred from growth medium to an acetate solution lacking nitrogen, the cells complete the mitotic cell cycle and embark on a program of differentiation, including meiosis and spore formation. The control of this process depends on both genetic and environmental conditions. For example, only diploids that are heterozygous for the **a** and α alleles of the mating-type (*MAT*) locus are able to initiate meiotic DNA synthesis; diploids homozygous for *MAT***a** or *MAT*α carry out many of the same biochemical functions as the *MAT***a**/*MAT*α cells but do not sporulate (8). The ability of diploid cells to sporulate also depends on the environment; sporulation is subject to both glucose and nitrogen repression (2).

In the past several years, we have been trying to understand how these various signals interact to initiate meiosis. We have found, for example, that the coordinate regulation of ribosomal protein synthesis is under different controls in sporulating cells than in a growing culture (6, 7). However, the differences in control are found both in *MAT***a**/*MAT*α and in *MAT*α/*MAT*α diploids and therefore appear to reflect a response to nitrogen starvation conditions rather than to the control of the mating-type system. Similarly, when we compared the 400 most abundantly labeled proteins in vegetative and sporulating cells, we found that nearly 10% of the spots were either turned off or turned on under sporulating conditions, but these same changes were found in *MAT*α/*MAT*α diploids as well (7).

In this paper we examine two other aspects of the regulatory signals that control the differentiation process. First, we wanted to make a distinction between proteins that were turned on or off merely by the absence of nitrogen and those that also depended on the proper carbon source (acetate instead of glucose) for this regulation. Second, we wanted to see whether there was a stage-specific regulation of various proteins, that is, whether the synthesis of some of these proteins depended on reaching a certain stage in the differentiation process. For these latter studies we relied on several temperature-sensitive sporulation (*spo*) mutants (1).

RESULTS AND DISCUSSION

Nitrogen and carbon source control of specific proteins. Because the same changes in protein spots were found in both **a**/α and α/α cells, it seemed that most of the changes in the abundant proteins are responses to the absence of nitrogen rather than to the genetic preconditions necessary to initiate meiosis and spore formation. We decided to distinguish more carefully between those proteins that were turned on or turned off simply by nitrogen starvation and those which responded only to specific sporulation conditions. We looked at changes in specific protein synthesis in defined media containing acetate or glucose, in the absence of nitrogen. Cells of strain AP1 were grown in AcII, a rich medium containing acetate as the carbon source (3), and then transferred in exponential phase to either 1% acetate or 1% glucose, in the presence or absence of 1.4% yeast nitrogen base (YNB), a source of nitrogen and vitamins. Cells were incubated for 4 h and then labeled for 5 min with [^{35}S]methionine. Not all of the 44 proteins that could be seen to change when acetate-grown cells were transferred to acetate sporulation medium could be accurately distinguished when they were shifted to glucose, because of significant changes in the abundance of a large number of spots. Nevertheless, 17 spots could be followed in each of these media (Table 1 and Fig. 1). We could distinguish three classes of changes in protein synthesis. First, 5 of the 17 spots showed very different levels of synthesis in the presence or absence of nitrogen. This was true whether acetate or glucose was the carbon source. These spots presumably represent sporulation proteins induced (or repressed) by the lack of nitrogen, regardless of carbon source. A second class of 7 proteins were more specifically controlled: they changed only in acetate medium. A third group of 5 proteins were not affected by the presence or absence of nitrogen, but were apparently responding to differences between the AcII growth medium and the much less supplemented 1% acetate plus YNB.

The changes we detected are illustrated in Fig. 1, an autoradiograph of major spots found in a truly sporulating culture (1% acetate, without nitrogen) labeled at T$_4$ (4 h after initiating sporulation). Those spots which are circled were synthesized to a greater extent when nitrogen was present; those indicated by arrows were depressed in the presence of nitrogen. The spots which responded to the differences in rich and minimal acetate growth media are in boxes. To illustrate the different responses more clearly, we selected one region from these gels and present enlargements of this section in Fig. 2. Spot 34 was maintained only in the presence of nitrogen, and spot 40 was depressed by nitrogen, regardless of carbon source. On the other hand, spot 42 depended on the presence of acetate. Spot 42 decreased (but was not absent) in acetate plus nitrogen but not in glucose plus nitrogen. (This spot also continued to be synthesized

TABLE 1. *Effect of carbon source and nitrogen starvation on the synthesis of some abundant proteins*[a]

Spot	AcII medium	Effect of YNB on presence of spot[b]			
		Ac + YNB	Ac	Gluc + YNB	Gluc
17	+	+	±	+	±
24	+	+	±	+	±
25	+	+	−	+	−
35	+	+	−	+	−
40	±	±	+	±	+
2	+	+	−	+	+
15	−	−	+	−	−
20	−	−	+	−	−
34	+	+	±	+	+
36	+	+	−	+	+
38	±	±	+	±	±
42	±	±	+	±	±
3	+	−	−	−	−
4	+	−	−	−	−
9	+	±	±	±	±
23	+	−	−	−	−

[a] Strain AP1 was grown in AcII growth medium and transferred to minimal growth medium containing either acetate or glucose plus YNB or else to solutions containing only 1% acetate or glucose. After 4 h, cells were pulse-labeled with [^{35}S]methionine for 5 min, and the labeled proteins were analyzed on two-dimensional gels (3, 5).

[b] Ac = acetate, Gluc = glucose, YNB = yeast nitrogen base. The 1% acetate medium (Ac) is the only medium which permits sporulation to occur. The relative abundance of each spot under different nutritional conditions is indicated. No indication of the abundance of these spots relative to each other is given. A ± symbol indicates that a trace amount was visible.

in a medium containing both glucose and acetate, plus nitrogen.) Clearly, the response to nitrogen starvation is quite complex and is modulated by carbon source as well.

Thus, among the 17 spots we examined, we could distinguish some proteins that changed in response to nitrogen starvation and others that were more "sporulation specific" in that they responded only to the conditions that actually promoted sporulation. There are clearly other signals operating here as well, as we also found proteins which responded to some unknown factor in the shift from AcII growth medium (containing yeast extract and peptone) to the minimal test medium containing only acetate and YNB.

Proteins affected by the *spo* mutations. Esposito et al. (1) identified 11 complementation groups of temperature-sensitive sporulation mutants that were blocked in various stages of

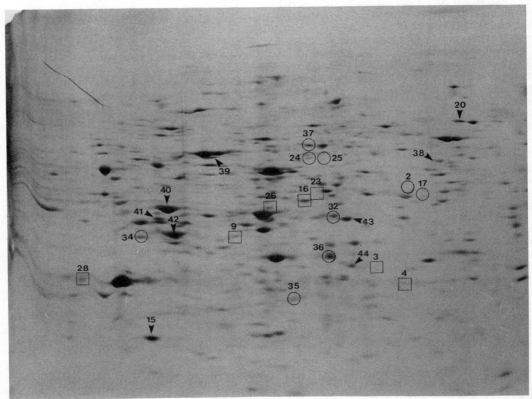

FIG. 1. *Some of the abundant proteins whose synthesis changed when strain AP1 was transferred from acetate growth medium (AcII) to sporulation medium (1% acetate buffered to pH 5.5) (3). The figure is an autoradiograph of proteins from a 4-h sporulating culture (T_4) which was pulse-labeled for 5 min with [^{35}S]methionine as previously described (3). Some of the 44 proteins that could be seen to change in their relative rates of synthesis in vegetative or sporulating cultures are indicated. Those which significantly increased in sporulation are noted by circles; those which were not synthesized or were synthesized to a much lesser degree during sporulation are shown by arrows. Those which changed when vegetative cells were shifted from a rich growth medium (AcII) to a minimal growth medium (see text) are marked by boxes.*

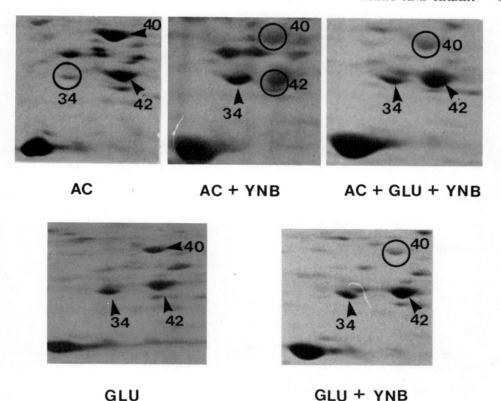

FIG. 2. *Changes in the abundance of several proteins under different nutritional conditions. Cells grown in AcII were transferred to minimal growth medium containing 1.4% YNB and either 1% acetate or glucose, or else they were transferred to acetate or glucose solutions lacking YNB. All media were buffered to pH 5.5 with succinic acid. After 4 h of incubation in the growth media or 8 h of incubation in the nitrogen-free media, cells were pulse-labeled with [^{35}S]methionine, and the labeled proteins were displayed on two-dimensional O'Farrell gels (3, 5).*

meiosis and sporulation. We selected 7 of these mutant strains to see if there were significant changes in the pattern of proteins synthesized at the restrictive temperature. We also wanted to know whether mutants blocked at the same stage of meiosis and spore formation would cause similar pleiotropic changes in the pattern of spots. We included two mutations (*spo9* and *spo11*) that block cells before meiotic DNA synthesis, two others (*spo1* and *spo2*) that block cells before the first meiotic division, and three that block cells at later stages (*spo3*, *spo4*, and *spo5*). The mutant strains were compared with the wild-type parent, S41, and with Spo+ siblings from backcrosses.

Before describing the results of this analysis, we must point out that during sporulation the strains carrying any of the *spo* mutations were not normal, even at 25°C, their permissive temperature. Protein synthesis, as measured by the

incorporation of radioactivity into acid-precipitable counts, was on average only about 30% of the level measured for their wild-type counterparts. This was true also at 34°C. There was, however, no such effect during growth, when the mutant and wild-type strains incorporated radioactivity equivalently. A further indication that these *spo* mutants were somewhat impaired at 25°C came from our observation that diploids homozygous for most pairs of these *spo* mutations were unable to sporulate even at 25°C. Diploids homozygous for *spo1 spo5*, *spo3 spo4*, *spo3 spo5*, *spo4 spo5*, or *spo4 spo11* did not sporulate normally at 25°C; only a *spo1 spo4* diploid could sporulate.

A summary of the effect of the different *spo* mutations on the pattern of labeled proteins displayed on an O'Farrell gel is given in Table 2 and in Fig. 3. In these strains, which are unrelated to strain AP1 used in previous studies (1),

TABLE 2. *Effect of* spo *mutations on the synthesis of proteins regulated during sporulation*[a]

Spot	Presence at 25°C		Presence at 34°C at T$_8$						
	Veg	Spo, T$_8$	Spo$^+$	spo1	spo2	spo3	spo5	spo9	spo11
Vegetative-specific:									
A	+	−	−	−	−	−	−	−	−
B	+	−	−	−	−	−	+	−	−
C	+	−	−	−	−	−	−	−	−
D	+	−	−	−	−	−	−	−	−
E	+	−	−	−	−	−	−	−	−
F	+	−	−	−	−	−	−	−	−
G	+	−	−	−	±	±	−	−	−
H	+	−	−	−	−	−	−	−	−
I	+	−	−	−	−	±	±	±	−
J	+	−	−	−	−	−	−	−	−
K	+	−	−	−	−	±	±	−	−
L	+	−	−	−	−	−	−	−	−
M	+	−	−	−	−	−	−	−	−
N	+	−	−	−	−	−	−	−	−
Sporulation-specific:									
O	−	+	+	−	−	−	+	−	−
P	−	+	+	+	+	+	+	−	+
Q	−	+	+	−	±	±	+	±	±
R	−	+	+	−	±	±	+	±	±
S	−	+	+	+	+	+	+	+	+
T	−	+	+	+	+	+	+	+	+
U	−	+	+	+	+	+	+	−	−
V	−	+	+	+	+	+	+	±	±
W	−	+	+	+	+	+	+	+	+

[a] The lettered spots are shown in Fig. 3A. Vegetative (Veg) and sporulating (Spo) cultures, at both 34 and 25°C, were pulse-labeled with [^{35}S]methionine for 5 min. The extracted proteins were then separated on O'Farrell gels (3), and the abundance of about 400 spots was compared.

we could resolve 23 proteins whose synthesis significantly changed when the wild-type strain S41 was transferred from growth to sporulation conditions (Fig. 3A). When the various mutant strains were labeled at 34°C during sporulation, after 8 h there were distinct differences in the synthesis of some of these proteins (Table 2). Of the 23 proteins, 10 were affected by one or more of these mutations. In a few cases, for example, spots O, Q, and R, most of the *spo* mutations affected the synthesis of the same spot.

Surprisingly, in most cases the changes in synthesis of a particular spot did not turn out to be temperature sensitive. For example, spot U was not synthesized under sporulation conditions in either *spo9* or *spo11* strains at either 25 or 34°C. In fact, among the sporulation-specific proteins, only spots O, Q, and R were actually temperature sensitive. These observations confirm that these *spo* mutants are nearly as impaired at 25°C as they are at 34°C. We should point out that all the temperature-sensitive and the temperature-independent changes in synthesis were found with the *spo* segregants and not with their wild-type siblings.

A further complication in the analysis of the

spo strains was that these mutations appeared to affect the synthesis of a large number of proteins which had not been identified as being sporulation specific. Many of these changes were indeed temperature sensitive, with spots appearing during sporulation at 25°C but not at 34°C, or vice versa. A representative sample of these additional spot changes is given in Fig. 3B. Here, a spot whose synthesis was temperature-sensitive is indicated by the number of the *spo* mutation that affected that spot. Again, this seems to suggest that the effects of the *spo* mutations are quite complex and that these mutations alter the levels of synthesis of a large number of different abundant proteins. None of these proteins, however, has been seen to change during the sporulation of a wild-type strain. Nevertheless, these changes have something to do with sporulation conditions, because their synthesis was not temperature sensitive when pulse-labeled proteins from vegetative cultures of these *spo* mutants were examined at 25 and 34°C.

From this analysis we can draw only guarded conclusions, because of the extensive pleiotropy of the *spo* mutants we examined. It does seem clear that the pattern of changes in protein

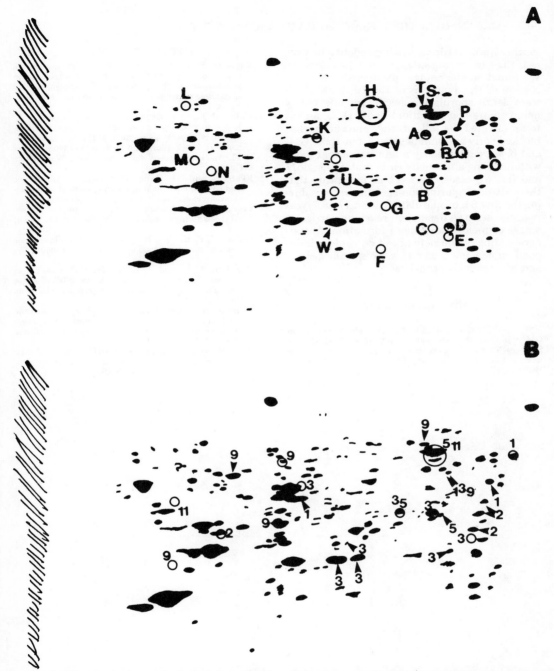

FIG. 3. *Changes in the synthesis during sporulation of abundant proteins which were affected by various* spo *mutations. (A) Spots which changed at 25°C when the wild-type strain S41 was shifted from growth to sporulation. Some of these spots were affected by certain* spo *mutations (see Table 2). The figure is a schematic representation of the pattern of abundant pulse-labeled radioactive spots from a sporulating culture labeled at T_8. The abundant proteins resolved in this experiment differ significantly from those shown in Fig. 1. This mostly reflects a significant difference in the unrelated strains S41 and AP1, but also reflects changes in the time of labeling and a slightly different isoelectric focusing range. (B) Changes in spots which did not change during sporulation of the wild-type strain, but whose synthesis was temperature sensitive in the presence of one or more* spo *mutations. The mutation which affected that particular spot is noted by the number adjacent to the arrow or circle. Those spots which are circled were synthesized to a greater degree at 34°C than at 25°C; those which are marked by an arrow were synthesized more at 25°C.*

303

synthesis is not ordered in a dependent pathway, as is the case during sporulation in *Bacillus* (4). We found no clear stage-specific pattern in the control of the synthesis of the 10 spots which were clearly sporulation specific and affected by one or more *spo* mutation. Spots O, Q, and R were affected by all of the mutations that blocked early and middle stages, but spots G, I, and K were affected by a late mutation (*spo3*) without being affected by early mutations *spo11* and (in two cases) *spo9*. Thus, it appears that the control of sporulation-specific proteins during the first 8 h of sporulation in *Saccharomyces* does not seem to depend on reaching particular stages of meiosis and spore formation. An understanding of the network of controls will depend on new mutants and more specific genetic and biochemical approaches.

ACKNOWLEDGMENTS

This work was supported by Public Health Service grant GM20056 from the National Institute of General Medical Sciences. This work was submitted by E.K. as partial fulfillment of the requirements for a Ph.D. in the Department of Biology.

LITERATURE CITED

1. **Esposito, R. E., N. Frink, P. Bernstein, and M. S. Esposito.** 1972. The genetic control of sporulation in *Saccharomyces*. II. Dominance and complementation of mutants of meiosis and spore formation. Mol. Gen. Genet. **114:**241–248.
2. **Haber, J. E., M. S. Esposito, P. T. Magee, and R. E. Esposito.** 1975. Current trends in genetic and biochemical study of yeast sporulation, p. 132–137. *In* P. Gerhardt, R. N. Costilow, and H. L. Sadoff (ed.), Spores VI. American Society for Microbiology, Washington, D.C.
3. **Kraig, E., and J. E. Haber.** 1980. Messenger ribonucleic acid and protein metabolism during sporulation of *Saccharomyces cerevisiae.* J. Bacteriol. **144:**1098–1112.
4. **Linn, T., and R. Losick.** 1976. The program of protein synthesis during sporulation in *Bacillus subtilis.* Cell **8:**103–114.
5. **O'Farrell, P. H.** 1975. High resolution two-dimensional electrophoresis of proteins. J. Biol. Chem. **250:**4007–4021.
6. **Pearson, N. J., and J. E. Haber.** 1977. Changes in regulation of ribosome synthesis during different stages of the life cycle of *Saccharomyces cerevisiae.* Mol. Gen. Genet. **158:**81–91.
7. **Pearson, N. J., and J. E. Haber.** 1980. Changes in the regulation of ribosomal protein synthesis during vegetative growth and sporulation in yeast. J. Bacteriol. **143:**1411–1419.
8. **Roth, R., and K. Lusnak.** 1970. DNA synthesis during yeast sporulation: genetic control of an early developmental event. Science **168:**493–494.

Role of the Mitochondrial Genome in Yeast Differentiation

RENEE SCHROEDER AND MICHAEL BREITENBACH

Institut fuer Allgemeine Biochemie and Ludwig Boltzmann-Forschungsstelle fuer Biochemie, University of Vienna, Vienna, Austria, and Genetisches Institut der Universitaet, Munich, Germany

In a recent publication (A. Hartig and M. Breitenbach, Curr. Genet. **1**:97–102, 1980) we have shown that sporulation is possible in a class of respiratory-deficient *Saccharomyces cerevisiae* mitochondrial mutants with mutations mapping in the central region of the *oxi3* gene. We report here the isolation, characterization, and mapping of a mitochondrial point mutation (V-17) in a mutant which is respiratory competent but germination deficient. Respiration and DNA synthesis during sporulation were quite normal, as was the morphology of the asci. The spores were, however, unable to germinate or to mate, as could be shown by micromanipulation and by experiments with pure spore preparations. It could be shown by cytoduction, by deletion mapping with a series of isogenic *rho*⁻ strains, and by restriction mapping that mutation V-17 was located on the mitochondrial genome between the loci *cob* and *oli2*. Experiments with the isogenic wild type showed that there were differentiation-specific products of mitochondrial transcription and translation.

The unicellular eucaryote *Saccharomyces cerevisiae* has been used by many investigators as a model species for eucaryotic cell differentiation and differential control of gene expression. Its life cycles contain three forms of differentiation: (i) conjugation, by which process two competent haploid cells form a zygote and, consequently, a stable diploid strain; (ii) sporulation, in which a diploid cell heterozygous for the mating-type locus, under conditions of starvation for nitrogen and for fermentable carbon sources, undergoes meiosis, four haploid spores being formed after the completion of meiosis; and (iii) germination, in which the spores, which are in a resting state, are induced to germinate and to grow out by media which would normally

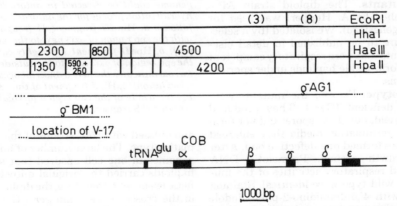

FIG. 1. *Physical map of the AG1/BM1-overlap. The map was established by use of the Southern (4) technique after cutting the mitochondrial DNAs of AG1 and BM1 with the restriction enzymes shown. The numbers in the HaeIII and HpaII fragments are lengths in base pairs. HaeIII fragment 2,300 and HpaII fragments 1,350, 590, and 250 lie totally within the overlap. Therefore, the minimum length of the overlap and the location of V-17 is given by the solid line below the lines symbolizing the AG1 and BM1 DNAs. The fractions of* cob *designated by Greek letters are the exons of* cob. *tRNA^glu indicates the gene for glutamic tRNA.*

sustain vegetative growth, forming stable haploid strains. We shall present evidence here for a specific involvement of the mitochondrial genome in sporulation and germination. Three differential experimental results support our hypothesis.

Sporulation of *oxi3 mit* strains. As rho^- and rho^0 cells (respiratory-deficient cells lacking large parts or all of the mitochondrial genome) do not sporulate (1), it was formerly believed that energy (ATP) production from acetate was the main function of mitochondria in sporulation. In a recent publication (2) we have shown that sporulation is possible at a very low level of respiration (below 1% of the respiration of the wild type) and that specific parts of the *oxi3* gene cluster (encoding subunit 1 of cytochrome *c* oxidase) seem to be necessary for sporulation. Cells with *mit*[-] mutations (respiratory deficient point mutations) located in the central part of the genetic map of the *oxi3* gene cluster are able to sporulate. Cells with mutations located to the right and left of the central spo^+ region are unable to sporulate (see Fig. 1 in reference 2). *oxi3* is now known to be a mosaic gene containing open reading frames within some of its introns. It is not known at present which of the genetic loci correspond to the exons (introns) of *oxi3*. The gene products or functions of the "genes within the introns" are also not known, but they offer a possible explanation for the effects described here.

Reversion or leakiness was excluded as the cause of sporulation in the spo^+ mutants.

Germination-deficient cells with mitochondrially inherited mutations. (i) Isolation of mutants. The diploid strain AP-3 (kindly supplied by A. Hopper) was used for $MnCl_2$ mutagenesis (3). We isolated (by a selection procedure to be published in detail elsewhere) mutants that were respiratory competent but did not form viable haploids under sporulation conditions.

(ii) Phenotype. Four of the mutants were germination deficient (Ger[-]). They produced asci in high yield, but these spores did not form colonies. On germination media they enlarged and sometimes formed one defective bud. After that, no growth ensued. Premeiotic DNA syntheses and respiratory activities of the mutant and the wild type were identical. Staining of the asci with 4',6-diamidino-2-phenylindole (5) showed that nuclear and mitochondrial DNAs segregated normally into the four spores of an ascus.

(iii) Genetic analysis. One of the mutants (V-17) was genetically analyzed in more detail. Haploid colonies, which occur at low frequency (10^{-6} to 10^{-7}) on selective medium, were picked

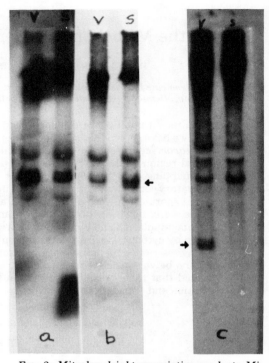

FIG. 2. *Mitochondrial transcription products. Mitochondrial RNAs of vegetative (V) and sporulating (S) cells were isolated and separated electrophoretically on 1.5% agarose gels containing 5 mM methyl mercuric hydroxide. The RNAs were transferred to diazotized paper and hybridized to ^{32}P-labeled restriction fragments of mitochondrial DNA ("Northern technique"). In this way, specific transcripts of comparatively small regions of the mitochondrial genome could be detected in autoradiograms. The hybridization probes for the autoradiograms were as follows: a, HindII fragment 11 (roughly correlated with the spo region of oxi3 next to the outside marker par); b, HpaII fragment 1 (roughly correlated with the spo region of oxi3 next to the outside marker oli2) (arrow indicates the new transcript occurring during sporulation); c, HhaI fragment of rho[-] AG1 comprising the whole of the cob region (arrow indicates the circular 10S transcript).*

and crossed with rho^0 strains of the appropriate mating type. The large number of mitochondrial genomes in one cell ensured that some of the haploids carried the original mitochondrial genetic lesion and, therefore, the diploids produced in the crosses were again ger[-]. Cytoduction of the haploid V-17/5 with a strain carrying the mutation *kar-1* showed that mutation V-17 was cytoplasmically inherited.

We crossed strain V-17/5 with a series of rho^- mutants derived from strain KL 14-4A. The rho^- mutations had been mapped against *mit*[-] and antibiotic resistance mutations, and their genetic lengths had been defined in this way. Every part of the mitochondrial genome was present in at least one member of the series. Only three (BM1, HB11, and AG1) out of 11 *rho⁻* mutants could restore wild-type germination properties when crossed with V-17/5 mutants. The three *rho⁻* strains had only one region of the mitochondrial genome in common, i.e., the region between the genes *cob* (coding for apocytochrome *b*) and *oli2* (coding for subunit 6 of adenosine triphosphatase). HB1 included a large part of this region, but AG1 and BM1 apparently had only a small overlap. Therefore, we undertook a more accurate localization by restriction mapping procedures. The results of these experiments are

shown in Fig. 1. The overlap defined the location of mutation V-17. It was outside any known gene, including the glutamic tRNA gene to the left of exon α of the *cob* region.

Differentiation-specific mitochondrial gene products in the wild type. (i) Transcription. The transcripts hybridizing to parts of the *cob* and *oxi3* regions of the mitochondrial genome are shown in the autoradiograms of Fig. 2. Some of the major differences between transcription in sporulating cells and that in vegetative cells are described below. The circular transcript of the first intron of *cob* was missing in sporulation. Processing of the *cob* transcripts was much more advanced in sporulating cells as

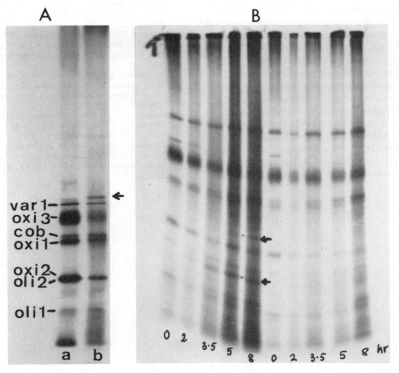

FIG. 3. *Analysis of mitochondrial products. Mitochondrially synthesized proteins were labeled with Na₂³⁵SO₄ in the presence of cycloheximide, and the proteins were separated on polyacrylamide gradient gels (10 to 15%) in the presence of sodium dodecyl sulfate. Equal amounts of radioactive material were applied to each slot of the slab gels. (A) Pulse-labeling of mitochondrial translation products of vegetative cells (a) and germinating spores (b). The arrow indicates a new polypeptide. Slot a shows the following polypeptides: var1 product (ribosome-associated protein), oxi3 product (cytochrome oxidase subunit 1), oxi1 product (cytochrome oxidase subunit 2), cob product (cytochrome b), oxi2 product (cytochrome c oxidase subunit 3), oli2 product (adenosine triphosphatase subunit 6, not resolved), and oli1 product (adenosine triphosphatase subunit 9). (B) Pulse-chase experiments with mitochondrial translation products. For the left five slots, mitochondrial proteins were pulse-labeled as above in the presporulation phase and then chased into sporulation with Na₂SO₄, cysteine, and methionine. Samples were taken at the indicated times (hours after transfer to sporulation medium). For the right five slots mitochondrial proteins were pulse-labeled as above, but the chase was performed under conditions of vegetative growth. Samples were taken at the indicated times (hours after transfer to fresh growth medium). Arrows indicate the two new polypeptides occurring under sporulation conditions. (Label was presumably incorporated from turnover of proteins labeled during the pulse.)*

compared with vegetative cells. A completely new transcript occurred in sporulation from the area of *oxi3* next to *oli2*. In the area of *oxi3* which lies next to the outside marker *par*, quantitative differences between the transcripts of sporulating and vegetative cells could be seen. However, there was no difference in transcription of the central part of *oxi3* (roughly correlated with the *spo*⁺ region) between vegetative and sporulating cells (autoradiograms not shown here). These results are in good agreement with the sporulation properties of *oxi3 mit⁻* mutants.

(ii) Translation. Mitochondrial products were analyzed (Fig. 3) in pulse-labeling studies during vegetative growth and germination and also in pulse-chase experiments during vegetative growth and sporulation. Pulsing in presporulation and chasing into sporulation seemed to us most meaningful, because sporulating cells normally take up only a small amount of amino acids or polypeptide precursors. Most of the mitochondrially made polypeptides were stable throughout sporulation, only var-1 (a component of the mitochondrial ribosomes) was degraded rapidly. Also during sporulation, two new polypeptides of molecular weights smaller than that of var-1 appeared, although they did not occur in pulse-chase experiments during vegetative growth. In pulse-labeling experiments during germination, a new polypeptide appeared

which had a molelcular weight in the same range as that of var-1.

We are presently trying to use peptide mapping techniques to establish a biogenetic relationship between var-1 and the new polypeptides.

Summarizing, we may state that differentiation-specific mitochondrial gene products which could be expected on the basis of the genetic experiments with *mit⁻* and *ger⁻* mutants have now been demonstrated directly.

LITERATURE CITED

1. **Ephrussi, B., and H. Hottinguer.** 1951. Cytoplasmic constituents of heredity. On an unstable cell state in yeast. Cold Spring Harbor Symp. Quant. Biol. **16:**75–84.
2. **Hartig, A., and M. Breitenbach.** 1980. Sporulation in mitochondrial *oxi3* mutants of *Saccharomyces cerevisiae*. A correlation with the genetic map. Curr. Genet. **1:**97–102.
3. **Putrament, A., H. Baranowska, and W. Prazmo.** 1973. Induction by manganese of mitochondrial antibiotic resistance mutations in yeast. Mol. Gen. Genet. **126:**357–366.
4. **Southern, E. M.** 1975. Detection of specific sequences among DNA fragments separated by gel electrophoresis. J. Mol. Biol. **98:**503–517.
5. **Williamson, D. H., and D. J. Fennel.** 1975. The use of fluorescent DNA-binding agent for detecting and separating yeast mitochondrial DNA. Methods Cell Biol. **12:**335–352.

Regulation of NADPH Levels in *Neurospora crassa* Conidia

JOSEPH C. SCHMIT, SHAO-YI SHEEN, AND R. LYLE CHRISTENSEN

School of Medicine and Department of Chemistry and Biochemistry, Southern Illinois University at Carbondale, Carbondale, Illinois 62901

The low NADPH levels in mature conidia were directly associated with a decrease in the water content of *Neurospora crassa* conidia. When conidia were rehydrated, the NADPH/NADP$^+$ ratio increased rapidly, as a result of both the reduction of NADP$^+$ and the net synthesis of NADP$^+$ or NADPH, or of both. The total NAD(H) level also increased when conidia were suspended in water, indicating that conidia were capable of the net synthesis of pyridine nucleotides from endogenous compounds during the early phases of germination. Dormant conidia contained sufficient levels of the pentose phosphate pathway enzymes to rapidly catalyze the reduction of NADP$^+$ to NADPH. The presence of high levels of NADP$^+$ in dormant conidia suggests that either these enzymes were inactivated as the water content of the conidia decreased or their substrates were not available until the initiation of germination.

Conidia are produced by budding from the tips of specialized aerial hyphae during the asexual developmental cycle of *Neurospora crassa*. Conidia have many of the properties of dormant spores (15) and, if dehydrated, are very heat stable (5). When conidia are suspended in liquid medium, they lose their heat resistance (15) and begin the process of germination by initiating the metabolism of endogenous storage compounds such as trehalose (6) and glutamic acid (14).

The formation of NADPH has been implicated in the activation of metabolism in dormant cells from a wide variety of organisms, including bacterial endospores (19), seeds (7; S. Brody, personal communication), and sea urchin eggs (3), and may be one of the earliest biochemical events that occur during conidial germination (16). Exogenous carbon sources were not required for the increase in NADPH during the germination of *Bacillus megaterium* endospores (19) or *N. crassa* conidia (this paper). The metabolism of endogenous storage compounds is sufficient to complete this biochemical change during the early phases of spore germination in both of these organisms.

This paper shows that the NADPH/NADP$^+$ ratio in dormant conidia is a consequence of the dehydration of the newly formed conidia. The activities of six NADP$^+$-specific dehydrogenases were measured in dormant and germinating conidia. Four of these enzymes were stored at sufficient levels in dormant conidia to catalyze the rapid conversion of NADP$^+$ to NADPH during germination.

RESULTS AND DISCUSSION

Effect of dehydration on pyridine nucleotide levels. Freshly harvested conidia contained 35 to 65% water (5; J. C. Schmit, unpublished data). The ratio of NADPH to NADP$^+$ in freshly harvested conidia varied considerably in different preparations of conidia. The water content of the conidia may have been the critical variable. The pyridine nucleotide content of mature conidia changed as the conidia were dehydrated (Fig. 1). The ratio of NADPH to NADP$^+$ decreased from about 3.5 to about 1.0, apparently as a result of the oxidation of NADPH during dehydration. There also was a steady decrease in NAD$^+$ levels (Fig. 1). The conidia remained viable throughout this experiment, as determined by their ability to form germ tubes on minimal glucose agar medium. The decrease in the NADPH/NADP$^+$ ratio that occurred when conidia were dehydrated suggests that the metabolic processes that oxidize NADPH could continue for a longer period of time than could those that regenerate this coenzyme.

Changes in pyridine nucleotide levels during incubation in distilled water. The ratio of NADPH to NADP$^+$ increased six- to eightfold when dehydrated conidia were incubated in distilled water (Fig. 2). The increase in the NADPH/NADP$^+$ ratio was due both to the

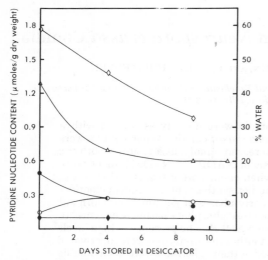

FIG. 1. *Effect of desiccation on pyridine nucleotide levels in conidia. Conidia from the mutant strain nada (13) were dry harvested and put into a desiccator with anhydrous CaSO₄. Samples were removed for pyridine nucleotide assays on the day shown, and dry weights were determined. Oxidized pyridine nucleotides were extracted from the conidia by pouring preheated (100° C) 0.05 N HCl directly on preweighed conidia. The conidial suspension was mixed vigorously with a Vortex mixer and incubated at 100° C for 3 min. This procedure completely destroyed the reduced coenzymes. The reduced coenzymes, NADH and NADPH, were extracted into 0.05 N KOH at 100° C by the procedures described for NAD⁺ and NADP⁺. NAD⁺ and NADH were measured by the cycling procedures of Kato et al. (8). NADP⁺ and NADPH were measured by the cycling procedure of Lowry et al. (11). The cycling assays were linear with respect to time and to the volume of extract added to the reaction mixture. No inhibitors were present in either the acidic or basic extracts that could interfere with the quantitative assay of the pyridine nucleotides. The percent water was measured as the difference in weight of conidia before and after they had been dried for 12 to 24 h at 100° C. Symbols: ◇, NAD⁺; ◆, NADH; ○, NADP⁺; ●, NADPH; △, percent water.*

reduction of $NADP^+$ to NADPH and to an increase in the total amount of these pyridine nucleotides (Fig. 2A). The NADPH content reached its maximum level within 10 min after the conidia were suspended in water and stabilized at a concentration of 0.7 to 0.9 μmol of NADPH per g of conidia. No further increase occurred, even after 3 h in distilled water.

When glucose was added to distilled water at a final concentration of 1% or when conidia were suspended in 10% Vogel minimal medium (20) with 0.2% glucose, a similar but more rapid increase in NADPH levels was observed. The NADPH content reached the same maximum

levels, but it took only 6 min rather than 10 min to reach that level.

The $NADH/NAD^+$ ratio also increased when conidia were incubated in distilled water, as did the total amount of NAD^+ and NADH in conidia (Fig. 2B). The observation that both total NAD(H) and NADP(H) levels increased when conidia were incubated in distilled water indicates that pyridine nucleotides can be synthesized de novo during germination from metabolites that are stored in the conidia during their formation.

Activity of $NADP^+$-specific dehydrogenases in dormant conidia. Dormant conidia contained high levels of the pentose phosphate pathway enzymes glucose-6-phosphate dehydrogenase and 6-phosphogluconate dehydrogenase (Table 1). The activities of these enzymes were sufficient to convert all of the oxidized $NADP^+$ in the dormant conidia to NADPH in less than 1 s. The increase in NADPH concentration dur-

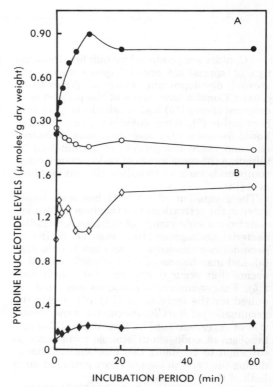

FIG. 2. *Changes in pyridine nucleotide levels during incubation in distilled water. Conidia that had been stored in a desiccator for 9 days were suspended in distilled water, and samples were removed at times indicated for pyridine nucleotide assays (see legend to Fig. 1). Symbols: ○, NADP⁺; ●, NADPH; ◇, NAD⁺; ◆, NADH.*

ing germination occurred much more slowly (Fig. 2A), indicating that the activity of these dehydrogenases might be dependent on water uptake itself or on the availability of substrates and coenzyme. Both of these dehydrogenases appear to be associated with intracellular membranes in *N. crassa* (9), and this association could provide a mechanism for their physical separation from their substrates.

Another explanation for the slow increase in NADPH levels could be that NADPH was reoxidized very rapidly during the early stages of germination. Oxidized glutathione accumulates in dormant conidia and is rapidly reduced to its sulfhydryl form in a reaction that requires NADPH during the early stages of conidial germination (4). Other reactions that oxidize NADPH probably also occur in germinating conidia since biosynthetic processes, including protein synthesis, begin as soon as conidia contact liquid medium (15).

Conidia contained moderate levels of two other NADP$^+$-specific dehydrogenases, isocitrate dehydrogenase and glutamate dehydrogenase (Table 1). Malic enzyme was not found in dormant conidia but did appear at about the time germ tubes were forming (Table 1). The absence of malic enzyme activity in conidia has been reported previously (22). There was no NADP$^+$-specific succinic semialdehyde dehydrogenase activity in *N. crassa* conidia (Table 1); activity was present, but it was specific for NAD$^+$ (Table 1). This enzyme is part of the γ-aminobutyric acid degradative pathway that has been implicated in the metabolism of the large pool of free glutamic acid during germination (2, 14). Dormant conidia also contained large amounts of malate dehydrogenase activity and had moderate amounts of NAD$^+$-specific glutamate dehydrogenase activity (Table 1).

During germination in minimal glucose medium, three different patterns of changes in enzyme activities were observed (Table 1). The pentose phosphate pathway enzymes and NADP-specific isocitrate dehydrogenase increased in activity during the first hour and then decreased to levels comparable to those in dormant conidia as germ tubes began to appear. Malic enzyme and NADP-specific glutamate dehydrogenase activities were absent or low in conidia and began to increase as germ tubes appeared. Other workers have reported similar changes in malic enzyme (22) and NADP-specific glutamate dehydrogenase (10) during conidial germination. The three NAD$^+$-specific dehydrogenases that were measured (Table 1) decreased in activity as conidia germinated.

In summary, *N. crassa* conidia contained high levels of glucose-6-phosphate dehydrogenase

TABLE 1. *Dehydrogenases in dormant and germinating conidia*[a]

Enzyme[b]	Sp act (nmol of NAD(P)H/ min per mg of protein)[c]			
	T_0[d]	T_1	T_3	T_6
A. NADP$^+$ specific:				
Glucose-6-phosphate dehydrogenase	364	521	314	329
6-Phosphogluconate dehydrogenase	263	410	304	263
Isocitrate dehydrogenase	24	57	30	22
Glutamate dehydrogenase	30	51	58	135
Malic enzyme	<1	5	14	29
Succinic semialdehyde dehydrogenase	<1	—	—	—
B. NAD$^+$ specific:				
Malate dehydrogenase	2,050	2,000	1,400	230
Succinic semialdehyde dehydrogenase	100	25	11	—
Glutamate dehydrogenase	20	7	—	—

[a] Conidia from the *nada* strain were obtained by dry harvesting 7- to 14-day-old slant cultures (2). The conidia were germinated in Vogel's minimal glucose medium (20) at 24°C on a rotary shaker (2).

[b] Glucose-6-phosphate dehydrogenase (17), 6-phosphogluconate dehydrogenase (18), NADP-specific isocitrate dehydrogenase (12), NADP-specific glutamate dehydrogenase (1), malic enzyme (22), succinic semialdehyde dehydrogenase, malate dehydrogenase (based on succinic semialdehyde dehydrogenase assay [21]), and NAD-specific glutamate dehydrogenase were assayed spectrophotometrically as described in the references given. The permeabilized conidia were added directly to the reaction mixture (2).

[c] Freshly harvested and germinating conidia were permeabilized with toluene and methanol (2). The protein content of the permeabilized cells was determined as described previously (2).

[d] The time, T, is given in hours after the conidia were first suspended in the minimal glucose medium.

and 6-phosphogluconate dehydrogenase activities and substantial amounts of NADP$^+$-specific glutamate dehydrogenase and NADP-specific isocitrate dehydrogenase activities. No malic enzyme or NADP$^+$-specific succinic semialdehyde dehydrogenase activity was found in dormant conidia. The inability of dormant conidia to maintain a high NADPH/NADP$^+$ ratio as they were dehydrating suggests that either these dehydrogenases were inactivated more rapidly than the enzymes that utilized NADPH or the substrates for these enzymes had become limiting.

ACKNOWLEDGMENT

This work was supported by National Science Foundation grant PCM 77-12551.

LITERATURE CITED

1. **Barratt, R. W.** 1963. Effect of environmental conditions on the NADP-specific glutamic acid dehydrogenase in *Neurospora crassa*. J. Gen. Microbiol. **33**:33–42.
2. **Christensen, R. L., and J. C. Schmit.** 1980. Regulation of glutamic acid decarboxylase during *Neurospora crassa* conidial germination. J. Bacteriol. **144**:983–990.
3. **Epel, D.** 1975. The program of and mechanism of fertilization in Echinoderm egg. Am. Zool. **15**:507–522.
4. **Fahey, R. C., S. Brody, and S. D. Mikolajczyk.** 1975. Changes in the glutathione thiol-disulfide status of *Neurospora crassa* conidia during germination and aging. J. Bacteriol. **121**:144–151.
5. **Fahey, R. C., S. D. Mikolajcyzk, and S. Brody.** 1978. Correlation of enzymatic activity and thermal resistance with hydration state in ungerminated *Neurospora* conidia. J. Bacteriol. **135**:868–875.
6. **Hanks, D. L., and A. S. Sussman.** 1969. The relationship between growth conidiation and trehalase activity in *Neurospora crassa*. Am. J. Bot. **56**:1152–1159.
7. **Hendricks, S. B., and R. B. Taylorson.** 1975. Breaking of seed dormancy by catalase inhibition. Proc. Natl. Acad. Sci. U.S.A. **72**:306–309.
8. **Kato, T., S. J. Berger, J. A. Carter, and O. H. Lowry.** 1973. An enzymatic cyclic method for nicotinamide-adenine dinucleotide. Anal. Biochem. **53**:86–97.
9. **Levia, S., N. Olea, and G. Pincheira.** 1979. Electron cytochemical detection of two dehydrogenase activities in *Neurospora crassa*. Cell. Mol. Biol. **24**:29–36.
10. **Loo, M.** 1976. Some required events in conidial germination of *Neurospora crassa*. Dev. Biol. **54**:201–213.
11. **Lowry, O. H., J. V. Passonneau, D. W. Schulz, and M. K. Rock.** 1961. The measurement of pyridine nucleotides by enzymatic cycling. J. Biol. Chem. **236**:2746–2759.

12. **Nealon, D. A., and R. A. Cook.** 1979. Purification and subunit structure of nicotinamide dinucleotide specific isocitrate dehydrogenase from *Neurospora crassa*. Biochemistry **18**:3616–3622.
13. **Nelson, R. E., C. P. Selitrennikoff, and R. W. Siegel.** 1975. Mutants of *Neurospora* deficient in nicotinamide adenine dinucleotide (phosphate) glycohydrolase. J. Bacteriol. **122**:695–709.
14. **Schmit, J. C., and S. Brody.** 1975. *Neurospora crassa* conidial germination: role of endogenous amino acid pools. J. Bacteriol. **124**:232–242.
15. **Schmit, J. C., and S. Brody.** 1976. Biochemical genetics of *Neurospora crassa* conidial germination. Bacteriol. Rev. **40**:1–41.
16. **Schmit, J. C., R. C. Fahey, and S. Brody.** 1975. Initial biochemical events in germination of *Neurospora crassa* conidia, p. 112–119. *In* P. Gerhardt, R. N. Costilow, and H. L. Sadoff (ed.), Spores VI. American Society for Microbiology, Washington, D.C.
17. **Scott, W. A.** 1975. Glucose 6-phosphate dehydrogenase from *Neurospora crassa*. Methods Enzymol. **41**:177–182.
18. **Scott, W. A., and T. Abramskey.** 1975. 6-Phosphogluconate dehydrogenase from *Neurospora crassa*. Methods Enzymol. **41**:227–231.
19. **Setlow, B., and P. Setlow.** 1977. Levels of oxidized and reduced pyridine nucleotides in dormant spores and during growth, sporulation, and spore germination in *Bacillus megaterium*. J. Bacteriol. **129**:857–865.
20. **Vogel, H. J.** 1964. Distribution of lysine pathways among fungi: evolutionary implications. Am. Nat. **98**:435–446.
21. **Walsh, J. M., and J. B. Clark.** 1976. Studies on the control of 4-aminobutyrate (GABA) metabolism in "synaptosemal" and free rat brain mitochondria. Biochem. J. **160**:147–157.
22. **Zink, M. W.** 1967. Regulation of "malic" enzyme in *Neurospora crassa*. Can. J. Microbiol. **13**:1211–1221.

Author Index

Subject Index